AF751605

Birds of Western Australia

2ND EDITION

Woodslane Press Pty Ltd
10 Apollo Street
Warriewood, NSW 2102
Email: info@woodslane.com.au
Tel: 02 8445 2300 Website: woodslanepress.com.au

First published in Australia in 2013 by Simon Nevill Publications
Re-issued edition published in Australia in 2018 by Woodslane Press
Reprinted 2021, 2022, 2025
This 2nd edition published in Australia in 2024 by Woodslane Press
© 2024 Woodslane Press, text and images © 2013 Simon Nevill

This work is copyright. All rights reserved. Apart from any fair dealing for the purposes of study, research or review, as permitted under Australian copyright law, no part of this publication may be reproduced, distributed, or transmitted in any other form or by any means, including photocopying, recording, or other electronic or mechanical methods, without the prior written permission of the publisher. For permission requests, write to the publisher, addressed "Attention: Permissions Coordinator", at the address above.

The information in this publication is based upon the current state of commercial and industry practice and the general circumstances as at the date of publication. Every effort has been made to obtain permissions relating to information reproduced in this publication. The publisher makes no representations as to the accuracy, reliability or completeness of the information contained in this publication. To the extent permitted by law, the publisher excludes all conditions, warranties and other obligations in relation to the supply of this publication and otherwise limits its liability to the recommended retail price. In no circumstances will the publisher be liable to any third party for any consequential loss or damage suffered by any person resulting in any way from the use or reliance on this publication or any part of it. Any opinions and advice contained in the publication are offered solely in pursuance of the author's and publisher's intention to provide information, and have not been specifically sought.

NATIONAL LIBRARY OF AUSTRALIA

A catalogue record for this book is available from the National Library of Australia

Printed in China by SC (Sang Choy) International Pte Ltd
Original book design by: Leanne Quince
Subsequent edition design: Graphic Source
Cover photograph: Western Rosella
Back cover photograph: Crested Shrike-tit

FSC www.fsc.org MIX Paper | Supporting responsible forestry FSC® C006474

CONTENTS

INTRODUCTION

This guide book is an expanded publication of my first bird book, a 'Guide to the Birds of the Greater South West'. It includes an additional 189 species and now covers the vast majority of species that are either resident or are regular migrants throughout Western Australia and Christmas Island.

The guide-book attempts to cover the needs of both those beginning birdwatching and those who have a more extensive knowledge of birding. In chapter I, the emphasis is on assisting the novice birdwatcher with general background information on birds and their chosen habitats. Chapter II contains photographs of 492 species, of which 26 are endemic to Western Australia and Christmas Island. The total species for the State at this present time stands at 629 species based on the IOC bird list (International Ornithologists' Union). However it should be noted that within this list, many are very rare vagrants some of which have only been sighted once in the State and so have not been included.

Most of the more common Pelagic (of the open sea) birds have been included. Pelagic birding is a very specialised hobby but a very exciting and challenging one, that requires intense skills in identification, particularly with the many similar species that occur on the high seas such as the petrels and prions.

There is a vast array of birding books that have been published throughout the world, no more so than in Australia, and most are of a high standard. The objective of this book is to give more detailed local information pertaining to those species found in Western Australia. Photographs have been used rather than paintings to illustrate the birds. There are advantages to both forms of presentation. Photographs are simply what one sees, the actual bird as seen by the observer in its natural environment. The majority of photographs were taken by the author, some 454 plus species but other photographers have kindly helped with the balance required.

I have attempted to write in a more personal and direct manner unlike many textbooks, particularly as the vast majority of information imparted in this book is gleaned from personal experience over the last forty five years of bird watching in Australia.

Bird watching can be a very rewarding hobby that requires little capital expenditure unless you are one of those who choose to travel the world looking for new species to add to your world list. Here in Western Australia, we are blessed with some wonderful parks and reserves that contain a great variety of bird species; it's a wonderful hobby and whether you're a lister or just simply enjoy observing the beauty of birds and nature in general, I really hope this publication may be of some use to you.

CHAPTER I

THE REGIONAL AREA OF WESTERN AUSTRALIA

The major regional zones described in the Guide to the Birds of Western Australia

The Swan Coastal Plain: Describes the lands that run as a lateral belt from just north of Perth to the Busselton area in the south and is flanked by the Indian Ocean on the west and the Darling Range on the east. It consists primarily of limestone and alluvial sands with many permanent wetlands being an important area for waterbirds.

The Jarrah Forest: Describes the dominant eucalypt that grows on laterite soils mainly on the Darling Range, although in vegetation terms it should truly be classed as the Jarrah-Marri Forest with Marri being the equally dominant eucalypt. It is not rich in bird species but nevertheless, is a very important forest belt that contains the highest numbers of certain species found in the state.

Karri-Tingle Forest: Describes the area of the restricted tall forest area of the Karri and Tingle trees. It tends to have a slightly higher variety of birds than that of the Jarrah Forest.

The Southern Coastal Belt: Describes all lands within ten to thirty kilometres of the southern coast and is a region that not only has many seabirds and shore birds but is also home to some of our rarest passerines that occur in the coastal heaths and wetter vegetation gullies. The area loosely described covers a narrow belt from Cape Naturaliste in the west to Cape Arid in the east.

The Wheatbelt Region: Describes those lands that occur within the farming region. Although much of the natural vegetation has been cleared, the region supports a higher number of bird species than the thick Jarrah Forest. The whole area is interspersed with fragmented remnant reserves; some are extremely flora species rich and accordingly have a greater variety of birds. Within this region there are several macro habitat types including granite outcrops, salt lake and samphire systems. 'breakaway' country, Mallee Woodland, Salmon Gum Woodland and Kwongan Heath.

Transitional Eucalypt Woodland: Describes the eucalypt belt that is the natural transition zone between the cleared lands of the Wheatbelt and the semi-arid lands of the Mulga Region. It contains many species of eucalypt, particularly between Hyden and Norseman and in the Kalgoorlie region. It supports many species of birds that have been driven out of the Wheatbelt area by land clearing. It is also an important area for the nesting sites of several members of the parrot family.

The Mulga Region: Describes a vast region that extends north of the Transitional Woodlands. Its primary vegetation type is the acacia family, which is a huge plant family, in fact the largest plant genus in Australia with well over two thousand plant species. One of the dominant acacias is *Acacia aneura* simply known as 'mulga', a collective term loosely applied to describe most of the acacias in the semi-arid and arid regions although the true mulga *(Acacia aneura)* only touches the coast near Carnarvon and north at Karratha. It is home to a wealth of bird species some of them being very uncommon or even rare.

Southern Coastal Belt: Refers to a narrow band of country that follows the southern coastline from Cape Leeuwin in the west to Cape le Grand in the East. It mostly contains dense coastal heaths, with occasional river gullies that run down to the sea. There are some interesting geological changes along this band and this influences the varieties of plant life that occur along this coast and in turn the bird species that occupy that habitat.

The Pilbara: The region in the north west that covers a vast area with the Hamersley Range as its core centre. The area refers to lands from north of the Gascoyne River to the De Grey River in the north and roughly terminates in the eat on the borders of the Great Sandy Desert and Little Sandy Desert.

The Deserts: As the name implies these are the dry arid deserts that receive little or no rain through the year except the occasional cyclonic rain or in the south winter rains. We have several deserts each with some variation in habitat type. The deserts are The Victoria. The Gibson. The Little Sandy and the Great Sandy Desert. Their habitats vary including open stony plains, sand dunes, breakaway country, salt lakes, desert oak woodland and mulga woodland.

The Central Ranges: Includes several ranges of hills that abut the border of South Australia and the Northern Territory. They include the Jamieson Range. Cavanagh Range. Tomkinson Range. Rawlinson Range. Petermann Range and the Schwerin Mural Crescent Range. These are rugged stony hills with mixture of spinfex and mallee.

The Nullarbor: A vast limestone karst plain stretching from east of Kalgoorlie across the South Australian Border.

The Kimberley: A magnificent region with a diverse array of habitat from deep rocky gorges with permanent freshwater pools many holding water all year round. Grass plains. Wetlands. Rugged ranges, Mangroves. A myriad of islands. Rainforests and small and large river systems.

Maps on the following pages are a guide only and official maps should be referred to for travel, and more concise detail.

MAPS OF WESTERN AUSTRALIA

General birding zones used within this guide book.

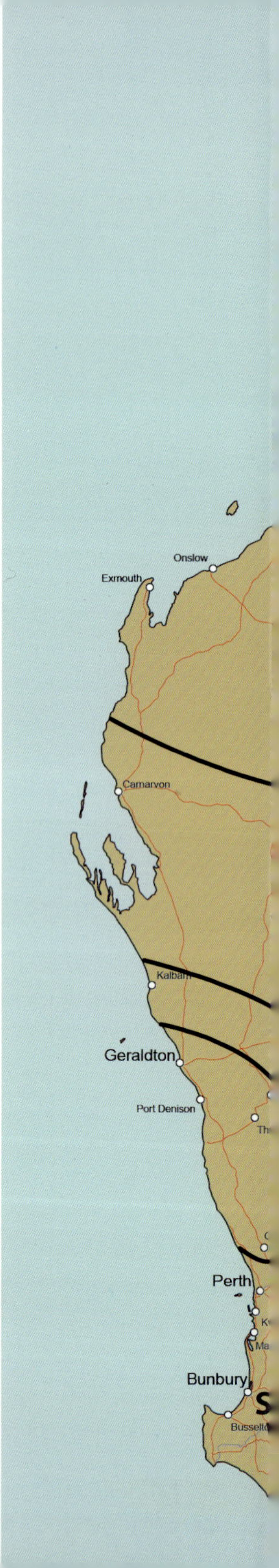

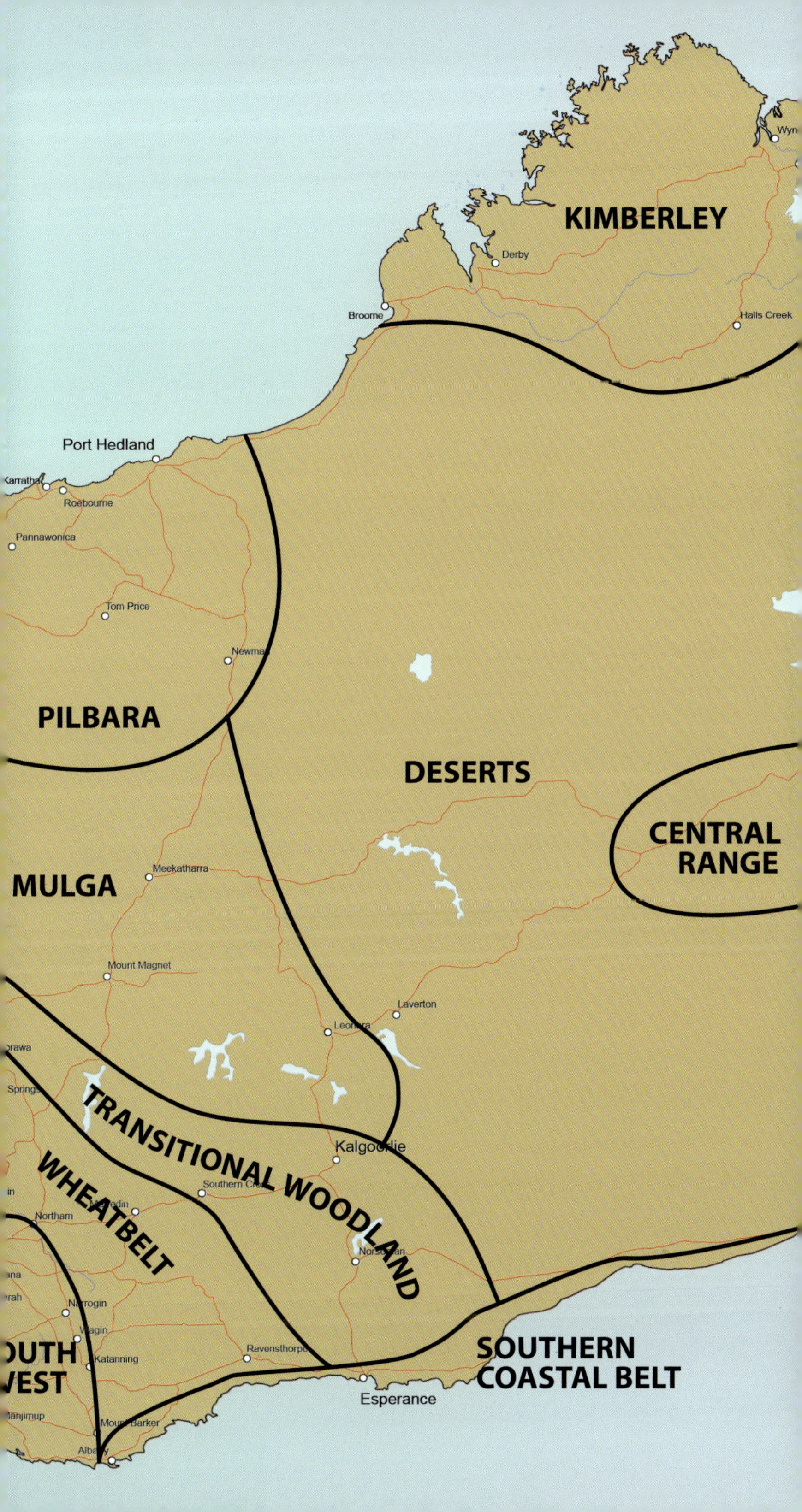
KIMBERLEY
Derby
Broome
Halls Creek
Port Hedland
Roebourne
Pannawonica
Tom Price
PILBARA
DESERTS
CENTRAL
RANGE
Meekatharra
MULGA
Mount Magnet
Laverton
TRANSITIONAL WOODLAND
Kalgoorlie
WHEATBELT
Northam
Narrogin
Wagin
Katanning
Ravensthorpe
SOUTHERN
COASTAL BELT
Esperance
OUTH
VEST

PERTH REGION

1. ERIC SINGLETON SANCTUARY
2. HERRISON ISLAND
3. KINGS PARK
4. LAKE MONGER
5. HERDSMAN LAKE
6. BOLD PARK
7. FLOREAT BEACH RES.
8. LAKE GWELUP
9. BIG CARINE SWAMP RES.
10. STAR SWAMP RES.
11. WHITEMAN PARK
12. GNANGARA
13. LAKE GOOLLELAL
14. LAKE JANDABUP
15. LAKE JOONDALUP
16. JOHN FORREST NP
17. GOOSEBERRY HILL
18. KALAMUNDA NP
19. MUNDARING WEIR
20. BICKLEY BROOK
21. ELLIS BROOK
22. VICTORIA RESERVOIR
23. BUNGENDORE RES.
24. WUNGONG BROOK
25. JANDAKOT RP
26. FORRESTDALE LAKE
27. THOMSONS LAKE
28. KOGALUP LAKE
29. YANGEBUP LAKE
30. BIBRA LAKE
31. MANNING LAKE
32. WOODMANS POINT
33. NORTH MOLE
34. BOORAGOON LAKE
35. BLUE GUM LAKE
36. ALFRED COVE
37. POINT WALTER
38. ROTTNEST ISLAND

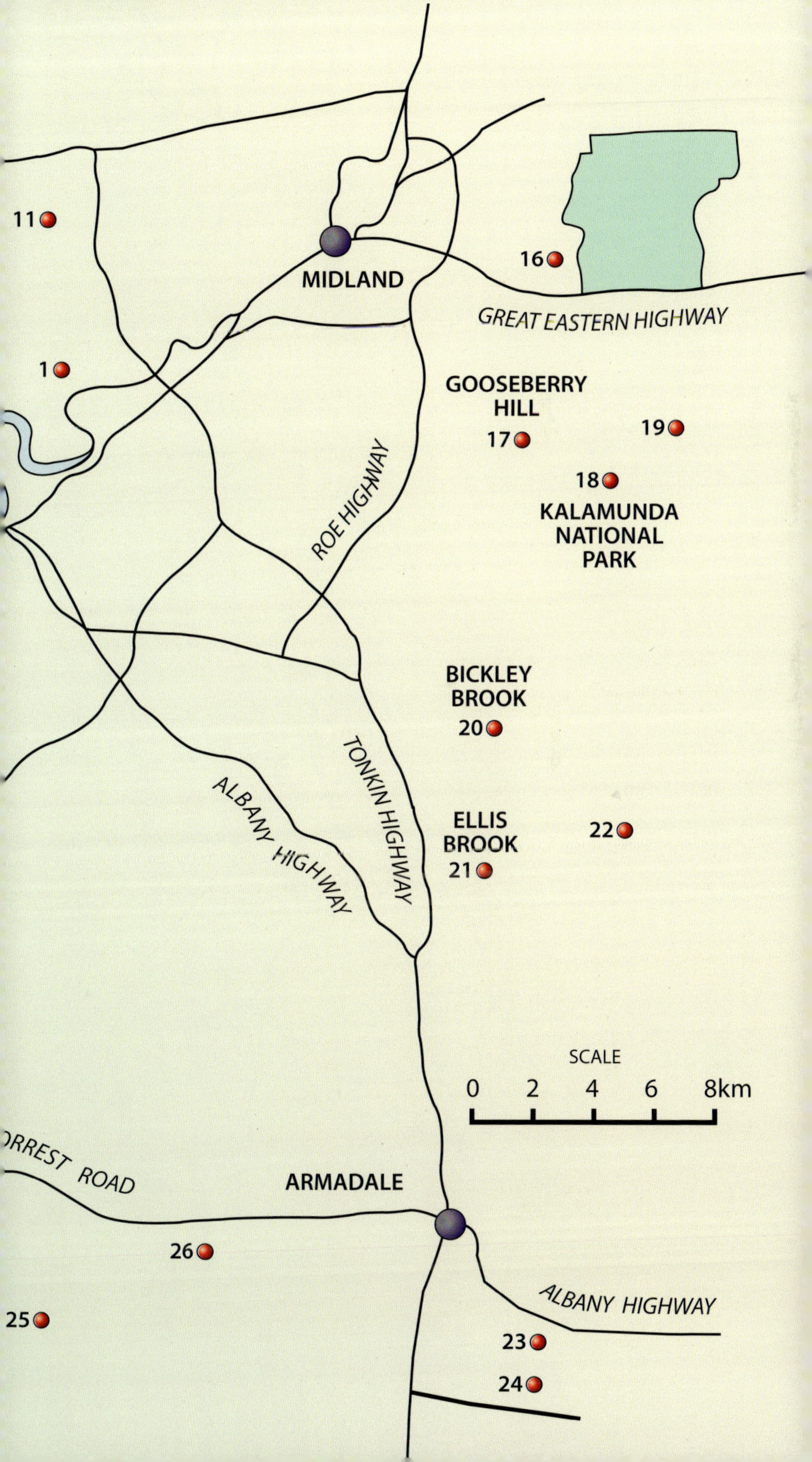

11
MIDLAND
16
GREAT EASTERN HIGHWAY
1
GOOSEBERRY HILL
19
17
18
KALAMUNDA NATIONAL PARK
ROE HIGHWAY
BICKLEY BROOK
20
TONKIN HIGHWAY
ALBANY HIGHWAY
ELLIS BROOK
22
21
SCALE
0 2 4 6 8km
ORREST ROAD
ARMADALE
26
ALBANY HIGHWAY
25
23
24

SOUTH WEST REGION

1. BOYAGIN ROCK NR
2. DRYANDRA WOODLAND NR
3. TUTANNING NR
4. DONGOLOCKING NR
5. TARIN ROCK NR
6. NORTH TARIN ROCK NR
7. CORNEECUP NR
8. COYRECUP NR
9. JINGALUP NR
10. NARLINGUP NR
11. CHINOCUP NR
12. DUMBLEYUNG LAKE
13. COBLININE NR
14. TOOLIBIN NR
15. NORTH YILLIMINNING NR
16. EAST YORNANING NR
17. CORRIGIN WATER RES.
18. JILAKIN ROCK
19. BENDERING NR
20. NORTH KARLGARIN NR
21. LUPTON CON. RES.
22. WANDOO NP
23. BOOLANOOLING NR
24. NANGEEN NR
25. CHARLES GARDNER RES.
26. STIRLING RANGE NP
27. CAMEL LAKE NR
28. UNICUP LAKE NR
29. KULUNILUP LAKE NR
30. LAKE MUIR NR
31. PORONGURUP NP
32. TORNDIRRUP NP
33. TWO PEOPLES BAY NR
34. WAYCHINICUP NR
35. LAKE PLEASANT VIEW NR
36. WEST CAPE HOWE NP
37. WILLIAM BAY NP
38. WALPOLE-NORNALUP NP
39. MT FRANKLAND NP
40. D'ENTRECASTEAUX NP
41. SCOTT NP
42. LEEUWIN-NATURALISTE NP
43. TUART FOREST NP
44. LESCHENAULT PENINSULA CP
45. YALGORUP NP
46. LANE POOLE RES.
47. LAKE Mc LARTY NR
48. PEEL INLET

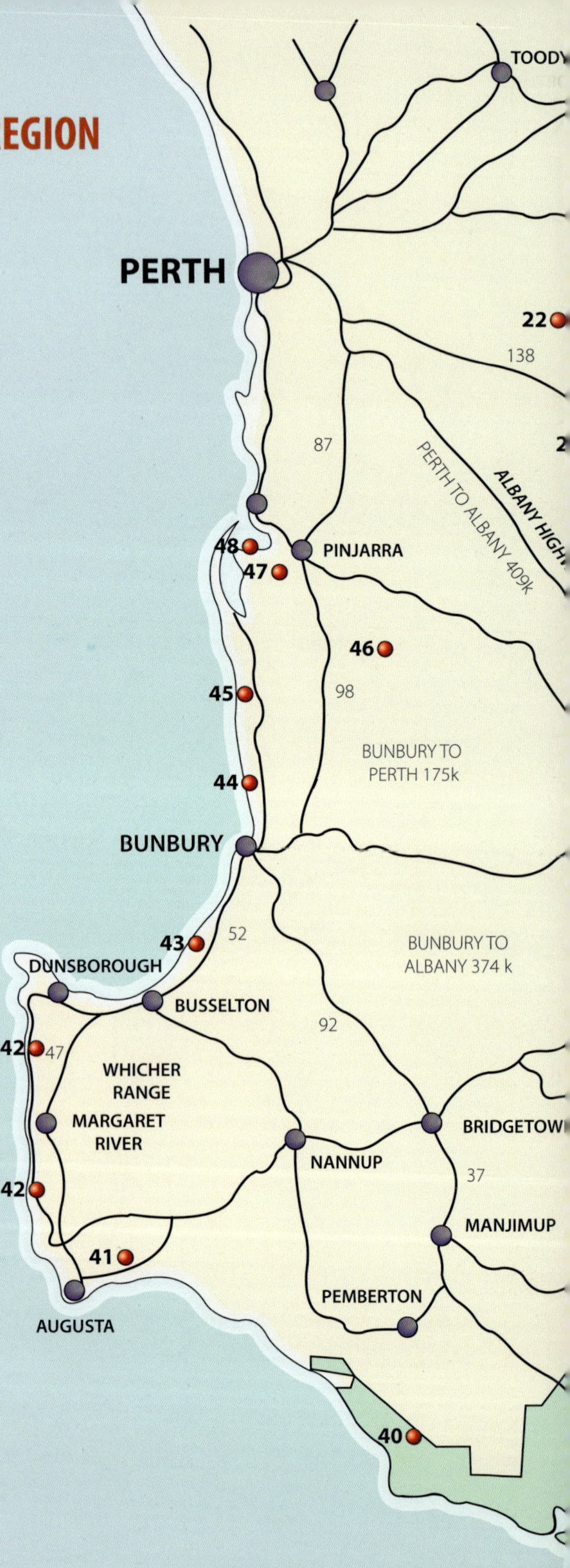

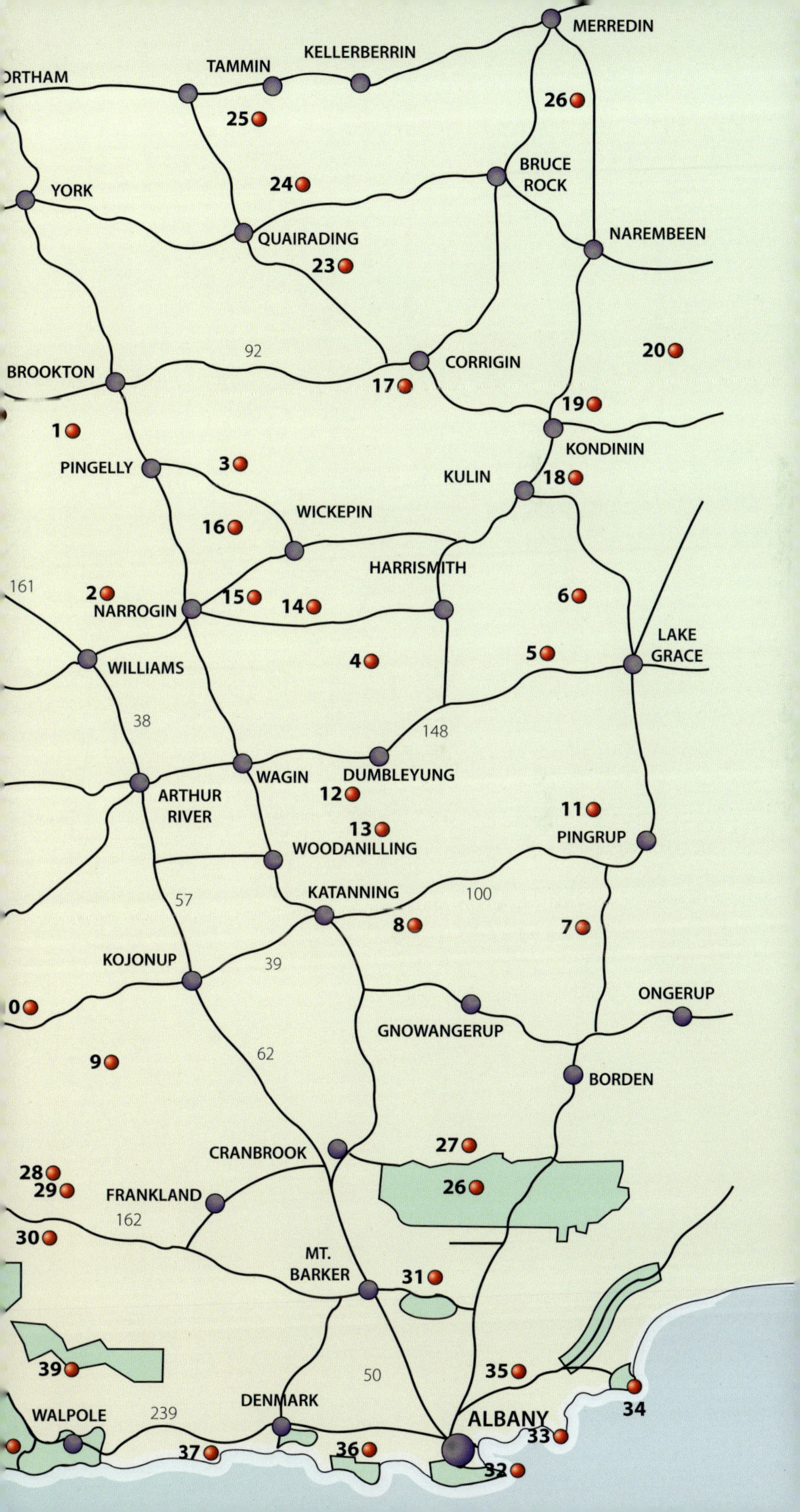

MERREDIN
KELLERBERRIN
TAMMIN
ORTHAM
26
25
24
BRUCE ROCK
YORK
QUAIRADING
NAREMBEEN
23
92
20
CORRIGIN
BROOKTON
17
19
KONDININ
1
PINGELLY
3
KULIN
18
WICKEPIN
16
HARRISMITH
161
2
15
14
6
NARROGIN
LAKE GRACE
5
4
WILLIAMS
38
148
WAGIN
DUMBLEYUNG
ARTHUR RIVER
12
11
13
PINGRUP
WOODANILLING
KATANNING
100
57
8
7
KOJONUP
39
ONGERUP
0
GNOWANGERUP
9
62
BORDEN
27
CRANBROOK
28
26
29
FRANKLAND
162
30
MT. BARKER
31
39
35
50
34
DENMARK
WALPOLE
239
ALBANY
33
37
36
32

PERTH ESPERANCE REGION

1. CHIDDARCOOPING NR
2. WALYAMOHNING NR
3. JAURDI NR
4. ROWLES LAGOON
5. SANDFORD ROCKS NR
6. LAKE CAMPION NR
7. DUROKOPIN NR
8. CHARLES GARDNER RES.
9. MAUGHAN NR
10. DRAGON ROCKS
11. THE HUMPS NR
12. LAKE CRONIN NR
13. VICTORIA ROCK
14. DUNDAS NR
15. FRANK HANN NR
16. PEAK CHARLES NR
17. JILBADJI NR
18. CAPE LE GRAND NP
19. CAPE ARID NP
20. LAKE MAGENTA NR
21. DUNN ROCK
22. FITZGERALD RIVER NP
23. STOKES NP
24. LAKE GOUNTER NR

PERTH to the MIDWEST

1. BUNTINE RES.
2. JIBBERDING NR
3. WHITE WELLS NOW CHARLES DARWIN NR
4. GOODLANDS NR
5. PETRUDOR ROCKS
6. EAST NUGADONG
7. L HINDS
8. MANMANNING NR
9. BOONARING NR
10. MOORE RIVER NP
11. NAMBUNG NP
12. BADGINGARRA NP
13. WATHEROO NP
14. LESUEUR NP
15. HILL RIVER
16. ALEXANDER MORRISON
17. TATHRA NP
18. COALSEAM NR
19. ELLENDALE POOL
20. WANDANA NR
21. KALBARRI NP
22. TOOLONGA NR
23. FRANCOIS PERON NP
24. DIRK HARTOG ISLAND
25. ROCKY POOL

The stations are listed as guides to localities but most do not have tourist facilities although several sheep stations have accommodation for travellers. Check with the various tourist centres.

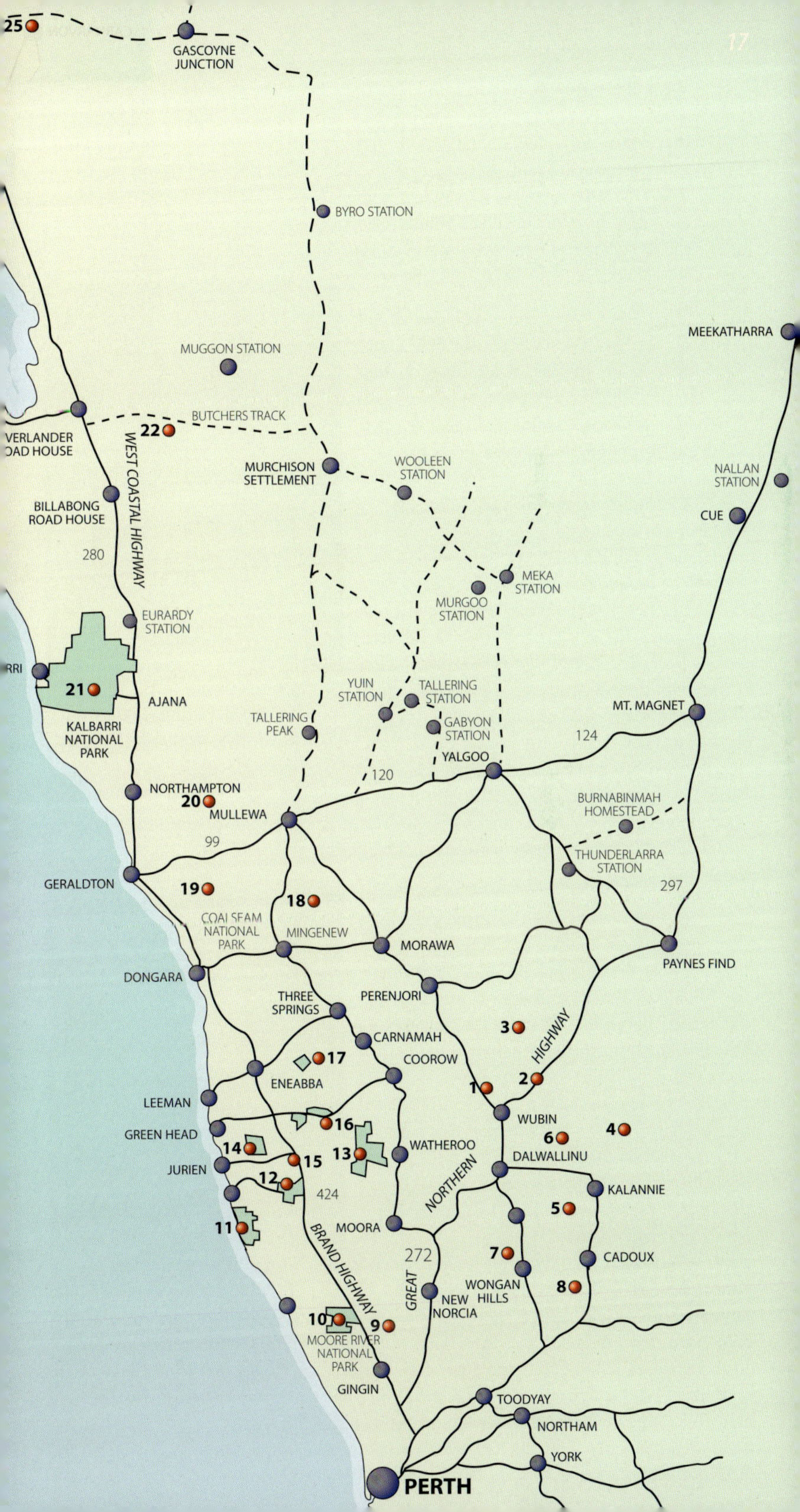
25
GASCOYNE JUNCTION
BYRO STATION
MUGGON STATION
BUTCHERS TRACK
22
VERLANDER
OAD HOUSE
WEST COASTAL HIGHWAY
MURCHISON SETTLEMENT
WOOLEEN STATION
MEEKATHARRA
NALLAN STATION
CUE
BILLABONG ROAD HOUSE
280
MEKA STATION
MURGOO STATION
EURARDY STATION
RRI
21
AJANA
KALBARRI NATIONAL PARK
YUIN STATION
TALLERING STATION
TALLERING PEAK
GABYON STATION
MT. MAGNET
124
YALGOO
120
NORTHAMPTON
20
MULLEWA
BURNABINMAH HOMESTEAD
99
THUNDERLARRA STATION
297
GERALDTON
19
18
COALSEAM NATIONAL PARK
MINGENEW
MORAWA
PAYNES FIND
DONGARA
THREE SPRINGS
PERENJORI
3
HIGHWAY
CARNAMAH
17
COOROW
ENEABBA
1
2
LEEMAN
16
GREEN HEAD
14
WUBIN
6
4
15
13
WATHEROO
DALWALLINU
JURIEN
12
NORTHERN
424
KALANNIE
5
11
MOORA
BRAND HIGHWAY
272
7
CADOUX
GREAT
WONGAN HILLS
8
NEW NORCIA
10
9
MOORE RIVER NATIONAL PARK
GINGIN
TOODYAY
NORTHAM
YORK
PERTH

Habitat types and the associated bird species

The type of vegetation habitat is the primary factor that influences where a certain species will reside. To aid the visitor and also the resident birdwatchers of this State, some of the typical species of birds to be found in certain habitat areas are given below. Obviously, many of the birds listed can be sighted in a variety of habitats but those listed are typical of that habitat.

Karri Forest

In the deep South West and along part of the southern coastline, lies an extensive belt of tall Karri trees *(E. diversicolor)*. Below these massive trees grows a thick understorey of Karri Hazel *(Trymaliun spathuatum)* with its distinctive smell, which always reminds one of the Karri Forest.

This is one area where the two vegetation strata have you either straining your neck to see birds in the canopy or looking through the lower bushes for ground frequenting birds.

Birds to be found: White-naped Honeyeater. New Holland Honeyeater. Tree Martin. Splendid Fairy-wren. Scarlet Robin. White-browed Scrubwren. Rufous Treecreeper. Purple-crowned Lorikeet. Grey Currawong. Red-tailed Black Cockatoo. Baudin's Black Cockatoo. Western Rosella. Inland Thornbill. Golden Whistler. Dusky Woodswallow. Grey Fantail. Shining Bronze-Cuckoo. Fantail Cuckoo. The majority of deep South West raptors particularly the Peregrine Falcon. Square-tailed Kite and Collared Sparrow Hawk.

Birds for listers: White-breasted Robin. Red-winged Fairy-wren. Crested Shrike-tit (the uncommon western subspecie *leucagaster*). Red-eared Firetail. All three except the Shrike-tit are found in the lowest stratum particularly along stream gullies.

Jarrah Forest

Even though the Jarrah Forest is far more extensive than both the Karri and Wandoo Woodland, it does not carry quite the variety of species that particularly the Wandoo Woodland holds. However it has the greatest number of Western Thornbills, not a rare bird but surprisingly missed by some visiting birdwatchers on short trips to the state.

It is also the primary habitat for the subspecies of the Red-tailed Black Cockatoo *naso* that feeds on the large Marri and smaller Jarrah fruits. The largest population of Baudin's Black Cockatoos also occurs here as well as the few Masked Owls in the deeper South West Jarrah Forest.

Birds to be found: Common Bronzewing. Varied Sittella. Australian Ringneck. Red-capped Parrot. Western Rosella. Weebill. Baudin's and Carnaby's Black Cockatoo (in this region more so the Baudin's). Inland Thornbill. Western Gerygone. Grey Shrike-thrush. Splendid Fairy-wren. Western Spinebill. Scarlet Robin. Grey Butcherbird. Grey Currawong.

In the northern Jarrah Forest and around the Perth region if you wish to see three of the endemics, Red-eared Firetail. White-breasted Robin and Red-winged Wren, it is best to walk along the riverine habitats that hug the streams on the Darling Scarp.

Be aware when looking for the Red-eared Firetail that the introduced east coast species the Red-browed Finch also occurs in a few isolated gullies adjacent to Perth.

South West beaches

The remoter beaches are by far the best locations to visit as human disturbance certainly affects the number of birds one can see.

Birds to be found: Besides the common species Silver Gull. Pacific Gull. Crested Tern. Red-capped Plover. Pied Oystercatcher, less common species occur, namely Hooded Plover. Rock Parrot. White-bellied Sea Eagle. Sanderling. Sooty Oystercatcher. Fairy Tern.

Off shore: Australian Gannet. Yellow-nosed Albatross. Black-browed Albatross. Occasionally Shy Albatross and Wandering Albatross. Wedge-tailed Shearwater (mostly on the west coast). Fleshy-footed Shearwater (common off the south coast, less so off the west).

Occasionally Short-tailed Shearwater and Great-winged Petrel.

If you wish to find Rock Parrots don't just walk along the beach but climb up over the sand dunes as they often feed in the swales of the sand dunes on the fruits and seeds of succulent plants.

Coastal Heath

The flora rich belt of heaths that abuts the coastal belt of the South West Region.

Birds to be found: In the habitat shown here at Waychinicup, three uncommon birds occur: Noisy Scrub-bird. Black-throated Whipbird and Western Bristlebird. Also many of the other endemics are here including White-breasted Robin. Red-eared Firetail. Red-winged Fairy-wren. Western Rosella. Red-capped Parrot. Western Wattlebird and Western Spinebill.

The low heath in the foreground shown in picture below is the territory of the Western Bristlebird. The vegetation is often no higher than 60-80 cm. On the hill side in the distance resides the Black-throated Whipbird in the taller mallee trees known as Tallerack with their glaucous leaves. In the gullies lives the elusive Noisy Scrub-bird. The author maintains this is one of the hardest birds to see and as a comparison finds it much easier to sight the east coast Rufous Scrub-bird. Although people at Cheynes Beach often get good sightings and some birdwatchers get lucky and see all three, I can assure you that is not the norm. This area is also ideal country to see Southern Emu-wren. Brush Bronzewing. White-cheeked Honeyeater. Western Wattlebird and Swamp Harrier.

Kwongan Heath

Kwongan is one of the few Aboriginal names to be used for a type of vegetation habitat that occurs in the South West Region, particularly common in the Wheatbelt region. It is one the richest flora regions in the world. The photograph shows a typical small patch of Kwongan Heath in Dryandra Woodland Reserve.

Birds to be found: The varying flowering species attract several honeyeaters including White-cheeked Honeyeater. Brown Honeyeater. New Holland Honeyeater. White-eared Honeyeater. Brown-headed Honeyeater. Red Wattlebird and Little Wattlebird and Western Spinebill. White-browed Scrubwren. Blue-breasted Fairy-wren. Splendid Fairy-wren and Silvereye also frequent the heath.

Wandoo Woodland

Wandoo Woodland carries a greater variety of habitat within its range than most other vegetation zones including the flora rich Kwongan Heath.

Birds to be found: In a rich Wandoo reserve like Dryandra Woodland Nature Reserve you may find Common Bronzewing. Brush Bronzewing. Yellow-plumed Honeyeater. White-naped Honeyeater. Brown-headed Honeyeater. White-cheeked Honeyeater. New Holland Honeyeater. Tawny-crowned Honeyeater. White-eared Honeyeater and Brown Honeyeater. Western Spinebill. Splendid Fairywren. Western Yellow Robin. Red-capped Robin. Scarlet Robin. Golden Whistler. Rufous Whistler. Western Gerygone. Western Thornbill. Weebill. Inland Thornbill. White-browed Scrubwren. Striated Pardolote. Spotted Pardalote. Dusky Wood Swallow. Welcome Swallow. Tree Martin. Rainbow Bee-eater. Grey Shrike-thrush. Black-faced Cuckoo-shrike. Grey Currawong. Restless Flycatcher. Shining Bronze-Cuckoo. Fantail Cuckoo. Pallid Cuckoo. White-winged Triller. Sacred Kingfisher. Collared Sparrowhawk. Wedge-tailed Eagle. Little Eagle. Whistling Kite. Peregrine Falcon. Australian Hobby. Western Rosella. Australian Ringneck. Elegant Parrot. Purple-crowned Lorikeet. Baudin's Black Cockatoo. Varied Sittella. Jacky Winter. Rufous Songlark (not every season). Tawny Frogmouth. Boobook Owl. Barn Owl. Owlet Nightjar.

Birds for listers: Blue-breasted Fairywren. Hooded Robin (getting scarce here). Crested Shrike-tit. Mallee Fowl. Bush Stone-curlew. Regent Parrot. Square-tailed Kite. Painted Button-quail. Masked Owl. The author has seen Black Honeyeater several times (rare in Dryandra), the last of the Crested Bellbirds (now gone) and Horsfield's Bronze Cuckoo. This is obviously not a complete list.

Noisy Scrub-bird habitat

Western Bristlebird Photo: Simon Nevill

Female Noisy Scrub-bird Photo: Graeme Chapman

Black-throated Whipbird Photo: Simon Nevill

Mallee Woodland

A vegetation type that stretches from the west just east of the Stirling Range as a narrow band along the southern coast to the southern edge of much of the Nullarbor and across to eastern Australia. Mallee can sometimes be very extensive as illustrated. Much of the Mallee belt has been cleared to make way for the Wheatbelt farms and only remnant patches remain. It is well known that this is the prime habitat for Mallee Fowl, which still survive in several inner Wheatbelt reserves.

Birds to be found: In the southern Mallee: Mallee Fowl. Purple-gaped Honeyeater. White-eared Honeyeater. Common Bronzewing. Brush Bronzewing. Blue-breasted Wren. Striated Pardalote. Southern Scrub-robin. Blue-breasted Wren. Yellow-rumped Thornbill. White-browed Babbler.

Birds for listers: Spotted Pardalote. (Yellow-rumped form *xanthopygus*). Shy Heathwren. Spotted Nightjar. Painted Button-quail.

Transitional Woodland

This tall eucalypt belt runs as a narrow band from north of the Murchison River at Kalbarri in a south east direction all the way to east of Hyden and down to the southern coast. In the north and middle of this band grows the dominant eucalypt York Gum (*E. loxophleba subsp loxophleba*). When one leaves the northern and eastern Wheatbelt you will enter this woodland. It is an important region for the parrot family with Major Mitchell's Cockatoo or Pink Cockatoo. Red-tailed Black Cockatoo (the arid sub sp *samueli*). Regent Parrot. Western Corella. Little Corella and Galah.

Birds to be found: Rufous Treecreeper. Red-capped Robin. Weebill. Rufous Whistler. White-browed Babbler. Redthroat. Chestnut-rumped Thornbill. Splendid Wren.

At the very northern part the Transitional Woodland, I have seen Black-breasted Buzzard, uncommon that far south although he has had an even lower sighting just before Johnston Lakes east of Hyden a long way out of its normal range. In the transitional woodland east of Hyden shown illustrated, you will come across woodland dominated by Salmon Gum

(*E. salmonophloia*) with an understorey of low trees called Goldfields Teatree (*Melaleuca pauperiflora*) and Bluebush (*Maireana sedifolia*). This is the ideal habitat for Gilbert's Whistler and Chestnut Quail-thrush.

Birds to be found: Western Yellow Robin. Regent Parrot. Dusky Woodswallow. Spiny-cheeked Honeyeater. White-fronted Honeyeater. White-eared Honeyeater. Restless Flycatcher. Jacky Winter. Crested Bellbird. Purple-crowned Lorikeet. Western Rosella (drier region subsp. *xanthogenys* no green on back). Rufous Treecreeper. Grey Currawong.

I have seen flocks of Masked Woodswallows on the Hyden-Norseman road. The southern Transitional Woodland is also an important stronghold for nesting Square-tailed Kite, maybe more so on the western side of this southern band of eucalypts adjacent to where extensive heaths occur 60-100 km east of Hyden.

Mulga Woodland

Much of the mulga in recent years has experienced a prolonged drought. Some of my station friends have not had decent rains for the last seven years. Obviously birds still survive in bad times even in the arid mulga. However when good rains fall the countryside can become a blaze of colour with the various everlasting species, and bird activity and the number of species increases greatly. At Wooleen Station one year, I found a particularly active breeding area, in about a 300 square metre block where I set up my hides. The following species were found breeding: Red-capped Robin. Southern Whiteface. White-browed Treecreeper (not common throughout the mulga). Crimson Chat (4 nests). Bourke's Parrot. Hooded Robin. Pied Honeyeater (at least 3). Zebra Finch. Rufous Songlark. Chestnut-rumped Thornbill. Spiny-cheeked Honeyeater and the not so common Ground Cuckoo-shrike.

Birds to be found: All the above as well as Orange Chat. Masked Woodswallow. Little Woodswallow particularly near granite outcrops. Crested Bellbird. White-browed Babbler with Grey-crowned overlapping above the Murchison. White-winged Triller. White-plumed Honeyeater. Singing Honeyeater. Grey-fronted Honeyeater. Black Honeyeater. Rufous Whistler. Little Button-quail (mainly in the northern mulga grasslands and into the deserts). Spotted Harrier. Little Eagle. Wedge-tailed Eagle. Whistling Kite. Brown Falcon. Australian Bustard. Banded Lapwing. Bushstone Curlew. Common Bronzewing. Crested Pigeon. Australian Ringneck (the drier region subsp. *zonarius* with the yellow instead of green underbelly, also markedly smaller than in the northern regions). Mulga Parrot. Budgerigar. Cockatiel. Little Corella. Pallid Cuckoo. Horsefield's Bronze-Cuckoo. Black-eared Cuckoo. Brown Songlark. Variegated Fairy-wren. Red-backed Kingfisher. Sacred Kingfisher. Western Bowerbird. Western Gerygone. Inland Thornbill. Mistletoe Bird. Pied Butcherbird. Little Crow.

Birds for listers: Western Quail-thrush. Banded Whiteface. Inland Dotterel. Black-breasted Buzzard. Grey Honeyeater. Slaty-backed Thornbill. Samphire Thornbill. Remote possibility of Black Falcon and Grey Falcon. I have seen Black Falcon flying over Lake Annean. White-browed Woodswallow with Masked Woodswallow occasionally.

Granite outcrops

The South West Region is one of the few areas in the world that has remained fairly stable during the geological and topographical changes throughout the world. One of the larger geological areas in this region is known as the Yilgarn Craton and stretches from the Darling Range in the west to east of Kalgoorlie. The Wheatbelt Region covers a greater part of the Yilgarn Craton and throughout the region there are numerous intrusions of the ancient underlying bedrock that consists of granite and gneisses and the second oldest rocks in the world occur here. Often they form huge domed features in the landscape as in Boyagin Rock Nature Reserve.

With the additional water runoff from these outcrops, thicker vegetation often occurs at their base and they are good areas to bird watch. In the Transitional Woodland zone at the base of these rock outcrops there are often Purple-gaped Honeyeaters. In the northern Wheatbelt Redthroat are often seen near the outcrops and further into the Mulga, Little Woodswallows frequent some of them. A few specialized mistletoes parasitise certain acacias at the base of the rocks and when in bloom attract Spiny-cheeked Honeyeater, White-fronted Honeyeater, Mistletoebird and even Grey Honeyeater.

Inland river systems

The country that abuts the inland rivers and creeks, often with tall stands of River Red Gum *(E. camaldulensis)* support a higher density of birds than the surrounding open country, particularly raptors, parrots and nocturnal birds, as tall trees supply good nesting sites.

Birds to be found: White-plumed Honeyeater. Spiny-cheeked Honeyeater. Striated Pardalote. Red-browed Pardalote (southern limit). Rufous Songlark. Owlet Nightjar. Boobook Owl. White-browed Babbler. Grey-crowned Babbler. Blue-winged Kookaburra (southern limit). Red-backed Kingfisher. Mulga Parrot. Budgerigar. Little Eagle. Black-breasted Buzzard. Whistling Kite. Grey Falcon (rare). Australian Hobby.

Breakaway country

In the picture taken in the Tallering Peak area north of Mullewa are stony flat tops which support populations of Western Quail-thrush. Nesting on the cliff face are a resident pair of Peregrine Falcons. The male thinks nothing of taking Ringneck Parrots in the valley below. Little Woodswallows nest on the small rock crevices. The mulga in the valleys supports numerous other birds.

Stony plains

In many areas in the Mulga Region one can come across areas of open stony plains. In eastern Australia they are called Gibber Plains and are often quite extensive, particularly in northern Southern Australia and south-east Queensland. Here in Western Australia, the Arabic term 'hamada' plain is given to these sparse stony areas, although they are rarely devoid of some form of plant life unlike their counterpart in eastern Australia. Where the photograph was taken, I decided to camp in a very exposed area of stony country. The following morning, scanning for Banded Whiteface and possibly Inland Dotterel, I sighted a bird that I knew was in the region but thought this particular location was far too exposed for them, but from under one of the low bushes in the middle of the photograph walked out a Western Quail-thrush with its characteristic jerky head movement, uttering the typical high pitched Quail-thrush call when it stopped walking. It demonstrated that you could never know everything about what bird will reside where. The country is typical of that required by Inland Dotterel and Banded Whiteface, although Banded Whiteface exist in a variety of areas but the one common factor is that it is generally very open country, unless one is in the desert regions where they often are found in the swales of sand dunes. Open stony country is also favoured by Ground Cuckoo-shrike and Banded Plover.

WETLANDS

Perth wetlands

Before European settlement, the Swan Coastal Plain had an extensive wetland system. Many of the water bodies particularly the smaller ephemeral wetlands have been filled in, which has made the feeding areas for uncommon birds like Long-toed Stint, even more difficult to find. In my own lifetime here, areas that I would visit to see this species are now gone. Luckily a few lakes still remain. The majority of all the *Anatidae* family, which includes Ducks, Swans and Geese, can be found in Perth including Hardhead. Black Duck. Grey Teal. Australasian Shoveler. Pink-eared Duck. Australian Shelduck. Blue-billed Duck. Australian Wood Duck. Musk Duck and occasionally Chestnut Teal and less so our rare duck, the Freckled Duck. Other wetland birds to be found in Perth include Little Pied Cormorant. Little Black Cormorant and Great Black Cormorant. Yellow Spoonbill. Occasionally Royal Spoonbill. Australasian Grebe. Hoary-headed Grebe. Great Crested Grebe. White-necked Heron. Nankeen Night-heron. Great Egret and to a lesser degree Intermediate Egret. Straw-necked Ibis. Glossy Ibis. Australian White Ibis. In the reed beds can be found Baillon's Crake. Spotless Crake. Australasian Spotted Crake. Buff-banded Rail. Three Bitterns are threatened in the South West. Two have occurred in the Perth region, Little Bittern and Australasian Bittern. As they are very cryptic and elusive this makes them harder to see. In the mid 1980s they were in greater numbers than we think they are now. Swamp Harriers still frequent some of our larger wetlands particularly Herdsman Lake. On the muddy shorelines of lakes like Thompson Lake. Forrestdale Lake and Joondalup Lake, you may find Red-need Dotterel. Red-capped Plover. Marsh Sandpiper and Sharp-tailed Sandpiper may be found. In late spring, early summer Thompson Lake can still find Long-toed Stints and the occasional Pectoral Sandpiper and rarely, Ruff.

Paperbark swamps

For those willing to wade or use a small canoe, some of our dense paperbark swamps support the occasional Freckled Duck. Also if there are reed beds Crakes will allow closer viewing if you are actually in the water rather than approaching by land, a trick that a fellow birdwatcher told the author. Many duck species breed in paperbark swamps including Grey Teal. Black Duck. Hardhead. Wood Duck as well as cormorants, egrets, spoonbills and herons that make their rookeries in them.

South West wetlands

Wetlands like the Vasse-Wonnerup Estuary illustrated, support huge numbers of water-birds far more than the wetlands in Perth. In one count in January 1986, 33,000 birds were counted comprising 60 species and ducks and swans accounted for 12,500 of that number. These are very large numbers and show how important the South West wetlands are but there is immense pressure to develop some of the foreshores of these wetlands, which would greatly affect the number of species visiting this region.

Inland salt lake systems and surrounding samphire flats

After heavy rains, these inland playa lakes systems can fill often for several kilometres in some locations. When they do, literally thousands of birds will take advantage of these newly created wetlands. One does not always associate Black Swans with these inland areas but the photograph shows, there were many hundreds of swans breeding as well as numerous pelicans in the distance. The lake is Lake Annean south of Meekatharra. The surrounding samphire flats are the nesting locations for two specialised birds, the Samphire Thornbill and Orange Chat, as well as White-winged Wrens and Brown Songlark. To see Orange Chats, just sit patiently with binoculars and scan the tops of the samphire. Soon enough, one will pop up on top of the samphire showing its bright orange plumage.

Ephemeral lakes

These inland freshwater lakes can as the name implies, be short lived but like the salt lakes discussed above they can attract many species of waterbirds including Freckled Duck.

I have camped at one such lake on the Gary Highway called Lake Cohen. Several species of parrot took advantage of this additional water to nest nearby (parrots have to drink regularly) including Mulga Parrots, Major Mitchells and Budgerigars. Black-tailed Native-hen and several species of duck also were nesting there. If you check the rainfall on the Internet regularly, you get to know where certain areas will receive greater rains and in turn create these ephemeral lakes.

Mangroves

Mangroves stretch from as far south as Bunbury and then various locations all the way to the Northern Territory border. There are several mangrove frequenting species that occur in Western Australia's mangroves and they represent the vast majority of the Australian species that frequent mangroves including: White-breasted Whistler. Mangrove Golden Whistler. Red-headed Honeyeater. Canary White-eye. Little Bronze Cuckoo. The endemic Dusky Gerygone. Mangrove Gerygone. Mangrove Fantail. Striated Heron. Collared Kingfisher. Mangrove Robin. Shining Flycatcher and Black Butcherbird.

Reed swamps

In the South West, particularly close to the south coast are many wetland areas with extensive reed beds like the one illustrated. These swamps are almost permanently waterlogged. The swamp illustrated here is typical of dense reed bed that the Australian Bittern requires. Also the low heath that abuts the swamp has been the territory of one of our rarest land birds, the Western Australian species of the Western Ground Parrot, whose population is threatened.

Offshore islands

Western Australia is blessed with numerous islands stretching from the Recherche Archipelago in the south east, to the Kimberley and Northern Territory border in the north. There are in fact over 200 islands larger than 30 acres. On the south coast, some of the islands can be visited by tourist boats. One such island is Woody Island, just off the coast of Esperance. On a boat trip you can normally see Black-faced Cormorant, uncommon in Western Australia. The island is a good place to see Rock Parrot. Red-eared Firetail. Golden Whistler. Brush Bronzewing. Spotted Pardalote. Brown Quail and Sea Eagle. In winter and early spring Shy Albatross. Black-browed Albatross and Yellow-nosed Albatross can be sighted; the tourist boats do not always go out in the winter months, so do check.

Closer to Perth is Rottnest Island. Here, besides the numerous small sandy bays that occur, there are several saline lakes that can be very productive for birds, particularly Government House Lake. Lake Herschell and Lake Serpentine. Several migratory birds visit in late spring mostly common species like Ruddy Turnstone. Grey Plover. Red-necked Stint. Curlew Sandpiper. Grey Plover. Sanderling. One bird that is rare in Australia but has been seen over many years is the Red-necked Phalarope with its distinct high raised neck, its body twisting and turning as it takes insects from the surface of the water. It is also a very good place to see non-breeding Banded Stilt both in summer and winter, although numbers in winter are far lower and most are immature birds, the adults having left to nest in some far inland salt lake where rains have fallen. Often Red-necked Avocet are on the lakes and a small number occasionally nest on the island. Both the Avocet and Stilt feed on the brine shrimps in the lakes. There are a few pairs of resident Eastern Osprey that nest on the offshore limestone stacks. A few Reef Egrets are here but not common in the Perth region. Bridled Tern. Crested Tern and Fairy Tern have nested on the island. It's a good place to look for Rock Parrot and White-fronted Chat.

The Houtman Abrolhos archipelago off the Geraldton coast is one of the gems of Western Australia. Here the only breeding population of Lesser Noddies breed. The other subspecies nest thousands of miles away on the Seychelles and Maldives Islands in the Indian Ocean. There are some relatively large islands in the Abrolhos group, including Woody, North,

East and West Wallabi and Pelsaert Island. The author has visited East Wallabi, Beacon and Long Island but has concentrated most visits on the long island of Pelsaert, out of the main breeding season in April but mostly in the breeding season in November, when there are thousands of breeding Common Noddy and Lesser Noddy. The Common Noddy nests in the samphire and saltbush. The Lesser Noddy requires mangroves. To see thousands of birds that do not fear man is a real joy and privilege in these hectic times. On the island there are several other species to be seen including Roseate Tern. Bridled Tern. Caspian Tern. Crested Tern. Fairy Tern. Sooty Tern. The Roseate Terns nest on a small islet at the northern end of Pelsaert. Other birds seen include Red-tailed Tropicbird. Pacific Gull. Osprey. White-breasted Sea-Eagle. Pied Cormorant. Buff-banded Rail. Curlew Sandpiper. Grey-tailed Tattler. Ruddy Turnstone. Greater Sand Plover. Pied Oystercatcher.. On Dirk Hartog Island there is accommodation available at the station and you can visit the island from the township of Denham. On the island are two small passerines that are very interesting. The black and white form of the White-winged Wren subsp. leucopteris is truly black in colour the other specie of interest is the remnant population of Southern Emu-wren sub species *hartogi*. Further studies regarding this specie would not go amiss as the nominate race is nearly 600 km further south and occur in a totally different vegetation habitat with no interbreeding occurring. There are many more islands in the north of the state and ofcourse the Cocas Keeling Islands and Christmas Island south of Indonesia in the Timor Sea both great destinations to visit with several endemic birds to see as a bonus, discussed further in 'Where to watch birds' section.

Mountain and plateau country

The central ranges between Warburton and Docker River hold a variety of species that do not occur in the deserts that surround them. It is an area where Dusky Grasswren occur, also Black-chinned Honeyeater. Pink Cockatoo (Major Mitchell). Spinifexbird and Peregrine Falcon. The Hamersley Ranges in the Pilbara have the attractive form of the Striated Grasswren subspecies whitei. They tend to hug the rocky dry creek gullies and stony boulder strewn hills sides. Rufous-crowned Emu-wrens occur both on the lower plains as well as on the very top of the ranges. Hooded Robins are common in certain locations. Grey Honeyeaters have a preference for the mulga woodland and although not common can be found more easily here than many other regions. The hills and gorges of the north-western Kimberley are the realm of another grasswren, the Black Grasswren. The attractive race of the Variegated Fairy-wren subspecies *rogersi* shares the same habitat along with White-quilled Rock Pigeon. In the valleys between the hills are large permanent freshwater pools: Red Goshawk and the 'Silver backed' form of the Grey Butcherbird and Black Bittern all frequent these pristine locations, while the grassy slopes support populations of the now diminishing yellow faced form of the Partridge Pigeon.

Rainforests

The Kimberley holds the last remnant rainforests of Western Australia that once covered far more of the region but still they manage to hold typical rainforest frequenting birds such as Rufous Owl. Cicadabird. Rainbow Pitta. Rose-crowned Fruit Dove. Varied Triller. The lush Vine thickets that typically hug south facing hillsides hold small populations of Green backed Gerygone. Buff-sided Robin and the Kimberley endemic, the Kimberley Honeyeater.

MACRO HABITATS

Over the previous pages, larger general habitats have been discussed but what is of interest to some birdwatchers is that within a general vegetation zone, such as the Wandoo Woodland Region, there are differing micro habitats. The picture shows the typical dominant habitat of open Wandoo Woodland showing the Sandplain Poison (*Gastrolobium microcarpum*) in the foreground and Wandoo trees (*E. wandoo*).

If you walk through this country you may leave the taller Wandoo trees and come to a lower stratum of closed shrubheath with a few interspersed mallees as in the photograph. The experienced birdwatcher knows that possibly Southern Scrub Robin and Shy Heathwren may occur here and where the photograph was taken in Dongoloking Reserve they certainly did.

Walking only 30-40 metres from this point one looks over a large area of low heath as illustrated in the photograph. Besides several honeyeaters visiting the area like Tawny-crowned Honeyeater. White-cheeked Honeyeater. Western Spinebill and on the perimeter White-eared Honeyeater, there is another elusive ground-dwelling bird very similar to the Shy Heathwren, namely the Rufous Fieldwren. For many it is a real joy to know that such birds still survive within the Wheatbelt with all the environmental pressures of fire, fox and cat predation and loss of habitat by clearing.

Walking back into the open Wandoo one may come across a dense concentration of young Wandoo trees. This is often a favoured habitat for the Western Yellow Robin that loves to cling to the sides of the sapling in typical robin fashion, looking for moving insects on the ground.

Further on one will come into a very dense stand of sheoak growing on the shallow soils that surround granite outcrops. You rarely find Yellow Robin in sheoak thickets but you are more likely to get Red-capped Robin along with Grey Fantail, and on the edge of the sheoak is a favourite location for roosting Tawny Frogmouth. You will not find a great variety of bird life in here although you will see a good variety of orchids which love the acidic soil.

Leaving the Sheoak you may come to the long fence line that stops abruptly where the cleared country occurs. At first glance you may think there is nothing on the edge but it's a favoured area for the Jacky Winter, Red-capped Robin and Yellow-rumped Thornbill, all taking advantage of the open space but having the added benefit and safety of the sheoak and wandoo to return to. At night Barn Owls and Tawny Frogmouth use the fence posts as launching points to retrieve prey.

The point of the above, is to highlight the importance of knowing a little about what birds reside where and what habitat they require, as this helps you find the birds you know as well as birds that you may not have found before.

One of the joys of bird watching is to see a bird species you have never seen before. However when you first start seriously looking at birds you may find that what you thought was a Weebill was in fact a Western Thornbill, or the Scarlet Robin was a Red-capped Robin, or the White-naped Honeyeater was a Brown-headed Honeyeater, but you will soon learn to pick up the small features that differentiate one bird from another.

Then you reach another level of difficulty. Was the wren a Blue-breasted Fairy-wren or Variegated Fairy-wren, or was the bird you just got a glimpse of a Western Gerygone or the uncommon Grey Honeyeater? It really does not matter as long as you keep learning and building on your own knowledge base.

Equipment

Firstly, it is essential to purchase a pair of binoculars and there is a good reason for this. Birds have a tolerance distance. What is meant by that, is that birds will go about their business relaxed and unperturbed even when you are walking towards them, but suddenly they become alert and nervous when you get too close. The reason is that you are now a potential threat and they will either walk away or fly off. Here lies the problem for you. What was the colour of their throat or tail? Did they have markings on the wing? You couldn't answer these questions, as they were too far away to see such detail. Binoculars are the answer to this problem but what size to get is the real question. Binoculars come in various sizes and types. Modern binoculars have internal roof prisms, which makes them lighter and slimmer and easier to handle. They come in various sizes 7 x 30. 8 x 35. 8 x 40 or 10 x 40. The first smaller number is the magnification, the second and larger number is the field of vision. Generally speaking the 7 x 30s are good for backpacking and just general identification where it not vital if you miss birds; they are better than nothing, similar to those that you would take to a theatre. The minimum magnification is ideally 8 but the field of vision is very important. If you select say 8 x 30s you may have trouble getting onto a bird as you literally don't have much area to look at and it can be like looking through a keyhole rather than a small window.

So what is the ideal size?. Well this depends on a few things. Some people will always have problems getting onto a bird with binoculars, as the author knows from years of guiding.

If this is the case, it is better to cut down on the magnification to 8 but retain a good field of vision namely 40. I use 10 x 40 as I don't have a problem getting onto the bird but need reasonable magnification which is the '10' size .

How much do I have to pay? Well you can get cheap binoculars, but there is one thing you cannot do with them and that is to drop them or even just knock them as the prisms inside will almost certainly go out of alignment, although surprisingly your eyes will in fact try and adapt to this misalignment if it is not too great but it certainly is not good for your eyesight. Even the best binoculars will be damaged if you drop them badly but generally if they are top quality they will take reasonably rough wear and tear.

Obviously when you are first starting, don't waste money on the best equipment as you may not enjoy the hobby. Serious birdwatchers spend up to one and half to two thousand dollars but that may be the only major purchase they may ever have to make to enjoy this wonderful hobby.

Telescopes

Telescopes are used mostly by serious birdwatchers who are often particularly interested in wader and waterbird watching, although people also use them in rainforest birding or take them on trips when driving to aid viewing some distant bird. The telescope will more than triple the magnification of most binoculars, sometimes as much as five times.

If you do buy a scope, I find a 45-degree eyepiece far better than the horizontal type. This makes viewing up into tree tops much easier and better even for horizontal viewing, say for waders, as you don't have to lean down straining your back. It takes a while to get used to this, but once mastered is less tiring on the neck and back.

There is also the option of having a zoom eyepiece, say 20 to 40 times magnification. This can be useful if the birds are so far that you need every possible magnification you can get, although there is sometimes a degree of loss in sharpness at the larger magnification end of the zoom.

Clothing

We know birds can see colour, in fact studies have shown that some species can actually see into the ultraviolet part of the spectrum; that is to say they can see some colours that many mammals cannot see, including man. Colour aids identification of other birds not only for us but also for them. A Red-capped Robin can see at a glance that the head of a Scarlet Robin has a white dot, immediately warning him of the presence of that species; the Red-capped Robin will normally make a hasty retreat as Scarlet Robin dominates when territories overlap which is not often but does happen. Colour has aided that recognition.

If we wear bright colours like red or yellow shirts, we are bound to be seen sooner than if we wear sombre colours that match the bush, i.e. browns and greens. The clothes will not win you adulation on a catwalk, but will certainly help you blend into the background and allow you a much closer approach to many birds.

Identifying Birds

Identifying birds can become a daunting affair; for example even in this small publication there are 489 species illustrated, but with sharing knowledge, fieldtrips and perseverance you will slowly build up your knowledge and species list and, for many, the challenge is very rewarding. However, before you start travelling, it pays to read as much as possible and check the birds in your guidebook to build up pre-knowledge.

There are some important things to do when you first sight a bird you don't know.

1. First and foremost, get the colouring of the bird sorted out quickly. Where are the colours i.e. what is the upperbody colour? What is the head colour? What is the underbody colour? Then look for colour features, i.e. red on the throat or a white neckband. Are there markings on the wing? Often birds have wing bars called 'chevrons'.
2. What is the size of the bird? Sounds obvious but it eliminates the bulk of other birds if you gauge the size. 'How do I gauge the size'? When you first start looking at birds in your garden, you may notice that a Singing Honeyeater is smaller than a Red Wattlebird but is larger than a Brown Honeyeater. You see, you have already established the foundation of comparative size. So from your humble garden, you start the long process of establishing comparative sizes for all birds you see.
3. What is the shape of a bird? This is very important. Pigeons in general are thick bodied. All Fairy-wrens have small rounded bodies and long tails. Most Honeyeaters have a similar shape (not all).

When raptors are in flight, it is predominantly the shape and angle of wings that tells the experienced birdwatcher what species it is, particularly if it is flying against the light showing no colour or flying extremely fast.

When you see a bird very quickly flying by, you don't have time to get all the colour features and when you are starting you may not be able to identify the bird; but with experience and further sightings, next time when the bird is flying longer and closer, you may say 'that was the bird I saw the other day', confirming what you have seen. For experienced birdwatchers, shape is vital and often the term 'jizz' is used. It actually should be written 'gis' as it comes from the abbreviation for the words 'general identifying shape' which is a phrase used by fighter pilots who have been trained to identify the type of plane flying at them by its shape, and so be aware of its capabilities. So too will knowing the shape of the bird aid you with your identification.

4. What is the bird doing, how is it behaving? Is it climbing up a trunk like a treecreeper? Is it running along the ground not taking flight like a quail? This will help you when checking the guidebook.
5. What is the beak shape and how is it using it? If it is fine and curved it will most probably be a honeyeater. You may see a beautiful uncommon bird with a thick wide beak, prising bark off a Wandoo tree. The colouring is similar to the lovely Golden Whistler that you have become familiar with but this bird has a thick bill. What is it? Turning to the book will show you it is a Crested Shrike-tit and it was the beak that helped confirm what you had seen.
6. What is the length of the legs? This helps with all the plovers and waders; it is basic information but it lays the foundation to your early sightings.

All this sounds a lot to take in at the first sighting of a new bird. Some birdwatchers recommend taking a notepad and sketching and noting the detail. I prefer to use a mini tape recorder so I can still look at the bird not lose sight of it and can record on a tape in seconds what would take minutes by hand. However, I will never have the lovely art piece of a notepad that becomes the record for some good birdwatchers with beautiful sketches included to highlight various features.

The tape is mainly used while driving, obviously in remoter country where it is safe to do so with one hand firmly on the wheel. It is used when recording birds in certain locations when I don't wish to stop. It is also used to describe locations or features that I wish to revisit. At the end of the day, the diary is brought out and the daily notes are transcribed. It has been a wonderful aid for over thirty years.

When you have identified a bird, say in a tree and wish to show it to your friend, it can be frustrating if they can't even see it, let alone know what it is.

Time is critical in this situation as the bird may fly off. Having led large groups for years, the author finds by far the best way to get people to see a bird is to be specific and positive. If you just say 'it's in that tree over there', they may ask 'What tree'. The best way is to give quick instructions: 'See the broken off white dead tree ahead, go to the first tree on the right, go in about one metre from ten o'clock of the tree and there is a small gap, the bird is sitting on a branch in that gap'. So rather than refer just to a tree, as there may be many, its best to use some feature which stands out from everything else. It may be a large rock or a dead white tree trunk, which is surrounded by lots of green trees. This immediately locates the distance and area of view and its then easy to go to one of the trees near it. The face of a clock is used by many tour group leaders to describe the location of a bird within the profile of a tree. Giving the time location, and then distance in, helps tremendously, particularly if the tree is big and close, so 'big tree 30m along side fallen log (pointing), 10.00 o'clock, 3m in' end of statement. All this sounds so basic (remember twitchers, you were there once) but I have seen and witnessed people trying to direct other people to a bird and the bird has flown off long before their complex instructions were finished. Keep it simple and quick. If you are close to a skulking bird in low bushes and helping people, make sure people are still and silent and not throwing hands everywhere, then lower your voice and describe where a bird is. Just slowly point with a clear description and normally they will get it. Silence is so important when birding as sound is vital to birds and they may hear you before sighting you.

How to approach birds and where to find them

Try not to walk straight towards birds if you are in an open area but walk on an angle as birds find this less threatening. If there are trees or bushes to block your view make your way towards the bird by using them as a hide or cover.

One of the best methods to see birds is actually just to sit quietly somewhere and it is surprising how close birds will come if you are not moving and blending in with the scenery.

Most birds must come to drink somewhere particularly in warmer weather. If you sit near any small body of water well away, you will increase your chances of getting birds. Visiting a lake system will normally guarantee you some waterbirds. If you are really keen, you can walk with boots into areas of reed beds and you will be amazed how accepting of your presence birds will be particularly if you take a container and just sit; many of the waterbirds will become quite tame, particularly the crakes and rails and literally walk around you. (thanks Tony K).

Checking weather on the Internet before you plan a trip will let you know where the heavier rains have fallen and, in the appropriate season, birds will breed in profusion and make a trip most rewarding. If you don't have Internet connection try to find out where the rains have fallen in the previous months.

Try to know the seasons when birds breed as this will increase your chances of seeing more species, and also the birds are far more active and will often allow a closer approach as they are so intent on defending territory and being involved in the breeding process. In the deep South West the majority of birds will breed from September right through to December but in the mid west they will normally breed earlier from August through to October. However in the drier regions birds may breed in the middle of winter or even in summer after heavy cyclonic storms turn the brown spinifex green and the insect life multiplies even though the temperatures will be high.

Early morning is by far the best time to go bird watching and the first 3 hours or so will often give you twice to three times the number of species as in the mid afternoon particularly in hotter weather.

Bird songs as a means to identification

If the author had any special skills in bird watching at all, it may be his knowledge of bird calls. This has proved invaluable throughout Australia. All birds sing or make calls of some kind, it is their way of communicating and if you can build up knowledge of their calls over time, it is of great benefit. If I had any skills at all, it was the capacity to retain bird calls in my head having a musical ear I guess, so when getting off a plane in Cairns say with a new tour group, suddenly the memory starts kicking in and I start sorting out Graceful, Lewins and Yellow-spotted Honeyeaters for the group, so essential for species that look alike. Where it really helps is when you hear a call you're not sure of and that will often lead you to a species you may not have seen for a long time and often it's a rare bird. I well remember driving for the first time in the Iron Range when the calls of about eleven birds were well programmed into the brain from listening to tapes; when suddenly I heard a call, the car was brought to a halt fast and everyone leapt out. There above in a large tree were at least eight Palm Cockatoos, a wonderful new bird for all to see.

Identifying waders

This may not be the first group of birds that you will start to view, mainly because getting close to waders is not easy and those that do watch waders seriously tend to purchase telescopes, which is not an essential purchase until you really start to enjoy this hobby. However when you do start, the identification of waders is still not easy, particularly as most of our waders are seen in their non-breeding plumage and at best in their eclipse plumage. Also there are subtle changes in moult. A first year non-breeding bird retains its juvenile wing feathers before it develops full adult plumage. So in simple terms most of them end up being dull grey brown in overall appearance making the task even more difficult. Waders test even the most experienced birdwatchers but there are some points that may assist with the process of identification.

When there is a group of waders all together don't get daunted. It's actually better that way as you can compare sizes and shape.

First select the larger or more obvious waders like curlew or the godwits. This gives you a base from which to start. Size is the first important factor. How does it compare say with a godwit? There may be very small waders in the group, most probably Red-necked Stints, but there are some dumpy birds larger than the stint but smaller than the godwit. They could be Knot. It's not important when you start for the first time whether the godwits are Bar-tailed or Black-tailed or that the Knot may be a Red or a Great Knot. You are already recognising the genus to which the birds belong.

Now it is time to look at the guidebook. There are two Godwits together but one has a slightly upturned bill (not its best diagnostic feature although easy to see) but more importantly one has a darker duller upperbody with less white on the margins of the wing feathers (coverts and scapulas). It's not clear but it is just showing solid black where as the other is showing barring on the rump. Ah, the duller plumaged bird has just stretched its wings showing clearly a white rump and black to the end of the tail. That confirms it, this is the Black-tailed Godwit but what of the dumpy birds alongside, they seem to be two sizes but it's not as easy as the godwits. After looking at the book you decide they are Knots but which one is it ?. One is slightly larger and the bill is one and half times the length of its head (nape to forehead) whereas the other is certainly smaller and the bill is the same length as the head, so the smaller one must be the Red Knot.

You have now started the skilful process of identification, comparing one bird to another. It is much better to start with a group of waders first than a single bird. Also the more you view waders the easier it becomes. Sometimes I have not looked at waders for a long time, even though I have been viewing waders for many years, it takes a while getting back into it and recalling all the important features one knows.

How best to view waders

Most importantly try to view waders with the sunlight falling on them and from behind you and the lower the angle of light the better the view you will get.

Waders will disperse at very low tide but tend to congregate at high tide. Waders feed most on the receding tide soon after high tide and following the tide out.

Waders can be nervous birds with people getting to close. If you posses a light stool/chair take it with you and sit yourself down not to far from the waters edge and you will be amazed how accepting they can become. Sitting is always less threatening to both shore and land birds.

To summarise, knowledge of bird calls and knowing the jizz of a bird are the main tools that I could not do without. Developing a memory for what a bird looks like, particularly in flight will help you go a long way to identifying species particularly the difficult birds. It all sounds so easy forty years on from my first birds but with perseverance your knowledge will grow ten fold so never give up.

BUSH SAFETY AND AWARENESS

Many areas mentioned in this book, particularly in the Mulga Region, are fairly remote and one must be extremely careful to avoid getting lost or disorientated. It is tempting when seeing a bird briefly while driving, to suddenly stop and quickly get out of the car and start following it into the bush. It is at this point that you may get into trouble. However tempting it is, you must first take bearings of where you are and where you intend to go.

We live in the southern hemisphere, so familiarise yourself by knowing that the midday sun will be in the north. In winter it is easier to relate to north as the angle is lower but in the summer the angle is higher and it is not so easy. Knowing where north is will certainly help with knowing where you are. However it's meaningless if you have not registered where you entered the bush. Did you enter an area walking south or north-east or west?

However experienced you are, it is essential to take along a compass. Even though I have been walking in the bush for nearly 40 years, I still check a few facts before setting off in remoter country. Get out your compass and check in what direction the road you are on runs. Take a bearing on the road. To make it easier to assess where you are. If it runs east – west you will either walk north of the road or south. If the road is running in a north-south direction register if you will walk east of the road or west. You then can set off in a much more relaxed manner knowing that you can keep walking even in the thickest of woodland, and when it is time to return, just check your compass and then head back. A word of warning, it's easy to be too relaxed and you can forget - did I enter the mulga woodland from the east or the west? If the sun is out it helps but it still does not help if you forgot what direction you walked in. Some of us are lucky and have a built-in memory map of all the main tracks and where they lead to, but not everyone has this. I carry a GPS which of course makes it so easy to return to your original starting point. I rarely use it to guide me back, more often for registering the location of a bird, but it certainly helps when the direction you walked in is not a simple north-south or east west. Even if you possess a GPS, always carry a couple of batteries as a back up and pocket compass as back up.

A simple word about walking at night in the bush: sure if the sky is clear and you're one of those few who know the stars and where they are at anytime of the year, then it may be all right; but if you're not and it is an overcast night, be careful. Most of you would never walk in the bush in the dark anyway. Even leaving the tent with a torch to go to the toilet in the night can lead to disorientation, when there is no full moon or an overcast sky. Ideally leave a light on near your tent to go to the toilet in the night. With no full moon or an overcast sky, even with a torch, it's easy to get disorientated, so be very careful when it's pitch black and ideally leave a spare light somewhere near the tent.

So please be careful when travelling to remoter areas.

LOCATIONS TO BIRD WATCH IN WESTERN AUSTRALIA

Having run bird tours for 20 plus years, one certainly gets to know where most of the best locations to view birds occur. I am often reluctant to mention them all as people have differing needs or likes but some of the better areas are described here in general terms only. As much as I would like to assist in giving specific locations particularly for rare birds I have found the over use of tapes or the constant visits to a specific areas can have a drastic effect on the population on specific birds. Some birds like Noisy Scrub-bird or Rufous Scrub-bird in popular localities have been 'taped out' or the vegetation down trodden and compromised.

Some 'Where to watch' guide books have written so specifically that they state 'walk up to this particular tree and you will find such and such a bird'. Imagine what that does to a rare bird particularly the nocturnal species that may have to fly off and then be harassed by every diurnal bird in the area. I'm not trying to denigrate the writers, it is more about changing the ethics of twitchers to show respect for our fragile environment and the birds' welfare. So, alas, you will only get general areas from me but it makes finding a rare bird by working at it far more rewarding. To the listers who use the services of professional guides, go easy on them, most are trying their very best. I well remember a tour where we had seen Red Goshawk, Black Grasswren, Yellow Chat, most of the mangrove birds in just over an hour and days later many other Kimberley specials. It was then time to now go down the Canning Stock Route. We had several separate sightings of Princess Parrots (our target bird). Rufous-crowned Emu-wren. Orange Chat. Banded Whiteface. Grey Falcon. Black-breasted Buzzard. Major Mitchell. Letter-winged Kite. This was not a bad selection for the arid zone that does not have the species count as the Strezlecki or Birdsville Tracks. On one of the last days around the campfire someone was complaining that they had not seen a specific species after all the wonderful special birds we had seen. Luckily for me a lovely lass leaned over my shoulder and said 'Simon, there is no pleasing some people'. It put it all into perspective for me, so relax and just enjoy this wonderful hobby of ours.

Perth to Dryandra Nature Reserve and the Stirling Range

If you are a visitor to this state and your time is limited and your primary objective is to see most of the endemic birds of the WA, then you must travel down to the south coast east of Albany. A typical seven day journey from Perth could take you via Dryandra Woodland Reserve located between the towns of Williams and Narrogin. There is basic accommodation in the old foresters' houses within the reserve itself. For more upmarket facilities one would have to stay overnight in Narrogin. There are some farm stays adjacent to the reserve as well. Several endemic birds and near endemic birds can be found, as well as a multitude of other species. The endemics include Western Thornbill. Western Spinebill. Short-billed Black Cockatoo. Western Rosella and the near endemics, like the Blue-breasted Fairy-wren. Rufous Treecreeper and Western Yellow Robin and the uncommon subspecies of the Crested Shrike-tit *leucogaster*.

The next area to visit on your way south could be in the Stirling Range, where a very good camping and accommodation facility is located, namely the Stirling Range Retreat. The staff are most accommodating and know where many of the bird species occur. In birding terms it is a reasonable area and many occur within the camping and accommodation grounds; Floristically this region is one of the flora hot spots of the whole state, being incredibly rich particularly in regards to the quantity of terrestrial orchids. Possible birds to see: Elegant Parrot. Rufous Treecreeper. Varied Sittella. Square-tailed Kite. Little Eagle. Western Whipbird. Southern Emu-wren. Peregrine Falcon. Crested Shrike-tit. Western Rosella. Regent Parrot and Owlet Nightjar.

Stirling Range to Cheyne Beach and Albany

Not far from the Stirling Range is the Porongurup Range with a totally different suite of flora including some impressive Karri trees in one of the most eastern populations of this wonderful tall tree. There are bird species here accustomed to the wetter South West forests. You may go onto Cheyne Beach or decide to stay here. There are camping and accommodation facilities and tearooms in the area. From Porongurups or the Stirling Range you can travel down to Cheyne Beach where there is camping and basic accommodation in the form of self contained chalets. Here several of the rare endemics are located, including Western Bristlebird. Noisy Scrub-bird also other endemics like Red-eared Firetail. White-breasted Robin. Red-winged Fairy-wren and the near endemic Western Whipbird. Sadly, we think that the Western Ground Parrot has gone from the Waychinicup area near Cheyne Beach but it still may remain in some localities nearby.

Not far from the beach township of Cheyne Beach is the large town of Albany; near here is Two Peoples Bay and it has all of the birds listed above for the Cheyne Beach and Waychinicup area. Albany itself has some good birdwatching localities, particularly some wader watching areas such as at Emu Point. The walk along Middleton Beach to Emu Point; is a good site for Southern Emu-wren and Rock Parrot. Not far from here is Lake Seppings, which is a very good place for Red-winged Fairy-wren. Western Rosella and White-breasted Robin. I have had a species of snipe here, most probably Swinhoe's Snipe. Except in the Kimberley, Snipe are very uncommon in Western Australia unlike the eastern states. Musk and Blue-billed Duck occur on the lake. Across the road is the golf course and Rock Parrots often feed on the grasses here but do seek permission to walk around the perimeter to avoid golfers.

Hyden and beyond

If your visit to the state was brief, then you will have to return to Perth hopefully after seeing many of the endemics; but if you have more time, you can combine what you have just done over the last few days and, instead of returning to Perth, you can travel to Hyden, either along the coast via the magnificent Fitzgerald River National Park with its 1800 plus species of plant life or head directly north east, wandering through the Wheatbelt Region to Hyden where there are camping, chalets and also motel facilities. Due east of Hyden past the Vermin Fence are some interesting habitat zones. The open heath has Blue-breasted Fairy-wren. Rufous Fieldwren. Tawny-crowned Honeyeater and occasionally Black Honeyeater. In the sand pits White-backed Swallow. Further into the Salmon Gum Woodland with melaleuca understorey there are Chestnut-backed Quailthrush and Gilbert Whistler. In the denser vegetation, Shy Heathwren and Southern Scrub-robin. Other birds include Regent Parrot. Purple-crowned Lorikeet and Square-tailed Kite (spring–early summer). None of these birds is rare but the habitat is more pristine than the Wheatbelt although, sadly mining is moving into many more areas in a big way now and it is losing that remote feel.

The Kalgoorlie region particularly north to Goongarrie Nature Reserve has lots to offer; you are now entering the marginal areas that abut the true desert regions. Goongarrie Nature Reserve is a good place to see White-browed Treecreepers. In the samphire after good rains Orange Chats can be seen. It is not uncommon to start seeing Budgerigars and Cockatiels here. Keep your eyes out for Scarlet-chested Parrot as they have been sighted in this part of the far eastern Goldfields. Heading north to Laverton and the deserts takes you to wonderful country but refer to the section on the 'Deserts'. North-west of Coolgardie is the nature reserve of Rowles Lagoon. It is a good place to see many inland species of duck and it is one of the few wetlands where our rarest duck is seen regularly, the Freckled Duck. The eucalypts of the Goldfields region are exceptional and more species of eucalypt can be found around the Kalgoorlie area than anywhere else in Australia. Both south and north of the Great Eastern Highway near Boorabbin, many areas have Gilbert's Whistler and Chestnut-backed Quail-thrush. Purple-crowned Lorikeet and Regent parrots.

South of Kalgoorlie via Norseman is the Esperance region with its stunning beaches and rich flora heath. There are some very good wetlands namely Lake Wheatfield, Lake Warden and Lake Windabout. Some of the greatest numbers of Hooded Dotterel in the state and in fact Australia are found here, particularly at Lake Warden. Near the car park at Lake Wheatfield there are often many Western Wattlebirds feeding on the Showy Banksia (*B. speciosa*). Cape Barren Goose can be found on the grasslands on the small farmlets on the edge of town as well as the Esperance Golf Course. The islands off the township of Esperance have several species of birds including Sooty Oystercatcher. Cape Barren Goose. Black-faced Cormorant. Rock Parrot. White-breasted Sea-Eagle, even Red-eared Firetails. Golden Whistler and Brush Bronzewing on some islands such as Woody Island.

Hyden to Wubin

If you choose not to venture east of Hyden, you can head due north-west making tracks to any of the lower mid west towns such as Wongan Hills or Dalwallinu. There are several reserves in this region and Western Corella subspecies *derbyi* is found here. Some of the populations of Redthroat closest to Perth can be found in the Wongan Hills area. Floristically, it is very rich even though the land is predominantly cleared. The roadside verges just north of Wubin travelling to Perenjori have some of the highest species counts in Australia and are best seen in late August to early October. The New Norcia area is not only worth visiting for historical reasons but it also has a population of Western Corella, so too has Moora.

Wubin north-east along the inland highway

From the mid west area you can leave the Wheatbelt Region and pass through the Transitional Woodland area north east of Wubin on the Great Northern Highway that eventually goes all the way to Newman and then to Port Hedland via the inland route.

Approximately 40 km from Wubin you pass through patches of Transitional Woodland and also thick melaleuca heath in between each area of open woodland. Stop in the woodland off the main road and several species of parrot occur including Regent Parrot. Red-tailed Black Cockatoo (subspecies *samueli*). Purple-crowned Lorikeet, and best of all, Pink Cockatoo, always a stunning sight (you obviously may not always get these birds when there). Leaving the Transitional Woodland just north of the Ninghan Station turn off, you enter the vast region of the Mulga Woodland; obviously the region is not restricted to the core vegetation type of mulga but has other diverse habitats, including breakaway country, salt lake systems with samphire as well as open stony plains.

In this region, a wealth of bird life awaits you. Common birds like Chestnut-rumped, Yellow-rumped Thornbill. Slaty-backed Thornbill (not as uncommon as people think in some areas). Inland Thornbill. Weebill. Western Gerygone. Crested Bellbird. Spiny-cheeked Honeyeater. White-plumed Honeyeater. Bourke's Parrot. Budgerigar. Mulga Parrot. Torresian and Little Crow. Then less common birds such as Redthroat. Spotted Bowerbird. The further north one travels, other species become more common such as Diamond Dove. Cockatiel. Chiming Wedgebill and less common species like Banded Whiteface (Nallan Station southern boundary). Inland Dotterel. Grey Honeyeater (Paynes Find area) and White-browed Treecreeper (Before Pindabunna T/O). The raptors may become more plentiful with Whistling Kite. Little Eagle. Brown Falcon. Kestrel and Wedge-tailed Eagle being some of the more common and then Black-breasted Buzzard and Spotted Harrier less so. When the Mulga is ablaze with colour from the everlastings, Crimson Chat. Pied Honeyeater. Black Honeyeater breed in profusion. Budgerigars can turn up everywhere and the cuckoos such as Pallid Cuckoo, Black-eared Cuckoo and Horsfield's Bronze-Cuckoo centre on these areas to parasitise many of the passerine species.

Many of the stations in this region offer basic facilities and I would far rather share an evening meal any day with a station owner, than be stuck in a five star plastic hotel in a city. The bonus of staying on a station is that they will sometimes (mostly) allow you to visit some of the interesting areas on their property such as the granite outcrops or wetlands (often rare in this region) which you would otherwise never have been able to see or visit. Obviously be 'bush wise' about where you are, as station owners have one major fear and that is you getting lost in the often featureless sea of mulga; so do register the tracks indicated and most station owners will give you a small printed map of their station or at least sketch locations. One such station is Nallan Station just north of Cue where the people are very friendly and the mulga bird life is very good. It is almost one of the southern limits of Banded Whiteface. Also it is a good location for Grey Honeyeaters and Slaty-backed Thornbill. Around the homestead, Western Bowerbird is common and some of the nearby water tanks have Bourke's Parrots coming to the troughs just after sunset and just at first light before sunrise.

Wubin to Mullewa

If you have chosen not to travel this time on the inland highway, you may have decided to go north on the Dalwallinu, Wubin, Perenjori, Morawa and Mullewa road or the New Norcia, Moora, Three Springs road. As mentioned before, the road between Wubin and Morawa is one of the richest flora drives, with many small towns that one can stay in. From Mullewa due east is the small town of Yalgoo. In the station country here is a good place for Grey Honeyeater and Slaty-backed Thornbill. On the samphire flats south east of the townsite, Orange Chat and Brown Songlark can be found when the season is good; and occasionally the uncommon Samphire Thornbill has been sighted here although I have not seen it in this particular locality. There are a few sheep stations particularly north of the Mullewa-Yalgoo road that have tourist accommodation on their property.

Mullewa to Murchison Settlement and Gascoyne Junction

Another wonderful drive takes you through the small Murchison Settlement on the road to Gascoyne Junction. In the winter months, although cold at times particularly at night, the weather can be glorious during the day with clear skies and temperatures in the low twenties. The best time to visit is mid August (still cool) through to the second week of October. The main breeding season occurs from July to October. Areas like Tallering Peak (may be no access now-mining) have Western Quail-thrush. Peregrine Falcon and Little Woodswallow.

The Quail-thrush becomes quite common in the area around the turn off where the Butchers Track commences to go to the Coastal Highway. Wooleen Station has very good accommodation facilities in the form of quality rammed earth chalets. Wooleen is perhaps one of the finest birding areas in the mid west, as it has such a diverse range of habitat, not least a wonderful wetland, Wooleen Lake with its beautiful white barked Coolibah trees (*E. victrix*). Most years there is some permanent water there and the adjacent samphire has populations of Orange Chat and lots of Brown Songlark. All the inland duck species have been recorded here. The Mulga Woodland in a good season is wonderful and White-browed Treecreepers and Ground Cuckoo-shrikes breed here.

Perth to Geraldton

On this route, there are countless areas you can visit and some of the major reserves have been located on the northern map.

On the coastal belt just north of Perth, around the Mt Lesueur area there is a cross-over region for Blue-breasted Fairy-wren and Variegated Fairy-wren so be careful of too hasty an identification. In the small wetter vegetated gullies around Hill River there still occur small populations of White-breasted Robin. The area has also several populations of Western Corella near Mt Lesueur. You may combine your birding with viewing the Pinnacles in Nambung National Park. On the coastal dunes Southern Emu-wren and Rufous Fieldwren occur. White-backed Swallows can be found in some of the sandpit areas on the road to Jurien. Fairy Terns and Rock Parrots occur all the way from Lancelin up to Leeman on some of the beaches.

Ellendale Pool about 20 km north-east of Walkaway (gets a little overcrowded at the weekend so try midweek) has Peregrine Falcon and Variegated Fairy-wren. A pleasanter and less frequented area is Coalseam Gorge about 27 km north east of Mingenew. Peregrine Falcon is resident there; it's a lovely place to camp and in early spring it is an oasis for everlastings in the centre of a cleared farming region and several mulga frequenting species of bird manage to survive here.

Out of Geraldton is the wonderful Abrolhos Islands and if you contact the tourist centre in Geraldton, they may be able to put you onto a charter boat that may be going out to one of the islands, but do this well ahead of your travels as they are not visited regularly. People pressure has to be avoided, so going with a legitimate charter group or tour is the only way to visit these wonderful islands.

Kalbarri – Denham

North of Geraldton is Kalbarri National Park and the township of Kalbarri with lots of tourist facilities and a tourist centre. The park has one of the richest floras in Western Australia and tends to have vibrant colours longer than many other areas, as the Banksias and some of the Verticordias are in bloom right up till Christmas, although the temperatures will be warm then. The park has lots of Rufous Fieldwren. On the top of the gorge face of the Murchison River at many of the lookouts, there is a surprising array of birds at times, more so than in the heath itself; birds like Black Honeyeaters, Southern Scrub-robin, Redthroat and Black-eared Cuckoo and Horsfield's Bronze-Cuckoo have been seen by me close to the top of the gorge.

Leaving Kalbarri National Park and only 14 km along the North West Highway, one crosses over the Murchison River. It is not a favoured area of mine as there can be a lot of overnight stayers there, but it is worth a stop for a while as there is almost always some permanent water in the pools just upstream from the bridge. This is a major attraction for birds in the semi-arid region and even Grey Falcon has been sighted here, one of the most southerly sightings for this species. You may also see your first Diamond Dove or Peaceful Dove here on your journey travelling north.

Further north, you pass through some very rich flora heath with lots of Ashby's Banksia (*Banksia ashbyi*). Here in spring occasionally lots of White-cheeked Honeyeaters can be found, at their very northern limit; however a more interesting zone will soon be reached, just 40 km from the bridge near the turn off to Eurardy Station. This is part of the narrow northern band of the Transitional Woodland that extends all the way south-east to Norseman and beyond. It is home to a large population of Red-tailed Black Cockatoo subspecies *samueli*. You can stay on Eurardy station run by the Bush Heritage Council of Australia, either camping or staying in basic accommodation. The camping/accommodation area is a little exposed but it is a base from where you can visit parts of the Transitional Woodland; lots of semi arid species can be found here but the big flocks of Red-tails are the real bonus along with reasonable chances of Regent Parrot.

Further north starting south of Billabong Roadhouse and past Overlander Roadhouse opposite the turn off to Shark Bay, is one of the most spectacular regions to see everlastings in full bloom but it has to be a season when reasonable to good winter rains have fallen, this is normally between late July to mid September early October. In recent years this has not always been a regular occurrence. When it is, if you park well off the highway (do refer to the section on bush safety) or turn off the highway on the Butchers Track for a few kilometres, you will find a wealth of 'Mulga Region' birds particularly Pied Honeyeater. Crimson Chat. Rufous Songlark. Splendid Fairy-wren. Chiming Wedgebill. Crested Bellbird. Several of the Thornbills but less so Slaty-backed.

If you turn off to travel to the township of Denham in Shark Bay you begin to pass through some different country; firstly there will be a few kilometres of more open

Mulga, often with large clumps of one of the acacia loving mistletoes that parasitise the acacias and this is always an attraction for a variety of honeyeaters including Spiny-cheeked Honeyeater. Pied Honeyeater and Black Honeyeater. Soon you come to an expanse of samphire at the base of Hamelin Pool. If the samphire is lush after reasonable rains, Orange Chats may be here. If Monkey Mia is your destination, then you will be aware that this is the territory of Thick-billed Grasswren and around the car park and in the dunes you should have little trouble finding these little birds as well as Variegated Fairy-wren, White-winged Fairy-wren and Chiming Wedgebill. At the northern part of Peron Peninsula in the larger areas of samphire, Slender-billed Thornbill do occur here but be careful as these tracks are for 4WD drive and are particularly sandy in many areas and even tricky for experienced four wheel drivers.

Denham to Carnarvon

Back on the North West Highway heading north, closer to Carnarvon, you come to a turn off to New Beach and Bush Bay. Only 6-7 km down this dirt graded track you find a band of low mangroves running north and south. There is only one species of mangrove here, Avicennia marina, it is a low mangrove at most 4 metres high. Unlike the larger mangroves further north and in the Kimberley, like Avicennia eucalyptifolia and Rhizophora mucronata that hold a larger diversity of mangrove bird species, these mangroves have only four species specialising in feeding in mangroves, namely White-breasted Whistler. Mangrove Grey Fantail. Dusky Gerygone (Endemic) and Canary White-eye. Getting the White-breasted is not always easy here. Slender-billed Thornbill have been sighted in the samphire flats behind the mangroves, although in the last few years I have not managed to see them here but obviously it depends on the seasons as they do move.

Back on the highway, it's a short distance to Carnarvon and here one is truly aware that you have entered the more tropical regions of Australia and the bird life can mirror that. It is one of the most southerly towns that occasionally can get some of the rarer vagrant waders and uncommon waders like Little Curlew, Broad-billed Sandpiper. Also vagrants like Yellow Wagtail come as far south as Carnarvon. Carnarvon is one of the southern limits of White-breasted Woodswallow. Lesser-crested Tern and Brahminy Kite.

Carnarvon to Gascoyne Junction

Driving east from Carnarvon to Gascoyne Junction the road runs parallel with the large riverine system of the Gascoyne River. The River Red Gums that line the river are typical of trees along the bigger rivers of the mid west and Pilbara region. Lots of raptors, such as Little Eagle, Whistling Kite, Black-breasted Buzzard, Brown Goshawk, Collared Sparrowhawk and Wedge-tailed Eagle are regularly seen along this watercourse. There are no Laughing Kookaburras here; this is the natural habitat of the Blue-winged Kookaburra and they are common in this region. Taking side-tracks off the Carnarvon – Gascoyne Junction road and driving only a few kilometres south, say on the Yalbalgo North road (be careful this is reasonably remote country), you enter very good Mulga Woodland. In good seasons it is a good area for Black-eared Cuckoo and all of the common mulga species including the less common Slaty-backed Thornbill. You can decide to travel all the way around to Mullewa, travelling via Gascoyne Junction and Murchison Settlement. There is fuel at both these small townships. Up on the top of Kennedy Range is a disjunct population of Rufous-crowned Emu-wrens as the top plateau has sand dunes.

In the Cape Range NP, there are some isolated populations of Striated Grasswren, also Western Bowerbird subspecies carteri, Spinifex birds and also Rufous-crowned Emu-wren. Other northern species like Red-browed Pardalote and Painted Finch can be seen. Mangrove Robins have gone from the north-west Cape but are found in the mangroves at the base of Exmouth Gulf at Giralia Station and also at Bullara Station where you can stay, although not so easy to access the mangroves. Also White-breasted and Mangrove Golden Whistler and Dusky Gerygone.

The Hamersley Range is a spectacular mountain range and special birds like Spinifex bird. Rufous-crowned Emu-wren and Striated Grasswren occur here in greater numbers than the Cape Range. The Striated Grasswren can be seen in gullies near the mountain, the 'Governor' or the Weeli Wolli area, also on the hills not far west of Newman. Around Tom Price and the mulga woodlands just north of Auski Roadhouse are good places to look for the uncommon Grey Honeyeater. Hooded Robins are common at the base of the mountain the 'Governor'. Painted Finches are common in certain areas. At Millstream you can see Star Finches near the park head quarters sometimes. I have seen Black Bittern in the large paperbarks here. Grey Falcon is sighted regularly in the Pilbara and most are in the Marble Bar to Nullagine area and east to Carawine Gorge. Also seen near the Fortescue Marshes. Flock Bronzewing occur in the lower grass plains of the Maitland River and also just east of Roebourne around the T/O to Python Pool but this is very much a seasonal occurrence. Point Sampson is the most accessible area to see Mangrove birds. Painted Finch fairly common around Python Pool area and at the campsite, where they will come into the fire places and take carbon here. Striated Grasswren are in the area mainly in the gullies. Spotted Nightjar on the short gravel road into Python Pool some evenings.

The West Australian Deserts

Besides the Kimberley, perhaps the deserts are my favourite habitat type as I often choose to be where few people go. Where does one begin. The Victoria desert past Yeo Lakes may deliver you some Scarlet-chested Parrots, the bird photographed here was in the first sand dune country encountered after Bishop Riley's Pulpit on the Anne Beadell Hwy. In the Mulga and mallee Grey-fronted Honeyeater become one of the more prolific Honeyeaters as you enter the Great Victoria. In the mulga around Yeo Lakes Slaty-backed Thornbill occur. On the Connie Sue about 40 km north of Neale Junction you will start to enter sand dune country and Striated Grasswren and Rufous-crowned Emu-wren are here. Princess Parrots have been recorded east of Neale Junction many times.

Along the Gary Hwy you enter the Gibson Desert, around Charles Knob Western Quail-thrush can be seen as well as Slaty-backed Thornbill. Around Lake Cohen if there is water in it, many parrots will congregate including Pink Cockatoos and Mulga Parrots. White-browed Treecreepers at the T/O to Mc Pherson's Pillar (a lovely drive out there). North of Lake Cohen in the middle of desolate open plains with just a few low mallees, surprisingly there are a few Chiming Wedgebills.

In the Great Sandy Desert as many know, you have good chance of seeing Princess Parrot, particularly about 80 km due east of Gary Junction and then further on over the border near Jupiter Well. I actually am not a fan of the Canning Stock Route as so many people use it and obviously it's never graded and it punishes your 4WD big time, so I tend to use feeder tracks and those hardly used and I also come down from the north from the Breadon Hills always turning back at Lake Tobin. The Breaden Hills are great. Grey Falcon sometimes on the plains along side the peaks and also along Sturt Creek. Painted Finches in the hills and surprisingly Pheasant Coucal in the long grasses leading into the valley to Breaden Pool. Spinifexbird are relatively common in the plains north of Breaden Hills. Lake Tobin occasionally has Orange Chats. Princess Parrots over the years particularly in the 90's. The country from Well 45 down to Well 41 is classic Princess Parrot country with Desert Bloodwoods on the ridges and many desert seed producing plants in the wide swales. Masked Woodswallows can congregate in their hundreds here. Rufous-crowned Emu-wrens abound in the big domed Spinifex. In the occasional burnt country Banded Whiteface can be common.

FOOT NOTE:

I am sure you will be aware but travelling in deserts is not for the faint hearted. Vehicles should be well prepared and you most probably know but do travel with at least one other vehicle.

The Kimberley

A wonderful region alas becoming a very 'wanted' place to visit with many tour buses travelling through the central Gibb River Road. However there are ways to avoid the larger visited areas like Manning Gorge. The best thing to do is check the various stations that allow camping or accommodation. There are several benefits to this. One, the managers or owners generally know their areas well and can direct you to specific locations on their property where fewerpeople go. Remember about 40-50% of people will hurtle through the centre of the Kimberley not allowing themselves quality time to turn off into the many wonderful stations that exist off the Gibb River Road. So if you do venture into these stations, the numbers of tourists will be less.

Broome is an exceptional place. In the town there are some very good birding guides no more so than George Swann who knows the Kimberley far more than I ever will. Roebuck Bay is world famous to the wader fraternity. The numbers of wadersin the bay can be huge although alas some drastic falls in world populations are occurring right now as I write.

In the bay over 10,000 Great Knot can be seen in early summer. If you stay at Broome Bird Observatory even if just camping, book well in advance and do try and stay to your booking date in fairness to the staff. Just metres away you have waders on your doorstep. Not only the more common waders can be present but species such as Asian Dowitcher. Common Redshank (not a common bird in Australia) and Broad-billed Sandpiper can be seen. The range of waders is large. Not far inland wetlands like Lake Eda near the Roebuck Plains can produce some very uncommon species like Pectorial Sandpiper, the rare Spotted Redshank or uncommon Long-toed Stint also Pintailed and Swinhoe's Snipe. In winter there can be big numbers of Flock Pigeon. Also look out for Garganey but you really need to be up here late in the year to catch these rare vagrants. Little Curlew start arriving early October sometimes before and later Oriental Plover. Also Channel-billed Cuckoo and Koel are coming down. Lake Eda is on private land and must be visited with people who have approval to do so. If the Roebuck plains have water on them it's possible to see many Yellow Chats in some years. In the dry Australian Praticole can be found in large numbers. Behind the Observatory on the open plains you may be lucky to flush Chestnut-backed Button Quail. I have seen Letter-winged Kite on several occasions on the plains sometimes roosting in paperbark trees sometimes in the early evening hunting. With the onset of the hot months Oriental Pratincole start to make their arrival often in their hundreds. They will fly close to the tree tops along the water's edge mid day hawking for insects.

The Dampier Peninsula at the very top is good chance of Little Terns and even Rose-crowned Fruit Dove. Red-back Buttonquail are more common in many of the dry grasslands below the cape. Derby is always worth a visit either at the mangroves possible for Collared Kingfisher and Great-billed Heron. At the water treatment over flow for snipe and Long-toed Stint. Spotless Crake and Baillons Crake. Yellow Wagtails. Next possible area to visit would be Windjana Gorge, not only for its beauty but also to see Green-backed Gerygone and possibly Figbird but do get up very early before the tourists start walking in the gorge.

The real character of the Kimberley for me is when I pass through the limestone ranges of the Napier Range and start to climb up into the Leopold Ranges. If you head off to small side pools you may see your first Gouldian Finches. Further along the Gibb River Road Crimson Finches and Pictorella Mannikins may be sighted in the rank grasslands. A visit to Mornington Station may turn up Gouldian Finch anyway. At Annie Creek Purple-crowned Fairy-wren and Azure Kingfisher can be seen. Red-tailed Black Cockatoos are common in this area. Grey Falcon has been sighted on the drive into Mornington. Stop on the Gibb River Road when you see large Cadjeput trees, several species occur in these locations like Black Bittern and if there are flowering trees Banded Honeyeater. Bar-breasted Honeyeater and Rufous-throated Honeyeater can all feed together. If you are visiting Mt Elizabeth Station Northern Shrike-tit can occur in the woodland driving into the homestead. Grey-fronted Honeyeater are common.

In the future the Walcott Track may not be accessible but if the Laceys are still running Mt Elizabeth you can pay a small fee and gain entry to one of the last pristine areas in Australia. It's a rugged 4WD track but the rewards are Black Grasswrens. Partridge Pigeon. Possibly Rufous Owl and certainly Kimberley Honeyeater and if you are really fortunate you may get to find a Red Goshawk.

All these species can of course be found on the Mitchell Plateau including Rainbow Pitta and Grey Goshawk. If you do go to the Plateau check the pandanus lined creeks for Purple-crowned Fairy-wren between Drysdale Station and the T/O to the King Edward River.

The Kununurra region has so much to offer particularly in terms of waterbirds. White-browed Crakes are seen near Kona Caravan Park as well as resident Buff-sided Robins. Yellow Orioles and Shining Flycatcher. The grasslands near the town may produce Chestnut Mannikin and the far less common Yellow-rumped Mannikin.

Mirima (Hidden Valley) NP very close to town will certainly get you Sandstone Shrikethrush and a good chance of White-quilled Rock Pigeon. Parrys Lagoon north of Kununurra has lots to offer. Not only Golden-headed Cisticola but the less common Zitting Cisticola. If the plains have the right amount of water Yellow Chats occur here and I have had them breeding here. Black Falcon are seen here occasionally not a common raptor for WAY unlike the south east Australia. Most of the sub-tropical and tropical waterfowl occur here. In the woodlands not far from the lagoon Black-chinned Honeyeaters feed. Gouldians are also found in the Wyndham area.

Christmas Island

The island is a long way from Australia, but for keen birdwatchers it is certainly worth it. Flights leave from Perth and also Singapore. To see so many Frigatebirds and Tropicbirds flying around parts of the island is always a thrill. There is accommodation at the main town, simply known as the 'Settlement'. The island is easily traversed and you can hire a vehicle. Birdwatchers come for the 5 endemics: Abbot's Booby. Christmas Frigatebird. Christmas Island Imperial Pigeon. Christmas island Whiteye. Christmas island Hawk-Owl. Surprisingly the IOC has classified the Christmas Island Goshawk as a subspecies of the Brown Goshawk subspecies *natalis*, which I question. I think Christidis and Boles were closer in classifying it as a subspecies of the Variable Goshawk. To me the whole morphology and plumage detail and jizz suggest a full species status but we will see. For this publication I am referring to it as the Christmas Island Goshawk. There are also some interesting subspecies on the island, including the Common Emerald Dove subspecies *natalis*. The beautiful White-tailed Tropicbird race *fulvas* known as the Golden Bosunbird. Glossy Swiflet subspecies *natalis*. Island Thrush

The best area to view Frigatebirds and Tropic birds in great numbers is to take a small track above the golf course but it is not easy to find as you have to walk through an old mining area and take a faint track through woodland to the edge of a cliff face (be careful) but here with the uprising thermals the Frigate birds just cruse past at very close range. The location is accessed from a turning just before the entry to the airport on the left leaving the Settlement. The Imperial Pigeons can be found often around the offices of the Australian NP offices and the adjacent village area. Island thrushes are seen on the walk down to Lily Beach. Red-footed Booby are in this region. Abbots Booby nest in the centre of the island where the tallest trees occur particularly around Jacks Hill. Carrying on past Jacks Hill leads you past the Detention Centre down to the 'Dales' certainly not like the Dales in the UK. Here more Island Thrushes occur as well as Common Emerald Dove and crabs galore. Around the township the Settlement there is good viewing across the open seas watching the movement of Frigatebirds and

Tropicbirds. Walking on the coastal road to the golf course you will pass Glossy Swiftlet. People look for the endemic Christmas Island Hawk-Owl around the golf course. I dipped as very strong winds blew each night alas. There are good possibilities of rare vagrants being on island particularly in summer.

At the tourist centre there are people that may assist you with local guides.

As mentioned, this publication is not a 'Where to find birds' book so the above places to visit are only a brief guide to areas that may be of interest.

There is a small publication titled 'Birding Sites around Perth' written by the late Ron Van Delft (sadly passed away so young) which gives far more information on birding localities for Perth.

The Western Australia branch of Birds Australia has a useful web site http://www.birdswa.iinet.net.au/ giving detailed regional birding lists.

The keen bird watcher Frank O'Connor keeps a very detailed website http://members.iinet.net.au/-foconnor/ with more in depth details than in this book of some of the better birding localities in Western Australia.

Red-capped Robin Photo: Simon Nevill

BIRD PLUMAGE TOPOGRAPHY

The markings and features on a bird are known as bird 'topography'. As this publication is geared to ordinary birdwatching and not the study of true ornithology many avian biological terms have been avoided and the author does not wish to confuse or complicate the understanding of birds; however some specialised terms must be used as words like 'back' or 'head' can be too simplistic and not specific enough when describing the features of a bird so the following bird topography has been listed below to help the reader understand those features that may be described in the text.

The main features described in the body text are in bold lettering below. Little reference to the complex wing topography such as the wing coverts and scapulas have been used in the body text.

shaft
tail
rump
lesser
primary
coverts
upper tail
coverts
scapulas
mantle
crown
forehead
secondaries
carpal joint
lesser coverts
greater primary coverts
primaries

primaries
marginal
underwing
coverts
secondaries
greater
underwing
coverts
axillaries
lesser underwing
coverts
breast
belly
flank
vent

lores
median crown stripe
crest
eye stripe

plume over
ear coverts
gape

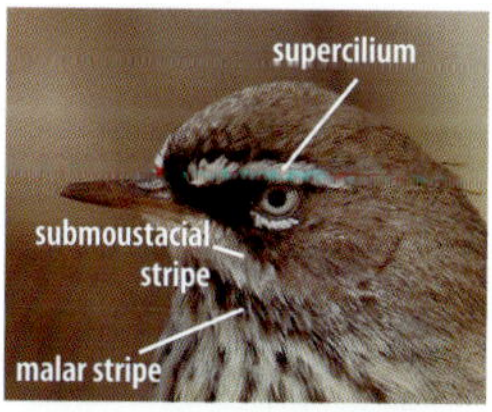
supercilium
submoustacial
stripe
malar stripe

facial
mask
or disc

eye ring

lores

plume

forehead

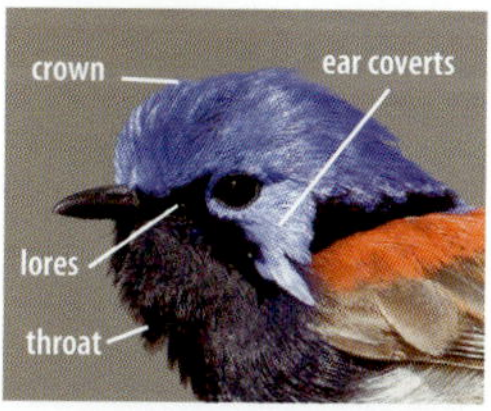
crown
ear coverts
lores
throat

Birds
of
Western Australia

CHAPTER II

Introduction to the photographic guide to the birds of the Western Australia

AN OVERVIEW OF CHAPTER II

As mentioned earlier in the guide, 489 species are illustrated but it does not include all the birds of the high seas (pelagic birds) or many of the rare vagrants but it does cover the vast majority of all land, sea and shore birds that are resident or migrate regularly to Western Australia.

My first bird guidebook was originally structured on the taxonomic revisions of Christidis and Boles in 2008. For this guide I have based the taxonomic nomenclature on the current IOC (International Ornithological Committee) world list.

The IOC is a worldwide body that attempts to give order to the complex taxonomy of world birds. The World Bird List is a work in progress and is constantly updated to conform to the current status of every species and subspecies, in terms of their true relationship in the world taxonomic order. It recommends the English names of each species, basing the nomenclature on an international basis rather than a localized naming system, so that species found in other parts of the world do not have conflicting or similar names. At present there are 10,488 extant species in the world with at least 149 now extinct. The IOC lists 20,984 subspecies.

This guide book is designed as quick-access guide for the reader and to meet that objective, I have altered the structural layout and re-arranged some of the taxonomic order of the species. This may not appeal to all but as mentioned before, it has been re-arranged to facilitate quick access and identification, mainly to aid the novice not necessarily the experienced biologist in finding species faster. For example, the waders and seabirds have been grouped together and some species like Skuas and Jaegars have been brought forward from their true taxonomic order and placed with other seabirds, so the reader does not need to flick through countless land birds to find these species. Similarly, Button-quail have been grouped with true 'Quail' and Frogmouths and Nightjar are lumped with other nocturnal species.

The book is simple in structure and layout and is certainly not a definitive work and the emphasis is on simple descriptive text. Personal observations have been used and hopefully this may add some interest. Along side the text is the matching photograph of each species. The distribution maps should be self explanatory, highlighting the main area of distribution in the darker green and the possible extreme range of distribution shown in a lighter yellow-green.

The complex procedure of grouping species to genus and genus to family and family to order is a never-ending process, no more so than in the last few years with the advent of DNA studies on several species, particularly in terms of hybridisation that helps to define a true singular species and its origins.

Most birdwatchers do not need to get bogged down in knowing about nominate species or subspecies but I cannot avoid describing those species that contain one or more subspecies. There is a good reason for this, as there is often a distinct variation in plumage colour with many subspecies and for some birdwatchers, it is important to know where they occur and their differences. In fact, since my last guidebook was published, many of the previous subspecies have now been given full species status.

To find a bird group quickly it is best to use the 'Fast Find Index' at the back of the book; for the more specific species name turn to the main index at the rear of the book.

The photographs

Several photographers have assisted with the balance of those species I did not have images for and their name is placed below the actual photograph and they have also been mentioned in the acknowledgments. In the case of my own photographs, photography at the nest was generally avoided, for obvious reasons, unless there was no fear of desertion by the parent birds. Several of the raptors and a few other species have been photographed in a captivity. The reason for this is that photographing for example raptors 40 to 50 feet up in a tree is not my favourite pastime and I certainly admire those like the father and son team of the Cuppers, or Michael Morcombe or the fearless John Young for doing so much photography at great heights. Although I know some who have tried and had serious falls, sustaining life long injuries, so its certainly not a pastime for the faint hearted. Some may question the ethics of having birds in captivity, even photographing them in that situation, but I feel that it is best to show the reader a good shot rather than a bad shot. One thing, if well kept, these birds bring so much joy to many who would otherwise never be able to see them in the wild and often they are injured birds that would otherwise have died in their natural environment.

Aesthetically I would have liked to show more background at times, to balance the photographs, but with such a small surface area to show each species it was felt best to crop and fill the frame, so the bird can be seen as clearly as possible to aid identification.

NONE OF THE PHOTOGRAPHS IN THIS PUBLICATION MAY BE REPRODUCED OR COPIED WITHOUT THE WRITTEN CONSENT OF THE PHOTOGRAPHERS. ALL PHOTOS BY THE AUTHOR ARE AVAILABLE.

The distribution maps

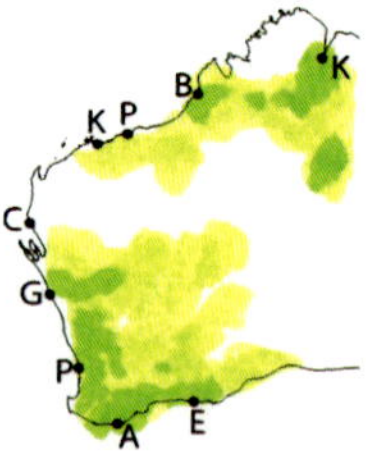

The small maps attached to the text for each species may have one or two colours. The darker green represents the main area where the species core population resides or visits on migration and will normally be the best area to see that bird. The lighter green areas show areas where either the population may occur but in low numbers or birds may occasionally expand their distribution. The maps are based mainly on the author's experience where species have been sighted in combination with the Birds Australia publication *The New Atlas of Australian Birds*, Barret et al. and Western Australian Birds Vol. 1 and 2 published by the Western Australian Museum. Ron Johnstone's work on past and current distribution is certainly the most accurate of species distribution in this state. Distribution maps are often open to debate and even the most authorative maps show major discrepancies. Also, it is not hard to find species way out of their normal range. Recently a Dollarbird was found in the middle of winter here in Perth and besides most probably feeling extremely cold for a subtropical species, its whole body clock and navigation aids would have been way out of kilter. The bird should have flown north to either Indonesia or another Asian country from the Kimberley rather than south to Perth, so do not be surprised sometimes when a bird is found way out of its normal range.

Species status

Various terms used to describe how abundant a species may be in the area covered by this book are as follows: Rare: simply rarely found in this region. Scarce: very low numbers in this region. Uncommon: means just that, uncommon. Common: means just that, common. Very common: the species is extremely common. Vagrant: means that a species rarely visits or is often accidental to the area. Irruptive: means that the species that will move to areas when their core centre of the habitat is experiencing severe weather conditions, normally drought, and they then move into areas where they would not normally occur. Endemic: in this case, refers to birds found only in the state of Western Australia (see below).

List of the endemic birds of Western Australia and Christmas Island

Fregata andrewsi	Christmas Frigatebird
Papasula abbotti	Abbott's Booby
Accipiter fasciatus natalis	Christmas Island Goshawk
Eopsaltria georgiana	White-breasted Robin
Ducula whartoni	Christmas Imperial Pigeon
Calyptorhynchus latirostris	Carnaby's Black Cockatoo
Calyptorhynchus baudinii	Baudin's Black Cockatoo
Cacatua pastinator	Western Corella
Purpureicephalus spurius	Red-capped Parrot
Platycercus icterotis	Western Rosella
Pezoporus flaviventris	Western Ground Parrot
Ninox natalis	Christmas Boobook
Atrichornis clamosus	Noisy Scrubbird
Malurus elegans	Red-winged Fairywren
Amytornis housei	Black Grasswren
Gerygone tenebrosa	Dusky Gerygone
Meliphaga fordiana	Kimberley Honeyeater
Melithreptus chloropsis	Gilbert's Honeyeater
Acanthorhynchus superciliosus	Western Spinebill
Anthochaera lunulata	Western Wattlebird
Dasyornis longirostris	Western Bristlebird
Acanthiza inornata	Western Thornbill
Psophodes nigrogularis	Black-throated Whipbird
Zosterops natalis	Christmas White-eye
Stagonopleura oculata	Red-eared Firetail

NOTE TO THE READER:

The IOC has made some major changes in the nomenclature of certain species. It has also established new species names that will confuse birdwatchers who were accustomed to previous common names, e.g. the Western Australian sub-species of the White-naped Honeyeater chloropsis has now been given full species status and will now be called the 'Gilbert's Honeyeater'. Other species such as the previously named Lemon-bellied Flycatcher have been given a new common name, Lemon-bellied Flyrobin. The genus being Microeca being different from Australian flycatchers, where the genus is Myiagra. I have placed many of the previous common names in brackets alongside the new.

Emu *Dromaius novaehollandiae* 1.6-2 m

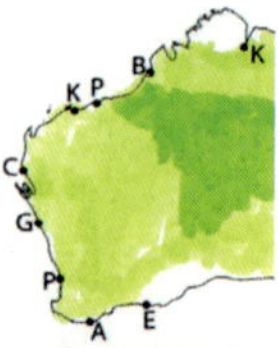

Generally a solitary bird although males will be seen with dependent young for up to 2 years. Will mass in larger groups occasionally. Adults will pair for about 6 months, the female laying in mid-winter in the mulga and slightly later in the lower south west. After laying the female

Immature bird

Photos: Simon Nevill

leaves the nest area when the last egg has been laid. The male takes on the full responsibility of raising the young, remaining on the nest for the full incubation period of 7 to 8 weeks without feeding or drinking through the first two weeks, and he will lose up to 8 kilos during that incubation period. After hatching the adult male remains with the young for a minimum of six months but in bad years will stay with the young for a year or even longer. Emus are omnivorous, eating fruits, seeds and insects, particularly grasshoppers. Voice: deep guttural grunts by both sexes. They are a most inquisitive bird and can be easily attracted by waving an item from, say, a car window. I once was banging tent pegs into hard ground and two adult birds came from behind to see what was happening, such is their curiosity. There is a natural tendency for emus to move from the drier regions in the summer, particularly in adverse seasons, to the lower south west. Tragically in drought years many hundreds die along the vermin fence trying to get to their natural south west migration route. **Nest:** No nest is made just a scrape on the ground or eggs laid on soft vegetation. **Similar Species:** None. **Status:** Uncommon in Jarrah Forest and Wheatbelt. Common in the drier pastoral country, particularly the mulga region. Small populations found as far north as Kalumburu but very uncommon in the Kimberley. **Where to find:** Perth near Lake McLarty near Pinjarra. Very easy to find in the Mulga Region beyond the Vermin Fence, providing country is not in a serious drought situation as they can move over 500 kms in a few months to find better feeding grounds. Avoids the sandy desert regions unless extensive mulga occurs in the area.

Malleefowl *Leipoa ocellata* 55-60 cm

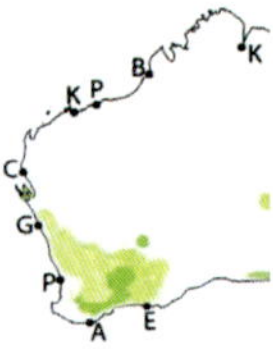

A member of the Megapod family, the 'mound builders' or 'incubator birds'. Malleefowl pair for life and are long lived birds, sometimes surviving 27 to 30 years. **Nest:** The eggs of the malleefowl are entirely incubated by the heat produced from the breakdown of organic matter within a huge mound acting like a compost heap. The mound is built by

The mound is nearly 2 metres high. Photos: Simon Nevill

both sexes but mostly by the male bird. In mid winter the mound is opened up and leaves and organic matter are placed in the open mound. With the first rains, the male bird becomes increasingly active pushing sand or gravel over the top of the leaves. The mound may be as high as 1.6 metres and 5 metres wide. At the end of spring in September, the male will call to the female letting her know the temperatures are correct and she can lay when she is ready. If the temperatures are not correct, he will dissuade her from laying. The female will lay anything from between 8-30 eggs but normally between 15-20 eggs. The male will control the temperature of the mound by opening up the mound and placing an open beak into the centre. The temperature must remain a constant between 32-34 degrees C. In summer when the leaf litter is almost dry, both the male and the female play an active role in controlling the temperature by opening or closing the centre of the mound. In doing so the male and the female move literally tons of earth in a season backwards and forwards off the mound. During the cool night, the earth is piled high to retain heat. With the first warm rays of the morning sun, the male with sometimes assistance from the female opens up the mound to warm the vegetation, staying close to the mound. He then checks the temperature and with the warming of the hot summer day, he will pile earth back on the mound. In Dryandra Woodland Reserve there are still a few old birds surviving. In 2006 there were two active mounds; the author had photographed one back in 1991 and it was active then. In another region, a small 138 acre bush block that was luckily spared from being cleared, thanks to an active conservation program and fox baiting by the farmer and the Malleefowl Preservation Group, has had as many as 5 active mounds in one season, which is a staggering quantity of nest mounds for such a small area. **Similar species:** None. **Status:** A threatened species, through persecution from fox and loss of primary habitat, the 'Mallee'. **Where to see:** In the southwest, the Wheatbelt: e.g. Dryandra Woodland NR. Buntine NR. Wongan Hills. Fitzgerald River NP. Dragon Rocks NR. and Magenta NR. They still survive in parts of the station country where dense stands of mallee and mulga exist. East of Hyden, particularly around the Holland Track area, there are still reasonable populations, alth ough their survival is always threatened by the introduced fox. Their northern range extends past Cue and across to Wiluna in the north and east right across to the northern Victoria Desert south of the Blackstone Ranges.

Orange-footed Scrubfowl *Megapodius reinwardt*

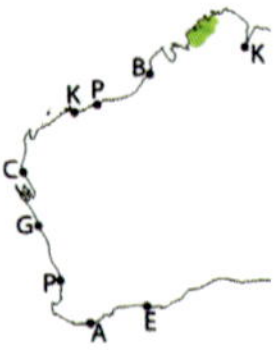

This species has the largest distribution of any member of the megapode family, with populations as far north as Sulawesi and much of the Sundas Islands. In Australia it is restricted to the tropical regions with the largest populations being in Queensland and Northern Territory. The subspecies *tumulus* is a very uncommon bird in WA, being restricted to few remnant monsoonal vine forests, mostly found on the southern parts of the larger north-west islands such as Bigge Island and the adjacent coastline. The country is hard to access as it is often found behind extensive mangrove systems. Small populations found inland as on the Mitchell Plateau but most monsoonal rainforest pockets inland are not large enough to support this specie. Feeds by turning over leaf litter. Generally a shy bird that will run off when approached or even take flight into the canopy of the forest. Its presence at dawn is always given away but its loud fowl like chuckles. Like other megapodes the Malleefowl and Australian Brush-turkey, the Scrubfowl roosts in trees at night, an instinctive habit for safety reasons. **Status:** Very uncommon in the Kimberley. **Nest:** The nest is the largest made by any megapode species, the mound having a diameter as wide as 12-13 m and as high as 2.5 m. However the clutch size is less than say a Malleefowl. Scubfowl lay on average 5-6 whereas Malleefowl lay on average 15 eggs. **Similar species:** Australian Brush-turkey not found in WA.

Photo: Simon Nevill

Where to find: A few islands on the north west coast, particularly in the Bonaparte Archipelago. Also monsoonal vine forests of the St George Basin and Prince Frederick Harbour and larger patches of rainforest such on the Mitchell Plateau. Easily seen in the Northern Territory such as in the city of Darwin. Common in tropical Queensland.

Stubble Quail *Coturnix pectoralis* 18-20 cm

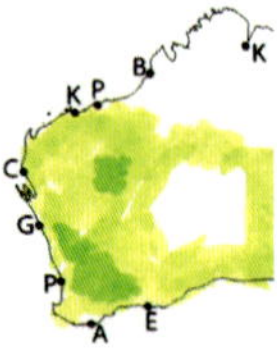

A highly nomadic quail, often travelling over 1000 km to find better feeding grounds. Heard more often than seen, mainly in the breeding season when the male makes his high pitched contact call, 'chewche-wit….chewche-wit'. Feeds mainly on seeds both native and introduced as well as insects. The male has an orange throat that is easily identified at close range. Seen mostly as a single bird or pair but out of the breeding season in small groups known as a 'covey' of quail. The Stubble Quail is one of the few birds that have benefited from the huge land clearing programs but this is a double-edged sword, as the breeding period can overlap with the harvest period and many are lost. **Nest:** Nests are placed on small bed of grasses in a cultivated crop or long grass, normally at the base of a thick clump. Normal clutch size 5 to 8. **Similar species:** Brown Quail, which is overall darker brown. Different patterning on flanks. Neither sex has an orange throat patch like the male Stubble Quail. Habitat should guide you to the correct species. Little Button-quail is much smaller with no heavy markings on the underbody. **Status:** Locally common otherwise generally uncommon. Nomadic. **Where to find:**. Mostly in the cropping region of WA. Found close to the Perth Region in areas such as Harvey and Pinjarra. Easiest to see in the breeding season on the edges of paddocks that abut reserves like northern boundary of the Fitzgerald River N.P. In the drier inland regions found in taller dry grasslands on floodplains, such as those on the Hope River floodplain some 40 km north west of Meekatharra. Also found on the Nullarbor Plain. Grasslands of the Pilbara such as north of the Fortescue River. Not the Kimberley.

Photo: Simon Nevill

Brown Quail *Coturnix ypsilophora Subspecies australis* 17-22 cm

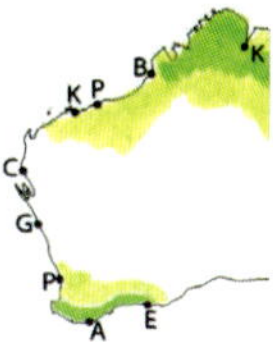

Found in wetter and denser habitat than the Stubble Quail particularly along the southern coastal belt and inland where lush grasses occur. Also throughout the Kimberley and parts of the Pilbara. Call is different from Stubble being a more mournful and drawn out two-noted ascending call of 'too-weeee'. When flushed, flies straight and upwards then dropping down quickly with breaking flight. **Nest:** Nest in thick rank grasses often near water. Normal clutch size 5 to 8. There are nine subspecies in Australasia and some south west pacific islands, two in Australia, *australis* on the mainland and the nominate subspecies *ypsilophora* in Tasmania. **Similar Species:** Stubble Quail, refer Stubble Quail notes. **Status:** Moderately common on the South Coastal Belt particularly adjacent to swamplands. Common in the grasslands of the Kimberley. **Where to find:** Hard to find although easiest at Cheyne Beach Caravan Park and Woody Island where they have become accustomed to people. In the Pilbara in the rank grasslands often near the coast between Wickham and Port Hedland, Pilbara birds are smaller and paler than those in the south west and also those in the Kimberley. In the Kimberley one of the more common quail. Can be found driving along Gibb River Road or Northern Highway where ever rank or thick dry grasslands occur near water.

Photo: Simon Nevill

King Quail *Coturnix chinensis*

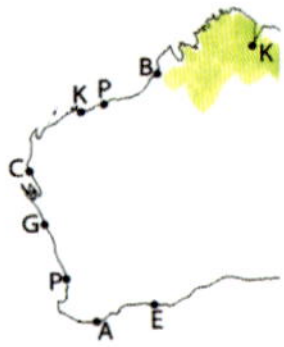

A quail of eastern India and south-east Asia. Uncommon to Rare in eastern Australia and certainly rare in the Kimberley. There are records of this species dating back to 1975 (Storr). Found in the Prince Regent area, also later seen at the Mitchell Plateau airport. Since then, a few sightings near Kununurra and near Mount House Station. The male is very attractive with unmistakable colour and markings. Has grey-blue breast and face with a contrasting white and black throat pattern. A rich chestnut under-body and vent, seen well on the rear flanks. The female heavy horizontal barring on breast with faint white throat pattern contrasting with chestnut upper breast and facial disc. Frequents tall grasslands and favours damp grasslands. **Nest:** Scrape on the ground with grasses and partial grassed hood. Clutch size 4-9. **Status:** Rare in WA. **Where to find:** Anywhere, where there are rank grasslands. Worth looking for as it could be over looked in WA but do remember this is a very small quail, even smaller than most button-quail and the female could be mistaken for a female Red-chested Button-Quail which has similar markings and same size.

Male

Female

Photo: Simon Nevill

Painted Button-quail *Turnix varia* 17-19 cm

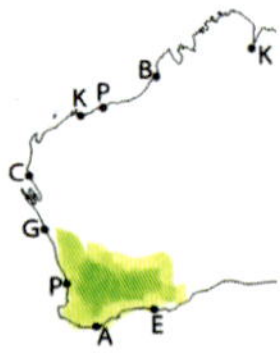

Much of the Painted Button-quail's original territory has been cleared. It prefers open dry woodland, particularly Wandoo and Salmon Gum woodland. It requires a reasonable cover of leaf litter on top of the base soils from which it gets most of its food. It feeds mostly in small family groups or in pairs, constantly turning in circles, using one foot and then another to kick away the leaf litter to expose the insects and seeds. You can see where button-quail have been feeding as small plate-sized areas, known as 'platelets', which are left on the ground. The Painted Button-quail is one of the larger Button-quails, not much smaller than the Brown Quail (unrelated). Sometimes you can catch them unawares. In open country they will try and walk away extremely slowly, rocking the body backwards and forwards. Most of the time they surprise the walker as suddenly, like all quail, they fly just in front of you with a loud 'brrrrrrrrrr' sound coming from the rapid wing beats as they fly off, normally landing about 30-40 metres away. They also give themselves away with their low soft booming call repeated every two or three seconds 'ooow…oooow…oooow'. There are three subspecies in Australasia, with two in Australia. Subspecies *varia* is found in the South West and in the eastern states. There is a small population of the subspecies *scintillans* on the Houtman Abrolhos Islands. **Nest:** No nest is made, just a scrape in the ground, although some dry grasses may be used as a base. Built normally below a bush or alongside a tuft of grass. Clutch size 3 to 4. **Similar species:** Nothing in their distribution range, the closest in terms of colour is the Little Button-quail which is smaller and a plainer lighter colour overall. The Painted Button-quail is more attractive with chestnut–brown flank and a dark mottled crown whereas the Little Button-quail has no chestnut coloured patch on flank. **Status:** Locally common in uncleared Wandoo, Salmon Gum and some mallee regions in the southwest.

Distribution stops past the Mulga/Eucalypt Line. **Where to find:** Most Wheatbelt reserves with reasonable acreages. Dryandra Woodland NR. Boyagin Rock NR. Dongalocking NR. Lake Magenta NR. Dunn Rock NR. Durokoppin NR. Billyacatting NR. Throughout the Transitional Woodland from Lake Cronin all the way east to past Dundas NR. North and east of Billyacatting, although one must be careful as it is possible to get Little Button-quail there. Little Button-quail have been recorded in much of the Wheatbelt as they have moved into the South West and have adapted to crops and stubble.

Photo: Simon Nevill

Little Button-quail *Turnix velox* 12-14 cm

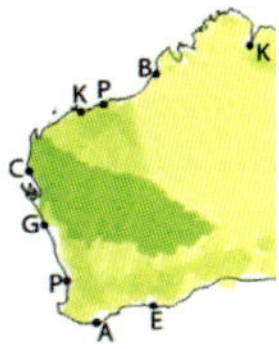

One of the smallest Button-quails along with the Red-backed Button-quail. You realize how small they are when you find dependent young chicks sitting motionless on the ground with heads outstretched, just slightly bigger than a 50c coin. Even the best birdwatchers have trouble with quail; I have seen this first hand as you don't get long to identify them sometimes, as they fly off at speed. The Little Button-quail looks paler when flying off, having an even coloured fawn back contrasting with its darker wings. Painted Button-quail have greyer brown but colours on their back are more mottled and darker. Definitely a quail of the semi-arid and arid regions, in fact perhaps the most common bird to be found in the desert regions. Feeds mostly on grass seeds and insects. Can often be missed when walking through tufts of grass, as it will scurry off at an incredible pace if it has not already flown. **Nest:** Similar to Painted Button-quail. Clutch size 3 to 4. **Similar species:** Refer to Painted Button-quail for

Photo: Simon Nevill

details. **Status:** Locally common where large tracks of grass plains occur or stubble between mulga. Throughout the drier areas. Does not like rank grass, wet conditions like Brown Quail. **Where to find:** Anywhere past the Vermin Fence where there is a good cover of native grasses. The further away from the South West Region the better your chances, particularly in deserts and the Nullarbor.

Red-backed Button-quail *Turnix maculosa* 12-15 cm

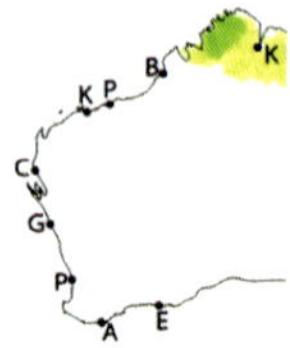

Button-quail although superficially looking like true quail *(Coturnix)*, belong to a different family namely *Turnicidae*. They posses no hind toe and no crop. So much is not known about the ecology of button-quail made even more challenging by the cryptic nature of these birds. They are often overlooked as when approached, they squat and freeze but if you are too close, will fly fast and direct away, sometimes not dropping for 100-200 m. A giveaway is the cooing of the female and that's when you freeze and slowly approach hoping to flush one. It's then that you get the chance to identify. Typical of all button-quail, the female is larger and brighter in colour particularly in the breeding season. She courts the male, selects and builds the nest but the male broods the eggs. They are a bird of the tropical grasslands **Nest:** Compact grasses often with a part dome covering mostly close to wetlands. Clutch size 3–4. **Similar species:** Red-chested Buttonquail has a heavier bill and streaked brown back rather than chestnut upper back as in Red-backed. Red-backed has darker crown and heavy dark barring on flank. Chestnut-backed Button-quail is slightly bigger than both above species. Has no chestnut colouring on underbody. Heavier bill. Legs yellowish tinge whereas Red-chested has cream coloured legs. Female Red-backed has the brightest rust red colouring on breast and neck to all the other Button-quail. **Status:** Generally uncommon. **Where to find:** Found in rank grasslands and black-soil plains. The grassland's behind Broome Bird Observatory and Roebuck Bay. Parry Lagoon. Easiest to see during and at the end of the wet season. The species is not seen south of the 20th parallel in WA.

Photo: Rohan Clarke

Red-chested Button-quail *Turnix pyrrhothorax* 12-16 cm

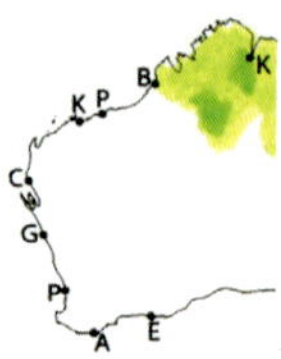

Not a common Button-quail in the Kimberley. Upper breast colouring a strong orange-brown, particularly distinct on the female. Legs pale pink not yellow. No heavy streaking on crown. Takes the place in the Kimberley of the more numerous Little Button-quail of the arid regions further south. Patchily distributed throughout the Kimberley from the

Roebuck Plains across to the grasslands around Kununurra. Tends to be more dispersive than most button-quails as numbers fluctuate considerably. **Nest:** Depression in ground with dried grasses and low hood over main nest. Clutch size 3-4. **Similar species:** Red-backed Button Quail which has richer rust red upper breast and neck collar and darker striped crown. **Where to find:** Roebuck Plains. Fitzroy flood plain. Grasslands and slopes around Kununurra.

Photo: Rohan Clarke

Chestnut-backed Button-quail *Turnix castanotus* 15-19 cm

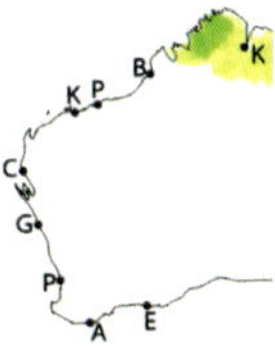

Largest of the Button-quails. Found only in the tropics of the NT and WA. In the Kimberley it is found in rank to dry grasslands out of the breeding season found in coveys of 5-20 birds. It takes the place of Buff-breasted and Painted Button-quail in the north west. If approached sticks head out straight and starts walking away slowly but if alarmed will take flight. Has no chestnut colouring on breast. Dark stripes on crown. Legs yellow. Shares the same habitat in parts of the Kimberley as Brown Quail, which is larger, dumpier and overall darker brown. **Nest:** Little known. Similar to most button-quails, being a scrape in the ground with woven grasses and partial dome. **Similar species:** Refer to Red-backed Button-quail. **Where to find:** Best place just south of Cape Leveque on the Dampier Peninsula. Rocky slopes with grass in the Mitchell Plateau region and most inland Kimberley rocky hillsides. Also on the way into Bachstein and the Calder River. Not easy to find, I have had more success in the Pine Creek in NT with this species.

Photo: Adrian Boyle

Red Junglefowl *Gallus gallus*

Introduced

Introduced to Christmas Island in the early 1900s. Population is now small around 20 birds mostly concentrated around the settlement and may now have interbred with domestic stock. **Status:** Stable. **Where to find:** As mentioned around settlement, also golf course and airport surrounds.

Photo: Simon Nevill

Common Pheasant *Phasianus colchicus*

Introduced

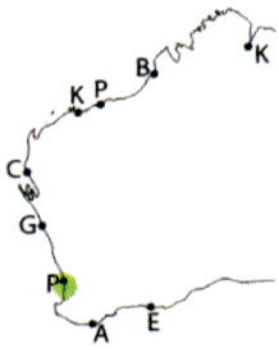

A bird originally native to Eurasia, there are at least 40 subspecies throughout the world. Introduced to much of Europe and now New Zealand and Australia. Feral populations occur in the wild on Rottnest Island and Islands in the Bass Straight. **Status:** Stable. **Where to find:** On Rottnest Island found mostly around the settlement, including the Barracks and Golf Course on Rottnest Island.

Photo: Simon Nevill

Magpie Goose *Anseranas semipalmate* 80-90 cm

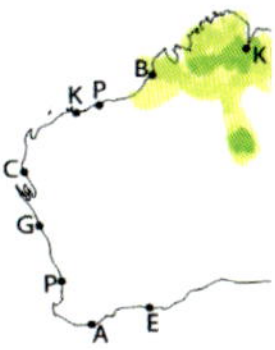

Found on the more extensive wetlands of the Kimberley particularly around Parry Lagoon near Wyndham and parts of the Fitzroy floodplain. They disperse far and wide during the 'Wet' seeking new areas recently flooded. They can be seen head and neck extended under water tearing up the succulent shoots of aquatic plants. When at rest they can be seen in trees roosting day or night. They fly in 'V' formation to new feeding areas. Their population in WA is far less than those of the vast wetlands of the Northern Territory, although they can be found in great numbers on Lake Gregory in times of extensive flooding. A bird primarily of northern and eastern Australia but occasionally coming all the way down to the Pilbara wetlands. **Nest:** Nests colonially in swamplands with both male and female attending the young. Has unusual feathered raised knob on crown. Make a continual honking noise in the larger colonies. Nest placed on trampled rushes often floating on top of water. Normal clutch size 4 to 8. **Similar species:** None. **Status:** Locally common in the north, although less so in the Kimberley than the Northern Territory. Rare vagrant in the South West Region as far as the Capel wetlands. **Where to find:** Parry Lagoon. Lake Eda (has breed there). Lake Argyle and occasionally the Fitzroy floodplain and Lake Gregory.

Photo: Simon Nevill

Plumed Whistling-Duck *Dendrocygna eytoni* 45-60 cm

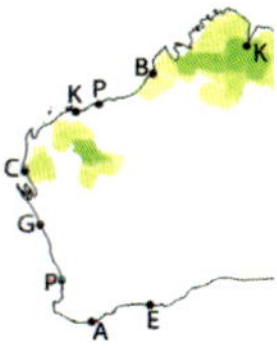

Found in greater numbers than Wandering Whistling-Duck in WA. Easily identified by its longer neck, extended cream coloured flank feathers, pink legs, pale crown and no dark coloured feathers to the nape and rear of neck like the Wandering. A bird of the tropics and temperate wetlands and grasslands. Grazes on land. Congregates in large numbers on tropical wetlands in the dry season and then disperses in smaller groups during the wet. Feeds on land during the night returning to roost two hours before dawn on wetlands and river banks. Occasionally will roost in trees. Will congregate with Wandering Whistling Duck. **Nest:** Nest on ground with gathered grasses placed in long grass above flood levels. Normal clutch size 5 to 8. **Similar species:** Wandering Whistling Duck, which is slightly smaller, with smaller side plumes and dark brown on crown rear of neck to the

Photo: Simon Nevill

mantle. **Status:** uncommon throughout the southern Kimberley and Fitzroy floodplain. Common in parts of the eastern Kimberley. Its stronghold is the huge floodplains of the NT and the open grasslands of central Queensland. However when exceptional rains have fallen and the Fitzroy floodplain is inundated huge numbers will disperse to this area. A vagrant to the South West Region. **Where to find:** Parry Lagoon. Lake Eda. Lake Argyle and occasionally the Fitzroy floodplain and Lake Gregory. Fortescue floodplains. Samphire floodplains huge numbers occasionally on the Mandora Marshes.

Wandering Whistling-Duck *Dendrocygna arcuata subspecies australis* 50-60 cm

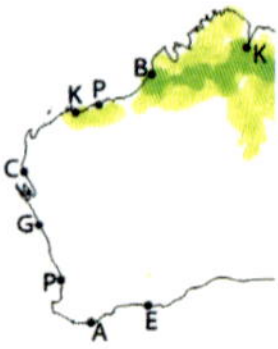

One of the two Whistling Duck in Australia there are 9 species in the World. Like the Plumed Whistling Duck, the Wandering Whistling Duck is a bird of the northern and eastern Australian wetlands but unlike the Plumed Whistling Duck it prefers to feed on open water taking more aquatic vegetation and insects. There are three subspecies in Australasia, the Philippines and Indonesia. Subspecies *australis* is on mainland Australia and New Guinea. **Nest:** Same as Plumed Whistling Duck. **Similar species:** Plumed Whistling Duck – refer Plumed Whistling Duck notes. **Status:** Generally uncommon in most areas of the Kimberley. Locally common in the east Kimberley. Rare vagrant to the South West. **Where to find:** Same localities as Plumed Whistling but far les numerous in WA particularly the Pilbara as this species prefers deep wetlands.

Photo: Simon Nevill

Black Swan *Cygnus atratus* 1-1.4 m

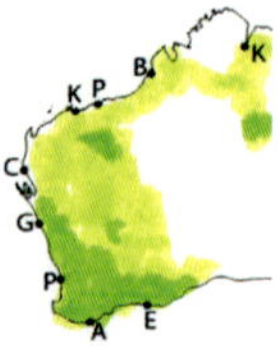

The unmistakable Western Australian icon. Found throughout the region in freshwater, brackish water and even saline waters. Feeds almost exclusively on aquatic vegetable matter, often seen tail up semi submerged grazing on the bottom of shallow wetlands. In the breeding season gathers in smaller groups but when in post-breeding moult, will gather in thousands on large safe permanent waters as they become flightless. Black Swans have complex territorial behaviours and social interactions leading to aggressive defence of territory or pair bonding between breeding pairs. **Nest:** Nest is either a mass of floating grasses or reeds up to one metre high or grasses gathered on islands. Normal clutch size 4 to 6. **Similar species:** None. **Status:** Very common in the southwest. Casual in dryer areas of the South West but will congregate in large numbers when unseasonable heavy rains fall in the pastoral region such as Lake Annean and Lake Austin and Wooleen Lake. Uncommon in the tropical parts of WA. **Where to find:** Throughout the South West wherever there are large bodies of water, particularly the brackish inlets along the south coast such as: Broke Inlet. Irwin Inlet. Wilson Inlet. Beaufort Inlet. Found throughout the state but far more irruptive, taking advantage of unseasonal inundations such as occur at Lake Gregory (4000 +) or the Fortescue floodplain and other inland wetlands.

Photo: Simon Nevill

Mute Swan *Cygnus olor* 1.3-1.6 m — Introduced

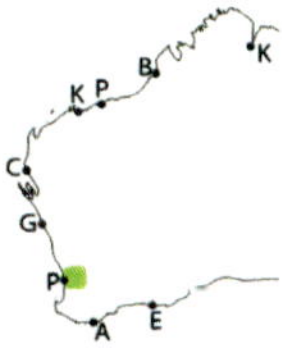

An introduced species native to Europe and central Asia, now established in Northam. In the early 1900s a pair was released on the Avon at Northam. The population peaked in the early 1960s and now there are around 40 individuals remaining stable over the last 20 years. **Status:** Locally common in the township of Northam. **Nest:** Nest similar to Black Swan. **Where to find:** On the Avon River within the township of Northam.

Photo: Simon Nevill

Cape Barren Goose *Cereopsis novaehollandiae* 78-90 cm

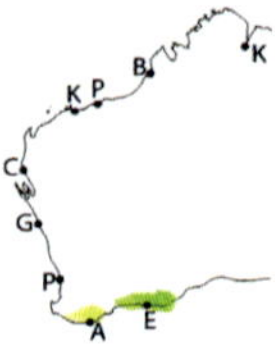

The only member of its genus and endemic to Australia. In Western Australia the subspecies *grisea* is found on the islands of the Recherche Archipelago in the Esperance region. This subspecies differs only slightly with the white crown touching the eye. Seen occasionally on beaches west as far as the Fitzgerald River National Park and east to Cape Arid, it is a herbivore, feeding on native and introduced grasses on both the islands and green pastures around the Esperance townsite. **Nest:** Nests on several of the islands in the Recherche. Many birds fly to the mainland to feed, particularly outside the breeding season. Feral goats on some islands are affecting the grazing areas for Cape Barren Goose. Nest is a gathered pile of grasses with hollow in the top where a quantity of down is used to cover eggs when unattended. Almost always nests on islands in Western Australia. Normal clutch size 4 to 5. **Similar species:** None. **Status:** Locally common on some islands and not endangered at present but overall WA population relatively small with approximately 1000 individuals. **Where to find:** On the mainland in the township of Esperance on small farmlets, particularly along Stearnes Road and on the Esperance Golf Course. To view them on the islands, there arc boat trips that visit some of the islands, including Woody Island, in the late spring till autumn.

Photo: Simon Nevill

Australian Shelduck *Tadorna tadornoides* 57-73 cm

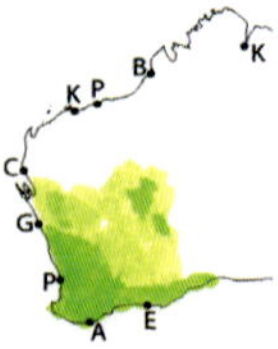

A grazing duck feeding on grasses on land or upending in shallow water to graze on aquatic plants. Formerly called Mountain Duck as on the east coast, it will graze on the high mountain grasslands. In general, it is far more opportunistic feeder than most other ducks many of whom have specialised feeding requirements. The most salt tolerant of all ducks. Like the Australian Wood Duck, the Australian Shelduck nests in tree hollows and like the Wood Duck has increased in numbers with the introduction of dams to the cleared Wheatbelt Region. Shelduck on Rottnest Island have adapted to nesting on the rock crevices by the sea, rather than high up in trees as on the mainland. The adult bird calls to the young from the sea whereupon they jump from the cliff making their way to the nearest beach and then walking to the nearest salt lakes on the island. The male lacks the white eye ring of the female. No white to the base of the bill and has a brighter orange-brown chest rather than the duller rust-brown chest of the female. **Nest:** Most nests are located in hollow trunks with a pile of grasses or leaves laid on wood debris. Tree hollow can be as low as 2 metres or as high as 15 metres or more. Normal clutch size 6-10. **Similar species:** None. **Status:** Common in the south west particularly the Wheatbelt. Less common in the drier regions however, will congregate in large numbers on inland salt lakes after good rains. Lake Gore in Esperance has had over 7000 in one count. Uncommon to rare in the north and desert regions of the state. Rare above latitude 20. **Where to find:** Any wetland in Perth or on or near Wheatbelt dams and wetlands along the south coast. Throughout the Mulga Region on large ephemeral lakes.

Male in foreground

Photo: Simon Nevill

Raja (Radjah) Shelduck *Tadorna radjah*

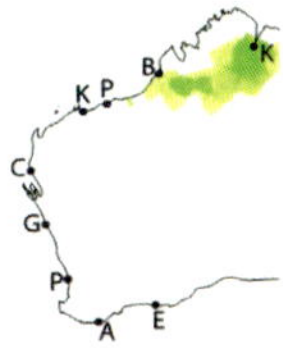

A shelduck of the tropical wetlands of Australia. Rare in west Kimberley found mostly in the wetlands of the east Kimberley and across into NT. Very numerous in Kakadu and Arnhem Land. In the east Kimberley found on lake Argyle particularly the southern end. Parrys Lagoon. Kununurra floodplains. Wetlands near Kingston Rest etc. Retreats to permanent wetlands in the dry season, dispersing to the floodplains of the Ord and Fitzroy and small freshwater ephemeral lakes in the wet. Often seen grazing on wet grasslands or upturned feeding on aquatic plants. Easily identified by size, white head with dark chestnut back and distinct breast band. Both male and female have distinct rattle nasal calls. **Nest:** Nests similar to Australasian Shelduck in the hollow of trees. Clutch size large 10-15 eggs. Parents coax young to jump and then guide to water. **Status:** uncommon. This species was far more widespread in the early 1900s but numbers now are far less. **Where to find:** Parry Lagoon often over 100 birds will congregate on the lagoons. Lake Argyle also can have large numbers. Ord Floodplain.

Photo: Simon Nevill

Hardhead (formerly White-eyed Duck)

Aythya australis 45-60 cm

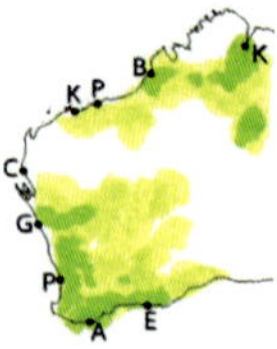

An irruptive species that will fly great distances to find new feeding grounds. Feeds on aquatic vegetation but also animals like freshwater shellfish. Prefers deep water. Can dive deeper than most ducks. Unlike some ducks, this species has declined in numbers since European settlement. The male is easier to identify with its white iris and dull white tip on the end of the bill, the female not so easy; she has a brown iris and no white on the end of the bill. **Nest:** Nest is a very large pile of reeds or other aquatic foliage with a large quantity of down used to cover eggs. Normal clutch size 6 to 8. **Similar species:** Female Blue-billed Duck which is smaller and has a more compact body. Bill shape on the Musk Duck is wedge shaped where as the Blue-bill has a typical slim duck bill not short and wedged. **Status:** Uncommon to common in the south west. Uncommon in the Pilbara and the arid regions but more prevalent in the Kimberley wetlands sometimes in large numbers as on Lake Gregory and Lake Argyle. **Where to find:** common in most large wetlands of Perth. Found throughout the South West Region where there are large bodies of freshwater. Many wetlands in the Kimberley but not common.

Photo: Simon Nevill

Musk Duck *Biziura lobata* 64-68 cm

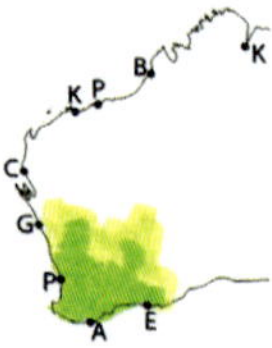

One of the stiff-tailed ducks. Male much larger than the female. Less nomadic than many other duck species, preferring the deeper permanent wetlands of the lower southwest corner of WA. Normally in small numbers but occasionally out of the breeding season in summer and autumn, large congregations can gather, such as on the Esperance wetlands or Peel Inlet; however it is an uncommon sight to see more than 200 birds on one wetland. The male has one of the most extraordinary courtship displays of any Australian duck. Under its bill is a leathery lobe that when not displaying remains as a loose appendage, but when it displays the lobe becomes more rigid and its cheeks inflated, while it extends the head upwards and back and draws the tail forward and over its back with the wings drawn upwards: then suddenly it will let out a far carrying high pitch 'pinging' call and draw its feet to the side creating a large splash either side of its body. **Nest:** Nests mostly on pile of aquatic vegetation, untidy with little or no soft lining. Normal clutch small 2 to 4. **Similar species:** Male none. Female may be confused with female Blue-billed Duck, which is smaller, has different bill shape and sits higher in the water. **Status:** Common locally but generally an uncommon Duck. **Where to find:** Common on most larger deep-water lakes in

Perth. In the Esperance region Lake Gore and Lake Warden. Lake Seppings in Albany. Most other wetlands have only small numbers. Found on some outer wetlands of the south west but not in the Pilbara. Desert or Kimberley regions.

Photo: Simon Nevill

Maned Duck (Australian Wood Duck)

Chenonetta jubata 48-58 cm

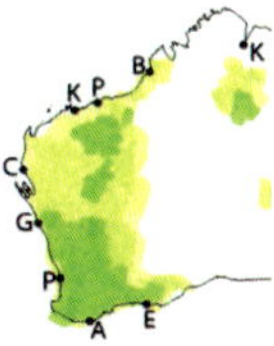

Often called Maned Duck in reference to the feathering on the lower nape. Even though this duck does not have brightly coloured feathers, when seen up close, both the male and the female have exquisite detail in their plumage. The male has greyer, less mottled flanks and darker chocolate head. This duck is primarily a grazing waterfowl, preferring to eat low grasses but in the Wheatbelt it has adapted to eating a higher proportion of grain seed. With the clearing of land and the availability of permanent water in the myriad of small dams that are strewn throughout the Wheatbelt, it has found a niche habitat. Also, the need to nest in tree hollows and have water nearby for their young has made this a common bird on the edge of many Wheatbelt Reserves. They nest in tree hollows as high as 15 metres from the ground. The parent bird calls constantly to the young to jump from the nest and then they freefall down to the ground extending their rudimentary wings and bounce like fluffy balls. It is an amazing sight. When all the chicks are out of the nest, the parent bird will walk them to water, often several hundred meters away. **Nest:** Nest normally 5 to 15 metres above ground mostly in live trees containing dead limbs with hollows, not far from water. Clutch size 6 to 8. **Similar species:** None. **Status:** Now common in the well watered South West but not in the thicker Jarrah Forrest unless abutting farmland. Uncommon in pastoral country. Uncommon in the Pilbara and a few wetlands in the northern desert regions and Kimberley but very uncommon. **Where to find:** Throughout the region but easiest in the deeper South West Region.

Photo: Simon Nevill

Freckled Duck *Stictonetta naevosa* 50-58 cm

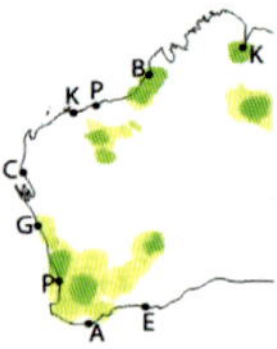

Australia's most uncommon duck other than the rare vagrants like Garganey, Northern Shoveler and Northern Pintail which visit mainly northern Australia. Seen up close, they have beautiful dark grey flecking on body. The female could be confused with a female Blue-billed Duck which is a smaller duck and does not have the same jizz, particularly the classic feature of the 'ski slope' on the forehead of the Freckled Duck that drops down to the bill and the slight crest on its the nape. Luckily, when the male is in breeding plumage the lores are bright red allowing easy identification but out of the breeding season can be very subdued. Perth is fortunate to have a good range of permanent wetlands giving species that are highly nomadic the opportunity to have somewhere to retreat to when the desert ephemeral wetland are dry. The bird illustrated was photographed on Herdsman Lake. One of their preferred habitats are wetlands that have dense stands of Freshwater Paperbark trees *(Melaleuca rhaphiophylia)*. Not only do they feed in these wetlands but also prefer to nest in paperbark trees in the south west, unlike the their other preferred site of nesting in tall lignum in large remote ephemeral lakes like Lake Gregory. **Nest:** A large structure of twigs or flood debris placed on water in heavily treed locations often in paperbark swamps in the South West. They sometimes use an old swan's nest. Normal clutch size 4 to 6. **Similar species:** Female Blue-billed Duck which is smaller and its neck is not as long and head is rounded with no steep forehead or crest on nape. **Status:** Uncommon to Rare in Southwest. **Where to find:** Can show up in Perth wetlands in any season. Found on wetlands such as Lake Dumbleyung (in the south east corner). Lake Tarblin. Lake Toolibin. Benger Swamp. Lake Chittering. Lake Warden. Lake Wannamal. If you are really keen to see this species in the South West, you will mostly need a canoe to explore many of the above wetlands, as they are often found in the central part of the lakes in dense stands of paperbark or sheoak thickets. Found regularly at Rowles Lagoon in the Goldfields. Is very irruptive and will fly hundreds of kilometres to find wetlands in the arid region when heavy rains have fallen, particularly locations like the Samphire floodplains off 80 mile beach. Lake Gregory can support the largest concentrations of Freckled Duck in the state when the lake is full.

Photo: Simon Nevill

Blue-billed Duck *Oxyura australis* 36-44 cm

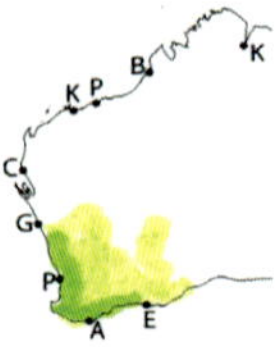

The bright blue bill of the Blue-billed Duck only lasts through the breeding period and by the autumn becomes a dull blue-grey. It is one of the two stiff-tailed ducks in Australia. It is almost totally reliant on a life in water. Does not like to venture far on land. When feeding, has the ability to dive to reasonable depths preferring deeper permanent wetlands. It prefers to dive when in danger and travel up to 30-40 metres just below the surface. Like the Musk Duck, it too has a spectacular mating display, which is varied and complex, including a

curious movement of pressing the head to the chest then extending the head into the water while blowing bubbles and then flicking water with its bill. An uncommon duck, although locally common in the Perth region and coastal belt down to Capel, mostly in small numbers but can reach large numbers in dry years in summer on wetlands like Peel inlet, with over a 1000 birds once counted. Does not like saline waters. **Nest:** A well constructed nest on water, sometimes as thick as a meter deep with hood over nest chamber lined with soft grasses and some down. Normal clutch size 4 to 6. **Similar species:** Female Musk Duck or female and eclipse plumaged male Freckled Duck. Refer those species notes. **Status:** Locally common on the Swan Coastal Plain but tends to be uncommon throughout the southern region and very uncommon past the northern Wheatbelt. **Where to find:** All the larger lakes in Perth particularly Herdsman Lake. Peel Inlet. Wannamal Lake. Vasse Estuary. Lake Warden Esperance. Rarely extends out side of the south west region.

Photo: Simon Nevill

Pacific Black Duck *Anas superciliosa* 50-60 cm

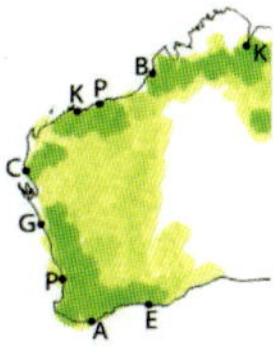

A duck that feeds in varied habitats and unlike many other species of duck, will feed on small bodies of water even small drainage channels. Has a preference for freshwater wetlands. It is one of the 'dabbling' ducks, taking food from the surface as well as upending to graze on aquatic plants. Normally sedentary where permanent bodies of water exist but many will migrate to inland areas after good rains. Males leave after the clutch has been laid joining other males in large numbers on some lakes. The female is left to attend the young until dispersal. Unlike many species of duck, the Pacific Black Duck will nest in a variety of locations including the tops of dead stumps, tree hollows and old nests and occasionally on the ground in vegetation. **Nest:** Nests vary, mostly located alongside water in grassed areas but sometimes placed in tree hollows. Normal clutch size 8 to 10. **Similar species:** Female Mallard. Female Garganey and eclipse male but far smaller 38-41 cm. Brown scalloped markings more darker and distinct. Some likeness at distance to Grey Teal, which is smaller with no facial striping. **Status:** Very common in South West Region. Uncommon in arid regions. Locally common on some wetlands in the Kimberley. **Where to find:** Any wetland area throughout the state.

Photo: Simon Nevill

Mallard *Anas platyrhnchos* 55-70 cm

Introduced

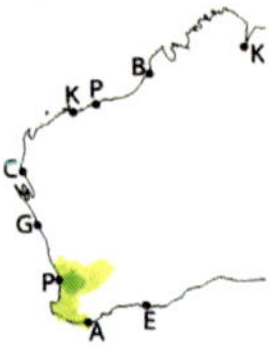

A hlarctic species, range circumnavigating the northern hemisphere. Some birds over winter in Africa and Asia.

Introduced to eastern Australia in the 1860s but were not introduced to WA until the early 1900s. There are still remnant populations scattered around the Perth region but many have hybridised with Pacific Black Duck. Male unmistakable with chestnut coloured breast and dark green head with fine white collar band. Female could be confused with Pacific Black Duck but supercilium stripe very pale and eye strip finer and less pronounced. Mallard legs are pinkish-orange where as Pacific Black Duck are grey-cream and overall colour paler. Mallard is slightly larger and thicker set. **Nest:** Consists grasses lined with down located in grasses or reeds near water. Occasionally in tree-hollows. **Status:** Uncommon. **Where to find:** Occasionally on Bibra Lake and Lake Monger.

Photo: Simon Nevill

Grey Teal *Anas gracilis* 42-46 cm

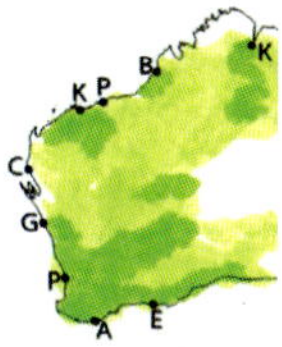

Found throughout Australia. This is one species that is extremely hard to differentiate between male and female, and colour of plumage is certainly not a help. Those who have studied the Grey Teal note that the male has a different standing posture with head held up while the female is lower, and that both have distinct calls. In the early nineteenth century, it was presumed that Grey Teal were the females of Chestnut Teal. Grey Teal can be found on fresh, brackish and saline waters. Highly nomadic. Any water body in the drier region has a good chance of having Grey Teal. This propensity to feed in so many areas throughout Australia may be due to the fact that a higher proportion of the diet is made up of invertebrates, the balance being vegetable matter. They breed in any season dependent on rain. **Nest:** Nests in tree hollows, bushes and on the ground but in the South West they favour nesting in tree hollows in flooded regions. Often roost in groups and will perch on logs, similar to Black Duck. Normal clutch size 5 to 10. **Similar species:** The female and eclipsed plumage male Chestnut Teal are both very similar in shape, size and coloration. The Grey Teal has a lighter throat but the main difference is a slight darker grey-brown coloration overall. **Status:** Common throughout the state, less so in the desert regions. **Where to find:** Any major in Perth lake, more so in the summer months. Any wetland throughout the whole state. Huge numbers on Lake Gregory when full. Also huge numbers at times on Parry Lagoon (20,000 +) and fairly large numbers (6000 +) also on Camballin Station.

Photo: Simon Nevill

Chestnut Teal *Anas platyrhynchos* 40-48 cm

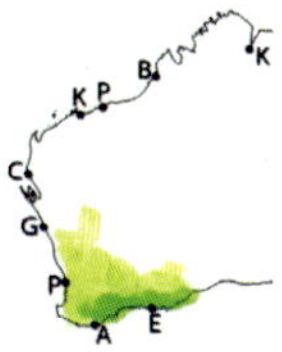

Although the Chestnut Teal is relatively common in the eastern states, particularly Tasmania, it is our second least common duck after the Freckled Duck, not including rare vagrants. It is one of the dabbling ducks gaining most food from the surface but also upending to get at aquatic vegetation. It is well adapted to living in waters with a high salinity level such as river estuaries, mainly on the south coast, increasing in numbers as one reaches the Esperance region. **Nest:** Similar to Grey Teal. Normal clutch size 5 to 10. **Similar species:** Grey Teal, the main two differences being that the Grey Teal has a whiter throat and the Chestnut Teal female is generally darker overall. **Status:** Locally common on some of the larger Esperance Lakes. Uncommon throughout the South West Region. **Where to find:** Occasionally in Perth but uncommon on the Swan Coastal Plain and along the coast to

Geraldton. Very uncommon in the deeper South West. Best seen on the southern inland and coastal rivers and inlets including: the lake systems that lie at the northern side of the Stirling Range, including small drainage channels. The estuaries of the Bremer River. The Gairdner River. Fitzgerald River inlet including all wetlands in the Fitzgerald River NP. All lakes inlets adjacent to the coast between Hopetoun and Israelite Bay with Lake Warden recording the greatest numbers. Uncommon outside the south west and south coast.

Photo: Simon Nevill

Australasian Shoveler *Anas rhynchotis* 47-53 cm

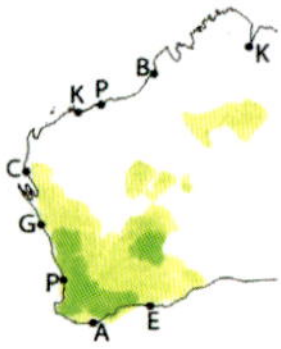

A very specialised feeder, one of the filter dabbling ducks, requiring wetlands with rich aquatic vegetation as bill is highly adapted to filter feeding. Found mostly on large deep-water lakes on the Swan Coastal Plain and the Southern region. More sedentary than most ducks, remaining where food sources are reliable but have been known to wander as far as Carnarvon and along the coast to Esperance. The male has brighter and more varied colours, with bright chestnut on belly and flanks, head is blue-grey rather than brown-grey. Male in eclipse plumage looks very similar to female but male flanks have chestnut tinge to the brown plumage, whereas the flanks of the adult female are plain brown. Also the male has a slightly darker head but these features are only noticeable when birds are close together. There are two subspecies in Australia, the nominate subspecies *rhynchotis* in Australia, the

Photo: Simon Nevill

other *variegata* in New Zealand. **Nest:** Nests close to water, nest lined with grasses and down. Normal clutch size 6 to 10. **Similar species:** None except the rare vagrant, the Northern Shoveler. The females are almost identical and very difficult to split. The Northern Shoveler female has a slightly lighter head and body but very difficult to note. With the male adults there is no confusion. **Status:** Uncommon to scarce throughout the region with the highest concentrations on the Swan Coastal Plain. Some authorities say the species is on the increase but the author questions this: some of the major lakes in Perth in the late 70s held far more than they do now. Jaensch et al between 1981 and 1985 recorded approx. 2000 on Lake Thompson and 2000 on Forrestdale Lake. Uncommon to rare outside the south west. Rare in the Kimberley. **Where to find:** Any of the larger lakes in Perth and the Swan Coastal Plain such as the Peel Inlet. Can be found on larger lakes out side the south west when dryer regions higher rainfall than normal like Lake Gregory but numbers in single figures.

Northern Shoveler *Anas clypeata* 47-54 cm — Vagrant

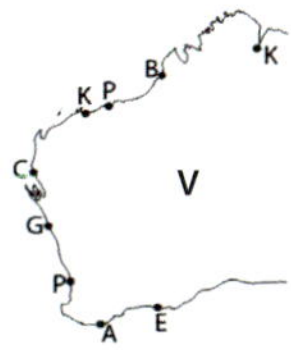

A northern hemisphere breeding bird from North America through Europe and to Alaska. Winters in Central America. Africa and Asia. Normal wintering range as far south as Malaysia but many flyas far as Indonesia with a few vagrants over wintering in Australasia. **Similar species:** Australasian Shoveler male in breeding plumage unmistakable. Lacks the contrasting white breast has scalloped flanks not plain chestnut colour. However in eclipse plumage the Northern Shoveler head is dull metallic green. Unfortunately the females of these two species are almost identical and virtually impossible to split. The bill is more orange coloured with faint brighter orange edge to lower beak with slightly more white in tail in flight on the Northern Shoveler female but it is difficult to pick these features out with a raft of ducks at distance. So let's hope it's a male even in eclipse plumage. **Status:** rare vagrant. **Where to find:** Possibly any freshwater wetland in the state, most records in south west.

Photo: Simon Nevill

Pink-eared Duck *Malacorhynchus membranaceus* 40-45 cm

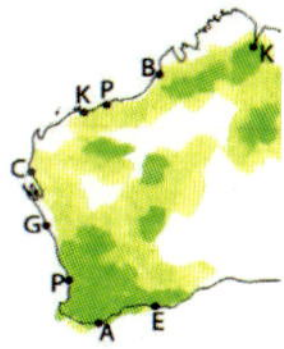

One of the most beautiful Australian ducks. A feature of the adult breeding birds not found on the juvenile is a small pink mark behind the eye, so small as to be indiscernible at distance. It has unique bill designed for specialised filter feeding on microscopic invertebrates and algae. A highly nomadic duck and unlike the majority of species of duck, wanders far into inland lakes and floodwaters. Has a preference for recently flooded country that is shallow with plenty of aquatic fauna and vegetation. Some ephemeral

Photo: Simon Nevill

wetlands just 100 km north of Perth have had numbers as high as 2500 birds but generally Pink-eared Ducks are uncommon on the Swan Coastal Plain. They have a peculiar feeding habit whereby two birds spin very fast in a tight circle only about 60-70cm apart, head to tail. This feeding technique stirs up invertebrates and is called 'vortexing'. At first sight it appears that it is a mating ritual but this is not the case. This form of co-operative feeding is unusual in the smaller waterbirds. **Nest:** Does not build nest but uses old nests or uses tree hollows. Large amounts of down are used and eggs can be totally covered. Normal clutch size 6 to 8. **Similar species:** None. **Status:** Generally an uncommon species but gathers in large numbers on ephemeral flooded wetlands often in remote regions. **Where to find:** In the Perth Region: Bayswater Sanctuary. Herdsman Lake. Lake Monger. Kogolup Lake. Lake Bibra. Forrestdale Lake. In the Wheatbelt: Dumbleyung Lake. Towerrinning Lake. Capamaura Lake. Coomelberrup Lake (excellent but only good after seasonal rains alas mostly dry). Nonalling Lake (a good wetland but often dry). In drier regions: Rowles Lagoon. Nallan Lake. Wooleen Lake. Fortescue floodplain. Any flooded area in the north west. Parry Lagoon can have large numbers. Lake Gregory (10,000 +) also Sandfire Flats can have high numbers.

Eurasian Widgeon *Anas penelope* 48 cm — Vagrant

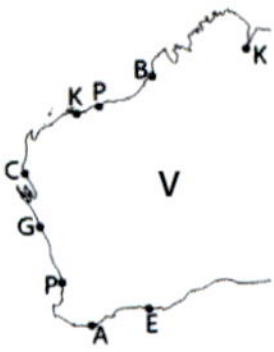

Extremely rare vagrant that breeds in the northern Paleartic. European birds over winter in Africa and those that breed in northern China over winter in parts of central East Asia including Thailand and the Philippines. So possibility of over passing to Australia. **Similar species:** Male none. Female possibly Grey Teal but far more rufous colouring on the flanks and head of the Widgeon. **Status:** Extremely Rare. **Where to find:** Any freshwater wetland.

Male

Female

Photos: Simon Nevill

Garganey *Anas querquedula* 36-41 cm

Vagrant

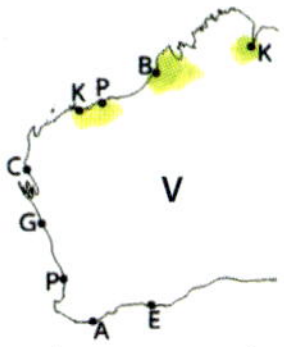

A northern hemisphere species breeding from the British Isles in the west to Japan in the east. Over winters in Africa and South East Asia and the extensive wetlands of southern New Guinea. A rare vagrant to Australia but every year there is a high possibility of this species being sighted particularly in WA, as the Kimberley is directly south of its wintering grounds. In WA it has been seen as far south as Yanchep National Park and more so the Pilbara. Most years it is sighted in the Kimberley from Lake Eda near Broome across to Parry Lagoon in the east Kimberley. The male in breeding plumage is fairly distinct having a clear white supercilium eye stripe running from the front of the eye to the rear of the nape. The white contrasts with a dark streaked head and scalloped breast. Also has extended long white scapular with black margins making it very distinct. In eclipse plumage the male white stripe on the side of the head is still clear. **Similar species:** The female can be confused with Pacific Black Duck but Garganey are a smaller duck, even slightly smaller than Grey Teal. In flight Garganey lack the bright green secondary coverts being a dull grey in both sexes. Check Pintail notes. **Where to find:** Possible to see on any Kimberley freshwater wetland most years. Has been sighted in many locations throughout Australia as far south as Perth in WA and in the Ballarat region of Victoria.

Photos: Simon Nevill

Northern Pintail *Anas acuta* 55-57 cm (not including extended tail)

Vagrant

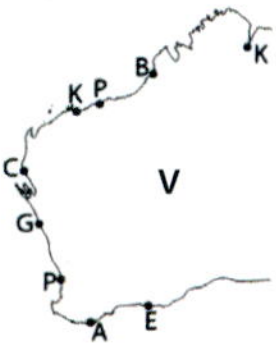

A rare vagrant from the Holarctic and north west China that may over pass from its wintering grounds in Asia. Has been sighted in WA and possible to be found on any wetland, not necessarily in northern WA wetlands as it is possible to find in any southern wetland like Vasse Inlet, as this is a duck of the temperate zones. The male in breeding plumage unmistakable but out of breeding very similar to female but back darker and flanks lighter brown. Importantly both sexes have longish tails the male's becomes much longer in breeding. **Similar species:** Both sexes of Garganey and

Photo: Simon Nevill

males in non-breeding plumage are smaller and have shorter neck and both have white eye markings. Both sexes have darker head than Garganey. The Grey Teal is closer to the size of Pintail but has a shorter neck and shorter tail. Both the male and female Pintail in eclipse plumage are much paler and have an overall creamy-brown rather than a grey-brown as in the Grey Teal and the dark browns of Pacific Black Duck and Garganey.

Photos: Simon Nevill

Green Pygmy-Goose *Nettapus pulchellus* 40-45 cm

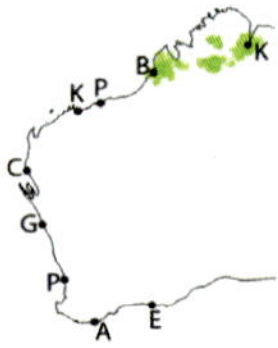

A waterbird of the tropical waters of Australia. In WA, as for most wetland birds this species is more prolific in the eastern Kimberley but possible to find on all wetlands across the Kimberley. Not only found on larger bodies of freshwater, this species is often found on the smallest of freshwater pools. Recorded far less in the eastern Kimberley than previous historical records. **Nest:** Conflicting reports but most records as for Cotton Pygmy-goose are to be found in hollows of trees. Clutch size quoted by 'North'. 8-12. **Similar species:** None. Cotton Pygmy-goose not found in WA. Has plain white flanks not black and white scalloped flanks. Has white neck and head. **Where to find:** Best Parry Lagoon. Lake Argyle. Camballin wetlands.

Photo: Simon Nevill

Great Crested Grebe *Podiceps cristatus* 48-61 cm

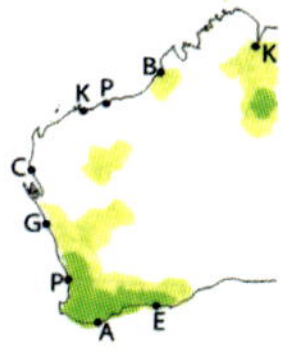

Native to Europe, much of southern Asia, parts of Africa and Australia. The Great Crested Grebe is noted for its spectacular courtship displays, in which the pair speed towards each other and meet face to face raising their necks high while leaning backwards and shaking their heads rapidly, then pull away. The most interesting display, which normally precedes mating, is the known as the 'Weed Dance', where the male dives and collects weed in his bill and swims towards the female shaking his head rapidly from side to side. They then both dive and gather weed and come together both shaking heads. If you are lucky in the breeding season (August–September) on the south west corner of Lake Monger or the south west corner of Herdsman Lake opposite Lakeside Drive, you may witness this unique performance. Great Crested Grebes mostly feed on small fish and hence are very strong swimmers and can dive with ease to reasonable depths. They require unpolluted fresh water and are normally a sign of a fairly healthy wetland although one may question this in the case of Lake Monger. There are three subspecies in the world, only one subspecies cristatus in Australia. **Nest:** Consists of a floating raft of reeds and aquatic grasses away from shoreline. Normal clutch size 4 to 5. **Similar species:** None. **Status:** Has expanded its range over the last century. It was first recorded nesting in Perth in the mid 1930's (Johnstone and Storr 1998). It now nests in several locations in Perth and the South West. **Where to find:** Lake Joondalup. Herdsman Lake. Lake Monger. Lake Joondalup. Wannamal Lake. Chittering Lake. Grassmere Lake. Has been recorded throughout much of the state including the Kimberley but is uncommon to rare outside of the south west region, although Lake Gregory has had over 70 sighted in one period.

Photo: Simon Nevill

Hoary-headed Grebe *Poliocephalus poliocephalus* 29-30 cm

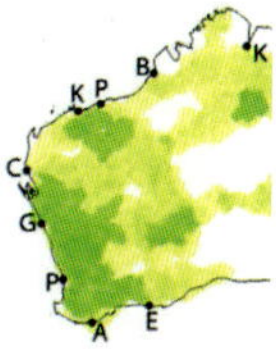

A nomadic Grebe that has a high tolerance to saline waters, hence it occurs throughout the arid regions as well as the south west corner. When claypans fill after heavy rains Hoary-headed Grebes may visit in large numbers feeding primarily on invertebrates and small fish. In the arid regions, they feed on brine shrimps. **Nest:** Nests are floating reeds or grasses often on remote lakes and sometimes in small colonies. Normal clutch size 4 to 6. **Similar species:** The non-breeding adult Australasian Grebe looks very similar to the non-breeding Hoary-headed Grebe and often requires the birds to be close to see the differences. The habitat tends to give you the first lead. The Hoary-headed Grebe has a longer neck. The author finds the plain grey flanks of the Hoary-headed Grebe the quickest identification detail; the Australasian Grebe has always a light chestnut-brown flank so tends to look a browner bird and the Hoary-headed a greyer bird. The dark crown meets the light cheek passing through the eye on the Australian Grebe and passes just below the eye on the Hoary-headed but these details are not easy to see unless very close. Colour shade on flanks is much easier. In breeding plumage, there is no problem differentiating both species. **Status:** Locally common on large open bodies of water of with higher saline levels than Australasian Grebe can tolerate. **Where to find:** In Perth: Lake Monger. Lake Thompson. Kogolup Lake. Forrestdale Lake (Thompson and Forrestdale dry out in early summer) and Bibra Lake can be good, particularly in summer and autumn. Found throughout the Wheatbelt particularly the Esperance Lakes. Found in the Mulga on inland lakes like Lake Annean and Lake Austin when heavy rains have fallen. Wetlands throughout the north both permanent and ephemeral wetlands.

Photo: Simon Nevill

Australasian Grebe *Tachybaptus novaehollandiae novaehollandiae* 23-25 cm

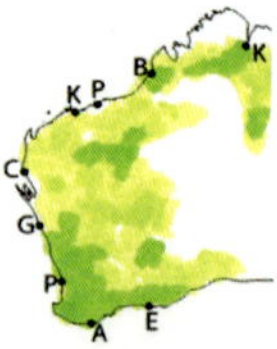

Although found together, the Australasian Grebe prefers wetlands containing fresh whereas the Hoary-headed Grebe can tolerate saline waters, which is why on the wheatbelt lakes you will get higher numbers of Hoary-headed Grebe. The Australian Grebe is found more in

Photo: Simon Nevill

the deeper South West and Swan Coastal Plain. The Australian Grebe will however travel to the arid regions in good seasons. Besides the bigger wetlands, they can be found on any small pool or dam. They eat small fish and aquatic invertebrates. Generally shy unless in urban areas. There are three subspecies in Australasia but only one *novaehollandiae* found in Australia and New Guinea. **Nest:** Nest is a floating pile of reeds with eggs covered over when bird leaves nest. Mostly close to shoreline but often concealed by reed beds. Normal clutch size 4 to 6. **Similar species:** Refer Hoary-headed Grebe. **Status:** Common in the southwest. Uncommon in the Station country except after good rains. **Where to find:** In Perth: Lake Joondalup. Lake Gwelup. Herdsman. Bibra Lake. Throughout the south west, including: Lake Chittering. Wannamal Lake. Vasse Estuary. Lake Toolibin and all freshwater lakes east to Esperance. Any wetland throughout the state but does not like wetlands with a high saline content. Found throughout the Kimberley.

Tricolored Grebe (formerly Little Grebe) *Tachyybaptus tricolor*

23-25 cm

Vagrant

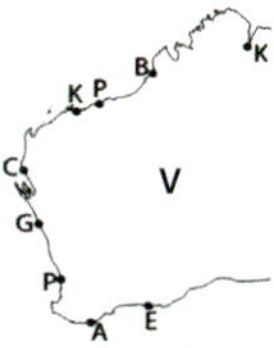

This species has only recently been split from the Eurasian bird the Little Grebe *Tachyybaptus ruficolis*. Also the subspecies *vulcanorum* has been lumped with *tricolor*. Range of this species: the Lesser Sunda Islands. Sulawesi. New Guinea and the Solomon Islands. This bird is often overlooked due to its close similarity to the Australasian Grebe and yet it is found close to north Australia being resident throughout much of south-east Asia as well as Java, Bali, Timor and New Guinea. Out of breeding plumage, this grebe would be difficult to split from Australasian Grebe, although in flight there is an important difference. Little Grebe has no white in the upper wings whereas Australasian Grebe has white on the secondaries. However, if a male Tricoloured Grebe is seen in full breeding plumage the extent of the rust colour covers the whole of the throat and is also a richer, brighter rust red and contrasts with a darker crown than Australasian Grebe. The rust colour on the Australasian Grebe is confined to the upper side of the neck and nape. **Status:** Rare vagrant, recorded only in the Northern Territory but there is no reason why it will not turn up on any Kimberley wetland. I feel this bird more than many species may be over looked and there are minor migration movements within Indonesia with this species. It is relatively common in Java and Bali and also New Guinea.

Photo: Simon Nevill

Australasian Darter *Anhinga melanogaster* subspecies *novaehollandiae* 85-90 cm

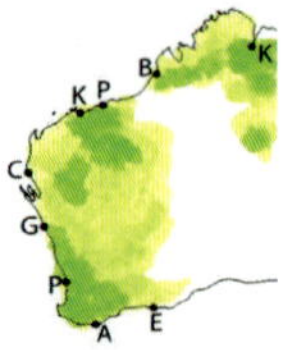

In the past, people gave the nickname Snake Bird to the Australian Darter, due to its ability to almost totally submerge its body leaving only its long neck exposed and giving it a snake-like appearance in the water. A fairly solitary feeding bird not forming feeding flocks like other wetland birds; however will congregate with other species when breeding within their colonies, particularly cormorants. Darters need to be solitary as their feeding technique requires stealth to catch their prey, mainly fish, by bending the long neck backwards in a snake-like fashion and then darting out the neck, spearing the fish with their sharp pointed bill. The permanent kink in the neck is thought to be part of their powerful mechanism to spring the neck forward at great speed. The male bird in full breeding plumage is a beautiful bird with a metallic black sheen to the plumage. It is identified from the female by having a chestnut throat, brighter when in breeding plumage, and is basically black all over, whereas the female has a cream underbody and no chestnut

Male

Photos: Simon Nevill

colour to the throat, and the upperbody and head is a lighter brown colour. It does get a little confusing with the immature male and the female as their upper and lower bodies are both a similar light brown colour but the immature male always retains a dark head with a chestnut coloured throat. The reason why darters are seen sitting with wings outstretched is that their feathers do not repel water and after constant feeding in water, they remain wet and must dry out. The absorption does however allow them to swim and feed for several hours semi-submerged. There are three subspecies in the world, only one subspecies, *novaehollandia* in Australia. Normally nests within colonies of cormorants. **Nest:** Large twig structure with hollow top lined with leaves or rushes placed on middle level branches, mostly in paperbark swamps. Clutch size 3 to 4. **Similar species:** None. **Status:** Moderately common throughout the wetlands of the state, but never in large numbers. **Where to find:** Throughout the South West Region where freshwater wetlands occur. In Perth: Oldham Park on the northern side of the Narrows Bridge. Booragoon Lake. All along the Swan foreshore around the Narrows and Causeway (occasionally Fremantle North Mole). Pilbara on the larger river pools. Throughout the Kimberley as in the Ord River system. Lake Argyle. Parry Lagoon. Parts of the Fitzroy River system. Lake Gregory when full.

Great Cormorant *Phalacrocorax carbo* subspecies *carboides*

80-85 cm

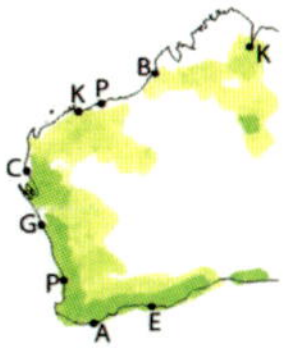

A cosmopolitan cormorant found in North America, Europe, Central and Eastern Asia, Africa and Australia. The Australian subspecies novaehollandiae is the largest cormorant. More of an inland species than the Great Pied Cormorant although can be found on coastal waters. Prefers big open bodies of water mostly in the South West Region. Dives well and can stay under water for nearly one minute. Feeds mostly on fish. A relatively solitary feeder unless food is abundant and even then it is uncommon to see more than 100 birds together. More nomadic than any of the cormorants flying hundreds of kilometres to better feeding sites. There are six subspecies in the world and only one in Australia. Nests mostly with other cormorants in colonies. **Nest:** A large twig nest is built, often being used year after year. Clutch size 2 to 4. **Similar species:** Little Black Cormorant, which is nearly 30 cm smaller, and has no white on the throat and yellow facial skin. **Status:** Moderately common in the southwest. Uncommon in the station country. Found in the Pilbara and the Kimberley. **Where to find:** In Perth on most of the larger lakes. Also Alfred Cove. Breeds at Booragoon Lake. In the South West Region: Namming Lake. Lake Dumbleyung. Wannamal. Guraga Lake. Harvey Estuary. Peel Inlet. Not so common on southern wetlands to Esperance. Larger pools in the Pilbara and Kimberley, also coastal estuaries on Pilbara coast. Dampier Peninsula and now on Lake Argyle.

Photo: Simon Nevill

Little Black Cormorant *Phalacrocorax sulcirostri* 55-58 cm

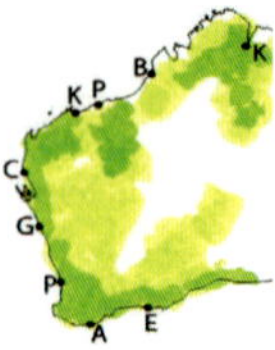

Found in greater numbers throughout the South West than the Great Black Cormorant. An all black cormorant with a dark green metallic sheen on the upper body, best when seen in full sunlight. Found on a diverse range of wetlands but also in large numbers on sheltered harbours and estuaries where rafts of cormorants will work together herding fish. Seen regularly flying up or down the Swan River in 'v' formation to new feeding grounds. **Nest:** A twig nest placed in trees; nests in large colonies, as at Booragoon Lake. Nest is smaller than most cormorants, placed at the extremity of branches mostly over water. Clutch size 3 to 5. **Similar species:** refer Great Black Cormorant. **Status:** Common throughout South West and mid west coastal regions. Scarce in the Wheatbelt and station country. Fairly common on pools in the Pilbara and Kimberley. **Where to find:** Throughout most of Perth wetlands and estuaries. Peel Inlet. Vasse. Lake Towerrinning (when flooded). Lake Toolibin (When flooded). Lake Warden. Fortescue River. Ashburton River. Fitzroy River. Ord River. Huge numbers on Lake Gregory (15,000 +) when sufficient water.

Photo: Simon Nevill

Australian Pied Cormorant *Phalacrocorax various* subspecies *hypoleucos* 72-80 cm

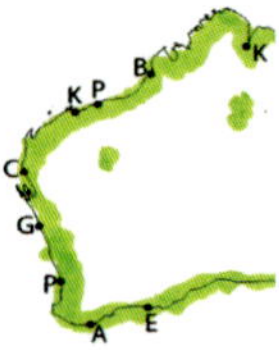

Primarily a marine species but does travel to some large inland lakes and estuaries. Mostly the adult birds stay within 30-40 km of their breeding area throughout the year. Feeds mostly on fish, some crustaceans taken. Breeds in colonies, sometimes in large numbers. Very numerous around the Denham and Geraldton regions (breeds on the Abrolhos Islands). Develops orange patch at base of bill in the breeding season. Needs to dry wings after long periods of diving. There are 2 subspecies, *hypoleucos* in Australia and *various* in New Zealand. **Nest:** A large twig structure with a shallow hollow centre. Nests in both trees and on the ground in colonies. Clutch size 2 to 3. **Similar species:** Black-faced Cormorant and to a lesser degree Little Pied Cormorant. Black-faced Cormorant has no yellow face or pink gape and no white

Photo: Simon Nevill

feathers near eye. Pied Cormorants in Western Australia are uncommon in the region around Esperance where Black-faced occur. Little Pied Cormorant is 18 cm smaller and except when juvenile does not have a black band on the flanks, and has no yellow facial patch. **Status:** Very common in Dirk Hartog Island and Denham region. Common along Geraldton Coast. Uncommon mid-west. Common off Perth all the way to Busselton. Not so common on the southern coast. **Where to find:** Perth coast and estuary. Alfred Cove. Rottnest Island. Vasse estuary. Peel Inlet. The entire coastline from the De Grey to Esperance and parts of the Kimberley coast. Huge numbers in the Shark Bay region. Established now on Lake Argyle.

Little Pied Cormorant *Microcarbo melanoleucos* 55-58 cm

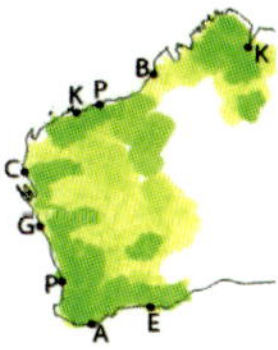

The smallest of all the cormorants. Almost a role reversal of the Pied Cormorant, it favours freshwater and inland waters but it too will feed on sheltered coastal waters though to a lesser degree. The most catholic in its choice of waters of all the cormorants. Prolific throughout most of the region, found in any small lake or dam. Not a popular bird with some farmers as they eat the yabbies from dams. However, their main diet is small fish. Feeds communally when food abundant, often with Little Black Cormorant. There are three subspecies of Little Pied Cormorant, only one in Australia namely *melanoleucos* which is also found in Indonesia, New Guinea and the Solomon Islands. **Nest:** Nests communally, often with other species of cormorant and Darters. Nest is small structure of twigs built in trees with small amount of leaves in central bowl. Clutch size 3 to 4. **Similar species:** Pied Cormorant, which is far larger and has black band on flanks and a yellow facial patch. **Status:** Very common throughout the region except inland in the drier regions. Less common in the northern part of the coastal belt gaining in numbers from Perth all the way around the South West coast to Esperance. Lake Warden can have up to 600 birds at one time. Throughout the Wheatbelt to Kalgoorlie in lower numbers but wherever there is water, there is the possibility of Little Pied Cormorants. Found in the Pilbara and Kimberley. **Where to find:** Throughout the south west but only casual in the inland regions. Found in the Pilbara and Kimberley wetlands and Lake Gregory but numbers far less than Little Black Cormorant.

Photo: Simon Nevill

Black-faced Cormorant *Phalacrocorax fuscescens* 62-70 cm

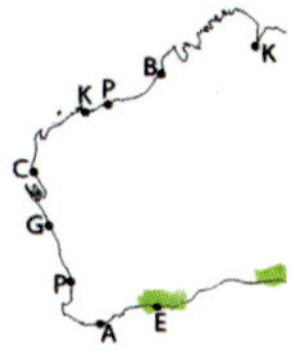

Unlike most cormorants, which may venture to inland waters, Black-faced Cormorants are the most sedentary birds of the cormorant group, not travelling far from their breeding and roosting sites in the southern seas. In Western Australia its range is restricted to the eastern southern coast of Western Australia. It nests on the islands off Esperance and those off Cape Arid National Park. Immature birds will fly to islands other than their original breeding site. **Nest:** Nests mostly on islands either directly on rock or sometimes large seaweed structures placed on ground. Clutch size 2 to 4. **Similar species:** Similar to

Photo: Simon Nevill

Pied Cormorant but lacks the yellow face and does not develop a pink gape during breeding. Black-faced has no blue orbital eye ring. The Little Pied is smaller again with no black band on the flanks and the eye is completely surrounded by white plumage on the face. **Status:** Locally common off the Esperance coast and further east to Daw Island. Does not come further west than the Hopetoun region. **Where to find:** The Esperance jetty often has Black-faced Cormorants at the end of the long jetty. A boat trip to Woody Island will pass by nesting sites on lone rock islets.

Australian Pelican *Pelecanus conspicillatus* 1.6-1.8 m

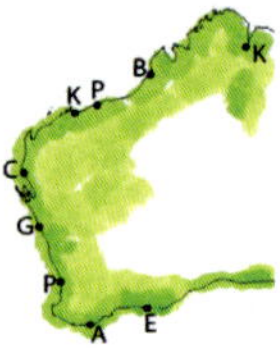

Of the 7 species of pelican in the world, this is the only member of its genus in Australia. Most people know that pelicans eat fish but there is a side to their behaviour, which is less well-known. They will manoeuvre close to ducks or other water birds, and particularly smaller birds such as ducklings, suddenly swing their necks and throw the body forward to try scoop up their prey. Luckily for the ducklings they often fail but it shows the extent of their carnivorous and opportunistic feeding behaviour. Mostly when not feeding alone, they work in groups to catch fish plunging their heads in unison. Found throughout most of the South West coastline and to a lesser degree on inland waters. Pelicans will fly as far as two thousand kilometres after heavy rains have fallen on inland salt lakes. They also fly regularly between the Swan Estuary and Peel Inlet by taking advantage of the thermals taking them to great heights whence they will literally glide down to their destination. Highly nomadic travelling many hundreds of kilometres to well watered areas, some banded birds being found nearly two thousand kilometres from their breeding site. **Nest:** Nests in colonies. Two of the main nesting sites in the south west are on the Peel Inlet and a small rocky island opposite Emu

Photo: Simon Nevill

Point in Albany. In the north, breeds on the Lacepede Islands. Lake Gregory. Nests mostly on islands on large bodies of water or sheltered inlets in dense colonies. Nest is scraped in sand or of small seaweed and debris on rock. Clutch size is normally 2. **Similar species:** None. **Status:** Common throughout the South West coastline. Uncommon along the Esperance/Cape Arid coastline. **Where to find:** Most large lakes in Perth, particularly Lake Monger (not guaranteed) and the Swan Estuary. All other major inlets along the coast to Bremer Bay. Most large estuaries and larger pools in the Pilbara and Kimberley will be visited by Australian Pelican but they will move to where fish are in reasonable numbers.

White-necked Heron *Ardea pacifica* 76-106 cm

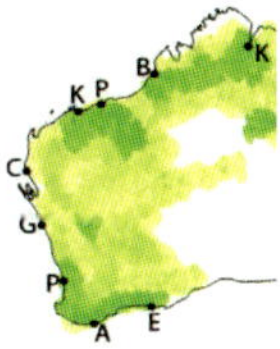

The largest heron in the South West. Unmistakable long white neck set against blue-grey body. Has a languid flight with the two white shoulder patches being visible in flight, also showing its kinked neck. Normally, a very nervous heron, which does not allow a close approach and soon takes flight, calling as it flies. Found on the edges of large lakes but has a preference for small ephemeral pools where it will clean out the trapped fish. Nests quite high in trees, sometimes solitary nester but other times in loose small colonies. Builds a broad flat twig nest. Clutch size 3 to 5. **Similar species:** None in this region. In the north of the state immature Pied Heron similar but far smaller. **Status:** Moderately common throughout WAY but not the desert regions and the Nullarbor. Locally common after good seasons in mid-west, particularly from Carnarvon to Gascoyne Junction. Moves to the South West when inland wetlands dry out in summer. Common on wetlands in both the Pilbara and the Kimberley but normally a solitary feeder and defends its feeding territory robustly. **Where to find:** Most of the larger freshwater lakes with shallow margins and dams particularly on the Swan Coastal Plain, also the south coast wetlands. Any larger water pools in the Pilbara and Kimberley but will also be found on small pools where it will remove the vast amount of fish remaining and move on.

Photo: Simon Nevill

White-faced Heron *Egretta novaehollandiae* 66-68 cm

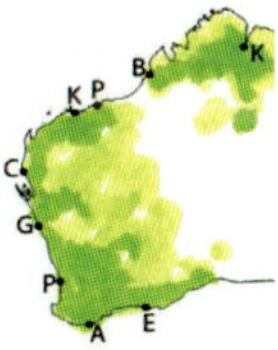

More numerous than White-necked Heron and seen on most small wetlands and small drainage channels. Feeds on fish, frogs, reptiles and invertebrates by stalking and waiting for its prey. It will gather in dry fields of stubble when grasshoppers are in high numbers, also well watered grass fields, feeding off insects, and occasionally found on beaches. Develops

White-faced Heron Photo: Simon Nevill

additional plumes on neck and breast in the breeding season. **Nest:** Flat structure of twigs between 5 and 15 metres above ground. Mostly a solitary nester. Clutch size 3 to 5. **Similar species:** Dark morph of the Eastern Reef Egret, which has no white on the face, a thicker bill and shorter legs. **Status:** Common throughout the whole state. **Where to find:** All types of wetland and, like the White-necked Heron will take advantage of ephemeral pools left over from heavy rains.

Great-billed Heron *Ardea sumatrana* 100-110 cm

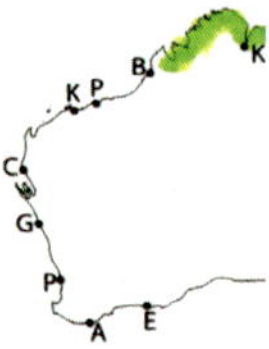

A large grey-brown heron with white underbody and white throat. In breeding white fine plumes on foreneck and small nape crest. Large stout blue-grey bill. Dark grey legs. Immature birds are all brown with cream underbody. Surprisingly not an uncommon bird in the Kimbcreley, as there are thousands of acres of relatively pristine mangroves that line the tidal creeks and coastline of the northwest and northern Kimberley. However, it is a shy bird that will take flight if approached too closely or simply walk back into the mangroves. Seen most often walking the margins of tidal mudflats at the edge of the mangrove belt.

Prefers feeding early or late in the day. Fairly territorial and remains in a favoured feeding area where it will also nest. It is a solitary feeder and does not nest or congregate in colonies. It has a loud far carrying call like a strong croak. **Nest:** Nest is very large located on horizontal branches of large mangroves. Apparently will use the nest year after year. **Similar species:** None. White-faced Heron far smaller and blue-grey not a brown-grey bird. **Status:** Relatively common in the remote mangrove regions of the Kimberley. Where to find:. Anywhere where there are exstensive mangroves.

Photo: Simon Nevill

Pied Heron *Egretta picata* 44-50 cm

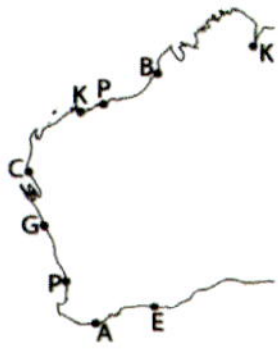

A heron of the tropics rarely found south of Broome. Requires classic tropical wetlands with wet grassy foreshores where it feeds. Often found in large groups. Its attractive plumage belies its scavenging habits although it does feed in normal heron fashion most of the time. Feeds on frogs and invertebrates mostly found in wet flooded grasslands alongside lagoons with head held forward waiting for movements of prey. Adult birds have a blue grey body, white neck and blue-grey crown that passes below the eye, pale cream legs. In breeding the crown feathers are extended and remain erect behind the nape and white breast feathers extend over lower chest. Legs pale yellow. Immature birds do not have a blue-grey crown but head and neck are totally white giving the appearance of a small White-necked Heron. **Nest:** Platform stick nest either in trees or mangroves, nest in colonies. Clutch 2–4. **Similar species:** None. White-faced Heron far smaller and blue-grey not a brown-grey bird. **Status:** Locally common in the Ord and Kununurra region uncommon in the western Kimberley. Where to find:. Occasionally found on the Roebuck Plains in the wet season and parts of the Fitzroy flood plain but largest concentrations are around Lake Argyle and Kununurra and Parry lagoon.

Photo: Simon Nevill

Grey Heron *Ardea cinerea* 90-100 cm Vagrant

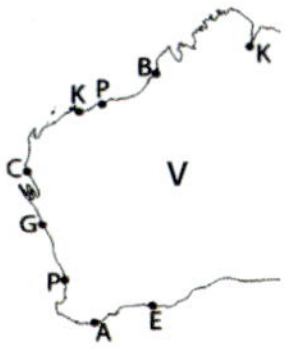

A rare vagrant to WA. but always possible to be sighted in the northern region as it is resident on the Indian Archipelago. The first sighting was in 2002 in Busselton WA.

Similar species: No other Heron similar in Australia.

Photo: Simon Nevill

Pacific Reef Heron *(Eastern Reef Egret) Egretta sacra* 60-65 cm

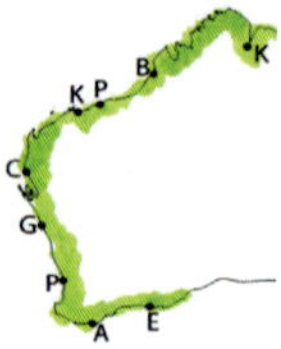

A bird found throughout Australasia and East Asia. There are 2 forms of Eastern Reef Egret: a white morph, which occurs more in the tropics and a dark blue-grey morph, which is more prevalent in the temperate regions of Australia. The 2 morphs do interbreed and not just at the transition zone as both occur in some areas together, although the white form is rare in the South West. They do not venture inland as other herons but remain on the coastal belt. They mainly feed on rocky parts of the coastline rather than sandy beaches, feeding from the small tidal pools left after the retreating tides. They eat fish, crustaceans and molluscs, easily breaking shells with their robust bill. **Nest:** Nests in the south mostly on islands, nest being a small collection of twigs placed on rock ledge. Clutch size 2 to 3. **Similar species:** The white morph is similar to Intermediate Egret but has slimmer body, finer all yellow bill, brown-black rather than pale yellow legs. Great Egret, far taller and longer necked. Little Egret, far smaller with fine black bill. It is rare for the white egrets in the South West to feed at tidal pools except for the Great Egret. Similar species to dark morph: White-faced Heron, which can feed on beaches but rarely does in the South West and does not like feeding on rocky coastlines as does the Reef Egret. The Reef Egret has no white on the face and has longer legs. **Status:** Uncommon to scarce on south coast with virtually no forms of the white morph. Moderately common on parts of the lower South West coast, increasing on the islands on the Abrolhos with a few white morphs. **Where to find:** Near Perth: on Rottnest Island, and the islands of Carnac, Penguin and Garden. On the south coast: Wilsons inlet. Cheyne Beach. Can be found where extensive rocky coastlines occur. Occurs along the full length of the West Australian coast mainly wherever there are rocky shores. Common on the Pilbara islands and the Kimberley coast.

Photo: Simon Nevill

Great Egret *Ardea alba* subspecies *modesta* 83-103 cm

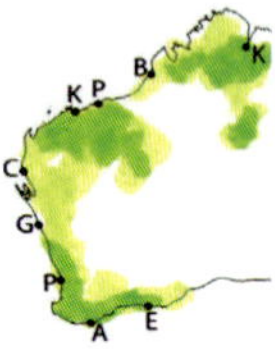

Tallest of all the egrets with very long 's' shaped neck, which at full stretch, is longer than the body length, a useful diagnostic feature. No other heron has that neck to body ratio, and it is the only egret where the gape passes the eye; also has the longest legs. People talk of bill colour as a main diagnostic feature but care must be exercised as the immature bird has an all-yellow bill, the non-breeding adult, a yellow bill with long black tip, and then in breeding plumage the bill becomes all black. In breeding plumage the facial skin becomes lime-green in colour. Found throughout much of the world, it was formerly a rare visitor to the South West but after the 1930s has increased in number. Will feed on inland wetlands, moist grasslands and coastal mud flats. Only one subspecies *modesta* in Australia. **Nest:** Nests mostly with other egrets and herons in small loose colonies. Flat twig nest placed in tree mostly over water. Clutch size 3 to 4. **Similar species:** Intermediate Egret has neck noticeably shorter than body length. Gape stops short just below eye and does not pass it. Legs in proportion to body shorter than Great Egret. You are unlikely to see Intermediate unless a vagrant comes to the Carnarvon region. **Status:** Common in most of the lower South West and particularly on the Swan Coastal Plain. Uncommon in the western part of the Wheatbelt and Southern Coastal Regions. Common throughout the Pilbara and Kimberley not only on the larger bodies of water but also the smallest. **Where to find:** Most large lakes in Perth. Vasse Inlet. Peel inlet. Australind. Throughout the Pilbara and Kimberley wetlands and river systems. Found as far inland as Lake Gregory.

Photo: Simon Nevill

Intermediate Egret *Egretta intermedia* 58 cm

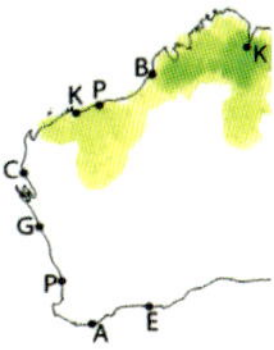

A species that has slightly increased in numbers over recent years, mainly due to the damming of the Ord River. It still is not a common bird in the far north east of the Kimberley. Found whereever large bodies of freshwater occur. On the east coast of Australia the species is far more common and is found as far south as Tasmania but in WA it is mainly restricted to the north west Kimberley wetlands. It occasionally comes down to the Pilbara region. In breeding plumage the male develops a reddish bill with a distinct emerald green cere. In non breeding has a yellow bill and cere. **Similar species:** Similar to all white egrets. The Great Egret is taller, bigger bird. Its head neck are twice that of its body length but in the Intermediate neck and head are approximately half the body length. Has black bill in breeding plumage. Little Egret always has a black bill. Much smaller in height. A very active feeder. In breeding plumage has two plumes leading from the nape. **Status:** Uncommon. **Where to find:** A Parry Lagoon. Ord River. Lake Argyle. In the irrigation channels close to Kununurra.

Photo: Simon Nevill

Little Egret *Egretta garzetta* subspecies *nigripes* 55-65 cm

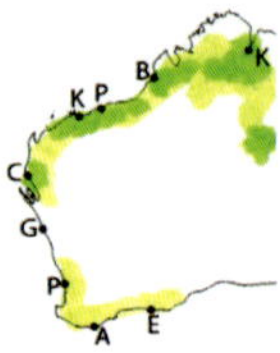

Native to Africa, Asia and Australasia, the Little Egret did not establish itself in the lower South West until the early 1900s. It is still scarce in the South West being centred on the Swan Coastal Plain. The smallest of all the egrets in Australia, with a fine bill that is always black, although the juvenile has just a slight yellow colouring to the lower mandible. Long legs for its size and as they lift their feet, occasionally the yellow soles can be seen from behind. Has two fine plumes coming from the nape during the breeding season. It is a more active feeder than the larger egrets, running and darting around to get prey. 2 subspecies in the world, only one *nigripes* in Australia. **Nest:** Nest and location similar to Great Egret. Clutch size 2 to 4. **Similar species:** None really: the Cattle Egret comes near it for size but it has shorter legs and a stocky build, having a shorter thicker neck, and a shorter and thicker bill that is yellow and not black. There is certainly no confusion when the Cattle Egret is in breeding plumage with its orange tinged feathers on the crown, back and chest. **Status:** Scarce in the South West and mainly on the Swan Coastal Plain. Numbers increase towards the Carnarvon region. Is more prevalent on the wetlands of the Pilbara but not common, however the species is more common in the Kimberley than those waters in the south. **Where to find:** Lake Jandabup (when water there). Lake Joondalup. Herdsman Lake. Nests in paperbark swamps near Pinjarra. One of the best places to see them is in the Peel Inlet and Australind area. Carnarvon is the next main area along the west coast. Ord River, parts of Lake Argyle. Found on many small wetlands particularly pools supporting water lilies, a sign of pristine pools that Little Egret prefers.

Photo: Simon Nevill

Eastern Cattle Egret *Bubulcus coromandus* 70 cm

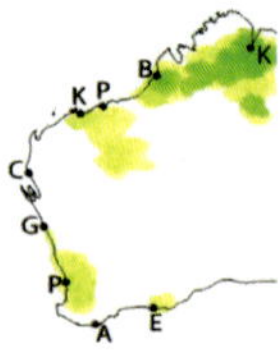

Photo: Simon Nevill

With the clearing of land worldwide and the introduction of cattle and water buffalo, the Cattle Egret has expanded its range coming to Australia in the late 1940s and first reported in the South West in 1952. Cattle Egrets have been recorded in several locations in the South West. As the name implies it is a bird that accompanies livestock, mainly cattle; the moving cattle stir up insects, which the Cattle Egret retrieves. This symbiotic relationship would have been going on for centuries with Brahmin cattle and water buffalo in Asia and Africa. When in breeding plumage, the bird is unmistakable with its orange head and chest, and bill that turns orange-red. When not in breeding plumage, its thickset jowls, short neck and short legs still make it an easy egret to identify. 2 subspecies in the world, only 1, *coromanda*, in Australia. **Nest:** Nest is small flat structure placed in tree or bush mostly over water and normally not more than 5 metres off the water level. Nests in colonies with other egrets, herons and cormorants. **Similar species:** Refer Little Egret. **Status:** Scarce in the well watered South West. Spasmodic distribution with greater numbers in parts of the Kimberley mainly around Lake Argyle but nowhere near the populations typically found in the eastern states.

Nankeen Night-Heron *Nycticorax caledonicus* subspecies *mannillensis* 55-65 cm

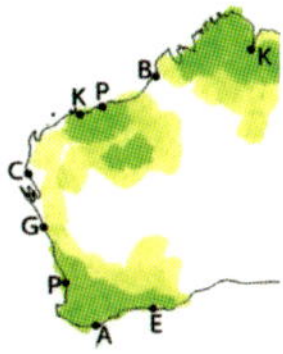

The name comes from the colour of the cotton cloth made in the city of Nanjing (once Nankeen) in China. Primarily a nocturnal feeder that roosts through the day although they will feed during daylight hours. They leave their well-sheltered roosting sites at dusk to find their feeding grounds for the night. They nest and roost in colonies such as the swamp at Australind. The adults do not develop their adult plumage until the end of the first year, which a rich chestnut colour on the upperbody and cream white underbody with no streaking. The crown is black with two white feathers leading from the nape of the head, called nuptial plumes. Above the eye and below the black crown is a white supercilium eye stripe. The immature birds lack the clean pure colours but have white and brown streaking on the breast and white and brown flecking on the upperbody. When very young, they have no black crown, just heavy white streaking. 6 subspecies in the world, only 1 *mannillensis* in Australia. **Nest:** Nest normally between 5-15 metres above water, mostly a solitary nester. Clutch size 2 to 3. **Similar species:** The Striated Heron has similarities to the immature Night Heron but the Night-Heron does not feed on exposed mud flats though it occasionally can be found on saline creeks. However, the Striated Heron does not occur south of the Shark Bay region. The Little Bittern is a far smaller bird with a chestnut face, no head plumes, darker back, and no white supercilium eye stripe, found mainly in long reed beds and very uncommon. **Status:** Relatively common on the Swan Coastal Plain and deep South West. Less common on the southern coast. Is found well into the Wheatbelt where there are heavy reed wetlands. Along the river systems of the Pilbara and Kimberley. Particularly common along pandanus-lined creeks in the Kimberley. **Where to find:** In Perth. Trees in the Perth Zoo. Narrows Bridge on the north side. Fremantle Esplanade opposite the Esplanade Hotel not so much in the pines but more in the dense cedar type trees. Herdsman Lake. Bibra Lake. Lake Chandala. Lake Toolibin (if there is water) and most wetlands in the south where there are stands of paperbark. On many of the larger pools in the Pilbara. Common on the Fitzroy, Mary, Durack, Pentecost and Ord Rivers. Numerous throughout the many freshwater pools throughout the Kimberley.

Immature

Photos: Simon Nevill

Striated Heron (Formerly Mangrove Heron) *Butorides stratus*

subspecies *rogersi* 40-44 cm

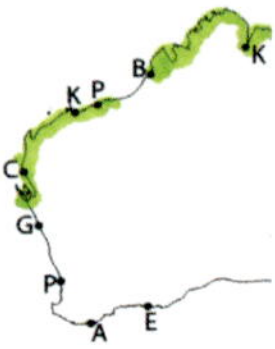

Found throughout much of Asia and Australasia. In Western Australia a bird of the northern region, feeding on exposed mudflats, and mud-lined tidal creeks often in mangroves along the coast. Feeds by lowering body just above the mud with neck withdrawn ready to strike, catching crabs and mud skippers. Immature birds have streaking on both the upperbody and underbody. There are 30 subspecies in the world with five in Australia, the only one in this region being *rogersi*. **Nest:** Nests almost entirely in mangroves. Clutch size 2 to 3. **Similar species:** Immature Striated Heron has similarities to Nankeen Night Heron but do not share the same territory. **Status:** Found from Shark Bay to the northern mangroves of the Kimberley. **Where to find:** On some of the mangrove-lined bays on northern Peron Peninsula. In some of the mangroves between New Beach and Carnarvon. Mangroves at Mangrove Point Carnarvon and Miaboolia just north of Carnarvon. Mangroves in Exmouth Gulf. Samson Point. De Grey estuary. Throughout all the mangrove areas on the Kimberley coast. Seen easily at Roebuck Bay.

Photo: Simon Nevill

Glossy Ibis *Plegadis falcinellus* 52-57 cm

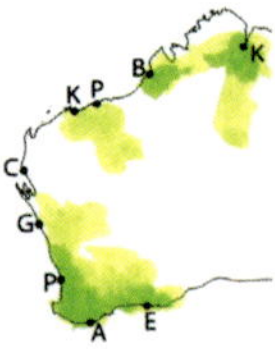

A bird found in parts of Africa, Asia and Australasia but is monotypic. The smallest Australian ibis. First recorded in Perth in 1906. Scarce to uncommon on the Swan Coastal Plain. Has nested at Chandala Lake near Chittering. Has a beautiful metallic bronze gloss when seen in full sunlight. Feeds on grassy edges of freshwater pools, has a preference for shallow waters. **Nest:** Nests either on flat rafts of reeds located in small swamps or up in paperbark trees on flat twig nests with leaves as base. Clutch size 3 to 4.

Photo: Simon Nevill

Similar species: None. **Status:** Uncommon. **Where to find:** In Perth: Not common but seen at Herdsman Lake regularly. Lake Joondalup also Thompson and Forrestdale Lakes when winter wet. On the Swan Coastal Plain: Wallering Swamp near Gingin. Benger Swamp. Lake McClarty when winter wet. Found on ephemeral lakes in the Pilbara. Samphire Flats after heavy rains. Wetlands at Roebuck Bay in the wet. Parry Lagoon. Taylors Lagoon. Lake Eda and Lake Argyle.

Australian White Ibis *Threskiornis molucca* 70 cm

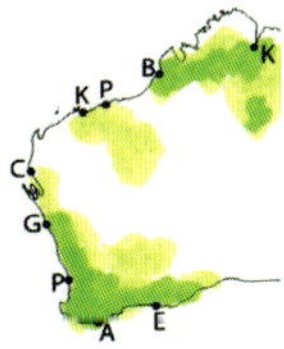

First recorded in the South West Region in the 1950s. Has expanded its population to become a very common bird. Alas, has become a scavenger at waste disposal sites and has also been displacing cormorants from their breeding sites. Seen in paddocks on the Swan Coastal Plain often with Straw-necked Ibis feeding on earthworms, insects and grasshoppers. They will also feed on freshwater mussels, fish and invertebrates. Gregarious, feeds in large groups. **Nest:** Nests both in reed beds and lignum but mostly in the South West in paperbark trees in dense colonies. Clutch size 2 to 3. **Similar species:** None. **Status:** Now very common on the Swan Coastal Plain expanding its range around the southern coast. Also travelling to station country when seasonal rains have inundated those areas. Locally common in parts of the Kimberley. Not so common in the Pilbara. **Where to find:** Throughout the Swan Coastal Plain and parts of the southern region such as the Kalgan River estuaries. Floodplains and fields in the Kununurra region.

Photo: Simon Nevill

Straw-necked Ibis *Threskiornis spinicollis* 75 cm

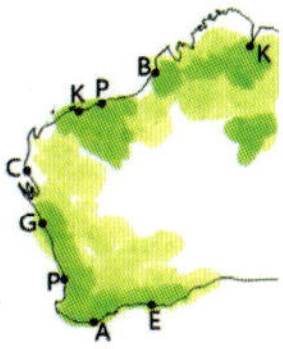

Slightly smaller than the White Ibis, the Straw-necked unlike the White Ibis is endemic to Australia. Often seen spiralling up hundreds of feet to new feeding grounds. Scarce in the South West until it became established in the mid 1950's, in fact all ibis have expanded their range in the South West due to the massive land clearing programs, with more grasslands being created. Seen often in large numbers in paddocks taking advantage of drying ground or simply feeding on grasshoppers and other invertabrates. They are quite nomadic and banded birds have been recovered several thousand kilometres away from the South West Region. **Nest:** They nest colonially. Location and clutch size similar to the Australian White Ibis, 2 to 3. **Similar species:** None. **Status:** Common in the South West Region, less so in the north towards the Carnarvon area. Numbers are increasing every year and in Esperance,

Photo: Simon Nevill

groups of up to 500 birds have been sighted. Not so common in the Pilbara but common in many parts of the Kimberley. Ventures further into arid regions than White Ibis, when conditions are favourable. **Where to find:** Carnarvon and inland along the lower Gascoyne catchment. South past Geraldton and all the way to Esperance. Common to very common on the Swan Coastal Plain. Common on parts of the Fitzroy flood plain as at Camballin. Ord River floodplain.

Yellow-billed Spoonbill *Platalea regia* 85 cm

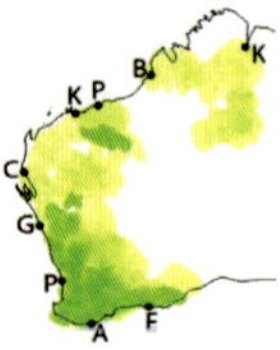

There are six species of spoonbill in the world, Australia having two of them; the Royal and the Yellow-billed. Spoonbills which are related to ibis but through evolution, rather than develop a probing method of feeding as ibis have done, spoonbills have developed a flattened bill with a spatulate end and become what is known as a 'tactile' feeder. By moving their open bill sideways back and forth through the water, they can feel the prey, they quickly clamp on and swallow it. The Yellow-billed Spoonbill is mostly a solitary feeder in Western Australia but will congregate with other spoonbills at times. In breeding plumage they develop nuptial plumes that grow from the lower throat and breast. They also develop a small red facial patch at the base of the bill. **Nest:** Nests in trees, making a flat stick nest lined with some small leaves or twigs but also nest on flattened reed beds in swamps, often in association with ibis. Clutch size: 2 to 3. **Similar species:** Royal Spoonbill which has a black rather than yellow bill. **Status:** Like many of the ibis and egret species, the Yellow-billed Spoonbill has increased in numbers from being very rare in the 1920s to now moderately common on some wetlands on the Swan Coastal Plain. Has spread as far as the Esperance lakes. Above Moora it is scarce and not until the Carnarvon region does it occur again in any reasonable numbers. Found on wetlands in the Pilbara more so than Royal Spoonbill. In the Kimberley, there are fewer Yellow-billed than Royal. **Where to find:** In Perth: Herdsman Lake. Lake Joondalup. Bibra Lake.

Photo: Simon Nevill

North Lake. Kogolup Lake. In the South West: Wannamal. Chittering. Toolibin (when water present). Kalgan River. McCarley's Swamp (Ludlow). Fortescue floodplain. Several of the larger rivers in the Pilbara and occasionally on the Fitzroy and Ord system.

Royal Spoonbill *Platalea flavipes* 74-81 cm

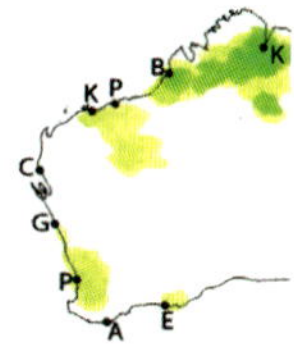

Uncommon in the South West. This species was not recorded in the South West Region until 1924 but it is still an uncommon bird. However, eight birds were recorded in one locality. It is very common on the east coast and across northern Australia. Feeds in shallow freshwaters in the South West but interestingly, will feed in shallow sea waters in the tropic regions of Australia. In breeding plumage they develop beautiful long nuptial plumes from the crown and nape. **Nest:** Similar to Yellow-billed Spoonbill. **Similar species:** Yellow-billed Spoonbill which has yellow bill, not black. **Status:** Uncommon in the South West, greater numbers in the Kimberley but still primarily a solitary or paired feeder. **Where to find:** Chance sightings may occur on the Swan Coastal Plain, including wetlands in Perth to Vasse and Peel Inlets area. They have however been sighted as far east as Esperance. Many remote wetlands as well as on the Ord and Fitzroy floodplain in the Kimberley.

Photo: Simon Nevill

Australasian Bittern *Botaurus poiciloptilus* 66-76 cm

One of the most secretive birds and one that is not easy to see. I have seen in my own lifetime the decline in numbers of this species. At Benger Swamp near Harvey there used to be several pairs in the late 1970s now you are lucky to get any. A combination of drying conditions for some wetlands combined with the desire to fill all wetlands has certainly contributed to their demise. Bitterns like reed beds and lots of it as they are not small birds and require a reasonable area of territory to feed. They also need wetlands that remain wet during the summer and that is getting harder to find, hence the small numbers. WA birds reside in the wetlands of the wettest regions of the south west. Australian Bittern is the largest bittern to be found in Australia, all brown with heavely streaked plumage a relatively short bill for the bird's size but a very robust bill as its diet is varied but needs the power to kill by pecking as well as grasping. Hunts with neck well forward in a heron like posture. It will take snakes, fish, but frogs are a major food source. They mostly feed during the later part of the day and into the night and early morning. To find or at least to know that they are in a particular wetland you have to know their deep booming call. This call is uttered by the male and increases in strength closer to the breeding season. The call is mainly given at dusk and into the early night.

Photo: Rohan Clarke

They bend forward and utter this strong booming call that eminates from the oesophagus. It is far carrying but not easy to locate as the boom has an echoing effect. When alarmed may stand motionless, with head and bill held vertically. **Nest:** Located on a bunched platform of broken reeds about 30 cm above water level. Clutch normally 3-4. **Similar species:** Could be confused with juvenile Nankeen Night-Heron other bitterns far smaller and different colour. **Status:** Scarce in WA and the population is tenuous. Far more common in Tasmania and particularly NZ. **Where to find:** Reed beds near Two Peoples Bay. Lakes near Mt Manypeaks. Lake Muir area.

Black Bittern *Ixobrychus flavicollis* 54-66 cm

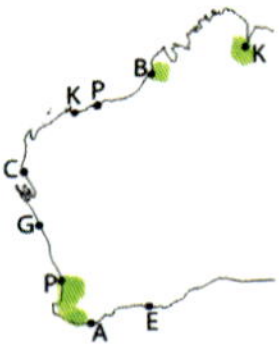

Found in Western Australia in two totally different climatic regions. The lower south west and most of the Kimberley region and parts of the western Pilbara region. An all dark brown Bittern with a prominent pale yellow streak running from under the bill down the sides of the neck to the wing shoulder. Also a series of fainter pale-yellow streaks from the throat down the chest. Stout bill. Another shy species that when approached may remain in cryptic vertical posture but more often will fly away. Many times I have approached small pools in the Kimberley, particularly early in the morning and have not noticed the bittern but it has seen me and off they would fly. Much of their feeding is done in the first hours of light

Immature

Photo: Martin Withers

and also late in the day. Tend to be found on smaller bodies of water particularly along creek lines. In the deep south west they are found on the quieter river systems often where there are large paperbarks as in the Warren. Donnelly and Shannon Rivers. Often seen from 5-10 m up in trees resting. **Nest:** Located from 2-15 m up in river or pool side trees, often paperbark. Flat small platform with finer twigs in centre. Clutch normally 3-5. **Similar species:** Not really likely to be confused with any other heron or bittern. **Status:** Scarce in the south west. Uncommon in the Pilbara. Relatively common in the Kimberley but does not like human disturbance. Where to find:. South West Margaret River. Donnelly. Warren. Gardner (Nr Northcliffe). Shannon and possibly the Deep River. Pilbara: At Millstream. Has been seen on the Fortescue and Maitland Rivers. Miaree Pool. Throughout the Kimberley but mostly north of Gibb River Road and on the Ord River system. Often in large paperbarks..

Black-backed Bittern (Australasian Little Bittern) *Ixobrychus dubius* 25-36 cm

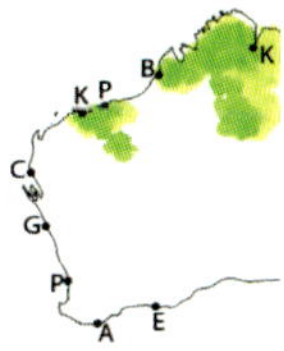

There are 4 subspecies of Yellow Bittern with populations in Europe. Africa. Northern India and the Australian subspecies *dubius* found in New Guinea and Australia. A very small elusive bittern. Very attractive. The male having an all black back and crown. Rich orange-chestnut face, nape with pale beige underbody with a distict dark brown stripe running vertically from throat to vent. Yellow bill that turns reddish at base of bill during the breeding season. The female is very similar but back is brown with cream coloured flecking. And neck and nape paler orange-chestnut. Legs pale cream. Hunting and postures similar to Australasian Bittern. Takes frogs but also takes small invertebrates. Occasionally seen flying across reed beds in the breeding season. They do not have a booming call but very low croaks that do not carry far. **Nest:** Located on a bunched small platform of broken reeds tighter bound than Australasian Bittern often woven around upright reeds. 30-40 cm above water level. Clutch normally 4-5. **Similar species:** Very similar to Yellow Bittern. Important feature Yellow Bittern has orange-brown back not black. Bill larger. No dark vertical dark brown band on underbody. Has a fine black facial line that passes through the eye. Yellow Bittern is rare in Australia but may be missed as it is common in Wallacea and Indonesia. **Status:** Uncommon in WA. **Where to find:** Lakes near Mt Manypeaks but often smaller than where Australasian Bittern occur. Lake Muir area but has a greater range as requires less extensive reed beds so can be found in lakes around Perth (very uncommon though) Jandabup. Joondalup. Sightings at Herdsman. Smaller lakes for example near Lake Muir. Lakes in the Esperance region. Possible further north but overlooked.

Photo: Adrian Boyle

Brolga *Grus rubicunda* 100-135cm

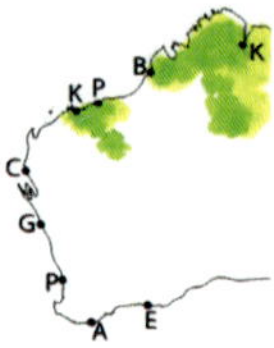

One of two species of crane to be found in Australia. A tall grey bird with a red head and a very long but stout blunt ended bill. Male slightly larger than female. Adult birds have black legs. The red plumage forms a band from the face around to the back of the nape, the crown is grey, has small grey patch within the red band over the ear. Importantly, the red stops at the jowl, has distinct drooping black-red dewlap hanging from chin. Juveniles are pale grey-brown with no red band but do have very pale red-brown crown. Fly in V formation often taking thermals together to gain height and then flying off together. Congregate in large numbers out of the breeding season. I have seen several hundred in just one scan of Lake Gregory and that was just one section of this vast lake. In breeding, they perform their well known dance-jumping display where both birds jump at varying intervals then throw their heads back while calling loudly. The female instigating calls and the male responding. They pair for life and pair bond develops in third year and breed in third or forth year. WA birds are found throughout the Kimberley. In the Pilbara small groups often arrive along the coastal belt in winter but return to the north with the onset of the wet. Brolga disperse and try to find quiet, small wetlands to nest, as they do not nest in large colonies. They are noted for their far carrying trumpeting call. **Nest:** Nest varies from platform built up on shallow water or on small islets directly on ground. Clutch 1-3. **Similar species:** Sarus Crane, which is slightly larger, red plumage covers face but most important difference is that the red plumage continues down the neck about 10 cm down below the jaw line and has no dewlap, legs are pink not black, grey crown less extensive. **Status:** Rare in WA. **Where to find:** Always possible in Parry Lagoon. Kununurra. Lake Argyle area.

Photo: Simon Nevill

Sarus Crane *Grus antigone* subspecies *gilliae*

Vagrant

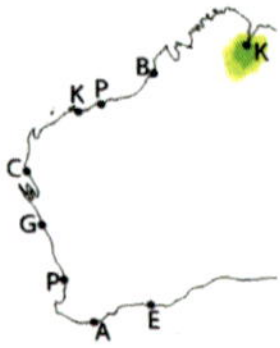

A rare vagrant to WA but slowly expanding range. Refer Brolga notes for difference between the two species.

Photo: Simon Nevill

Black-necked Stork *Ephippiorhynchus asiaticus* 110-137 cm

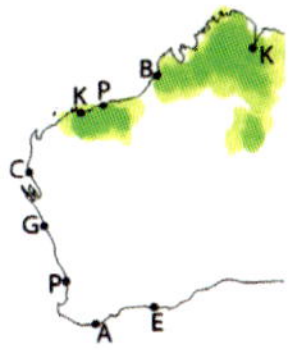

Of the 16 species of stork found in the World, not all are the same genus. The Australian species is found from eastern Parkistan to Australia and New Guinea, although rare in much of South-East Asia. A stork with distinct markings of black and white with bright red legs. Contrary to some speculations it is not an Aboriginal name but uses the South American common name of their stork *Jabiru mycteria*. Unmistakable in flight being slow and straight, with long trailing legs and black and white wings. The male has black iris whereas the female is yellow with the black pupil showing clearly against the yellow when good views. Does not nest colonially and generally a solitary or pair feeder. Diet is primarily fish and will handle fairly large fish. Will also scavenge on dead fish. Its distribution is interesting in WA. I have seen them on large wetlands like Lake Gregory although never many like Brolga or relatively small pools in the centre of the Kimberley miles from the nearest large wetlands. In the Pilbara I have found it on the edge of the desert at Carrowine Gorge or rivers such as the Yule River. These birds will move hundreds of kilometres to remote pools often far from their breeding area, so it is very much an opportunist feeder. **Nest:** Platform stick nest located in forks of very tall trees often dead trees. Some times in mangroves. Sometimes on dry reed beds. Clutch 2-4. **Similar species:** None. **Status:** Uncommon but wide spread. **Where to find:** Any wetland in the Kimberley. Along large river systems in the Pilbara often a long way inland. Occasionally found on the Roebuck Plains in the wet season and parts of the Fitzroy flood plain but largest concentrations are around Lake Argyle and Kununurra. Normally a few at Parry Lagoon most of the year.

Photo: Simon Nevill

Chestnut Rail *Eulabeornis castaneoventris* 48-52 cm

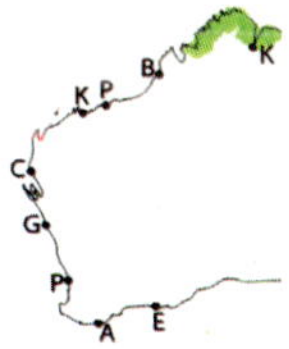

A very large, long billed rail with a grey head, white throat and olive-yellow legs. This is normally a very shy bird retreating back into mangroves even if you approach from a distance. The bird is often heard more than it is seen. Due to the many large expanses of pristine mangroves in the Kimberley, the population is fairly stable and most boat trips taken around the Kimberley coast will reveal this elusive bird for you. Normally a solitary bird but can be seen in pairs or a family parties feeding on the mudflats and on the mangrove line. Eats predominantly crustaceans. There is a variation in colouring between Chestnut Rails found

Photo: Simon Nevill

in NT and those found in WA, the dorsal colour in WA birds are olive-brown whereas the NT birds are chestnut-brown and there seems to be a gradual change across the north. **Nest:** Nest a platform of sticks placed in the crotch of a tree of a mangrove tree well above the high tide mark. Clutch 3-4. **Similar species:** None. **Status:** Moderately common throughout the mangrove areas in the Kimberley. **Where to find:** From Derby all around the Kimberley coast to the Cambridge Gulf through to the Gulf of Carpentaria.

Pale-vented Bush Hen *Amaurornis olivaceus* 24-28 cm

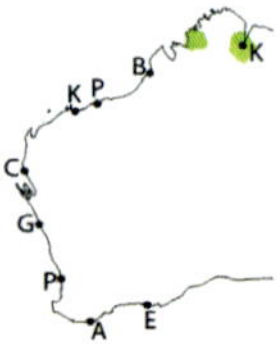

A very secretive bird although on the east coast in towns like Brisbane and Cairns this species will show itself but normally like most landrails, one more often hears it rather than sees it. It is smaller than the more easily seen Buff-banded Rail. The upper body is an olive-brown, throat, chest dark grey becoming darker on the belly, vent and under tail feathers chestnut, long dark legs, yellow stout bill with orange at base. For many years this species was thought to only occur on the sub-tropical and tropical parts of the east coast but sighting were later recorded in the NT, then in recent years the bird has been seen in the Kununurra region and then later George Swann of Kimberley Bird watching found one on the Walcott Inlet. This bird is generally hard to find, as it has a preference for tall rank grass. This type of habitat is not so prevalent in the Kimberley as it is in tropical north QLD. It pays to know the call, which it gives mainly during the breeding season as other times it remains silent. Its distribution and movements are not well known but it may in the dry season retreat to the wettest locations and then with the onset of the wet expand its range. **Nest:** A platform of grasses placed in long grasses often with a side hood of woven grasses. Clutch 4-7. **Similar species:** None in WA. **Status:** Scarce but may prove to be just uncommon as very elusive and hard to find. **Where to find:** Easiest locations to listen for this species would be rank grasslands around Kununurra region. Need to know call.

Photo: Ian Montgomery

Buff-banded Rail *Gallirallus philippensis* subspecies *melori* 30-33 cm

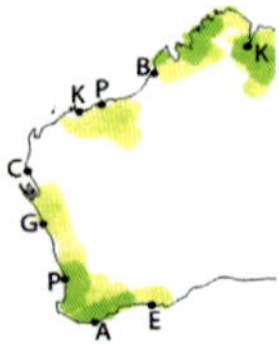

Crakes and rails can be difficult to see as they occur mostly in reed beds and thick moist undergrowth and generally are shy and elusive birds. The Buff-banded Rail can be an exception. It can tolerate a slightly drier environment and will come out into the open more than most other rails and crakes. Penguin Island is an example of where the birds venture onto open grassed areas where people gather. It is an extremely attractive rail, showing its orange breast band and barring on the underbody. It is taller and larger than all the Australian crakes and can tolerate brackish waters. It is a more opportunistic feeder than most crakes and rails, as it will eat carrion or take other birds' eggs but mainly feeds on insects, crustaceans, small reptiles and frogs. Calls mostly early morning or just after sunset. 10 subspecies have been identified throughout the world with only one, *melori*, on mainland Australia. There is a separate subspecies on the Cocos and Keeling Islands, *andrewsi.* **Nest:** It nests in a variety of habitats, mainly reed beds and long grass. The nest is a deep saucer shape, constructed of reeds or grasses often just above water level with grasses pulled over as a hide. In dry locations such as on islands, will gather a few grasses together and lay them on bare rock. Clutch size 4 to 5. **Similar species:** None really, the nearest is the Lewins Rail, presumed extinct in WA. **Status:** Moderately common in the south west, parts of the Pilbara and the Kimberley. **Where to find:** All larger lakes with reed beds, retreating to those that retain water through the summer. Alfred Cove in the reedy inlet. Penguin Island near the jetty. Bayswater Sanctuary. Canning River Wetlands. Herdsman Lake. Dams in the station country where permanent water runs from bores allowing reeds to grow. On islands such as Pelsaert Island in the Houtman Abrolhos Islands. Found in wetlands both in the Pilbara and Kimberley particularly in the Kununurra area.

Photo: Simon Nevill

Corn Crake *Crex Crex* 29 cm

Vagrant

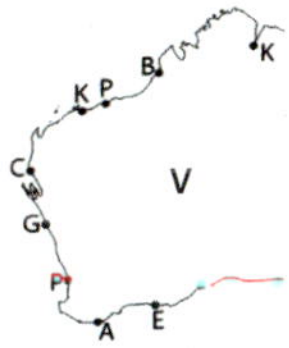

A rare vagrant landrail that has been sighted in WA previously and is on the WA bird list. A migratory bird that breeds in the northern Europe. Asia and southern Siberia and migrates primarily to Africa but also some to the Indian sub-continent. A strong migratory flyer and always possible for stray birds to fly off route and come to Australia. A skulking bird very hard to see normally just showing its head as it calls in the long grass. Unlikely to call if it does over-fly to Australia, as it would come in our summer being their non-breeding season. Coloration resembles a Baillons Crake but larger bird with a totally different jizz and stance. **Similar species:** None. **Status:** Very rare vagrant.

Photo: Simon Nevill

Baillon's Crake *Porzana pusilla* subspecies *palustris* 15-18 cm

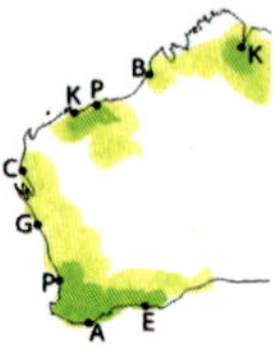

Crakes will travel far and wide to find suitable wetland habitat to feed and breed, so much so that the Asian crakes and rails occasionally, on their long sorties for better wetlands, end up in Australia. Baillon's Crake is found throughout Asia and parts of Africa, but it is presumed to be a resident bird in Australia. We still do not know where many of the crakes depart to when lakes are drying, as they will sometimes be in large numbers and then suddenly disperse from a wetland. They are more common in spring and early summer in the South West. The best time of year to see crakes with some degree of success is when waters are drying and the mud becomes exposed but is still wet and full of food matter; then the crakes will venture out from the reed beds into the open, particularly early morning and late afternoon. There are several subspecies of Baillon's Crake in the world from Europe through to Asia, Africa and the Far East, as well as Australasia, but there is only one subspecies in Australia, *palustris.* **Nest:** A very small deep saucer-shape with reeds or grasses pulled over as a canopy either above water or amongst reeds or alongside tufts of grass close to the edge of shallow water. Clutch size 3 to 5. **Similar species:** Australian Spotted Crake, which has a darker blue-grey breast, small red cere at the base of the bill; which is bright lime-green, the mottled back is darker and is larger. **Status:** Locally common in spring and early summer on the Swan Coastal Plain and several wetlands in the South West Region. **Where to find:** The Swan Coastal Plain, particularly Perth's wetlands: Lake Joondalup, Thomson Lake, Kogolup Lake. Bibra Lake, Wellard Wetlands, Namming Swamp. Found also in a few of the permanent wetlands in the South West, like Benger Swamp, Grassmere Lake, Unicup Lake and Kulinilup Lake. Found in wetlands in the Pilbara and Kimberley but often overlooked. Like all small crakes and rails, they are highly nomadic.

Photo: Simon Nevill

Australian Spotted Crake *Porzana fluminea* 22 cm

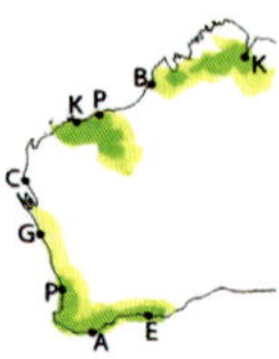

Largest of the crakes. Behaviour and habitat very similar to Baillon's Crake. Like most crakes, it moves when waters are drying in lakes. Another skulking bird, best seen in the early morning on the edge of reed beds. **Nest:** Similar to Baillon's Crake, placed at the base of or within lush grass tussocks. Clutch 3 to 4. **Similar species:** Baillon's Crake, which is smaller and has a brown mottled back, whereas the Australian Spotted Crake's back has white spots that can be clearly seen. Refer to Baillon's Crake for more details. **Status:** Locally common on Swan Coastal Plain and parts of South Coast. **Where to find:** The following wetlands in Perth can produce good sightings of all crakes:

Lake Joondalup. Thompson Lake. Kogolup Lake. Bibra Lake. Wellard Wetlands and Namming Swamp. Found in a few of the permanent wetlands in the South West. Some wetlands in the Wheatbelt in spring and early summer when water and reeds are present. Also in the Mulga Region where permanent bore water flows into reed beds. Rare in wetlands in the Pilbara and Kimberley but often overlooked. Like all small crakes and rails they are highly nomadic.

Photo: Simon Nevill

Spotless Crake *Porzana tabuensis tabuensis* 20 cm

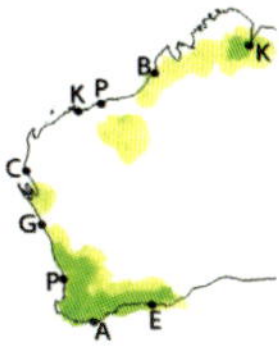

Slightly smaller than Spotted Crake. It is the easiest crake to identify, being dark all over with no spotting or flecking. When seen close up, its red eye, stands out against the dark purple-black head, also its red legs are a strong feature. Requires the same habitat as Baillon's Crake and Australian Spotted Crake. There are 3 subspecies in the world, 2 in Asia and the Far East and the one, *tabuensis*, in Australia. **Nest:** The nest is similar to other Australian crakes but is placed almost entirely over water and mostly in a higher location, with a small ramp of reeds leading up to it. Clutch size 3 to 4. **Similar species:** None. **Status:** Locally common in well-watered reed beds in any wetland in the South West. **Where to find:** The Swan Coastal Plain with Perth's wetlands have some of the best locations to see crakes. Lake Joondalup. Thompson Lake. Kogolup Lake. Bibra Lake. Wellard Wetlands. Lake Chandala. Namming Swamp. Found in a few of the permanent wetlands in the South West. Some wetlands in the Wheatbelt in spring and early summer when water and reeds are present. Casual further north.

Photo: Simon Nevill

White-browed Crake *Porzana cinereus* 18-20 cm

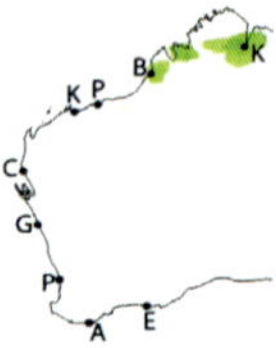

One of the less shy crakes found in the tropical wetlands of Australia. Easiest to recognize of crakes and rails, with its contrasting black and white stripes on the face. Has dark forehead, red cere, olive bill, brown back with black patterning, grey under body lightest on throat, black eye stripe contrasts with small white triangle at base of bill leading to eye, distinct white stripe below eye, red eye ring contrasts with black stripe, beige vent, cocks and flicks tail typically when alarmed like most crakes and rails. A very territorial crake before the onset of the wet season, defending its own section of a wetland. Have longer toes than most crakes, which assist them running over floating vegetation and lily pads. Most active in the first few hours of the day feeding in the open then retreat back into longer wet grasses during the heat of the day. **Nest:** A shallow cupped platform of grasses placed in long grasses or reeds with a hood of woven grasses. Clutch 4-7. **Similar species:** None really, Baillons Crake lacks contrasting stripes on face. Vent has distinct barring seen when agitated. Slightly smaller. **Status:** Common on large billabongs and wetlands. Most numerous on the wetlands surrounding Kununurra. **Where to find:** On the flooded inlets on the Ord River in Kununurra. Parry's Lagoon. Possible in any extensive wetland west or east Kimberley, particularly with surrounding long grasses and lilies.

Photo: Simon Nevill

Purple Swamphen *Porphyrio porphyrio* subspecies *bellus* 47 cm

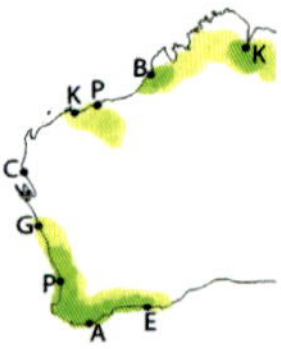

A massive rail also found in Africa, Asia and New Zealand, although not as massive as the close relative, the rare Takahe in New Zealand. The Swamphen can certainly fly and does when waters recede. In fact some birds banded in Australia have been found in New Guinea, such is the power of their flight, which is surprising considering their body weight to wing ratio. The Western Australian subspecies *bellus* is slightly larger than the eastern states subspecies and has a bluer, less purple breast colour. The bright blue-purple colour stands out on wetlands in contrast with its heavy red bill and red forehead shield. It has a tendency to be aggressive to other waterbirds. It will often corner small fledglings or other birds and kill them and is the bane of crakes and rails. Luckily most of its feeding is confined to aquatic plants, which it will hold parrot fashion in one its powerful feet, ripping the off the succulent part of root. There are six subspecies in the world, with two occurring in Australia,

Photo: Simon Nevill

bellus in the South West and *melanotus* in the east and northern Australia. **Nest:** Builds a broad rush platform amongst reeds. Clutch size 2 to 4. **Similar species:** None, except the Purple Gallinule not found as yet in Australia. **Status:** Common in the South West and uncommon parts of the Kimberley. **Where to find:** Most larger wetlands in Perth and the South West from Wannamal south east to the Albany region with an outlying population in Esperance. Rare in central Western Australia but found in the Kimberley but not common.

Dusky Moorhen *Gallinula tenebrosa* 35-40 cm

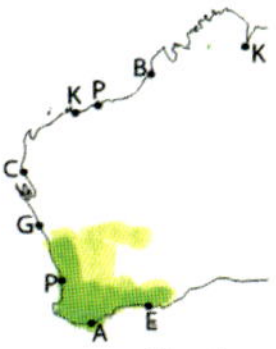

A medium-sized waterhen. Has adapted well to urban developments where there are grassed areas adjacent to wetlands, and it is common and far more confiding in city environments than birds found on remoter wetlands where they tend to be shy and at times elusive. Feeds on the edges of wetlands, mostly near reeds where it can seek refuge if threatened. Like many rails and waterhens, they have white feathers on the vent that one normally does not see, but when threatened will raise their tail, flicking it constantly, exposing the white as warning to other birds while at the same time giving its warning call. This cooperative early warning system is used by most rails and crakes and is a self-protective instinct aiding all waterbirds, mainly against raptors, particularly Swamp Harriers and Whistling Kites. Feeds on aquatic vegetable matter, insects and seeds. Its red forehead shield and yellow tip to its bill stand out from its dark body. Tends to swim more than most waterhens and will feed on the water surface for insects, and will also eat shoots. There are 3 subspecies in the world with the nominate race *tenebrosa* occurring in Australia. **Nest:** Similar to Purple Swamphen, but smaller and can be located up in a fork in a paperbark tree. Clutch size 5 to 7. **Similar species:** None. **Status:** Common in south west only rare further north than Moora. **Where to find:** Any water body in the South West. Most lakes in Perth. Not found outside the southwest region.

Photo: Simon Nevill

Black-tailed Native-hen *Tribonyx ventralis* 32-38 cm

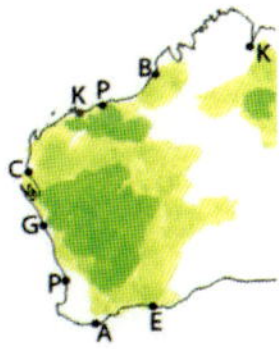

This is one of the few water hens that have a preference for bodies of water located in semi-arid and arid regions of inland Australia. It is a highly irruptive species that is often seen in large numbers, sometimes as many as 200-300 birds on large inland ephemeral wetlands, particularly where there are lignum swamps and wetlands like Lake Gregory, they can occur in their hundreds after good rains. When wetlands recede, they fall back on the station dams that may have permanent water from bores. Sometimes they will come to the northern edge of the metropolitan region of Perth. They feed on plants, insects and small reptiles. **Nest:** A small reed platform, with a hollow centre placed in reeds or lignum, often located on ephemeral lakes on claypans. Clutch size 5 to 6. **Similar species:** Tasmanian Native-hen, not in this region. Swamphen far bigger with blue underbody. Black-tailed Native-hen is an overall dull brown bird. **Status:** Locally can be very common in the arid zone, overall uncommon. May have suffered from loss of habitat in station country. **Where to find:** Permanent bores supplying large dams with reeds. Any large inland lake after receiving heavy rains, also in the Wheatbelt. Pilbara and lower Kimberley.

Black-tailed Native-hen Photo: Simon Nevill

Eurasian Coot *Fulica atra* 35-39 cm

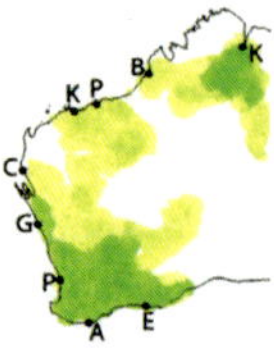

Found throughout Europe and Asia but not Africa or the Americas. An extremely common waterhen, seen on most wetlands through much of the South West. A pugnacious bird, if not chasing other coots in the breeding season they are chasing other birds from their feeding territory. Being an all-black bird with a white bill and white forehead shield makes it easy to identify. It is mainly herbivorous but also eats insects. **Nest:** A small neat platform with hollow dish centre, constructed of water weeds and placed over water, often within reed beds or at the base of a paperbark tree. Clutch size 4 to 6. **Similar species:** Dusky Moorhen is the same size but different in body colour and has a red bill and forehead shield making it easy to differentiate. **Status:** Extremely common. Wetlands through out the state. Nomadic at times and massive irruptions may occur as in Lake Gregory where several thousand can be recorded. Scarce in the arid zone. Seen at most large or small wetlands in Perth.

Photo: Simon Nevill

Australian Bustard *Ardeotis Australis* Male 110-120 cm. Female 80-90 cm

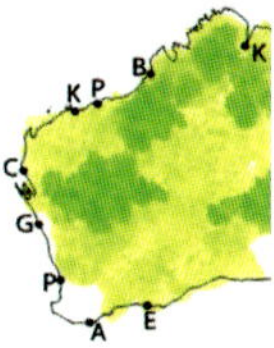

The Australian Bustard is one of the many bustard species found throughout the old world, most of them occurring in Africa. Australia has just the one species but it is one of the largest, in fact it is our third largest bird and our heaviest flying bird. When it is taking flight, it requires a reasonable distance over which to run before it can actually take off. It has always been one of the favoured foods of the desert Aborigine. Bustards are no longer common in the lower South West, but when the inland arid regions are experiencing severe droughts bustards will fly long distances to get to better feeding grounds and the South West in some years will receive several visiting birds. Farming friends have had them on their Tenterden farm several times over the last 20 years. When caught unawares, will stand with neck outstretched, bill to the sky, hoping the intruder will not see them; however, will soon take flight if they feel there is imminent danger. The male is much larger and his breeding display is very interesting. He will inflate his throat and loose neck skin into a huge bag that almost touches the ground. He then flicks the tail vertically with its white undertail coverts showing well and proceeds to utter deep booming grunts, while walking in circles. The female is attracted to this performance and they may mate. Other younger subordinate males may be present, witnessing and learning the complex procedure. Sadly, the bustard is still shot in many areas and it only remains common away from human disturbance. Its diet consists of small mammals and reptiles but when grasshopper plagues occur, it will gorge itself for several days. It also will come into areas that have been and are being burnt, to take advantage of the fleeing small animals. **Nest:** No nest is made but just a simple scrape on the ground, often located below a bush or mulga tree. Clutch size, mostly just one egg. **Similar species:** None in Australia. **Status:** Comes into the south west region occasionally, particularly if there have been droughts in the arid zone. Uncommon throughout the mid west. Northern Pilbara grasslands and the Kimberley. Locally common in the station country and desert regions. **Where to find:** Heavily grassed areas in the Mulga Region and open spinifex plains in the desert areas.

Photo: Simon Nevill

Bush Stone-curlew *Burhinus grallarius* 54-59 cm

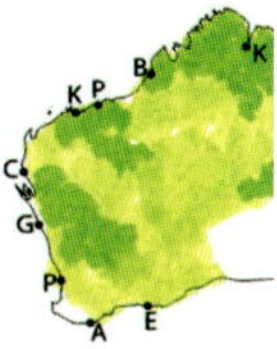

There are 5 thick-knees (Stone-curlews) in the world; Australia has two, the Bush Stone-curlew and the Beach Stone-curlew. Sadly, both have been driven out of much of their former territory The Bush Stone-curlew has certainly gone from much of the Wheatbelt. This is not only due to land clearing but also to persecution from the feral fox. Being a sedentary bird, once removed from an area they rarely return. The sound of the Bush Stone-curlew wailing at night to many is an eerie sound and those who have never heard their call often wonder who and what are making these screeching calls. They prefer dry open woodland areas and also mulga woodland. When approached will freeze on the ground with head and neck stretched horizontally. If the nest is approached they do their classic wing spreading defence posture. The chicks are well camouflaged and the streaked bodies make them difficult to see in the leaf litter or stones, where they have their nest. Bush Stone-curlews feed mostly on insects, some reptiles and seeds. They are very active at dusk and dawn before they go to their daytime roost. Feeding is mostly done at night, hence their large eyes aiding nocturnal vision. **Nest:** No nest is made, only a few twigs are drawn close to the nest but actually serve no purpose, almost a relic of an ancient instinct to create a nest. Will normally select a nest site, which has a clear surrounding view. Clutch size normally 2. **Similar species:** Only the Beach Stone-curlew which is far larger. Far larger bill. Sronger facial markings and black bar across the shoulder. **Status:** Scarce in the Wheatbelt. Uncommon on the edge of the Wheatbelt and into the mulga and York Gum and Salmon Gum Woodland. **Where to find:** Luckily at present, the small reserve known as Dryandra Woodland Reserve near Narrogin has a few families still surviving, one of the few populations in the western Wheatbelt. They are occasionally found in the eastern and northern Wheatbelt but more so in the Mulga Region and Transitional Woodland. Very uncommon in the desert regions. Found throughout much of the Pilbara and Kimberley.

Photo: Simon Nevill

Beach Stone-curlew *Esacus magnirostris* 54-56 cm

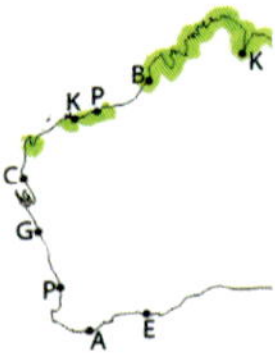

Thicknee or Stone-Curlew a contentious issue for some. Although Bush and Beach Thick-knee look alike they represent 2 genus. In the thick-knee complex there are a total of nine species, 7 species in the genus *Burhinus* of which the Bush Thick-knee is one, and 2 species in the genus *Esacus* of which the Beach Thick-knee is one, so Australia is represented in both genus. Beach Thick-knee is generally a shy bird requiring beaches away from populated areas. With the introduction of the fox and the presence of domestic dogs on many beaches, it has forced the Beach Thick-knee to retreat to remoter beaches, luckily in the Kimberley there are many such beaches. They occur wherever crustaceans are plentiful both on sandy beaches and more so tidal mudflats. Most of the larger islands off the north-west coast have populations of beach thick-knee. A well marked thicknee with distinct white stripe above eye contrasting with its black head, white throat with small black stripe leading from bill and yellow cere, massive black bill, dark brown upper body, on shoulder (scapulars) black-brown with distinct white bar, belly and vent pale grey, upper tail feathers black, long olive legs. In flight white on primaries contrasts with black on greater primary coverts giving a bulls-eye appearance. If encountered first it will slowly walk away and the if pursued will take flight. You can if there are two of you, one sit and the other skirt around the bird. It will often concentrate on the walker and the sitter will get good views as it walks past. Do not hassle the bird though and keep a good distance. **Nest:** A shallow cupped platform in sand with a few leaves and twigs encircled either on beaches or within mangrove areas above high water mark. Clutch 1-2. **Similar species:** Bush Thick-knee which has shorter neck, much smaller bill, heavy streaking on breast, lacks plain colouring of Beach Thick-knee rarely in same habitat. **Status:** Moderately common on remoter beaches. Beaches that I used to see this species north of Broome now with increased activity have left. **Where to find:** From the south heading north: Still on North West Cape, beaches just south of Onslow, south of Cape Preston, still on the Burrup away from the mining activities, Point Sampson, Breaker Inlet, still possible at Cape Keraudren but increased tourist activity is a problem, On 80 mile beach I have covered many, many kilometres but have not seen this species until Port Smith, so there may be a preference for inlets and beaches with mangroves or taller vegetation behind beaches where there is higher density of crustaceans. Found on beaches just north of Broome and all the way to Cape Leveque, then many remote beaches and mangroves all the way around the coast into the NT.

Photo: Simon Nevill

Pied Oystercatcher *Haematopus longirostris* 45-50 cm

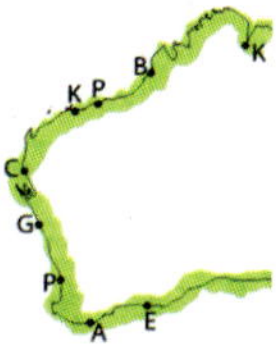

Oystercatchers are stocky birds with sturdy short legs and a strong thick bill, which they certainly need particularly when hammering on molluscs attached to rocks because the impact on the neck is immense. In the breeding season a pair can be very territorial, protecting their feeding and nesting area. It is interesting how Oystercatchers manage to feed on hard-shelled mussels and oysters. Shellfish have a strong adductor muscle that controls the opening and closing of the shell. The Pied Oystercatcher will stab at the shell and open it, quickly trying to strike the muscle and thus sever it, then eats the body parts. If it is not successful in opening the shell, it will peck hard into the side of the shell again, trying to cut the muscle. Both Pied and Sooty Oystercatcher use this technique. Pied Oystercatchers are more diverse in their choice of food than Sooty Oystercatchers. Pied Oystercatchers

Photo: Simon Nevill

mostly work the sandy beaches where they run backwards and forwards with the tides eating exposed molluscs, worms and small fish, whereas Sooty Oystercatchers feed mostly on rocks. **Nest:** Above high water mark, with no nest made except a scrape in the ground. Birds prefer to have locations where they can see approaching predators. Clutch size 2 to 3. **Similar species:** Sooty Oystercatcher, which instead of a pied plumaged body is all black. **Status:** Moderately common along much of the WA coastline. Can be found throughout the west coast all the way to the northern Kimberley. **Where to find:** Near Perth common on Rottnest Island. Common on the Abrolhos Islands. Common in the Albany region all the way to Esperance. Shark Bay. Cape Keraudren. Parts of Eighty Mile Beach. Cape Leveque.

Sooty Oystercatcher *Haematopus fuliginosus* 40-52 cm

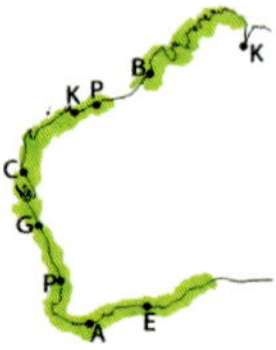

Slightly thicker-set than the Pied Oystercatcher. Has a preference for rocky coastlines, where it feeds off the molluscs on the rocks, either piercing the shell or hammering it on a rock or simply using a scissor action to prise open the shell. More solitary than Pied Oystercatchers and seen more often on remoter rocky coastlines.. There are two subspecies in Australia, *fuliginosus* on the southern coastline and *opthalmicus* on the northern coastline, the latter having a thicker bill and larger yellow-orange orbital eye ring. They overlap between Shark Bay and Ningaloo. **Nest:** No nest is made, just a scrap in ground but will draw debris close to nest site if available. Nest commonly placed between rocks and often on islands. Clutch normally 2. **Similar species:** Pied Oystercatcher, which has pied plumage. **Status:** Uncommon throughout the south west, locally common between Walpole and Bremer Bay, then to a lesser degree along the eastern coast to Cape Arid. **Where to find:** Mostly on the south coast and particularly on offshore islands in the Archipelago of the Recherche. They are found from Twilight Cove Bay all around the coast to the northern Kimberley but prefer rocky coastlines.

Photo: Simon Nevill

White-headed Stilt *Himantopus himantopus* subspecies *leucocephalus* 33-37 cm

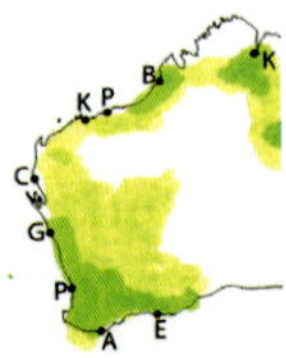

There are only four stilts in the world and Australia has two of them, the Black-winged and the Banded. Stilts are the longest-legged waders of all, with extremely long fine bills. The Black-winged Stilt delicately prods on the surface of the water for aquatic invertebrates that swim on the surface. If they don't surface-feed, they will select muddy areas where they can pick off small invertebrates and worms. Unlike the Banded Stilt, Black-winged Stilt will nest as a solitary pair on a small wetland, or more often just a small colony of 5-10 birds, occasionally as many as 100 but that is very uncommon in the west and nothing like the incredible numbers that the Banded Stilt colonies reach. Also unlike Banded Stilt, Black-winged Stilt prefer freshwater lakes and will feed on lake edges, swamps and samphire flats that have fresh water. There are five subspecies in the world with one in Australia, *leucocephalus*. **Nest:** Sites vary, sometimes a scrape in the ground if on sandy islets or if near or on water a pile of grasses are pulled together above the water mark. Will nest in small colonies. Clutch size 3-4. **Similar species:** Banded Stilt. Easy to differentiate when they are adult birds as the Banded has a chestnut band passing from the shoulder right around the belly to the other shoulder. Also has an all-white head, whereas the Black-winged Stilt adult has black on the back of the neck, stretching from the nape to the base of the neck. Juvenile Banded Stilt and Black-winged Stilt can be a little tricky, although it's rare for them to be feeding on their own without some adults nearby. The juvenile Black-winged Stilt has a white head like the Banded, with a faint black patch around the eye and on the crown. The main feature to note is that even if the Banded Stilt does not have a chestnut band it will still have a white back between the wings, whereas the Black-winged Stilt has a completely black back whether juvenile or adult. **Status:** Common on the Swan Coastal Plain. Can be in huge numbers after breeding. Over 1000 birds have been recorded at the Vasse Estuary and Peel Inlet. Locally common on some Wheatbelt lakes. Uncommon on the southern coast. Uncommon throughout the Pilbara but found in greater numbers when the desert regions receive extensive rains. **Where to find:** Many large lakes in Perth with shallow margins. Small Wheatbelt wetlands often with samphire edges may have only one or two birds at one time. Northern wetlands like Parry Lagoon have had numbers in excess of 1000.

Photo: Simon Nevill

Banded Stilt *Cladorhynchus leucocephalus* 35-43 cm

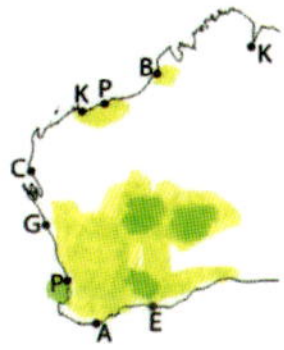

A bird that feeds on some of the most saline waters in Australia, wherever small crustaceans occur, particularly brine shrimps. They will wade in water picking off small invertebrates and crustaceans or swim in deeper water using the same pecking technique as Black-winged Stilt. Banded Stilt gather in some of the largest concentrations of any species, with numbers over 150,000. For years ornithologists did not know where Banded Stilt bred. It was not until 1930, when a large colony was found at Lake Grace, that observers realised how and where they breed. The colony found at Lake Ballard in Western

Australia had an estimated 179,000 nests. Because Banded Stilt are opportunistic breeders, they will fly long distances to remote salt lakes if sufficient water is present. For nesting they will normally select large islands, located in remote inland ephemeral lakes. Sometimes the water evaporates at such a fast rate that by the time many of the chicks have hatched the waters may have dried up and thousands of birds will perish. There is still so much that is not known about movements of this fascinating bird. We know that small groups can occur, say 50-200 birds, feeding on the odd salt lake or brackish waters, like Lake McClarty near Pinjarra. We also know that very large numbers can visit the salt lakes on Rottnest Island. Recently over 11,000 birds were gathered on one of the lakes on Rottnest for a brief period. It would be fascinating to know whether that massive group stayed together and also where they dispersed. **Nest:** No nest is made, just a scrape in the sand. They invariably nest where brine shrimps are available close by. Nest colonially in their thousands. Clutch size 2 to 4 eggs laid. Similar species. Refer Black-winged Stilt. **Status:** locally very common on selected salt lakes. Uncommon throughout most of the region. Always present on Rottnest, hence the ridiculous nickname of Rottnest Snipe was given to this specie in bygone days. Very uncommon north of 26th latitude as lakes are not saline enough. **Where to find:** Often on the Esperance if breeding has occurred in the remote Goldfields salt lakes. Rottnest Island is normally reliable for a few birds. Sometimes Peel Inlet. They can occur when not breeding on lakes in the Wheatbelt, like Lake Marmion near Dowerin or Lake Hinds near Wongan Hills. Lake Grace. Lake Cowcowing Lakes and Lake Campion. Lake Ballard. Lake Way.

Photo: Simon Nevill

Red–necked Avocet *Recurvirostra novaehollandiae* 40-48 cm

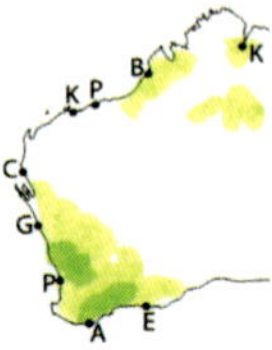

A beautiful Avocet, one of four in the world. It is endemic to Australia. Has the typical Avocet bill that it uses in a scything action backwards and forwards with bill slightly open so it can feel prey, just like the Spoonbills do, being another tactile feeder. Like the Banded Stilt, it can swim whilst feeding but mostly wades in water. Prefers to feed in saline or brackish waters, particularly ephemeral lakes. **Nest:** It nests in a variety of locations including islands in small salt lakes, freshwater swamps. Often nests on remote inland lakes as does the Banded Stilt but never in such large concentrations, normally in colonies of 6 to 20 birds but sometimes reaching larger numbers. Will nest with other birds like Black-winged Stilt. No nest is made, just a scrape on the ground in sand or earth but some soft material like grasses or samphire is often used as a base. Clutch size normally 4. **Similar species:** None. **Status:** Locally common, normally only individuals or small groups of 6 to 10 birds. Occasionally bigger groups of 30-100 at the most. **Where to find:** Rottnest Island. Joondalup.

Lake Thomson. In the Wheatbelt any salt lake after good rains. Station country like Lake Annean and Lake Austin. As they are nomadic birds they can be seen on any wetland in the Pilbara and southern Kimberley.

Photo: Simon Nevill

Pacific Golden Plover *Pluvialis fulva* 23-26 cm

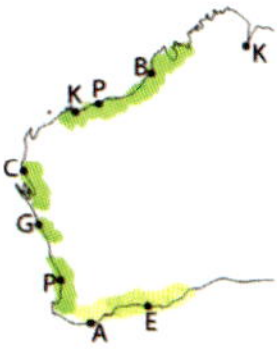

There are three Golden Plovers in the world: the American, Eurasian and Pacific Golden Plover. The species that primarily comes to the western shores is the Pacific Golden Plover. The Pacific Golden Plover is a slim upright wader of coastal beaches, estuaries and salt marshes. Breeds in the central and eastern Tundra and western part of Alaska. **Similar species:** None in breeding plumage but in non-breeding plumage the Golden Plover has similar features to the Grey Plover. If you manage to see the Grey Plover in flight (see illustration) it has a black 'armpit', the Golden has no black 'armpit'. When resting, the Golden has a rounded head less flattened on the crown (subtle difference but noticeable), and a slightly more upright stance. Grey Plover looks bulkier and is bigger, weighs a third more. Bill is larger and longer in the Grey. Pacific Golden Plover also has a golden tinge to

Photo: Simon Nevill

plumage, not as strong as when in breeding plumage but quite noticeable. Small brown facial patch to the rear of the ear covets. **Status:** Common on north west beaches. Broome region. Eighty Mile Beach. Shark Bay. Uncommon to locally common in the South West. Visitor from October to March. **Where to find:** Found on all major beaches and mudflats all along the WA coast. Broome region. Eighty Mile Beach. Sherlock Bay. Faure Island. Mangrove Bay, Carnarvon beaches to Gladstone. All the way down to Vasse Inlet. On the south coast Oyster Harbour near fishing harbour. Kalgan River. Occasionally seen in inland lakes not far from coast.

Grey Plover *Pluvialis squatarola* 27-31 cm

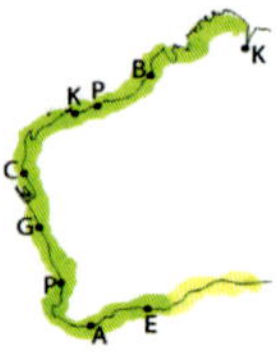

3 races recognised. The nominate *squatarola* comes to WA. Nests throughout the Siberian and the North American Tundra, almost circumpolar. More coastal than the Pacific Golden Plover and certainly in greater numbers on the Western Australian coast. Has a typical plover technique of feeding, that is look for prey while motionless, then run and pick and continue the process. Found on coastal beaches, particularly estuaries, coastal marshes. **Similar species:** None in breeding plumage. In non-breeding plumage Pacific Golden Plover. The Grey Plover stands more horizontally, has a thickset body, flattened crown, larger bill, black armpit, grey body rather than golden tinge. In flight bold white wing bar and shows white rump, Golden has no white on rump. **Status:** Common on north west beaches, particularly Broome. Uncommon to moderately common locally in the south west. Visitor from September to April but some over-winter. Normally solitary or in small numbers. **Where to find:** All suitable beaches and mudflats on the WA coast. Roebuck Bay. Eighty Mile Beach. Beaches north and south of Onslow. Shark Bay. Uncommon along the southern coastline. Oyster Harbour. Kalgan and King Rivers.

Showing black arm pits

Photos: Simon Nevill

Red-capped Plover *Charadrius ruficapillus* 14-16 cm

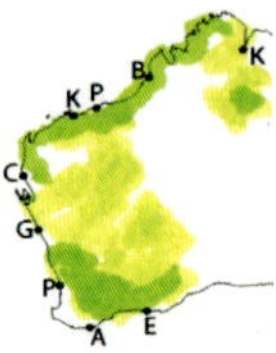

An endemic plover, very common and found in many habitats including sandy beaches, salt marshes, margins of freshwater lakes, inland salt lakes. A very small plover feeding with typical plover 'run and pick and stop' technique. In breeding the male cap is bright chestnut, the female a duller chestnut colour. The male cap becomes duller out of breeding plumage. The male has a black lateral band coming from the neck and stopping before the chest. Female has no black, only a shorter chestnut band. **Nest:** No nest is made, just a scrape in the sand or soil out in the open but often with tuft of grass or beach-washed piece of timber adjacent to nest. Clutch size 2 to 4. **Similar species:** Only one and extremely rare, the Kentish Plover seen by John Darnell at Leslie Saltworks. Very difficult to split but has broader black forehead band, is slightly larger but main feature is complete white collar, whereas Red-capped Plover has no white collar. Malaysian Plover is similar, is found on islands just north of Australia and although sedentary birds, there is the possibility of vagrants landing. **Status:** Very common throughout the coastal region of WA, particularly north west beaches. Most small lakes near the coast. Largest numbers on inland lakes. Recently the author had at least 1100 on Lake Thompson as it was drying out. On Jandabup 3000 have been recorded by others. **Where to find:** Rottnest. Lake McLarty particularly when water is receding in early summer. Lake Forrestdale. Most inland salt lakes in the Wheatbelt. Esperance lakes. Eighty Mile Beach and Roebuck Bay. Lake Gregory. Most inland larger lakes.

Photo: Simon Nevill

Ringed Plover *Charadrius hiaticula* 18-20 cm — Vagrant

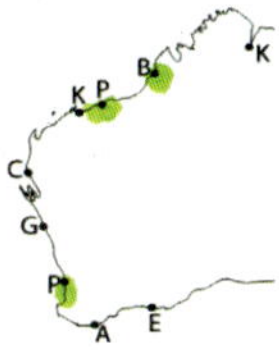

A rare vagrant to WA. This small sand plover has been sighted in several locations as far south as Peel Inlet and surprisingly, not in the Broome region where there are many skilled bird observers. Breeds in the Northern Hemisphere. There are two subspecies, the nominate *hiaticula* breeds in Europe, Greenland and north east Canada and the subspecies *tundrae* breeds on the far northern coasts of Lapland and Russia. **Similar species:** Easily confused with Little Ringed Plover. In flight easy as Little Ringed Plover lacks the white wing bar, which contrasts with darker trailing edge on wing and has pinkish legs, not yellow-orange, has brighter yellow eye ring not so easily seen in non-breeding plumage. In breeding look for markings on forehead; Little Ringed Plover shows two white bars, one above and one below black band whereas Ringed Plover has one wider white band above black. Ringed plover is slightly larger at 18-20 cm, Little Ringed 14-17 cm. They are not easy to split for the first time, so clear details should be taken. Semi-palmated *C. semipalmatus* is even harder to split but due to its rarity it will not be covered in this book. **Status:** Rare vagrant. **Where to find:** Any coastal beach or brackish inlet.

Photo: Simon Nevill

Little Ringed Plover *Charadrius dubius* 14-17 cm Vagrant

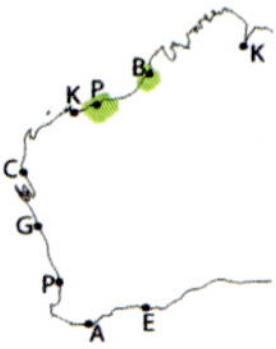

Not such a rare vagrant as Ringed Plover. Small slim sand plover with relatively long legs, distinct yellow eye ring more defused in non-breeding, pinkish not yellow-orange legs, no prominent dorsal white wing bar. There are 3 subspecies *dubius* from Philippines to NG, *jerdoni* from India to South East Asia, both are sedentary in their region but the subspecies *curonicus* is migratory. It breeds from Europe to Siberia. The Eurasian birds over-winter mostly in Africa and the Siberian birds over-winter in the Malay Peninsula and Indonesia but several will overshoot and land in Australia, mostly in the north. In their natural habitat they have a preference for stony shorelines around freshwater as well as brackish and coastal beaches, so it's possible to find these birds not only on beaches but also on freshwater lakes or perimeters of sewerage ponds. **Similar species:** Ringed Plover. Refer notes on Ringed Plover. **Status:** Uncommon but regular vagrant. **Where to find:** Seen regularly in the Broome region and Darwin. Found on any coastal beach, brackish Inlet or freshwater lagoon with better chances in northern Australia.

Photo: Martin Withers

Lesser Sand Plover *Charadrius mongolus* 18-21 cm

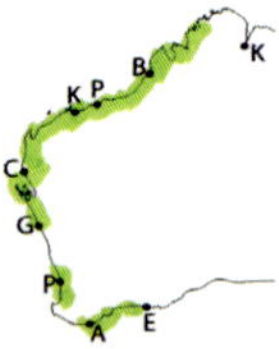

Breeds in Mongolia and parts of eastern Siberia, often as high as 5000 metres in the Himalayas on the western edge of Mongolia. Arrives in Australia mid September, moving slowly down the coast then returning slowly north by December and leaving March to April. Common in the north west of Western Australia. Scarce in the South West.

Photo: Simon Nevill

Similar species: Greater Sand Plover. Lesser Sand Plover is only slightly smaller but looks in proportion to other small plovers. The Greater Sand Plover head looks large compared to the body and the bill is larger and heavier than the Lesser. The legs are longer on the Greater. The head is flatter on the top, dropping steeply down to the bill on the Greater, whereas the Lesser has a more rounded head. If the birds are side by side you will soon see the differences. Sometimes you may get them in eclipse or even full breeding plumage, then the task of splitting them is made much easier. The Greater has a smaller orange-brown collar band than Lesser but the important thing is that the Lesser has a black edge band to the top of the breast band (in the male only). Having said all this, unfortunately it is very uncommon in the South West but far more common in the north west. There are 5 subspecies in the World with 3 that come to Australia's shores, mostly the subspecies *mongolus* but *stegmanni* and particularly *schaeferi* may come to Western Australian shores mostly to the north west. **Status:** Common in the north-west, uncommon in the South West. **Where to find:** Eighty Mile beach gets large numbers. Roebuck Bay.

Greater Sand Plover *Charadrius leschenultii* 22-25 cm

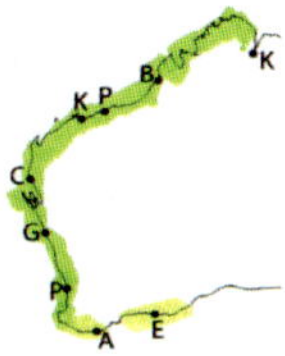

Breeds in central Asia in highland open plains, mainly the Gobi Desert in Mongolia. Arrives in Australia late August to mid September moving slowly down the coast or may remain in north west all summer. Some come down the coast as far as Busselton but numbers decrease from north to south. A few birds pass on down to the southern coast. Feeds almost entirely on the coast on estuaries and beaches. There are 3 subspecies in the World but only *leschenultii* comes to these shores. **Similar species:** Lesser Sand Plover and Oriental Plover, refer notes on Lesser Sand Plover. Oriental Plover is much easier to differentiate. Its neck is longer, has a finer bill, no dark eye band, legs longer and paler yellow, breast less marked. In breeding there is no confusion, breast colour configuration is different with black base to orange chest and white neck and face and no black eye stripe. **Status:** Moderately common in the northern part of this region. Uncommon from Geraldton down to Busselton and on the south coast. Locally common Peel Inlet. Oyster Harbour. **Where to find:** Roebuck Bay. Eighty Mile Beach. Onslow. Shark Bay. Carnarvon region. Mangrove Bay. New Beach. Denham. Alfred Cove. Peel Inlet. Oyster Harbour.

Photo: Simon Nevill

Inland Dotterel *Peltohyas australis* 19-23 cm

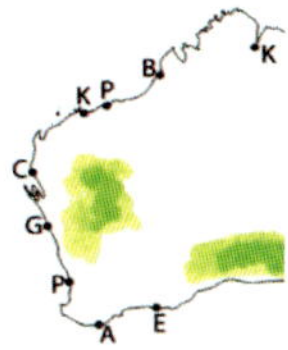

One of the most cryptic birds in Australia, with its transverse band running from the crown through the eye as well as a distinct 'v'-shaped breast band. Very hard to see in the stony country where it occurs as it blends so well with the red-brown soils. A bird of the central drier regions of WA including the northern Wheatbelt (scarce), Mulga, southern plains of the Pilbara. Found on open stony plains, samphire flats and claypans. It is one of the few birds that has encroached into the northern Wheatbelt after the massive land clearing program, finding suitable habitat in ploughed fields or fallow pastures. It is both a nocturnal and diurnal feeder having large eyes to aid in low light. In winter it can be seen far more in the daytime but in the hot summer months it does much of it feeding during the night. Feeds on insects and seeds. Often overlooked and best seen in the cold winter months when its can be more active during the day. Occasionally seen at night with one's headlights on the edge of remote outback roads. **Nest:** No nest is made, only a shallow depression, sometimes uses the base of a ploughed trough in field. Nest site can be located in fields, open samphire plain and more often open stony country. Clutch size usually 3. **Similar species:** None. Where to find. In the northern Wheatbelt. Mullewa. Koorda. Bonnie Rock. Better chance around in the Cue region on Austin Downs west of Cue. Stony country south of Gascoyne Junction. Mileura Station west of Meekatharra. Has come into the Swan Coastal Plain to Lake Clifton and Lake Wannamal. Also into the Southern Wheatbelt. For those that really wish to see it, best chances are on the Nullarbor Plain particularly south and just north of Rawlinna.

Photo: Simon Nevill

Black-fronted Dotterel *Elseyornis melanops* 16-18 cm

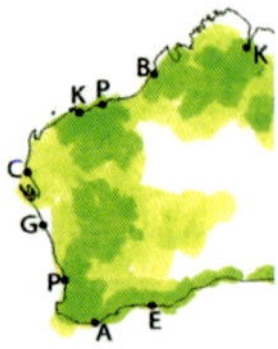

A common bird throughout much of WA except the driest deserts. Found on the margins of freshwater and brackish pools. Feeds on crustaceans and insects. Slightly larger than the Red-capped Dotterel but shares the same habitat. Tends to be found near freshwater pools rather than saline. The Red-capped Dotterel can tolerate more saline waters. **Nest:** No nest is made, nests in a variety of locations, in depressions in sand alongside water, on stony plains alongside creeks, even in gravel on roadside verges where streams are close by. Clutch size normally 3, sometimes 2. **Similar species:** Only the immature Red-capped Dotterel, which is smaller. Stands more upright and has black eye line. **Status:** Locally common, overall uncommon. **Where to find:** Most large lakes in Perth with grassed and muddy margins. Lake Kogolup. Lake Thompson. Herdsman Lake. North Lake and Bibra Lake. Lake Joondalup. Throughout the southern wetlands and the Swan Coastal Plain. Found alongside pools and rivers on south coast. Gairdner River. Kalgan River. In the northern region, common on pools remaining on larger rivers like the Gascoyne and Murchison. Found on station dams, particularly large permanent dams. Any small wetland or river in the Pilbara or Kimberley may have this bird.

Photo: Simon Nevill

Hooded Dotterel *Thinornis rubricollis* 19-23 cm

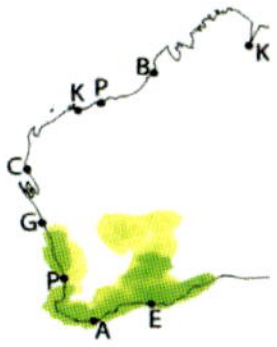

A species that has suffered since European settlement, having lost much of their original habitat and now confined to the more remote beaches where there is less interference. The Hooded Plover population in Western Australia is the only one in Australia that can be found on inland salt lakes as well as coastal beaches. They nest on remoter beaches, avoiding continual disturbance of their territory by human activities. After breeding many will migrate to larger salt lakes like Lake Clifton south of Mandurah or Lake Gore and Lake Warden in the Esperance region, sometimes gathering in reasonable numbers normally 40-50 but as many as 350 have been recorded, in the Esperance region. These are extremely large numbers for such an uncommon species. They feed by running along the beach stopping and starting, picking small molluscs and crustaceans, they will also walk along the base of sand dunes feeding on seed. Have a preference for wide sandy beaches. Their distribution extends from Horrocks near Northampton to Eyre on the Nullarbor. **Nest:** No nest is made just a scrape in sand or gravel, mostly very close to a waterline but commonly near fallen log or branch. Clutch size 2 to 3. **Similar species:** None. Unmistakable with an all black hood. Juvenile could be confused with juvenile Black-fronted Dotterel but it has a black eye stripe not black hood. **Status:** Locally common but generally uncommon. **Where to find:** Lake Clifton. Lake Preston. Beaches from Nornalup to Wilson Inlet. Bremer Bay (threatened there by beach vehicles). Beaches in the Fitzgerald River National Park. Largest numbers on the Esperance lakes. Inland on some of the smaller ephemeral salt lakes, particularly in the Salmon Gum woodlands north of Esperance and those north east of Hyden. Also the inland salt lakes between Hyden and Norseman.

Photo: Simon Nevill

Red-kneed Dotterel *Erythrogonys cinctus* 17.5-19.5 cm

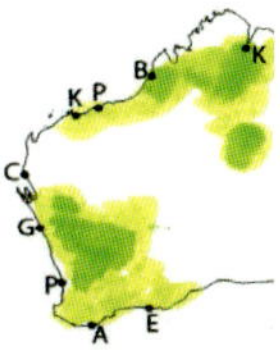

Often shares the same habitat as Black-fronted Dotterel, but Red-kneed Dotterel does not like coastal saline waters or gravel beaches which the former can occupy. Likes freshwater lake margins and swamps or inland large dams. An attractive Dotterel, having strong black and white markings and a chestnut flank, with red on the leg from the thigh to the knee, hence the name Red-kneed. Feeds on insects, molluscs and seeds. Is nomadic and will travel long distances to get good feeding sites. Found on many outback station dams. **Nest:** No nest is made, just a deep scrape but may put loose grasses or other debris in base of depression. Nests mostly close to waterline of inland lakes and swamps, particularly small islands. Clutch size 3 to 4. **Similar species:** None. **Status:** Moderately common throughout most of its range, increases near the coast in summer as inland freshwater pools dry up.

Where to find: In Perth, many of the large lakes. Herdsman Lake. Bibra Lake. Kogolup Lake. Lake Joondalup. Lake Claremont. Odd inland Wheatbelt lakes like Lake Coomelberrup. Nonalling Lake near Corrigin. Towerrinning but only after good winter/spring rains. Large numbers gather in the Cue region on various ephemeral wetlands and then disperse to station dams. Found on many wetlands in the north west particularly dams and freshwater pools in the Mulga region. Found on wetlands in the Kimberley.

Photo: Simon Nevill

Double-banded Plover *Charadrius bicinctus* 18-21 cm

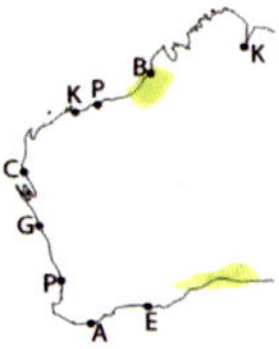

A sand plover that breeds in New Zealand and the remote Auckland Islands 450 km south of South Island NZ, There are two subspecies, *bicinctus* which breeds on the mainland islands of NZ and *exilis* on the Auckland Islands. Some of the population of Double-banded Plover migrate to the Australian mainland and Norfolk and Lord Howe Islands. The migratory patterns of this small sand plover are very interesting, those that do migrate to Australia are those birds that breed high in the central mountains of NZ, mainly over 500 m. Due to the cold conditions at the higher altitudes in winter limited food availability has forced these birds to migrate rather than join the sedentary birds that remain on the lowland beaches and wetlands. The migratory birds return faithfully to their original breeding sites in the mountains of NZ. The subspecies *exilis* that breeds on the Auckland Islands apparently remain there. Those birds that do come to Australia mostly come to the south-east corner of Australia, particularly Tasmania. Only a few birds migrate as far as WA and those that do, mainly come to the beaches on the Great Australian Bight, very few come to the deep south west and even less fewer around to the west coast. **Similar species:** In breeding plumage, unmistakable, the only plover with two bands on chest. In non-breeding could be confused with

Photo: Rohan Clarke

Lesser and Greater Sand Plover. Both are larger. Double-banded Plover even in non-breeding plumage normally shows two diffused bands, unless it is an immature bird, the jizz plays an important role here, Double-banded size, is really evident against Greater Sand Plover which is not only is bigger but has far longer legs, bill on Double-banded much smaller and finer than other two species, out of breeding breast bands on Lesser and Greater are always broader than Double-banded. Double-banded upper body is brown rather than grey-brown on Lesser and Greater. **Status:** Uncommon migrant to WA. **Where to find:** Beaches adjacent to Eyre Bird Observatory. Occasionally found off Perth coast. Seen at Lake McLarty near Mandurah.

Oriental Plover *Charadrius veredus* 21-25 cm

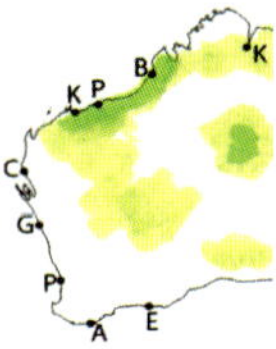

Breeds in a narrow band of habitat stretching from Mongolia to the Gobi Desert. They truly are a bird of open grassy and stony plains, as well as frequenting beaches in the north-west, they can be found miles inland in desolate arid plains. I remember well when I had a bird tour on the Nullarbor and there was a sand storm, as we drove across the plains there was a group of plovers, at first we wondered what we were looking at through the dust, until we stopped and could ID them, it just showed how far south the birds from the arid plains of the Gobi Desert could fly, so far south to more arid plains showing the need for habitat dependence such as these wind swept open plains. Sometimes they congregate in large numbers but most of the time in small groups. Oriental Plover is a slim, long-legged plover, has similarities to Greater Sand Plover but longer-necked and distinctly longer legged. In breeding plumage it is an attractive bird with rich chestnut breast, black chest band below, grey back, white face with grey tinge, diffused grey cap on rear of crown, long slimmer bill and longer legs than sand plovers, in flight reminiscent of pratincole but wings and tail not as long or sharp. In non-breeding (or breeding plumage) look for jizz, long neck, long legged, slim bill and importantly wide grey brown collar band from throat to lower breast, dark grey facial mask with surrounding white giving eye-patch look. **Similar species:** In breeding plumage it is very similar to the rare vagrant the Caspian Plover that breeds in Eurasia and Turkistan and migrates South to Africa. It is a smaller, shorter-necked bird with grey legs, not orange-pink, Clean-cut white face edged with distinct black line above breast, important feature is colour of legs, in flight ventral view shows contrasting white and dark-grey wings, whereas Oriental ventral view wings are all grey with dark grey tips. Greater Sand Plover, different jizz with short neck, proportionally larger head, thin grey breast band, stouter shorter bill, legs dark-grey. **Where to find:** Beaches near Broome and Roebuck Plains but mainly grassy and open stony plains south Cape Bossut to Samphire Roadhouse in the summer months. Found anywhere in the northern deserts but few travel these areas in the summer. Will come as far south as the Nullarbor, a strong flyer.

Photo: Rohan Clarke

Masked Lapwing *Vanellus miles* 35-38 cm

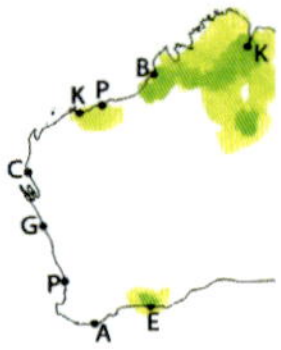

Even though this is a very common bird on the east coast, it is very scarce in the south-west with few sightings however the nominate subspecies *miles* is moderately common and found in the tropical parts of WA, mostly in the Kimberley. The subspecies miles is found from Eighty Mile Beach in the west across to the mid east coast, basically covering the tropical regions. The two subspecies are quite distinct. Subspecies *miles* is smaller with a slender body and proportionally longer legs, has larger wattles, and the most important visual difference is that there is only black on the crown. The subspecies *novaehollandiae* has a black on the crown that passes down the back of the neck and forms a black band on the side of the breast but does not meet in the front. The spurs on the wing shoulders are more pronounced. One of the few birds to have benefited from human expansion in Australia, becoming extremely common in grassed areas, particularly near water. Aggressive and noisy with intrusion into their breeding area where there are young, or if on eggs will do the 'broken wing' distraction display like many ground nesting birds do. **Similar species:** Possibly Banded Plover at distance but no wattles. White eye stripe, small red wattle at base of bill, black band surrounds white throat, smaller than Masked Lapwing. **Status:** Locally common on grassed areas around Broome and Derby, Wyndam and Kununurra, generally uncommon through most of the Kimberley, particularly central and northern parts. **Where to find:** Parks and gardens in Broome and on the Roebuck Plains, Wyndham, Derby and Kununurra towns, around fringes of Lake Argyle and when Lake Gregory has water reasonable numbers around fringes and islands. The few birds recorded in the south west are of the subspecies *novaehollandiae*.

Photo: Simon Nevill

Banded Lapwing *Vanellus tricolour* 25-29 cm

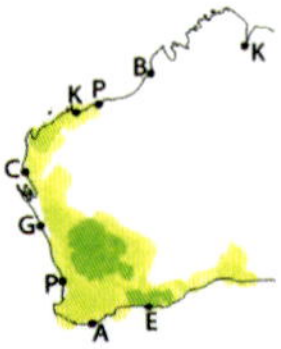

This attractive lapwing has a preference for open stony plains with short grasses, unlike the Masked Plover, which is happy to feed in longer lush grasslands. There are a few species like the Inland Dotterel and the Banded Lapwing that most probably have gained from the clearing of land in the northern Wheatbelt. We know that early historical records make little mention of Banded Plover in the South West. What has been recorded is the expansion into many areas by the 1930s, breeding as far south as Bridgetown. Banded Lapwing can be found throughout most of the South West except the deep South West forests and the Salmon Gum/ York Gum woodland. They feed on seeds and mostly insects and worms with the typical plover action of see, run and pick. Its preferred habitat and where it can easily be seen are the various small open stony or flat clay plans that occur throughout the Mulga Woodland. Like so many plovers, they will perform their broken wing act to distract you if you get too close to the nesting site. **Nest:** No nest is made just a hollow depression is scraped, with some dry materials used as a base for the nest. Mostly located on open plains with a good surrounding view. Clutch size 3 to 4. **Similar species:** Masked Plover which generally does not overlap in territory and found mostly in the subtropical and tropical parts of WA, very scarce vagrant in the South West. One should not have problems with these two species; the Banded Lapwing has a black band starting from the lores all the way around the belly, and a red cere. Masked Plover has no band on belly and a prominent yellow wattle. **Status:** Found throughout the southwest and much of the mulga region. Uncommon on the southern coastal regions and Swan Coastal Plain. Moderately common in certain areas of the Wheatbelt particularly the northern and north-eastern areas. Common in the Mulga Woodland and parts of the Pilbara and desert regions. Locally common on Rottnest Island.

Photo: Simon Nevill

Bar-tailed Godwit *Limosa lapponica* 37-39 cm

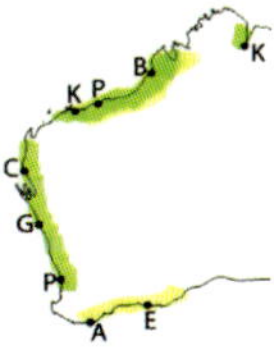

Of the world's 4 godwits, 2 migrate to Australia, the Black-tailed Godwit and the Bar-tailed Godwit. They breed in different regions, the Black-tailed in Europe, Iceland, Central Asia and western China, the Bar-tailed in Scandinavia, Russia and Western Alaska. Just over twice the number of Bar-tailed Godwits migrate to Australia than the Black-tailed Godwit and the population of Black-tailed Godwits is even less in the South West than Bar-tailed. There are three subspecies in the World and most probably the main migrant to Western Australia's shores would be the subspecies *menzbieri* as the majority of retrieved banded birds have been that subspecies. All godwits have long necks, long legs, small heads and long slim bills. They feed by probing into the mud with rapid probes searching for aquatic invertebrates, finding food by touch. They also feed in wet grassed areas, probing for worms. **Similar species:** Black-tailed Godwit. In non-breeding plumage the Bar-tailed has mottling on the wing covets (mostly

Bar-tailed showing barring on rump

Photo: Simon Nevill

on the upper body), looking overall a fawn-brown colour. The Black-tailed has a decidedly dull grey-brown look to the wing covets (mostly on the upper body). So in essence, one looks dull grey-brown, one lighter fawn-brown. The Bar-tailed has a slightly upturned bill and shorter legs than the Black-tailed and breast and flanks have slight flecking. Black-tailed has a fine straight bill, breast and flanks are plain light grey-brown. It is easily identified, is in flight. The Bar-tailed has, three thin dark black-brown bands on the tail, no white bar in the wing. The Black-tailed has a complete wide black band from the middle of tail to the base of the tail with a clear white rump above the black band, darker wing feathers with a distinct lateral bar running the full length of the wing. Hudsonian Godwit which has different underwing pattern with dark coverts and contrasting with white wing bar, where as Bar-tailed has mottled coverts and no white wing bar. Rare in north west. **Status:** Common in northern beaches getting less common towards southern coast. Arrives in late September to October departing end of March, early April. **Where to find:** Throughout much of the north west coast. Roebuck Bay. Onslow. Carnarvon estuary and Mangrove Bay south to Gladstone. The Abrolhos. Geraldton to Bunbury. Alfred Cove. Then a few estuaries on the south coast including Kalgan River mouth and Oyster Harbour.

Black-tailed Godwit *Limosa Limosa* subspecies *melanuroides*

40-44 cm

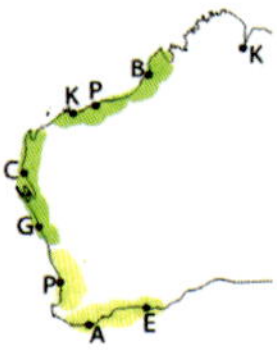

Less common than Bar-tailed in the South West but in the north west of the state you can see hundreds and even more in the Gulf of Carpentaria. Tends to feed in small groups. Migrates to Australian shores from late August onwards and leaves April-May. There are three subspecies in the world but the predominant one that comes to Australia is *melanuroides* which breeds in Mongolia and Siberia. **Similar species:** Bar-tailed Godwit, refer to notes. Hudsonian has black underwing coverts where as Black-tailed has fine black leading edge contrasting with wide white underwing coverts. **Status:** Common in the north of the South West. Uncommon down to Bunbury. Scarce in the South Coastal Region. **Where to find:** Roebuck Bay. Most estuaries in the north west. Carnarvon. New Beach. Denham. Alfred Cove. Black-tailed have a tendency to enter freshwater regions a little more than Bar-tailed and can be found occasionally on lakes like Thompson. Forrestdale. Lake McLarty and Benger Swamp.

Black-tailed showing white rump

Photo: Simon Nevill

Little Curlew *Numenius minutus* 28-31 cm

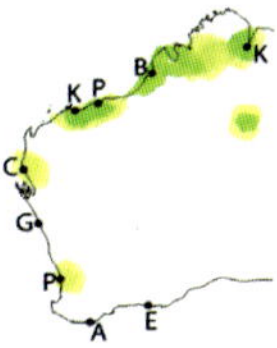

A bird that migrates from the Russian Tundra. Very rare in the South West, although I have seen this species at Woodmans Point with a tour group. There are chances of Little Curlew in the Carnarvon region but they rarely come past Port Hedland. The smallest curlew with a very short bill most unlike all other curlews, more like a Whimbrel but much smaller and the bill is shorter and straighter. Has an upright stance. Likes feeding on well-watered grassland, often seen on playing fields and parks in northern towns such as Port Hedland, Karratha, Broome and Derby. Also freshwater water wetlands. Found in the north west of the state in their hundreds, even more so in the north of the Northern Territory and the Barkly Tablelands. **Similar species:** Only the Whimbrel could possibly be confused but is a larger bird with a bill almost twice as long, and longer legs. **Status:** Rare in the South West uncommon on the Pilbara coast and common in parts of the west Kimberley particularly Roebuck Bay and Roebuck Plains. **Where to find:** Roebuck Bay and more so the Roebuck Plains. Port Hedland playing fields.

Photo: Simon Nevill

Whimbrel *Numenius phaceopus* subspecies *variegates* 40-45 cm

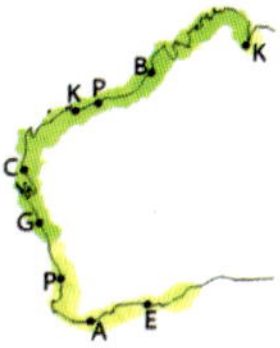

Breeds in eastern Siberia. A medium sized curlew with a long down-curved bill, dark crown with a slight eye stripe, short legs. Important feature is the white rump, in flight, looks wedge-shaped. Mainly a solitary wader or in small groups, feeding on beaches, mangrove mudflats and estuaries, although tends to prefer mangrove habitats where it will also roost. When it does roost it will come together with other waders. Eats small crabs, worms and shrimps. There are four subspecies in the world but most probably the only two that come to Australian shores would be *variegates* that breeds in Siberia and *hudsonicus* that breeds in Alaska and northern Canada; most Western Australian birds are *variegates*. **Similar species:** Eastern Curlew. Far larger with very long bill more down curved. No eye stripe. Main feature, has no white on the rump. Little Curlew is far smaller in size and the bill is, one third shorter and only slightly curved. **Status:** Common in the north west. Moderately common Carnarvon south to Gladstone and in Denham and Shark Bay region. Scarce in the south west. **Where to find:** Best in the north of the state.

Photo: Simon Nevill

Far Eastern Curlew *Numenius madagascariensis* 60-66 cm

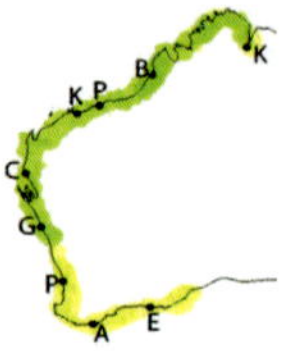

The largest wader to come to Australia, with a long neck, long legs, thickset body and very long bill, particularly the female which is larger than the male. Breeds in eastern Siberia in wet marshes. The call is familiar to those who wader watch regularly, sounding 'cur-loo'-like, although less so than the European Curlew. Feeds in a variety of habitats in marshes, beaches, salt-works, muddy creeks but prefers large muddy estuarine habitat. Feeds on crustaceans. The female is a larger bird with noticeably longer bill. **Similar species:** European Curlew *Numenius arquata* subspecies *orientalis*, which has a white rump and lower back, also white on the underwing at the axillaries and underwing coverts, and the bill is not as long as the Eastern Curlew. The Wimbrel is much smaller and has a much shorter bill. **Status:** Numbers of this species have declined over the last 30 years due to loss of habitat and persecution. Locally common at Roebuck Bay and some of the beaches in the Pilbara moderately common around Carnarvon from Miaboola Beach down to Long Point and Denham region. Uncommon further south. **Where to find:** Roebuck Bay. Estuaries in the Pilbara. Carnarvon estuary down to Long Point and Denham region. Alfred Cove. Peel Inlet. Albany region.

Photo: Simon Nevill

Eurasian Curlew *Numenious aequata* 50-60 cm Vagrant

A very rare vagrant to WA. There is only one accepted record from Eighty Mile Beach at present but it's possible the species could be overlooked with its initial similarities to Eurasian Curlew, Luckily the most likely location to see this species is in the Broome region where some of the most experienced wader observers in the country reside, and it is certainly the finest wader spot in Australia; other than walking down the Cairns Esplanade where the waders just ignore you. There are two races, the nominate *arquata* which breeds in Europe to the Urals in the east and *orientalis* that breeds between the Urals and eastern parts of Russia. Identification is difficult due to overlap of the two races.

Photos: Simon Nevill

Similar species: The main feature to look for is that Eurasian Curlew has a white rump whereas the Eastern Curlew has all-brown-mottled rump. Barring on tail of the Eurasian is dark brown and white which shows well, whereas Eastern has less contrast in the barring, being dark brown and pale brown. The Eurasian Curlew is markedly smaller with a shorter bill, although this is not a good feature to look for as the female in both species has a much longer bill. **Status:** Rare vagrant. **Where to find:** Possible from any beach south Broome to Carnarvon best chances Roebuck Bay to Cape Keraudren.

Marsh Sandpiper *Tringa stagnatilis* 22-26 cm

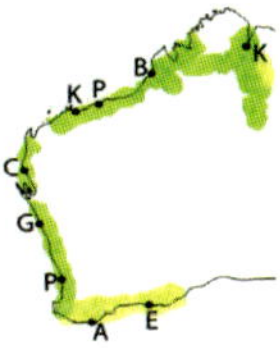

A medium sized wader, like looking a delicate miniature Greenshank. Migrating to Australian shores from September to March. The Marsh Sandpiper has a needle-like bill, small head and very long legs. Breeds in eastern Europe and central and far eastern Asia. This is one wader that has a preference for freshwater lakes and dams. It will feed on mudflats and beaches but not as much in the South West. It does mass in the north west of WA to migrate north for the breeding season. Feeds on aquatic invertebrates and crustaceans, mostly on the margins of small freshwater lakes but will venture deeper into lakes past the tarsus, even in water to the base of the belly, picking at swimming insects. **Similar species:** Greenshank, that is larger and much thicker-set, weighing more than twice the Marsh Sandpiper. Bill thicker and slightly upturned. Marsh Sandpiper has a fine and straight bill. **Status:** Common in the north and north west. Patchily distributed through the mulga on larger wetlands. Uncommon throughout most of the south west, locally common on a few freshwater lakes. **Where to find:** Lake Campion. Lake Eda. Roebuck Plains when surface water present. Parry Lagoon. Lake Argyle. Lake Gregory. Fortescue marshes. Lake MacLeod. Chinaman's Pool in Carnarvon. Inland bores with overflow drains between Carnarvon and Overlander Roadhouse. Pools on the Murchison River. Many freshwater pools inland after heavy rains have fallen. In Perth: Kogolup Lake. Bibra and North Lake. Thompson Lake. Herdsman Lake. Lake McLarty. Kalgan and Lower King River in Albany. Lake Gore in Esperance.

Photo: Simon Nevill

Common Greenshank *Tringa nebularia* 22-26 cm

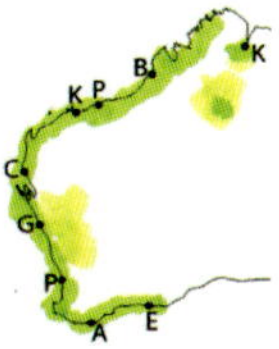

A large long legged wader, with a long slightly upturned bill. Its breeding range is vast, stretching from Britain right across Europe, northern Asia and as far as eastern Siberia. A visitor to Australia from August to April/May. Far more diverse in its range of feeding habitats than the Marsh Sandpiper, including mudflats, sandy beaches with some soft sand, salt marsh and mangroves and inland ephemeral lakes and permanent dams. Feeds on crustaceans, molluscs and small fish. Is the equivalent of the Sulphur-crested Cockatoo in the rainforest, in that it is the first bird to warn waders of imminent danger, giving its three note rapid 'teuu teuu teuu' call while normally flying off as it does, not always appreciated by birdwatchers. **Similar species:** Marsh Sandpiper. The Greenshank is a larger bird, thicker-set with heavier upturned bill, larger head and thicker neck. Unfortunately both have similar plumage features in flight with fine bars on the tail and white rumps looking wedge shaped. **Status:** Common in the north and north west. Moderately common throughout suitable habitat in the South West. **Where to find:** Roebuck Bay. Parts of Eighty Mile Beach. Sandfire Flats when in flood. De Grey estuary. Lake MacCleod. The Carnarvon mudflats but also the Gascoyne River pools. All mangrove-lined beaches from Carnarvon to Long Pont. Denham beaches. Wannamal Lake. Many of the Perth lakes. Alfred Cove. Peel Inlet. Vasse Estuary. Harvey Estuary. Albany region. Esperance lakes.

Common Redshank *Tringa nebularia* 29-34 cm

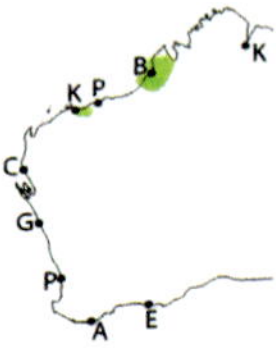

Common in Europe but certainly not common in WA, although always a regular visitor to the north west coast each summer particularly at Roebuck Bay Broome and Eighty Mile Beach. Common Redshank breeds from Iceland in the west across Russia to northeast China. There are 6 subspecies but it is uncertain which race comes to Australia. Mixes with other waders but is more nervous than most and when anxious, will start bobbing its head. Has bright orange-red legs, upper-body and crown grey-brown, under-body strong dark-brown streaking. In breeding under-body stronger contrasting streaking, head becomes darker, bill show stronger orange-red on base. White forehead supercilium stripe is lost in breeding. In flight unmistakable having all white secondaries and inner tips to primaries that contrast with dark-brown fore wing and sooty-black wing tips. No other wader has the white oval on back and trailing white band. Out of breeding plumage is duller and gone is the strong contrasting streaking on under-body and neck. **Similar species:** Spotted Redshank, which is a rare vagrant but has certainly been sighted a few times in Australia with at least two confirmed sightings in the Broome area one at Lake Eda, showing its natural preference for freshwater lagoons. Spotted Redshank in breeding plumage is unmistakable, all sooty-black head and body, upper-body and vent strongly streaked with white markings contrasting well with the black plumage. Out of breeding it gets very difficult; jizz is important here, Spotted is a larger bird and neck is slightly longer. and head looks proportionally smaller, bill is definitely longer, legs slightly longer all in all making this look a very elegant wader. More likely to be found near freshwater than Common Redshank (does frequent beaches). **Status:** Uncommon migrant. **Where to find:** Possible from any beach south of Cape Leveque to Carnarvon, best chances Roebuck Bay to Cape Keraudren.

Photo: Simon Nevill

Spotted Redshank *Tringa erythropus* 29-31 cm

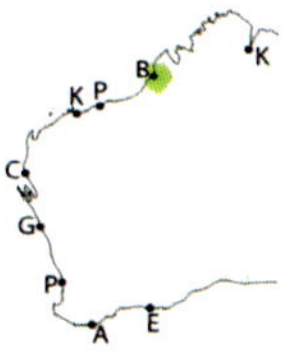

Refer some comments re this species under Common Redshank. Breeds in the pine forests of the Tundra and northern Scandinavia, interestingly the female leaves the nest after laying, leaving the male to raise the young, so the female can prepare for the long migration south to central Africa, the Indian subcontinent and south east Asia. A few passage migrants over pass to Australia. An elegant wader with a breeding colour combination of black body and red legs, its unlike any other wader except the Black Stilt, which is restricted to New Zealand. More likely to see in Australia in non-breeding plumage unless in late summer when breeding plumage can develop. In non-breeding has sharp black line on the lores which contrasts well with a white supercilium, far stronger in contrast than Common Redshanks markings. Elegant jizz is important, almost like looking at an oversized Marsh Sandpiper. Under-body whitish-grey. In flight resembles Common Redshank with white oval back but lacks white contrasting bar on rear of wings. Also legs protrude further past tail and bill longer. **Status:** Rare vagrant. **Where to find:** Possible on north west beaches but really look for this species on inland freshwater pools, if you're familiar with Marsh Sandpiper, this bird will stand out, much larger heavier-bodied bird.

Photo: Adrian Boyle

Wood Sandpiper *Tringa glareola* 19-23 cm

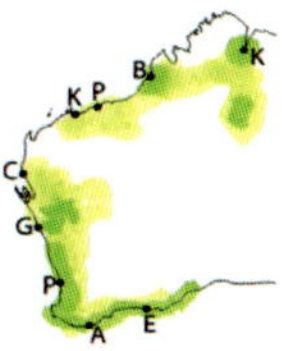

An attractive wader, breeding in Europe through to eastern Siberia. A medium-sized wader with a distinct darker back than most waders, although not as dark as a Common Sandpiper; the same size as a Sharp-tailed Sandpiper. Strong white markings to the edges of the scapulars and wing coverts, giving the appearance of a spotted upper-body. Long sharp straight bill, distinct supercilium eye stripe, and yellowish green legs. A feature of the Wood Sandpiper is that it bobs its tail constantly when on the move. Has a preference for freshwater, more so than most waders. Found on the margins of shallow freshwater lagoons and swamps, often with emergent vegetation. Will feed in flooded grass plains. **Similar species:** Sharp-tailed Sandpiper, which does not have the same coloured upper-body, being lighter plain brown; has bigger wing coverts and the supercilium eye stripe is not as prominent as the Wood Sandpiper. Sharp-tailed has a rufous crown and shorter legs. Common Sandpiper can share the same habitat but unlike the Wood Sandpiper feeds mostly on muddy estuaries and mangrove areas. Is a similar size but has shorter legs, a distinct collar and a strong green tinge to the back, which should not be confused with Wood Sandpiper. Green Sandpiper has been recorded in the NT and could be overlooked in the Kimberley. It has a darker upper body. Spotting is not as prominent as Wood Sanpiper. Feet extend further past tail with the Wood Sandpiper. **Status:** Locally common on suitable freshwater wetlands throughout the north and the Pilbara. Uncommon throughout the south west. **Where to find:** Most large freshwater lakes in the north and north west. Sometimes large numbers on Lake Gregory. Samphire Flats when in flood. Carnarvon region down to Long Point. Base of Shark Bay around Hamelin Bay. Found on some inland bores between Carnarvon and Overlander. Pools on the lower Murchison. Many of Perth's wetlands including drainage channels. Inland Wheatbelt ephemeral lakes but not if too saline. Uncommon on the south coast.

Photo: Simon Nevill

Terek Sandpiper *Xenus cinereus* 22-24 cm

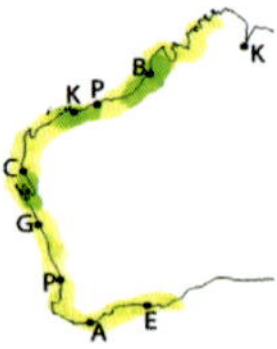

An unmistakable wader with bright orange legs and an upturned yellow-orange bill and a stocky body. The only species of its genus. Breeds in the higher regions of Asia and Siberia but not the Tundra and migrates down to Africa, the Indian sub-continent, east Asia and Australasia. A coastal bird of mud banks, estuaries and mangroves. Whenever it lands it bobs its tail and rocks its head. A fast feeder running erratically to various mud pools, digging its bill into mud or running with head down chasing small crabs. **Similar species:** None. **Status:** Not common throughout most of the north but can be locally common. Scarce in

Photo: Simon Nevill

the whole of the South West. **Where to find:** From Barred Creek south to Roebuck. Parts of Eighty Mile Beach. Shark Bay. Occasionally Lake McLeod. Mangrove Bay. New Beach. Alfred Cove. Peel Inlet and Mandurah Harbour. Can turn up at any muddy estuary in the South West. Found along the line of mangroves south of Carnarvon, as it loves small crabs.

Common Sandpiper *Actitis hypoleucos* 19-21 cm

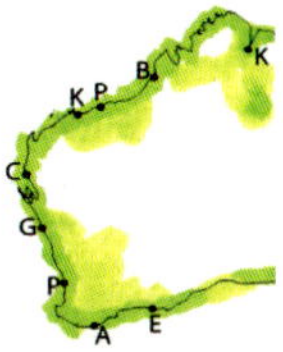

A long bodied small wader. Breeds over a broad band from Britain right across central Russia to the Far East. Found in a variety of habitats including estuaries, mangroves, beaches, rocky coastlines, lagoons, marshes, small inland ephemeral lakes as well as larger lakes and particularly lower pools on the large rivers like the Murchison and the Gascoyne Rivers. Like the Wood Sandpiper, can be found a long way inland. Is one of the most active feeders constantly running here and there chasing insects and aquatic crustaceans? Bobs its tail constantly, even when standing on rocks. Is often seen running over small boulders to snatch food in small tidal pools. Main diagnostic feature is the neck collar that passes around the chest, but there is a distinct white break from the wing shoulder and the dark green-brown collar. Has a long tail passing well past the wing tips. Has a prominent white eye ring. Green-brown tinge to all the upper body. In flight has a very noticeable thin white wing bar standing out from the dark wing. When approached, tends to fly low over water with stop-start wing beats with its wings drooping. **Similar species:** None that can be confused in the South West. **Status:** Moderately common in the northern region. Uncommon in the South West. Green Sndpiper has been recorded in the NT and there is obviously the possibility that it can come to the Kimberley. **Where to find:** Throughout the north west coast particularly behind mangroves and inland pools. Lake MacLeod. All mud banks in the Carnarvon area south to Long Point. Throughout Shark Bay region. Inland along the Gascoyne and Murchison Rivers. Also some inland pools and dams. On many of the islands including the Abrolhos group. Locally common on Rottnest Island. Penguin Island. Found along the south coast all the way to the Esperance region, including the inland lakes like Lake Warden.

Photo: Simon Nevill

Grey-tailed Tattler *Tringa brevipes* 25 cm

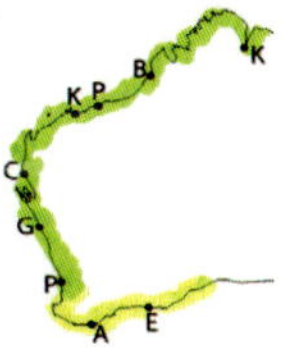

There are only 2 tattlers in the world, Grey-tailed Tattler and Wandering Tattler; both come to Australia, although the Wandering Tattler is far less common than the Grey-tailed and is almost entirely restricted to the eastern shores of Australia where it over-winters. We all marvel at the long flights that waders make to get to Australia but some of the habitat locations they breed in are also very interesting. The Grey-tailed nests in dry rocky stream beds but also occasionally, nests way up in trees in old birds' nests, high in the mountains of remote eastern Siberia. Apparently quite adept at perching on tree branches, something we don't see in Australia. It is a medium sized wader with short yellow legs, wings almost touching the tail when at rest. White supercilium eye stripe that passes just over the eye. Plain grey upper body and crown with greyish front to the breast, bleeding out to a white throat. Bill straight and slightly yellow. **Similar species:** Wandering Tattler. You will be lucky to see the Wandering Tattler in the west but if you do it is, perhaps of all the regular waders visiting these shores, one of the hardest to split from the similar Grey-tailed Tattler. All the described features are very subjective. It often gets down to 'jizz' as the Wandering is thicker in the body but only slightly. One of the main differences is the Wandering Tattler is definitely a darker grey but it's really best to have these birds side by side then you can get used to the shape and colour. Features like nasal grooves, supercilium eye stripe size, and primaries to the tail are variable and can be unreliable. Wandering has a preference for feeding on rocky coast lines and that certainly aids identification as they rarely feed on mud banks. In breeding plumage, it is much easier as the barring on the under-body on the Wandering is heavier and passes right under the body whereas the Grey-tailed has lighter barring that diminishes on the flanks to a white belly. **Status:** Moderately common in the northern region. Uncommon in the south west. **Where to find:** Any of the rocky shores of the north west. Carnarvon region down to Long Point also in the Shark Bay inlet and Peron Peninsula. Kalbarri down to the Vasse Estuary. On some of the western islands including Rottnest Island. Uncommon on the south coast.

Photo: Simon Nevill

Ruddy Turnstone *Arenaria interpres* 22-24 cm

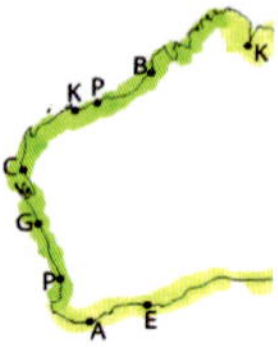

A small very stocky wader but its size certainly belies its capacity to travel. The Ruddy Turnstone nests as close to the Arctic Circle as you could possibly get. The breeding grounds in the Tundra stretch the full circle of the globe; there are five identified separate populations but they come together when flying south to their over-wintering grounds, which include Australia. They make this phenomenal journey leaving the Tundra in July and slowly make their way through Asia arriving in the north of Australia in August. So when you see these small waders, say on Rottnest Island salt lakes, spare a thought for the journey they have made to get here, losing up to one third of their body weight in the process. Turnstones cannot be confused with any other wader except the Black Turnstone, which would almost never fly to the west coast, possibly a remote possibility to the east coast. They are small, very dumpy waders with a black chest, partly broken by a white collar that is not clear cut until in breeding plumage. Dark brown coverts, turning to brighter rust-orange coverts in breeding plumage. Has short bright orange legs. A very short stout bill. Feeds in a variety of habitats but prefers scrambling over rocks feeding in small tidal pools. Also feeds on sandy and gravel beaches and estuaries. There are two subspecies in the world and both come to Australia but the main subspecies to Western Australia's shores is *interpres* that breeds in northern Europe and northern Siberia and part of Alaska. **Similar species:** None. Black Turnstone *Arenaria melanocephala* highly unlikely here, more so east coast. **Where to find:** Moderately common in the north-west and Carnarvon region down to New Beach. Also in the Shark Bay region. Common on all the larger islands south to Penguin Island. Most common on Rottnest Island. Uncommon from Augusta to Cape Arid.

Photo: Simon Nevill

Asian Dowitcher *Limnodromus semipalmatus* 33-36 cm

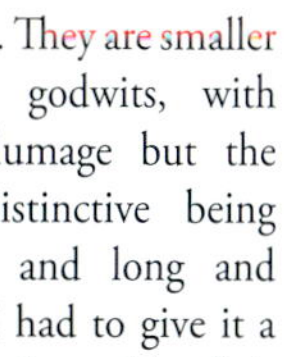

A similar wader to the Godwits but the bill is more reminiscent of a snipes bill. Uncommon in Australia. Some authorities put the estimated population to Australia at less than 500 birds, however Broome and Eighty Mile Beach receive reasonably large numbers in a season but still an uncommon wader. In 2005, 180 were seen at the Port Hedland Saltworks. Once you have seen a few Asian Dowitchers, you soon get to pick them. They are smaller than the godwits, with similar plumage but the bill is distinctive being snipe-like and long and heavy. If I had to give it a nickname, I would call it the 'sowing machine' bird as it prods in snipe fashion faster than most waders into the soft mud but will also stretch forward as it is walking looking for food. It requires soft sandy beaches and more so mud flats and estuaries. **Similar species:** Bar-shouldered Godwit, which is taller

Photo: Simon Nevill

with a larger body and thinner upturned bill. **Status:** Locally moderately common at Roebuck Bay. Scarce in the rest of the north west. Rare in South West. **Where to find:** Roebuck Bay Broome and Eighty Mile Beach always get some Asian Dowitchers. Seen in Albany region. Lake McClarty and Lake McLeod.

Great Knot *Calidris canutus* 26-28 cm

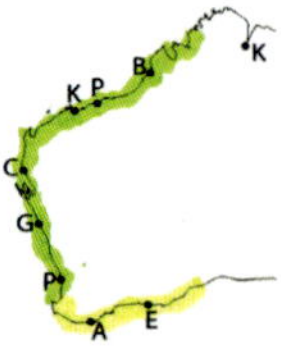

Great Knot breed in remote subarctic mountains in far eastern Siberia. For a bird that comes to these shores in greater numbers (approximately 350,000 +) than most waders, little is known of its breeding habits. What we do know in the last few years the world population is crashing fast which is a serious worry, mainly due to feeding area mudflats being reclaimed. There are two knots in the world and both migrate to Australia, the Great Knot and Red Knot. The Great Knot is a medium sized wader with a thick-set body, stout bill and relatively short legs in proportion to the size of its body. **Similar species:** Red Knot is smaller; bill is less than the depth of its head, whereas the Great Knots bill is larger than the depth of its head. The chest is more pronounced on the Great Knot, giving it a sturdier look. If in breeding plumage there is no problem identifying the two species. The scapulars on the back of the Great Knot are larger and appear more contrasting, like the back of a Bar-tailed Godwit compared with a Black-tailed Godwit in colour. **Status:** Uncommon to common in north west, numbers have drastically fallen in the last 25 years. Scarce to uncommon in the South West. **Where to find:** Roebuck Bay. 80 mile Beach. Shark Bay. Gascoyne estuary and Mangrove Bay. Less so down to New Beach. Beaches in the Denham region. Murchison and then further down the coast to Bunbury. In the Perth region. Alfred Cove. Woodmans Point. Peel Inlet. Scarce on the south coast.

Photo: Simon Nevill

Red Knot *Calidris canutus* 23-25 cm

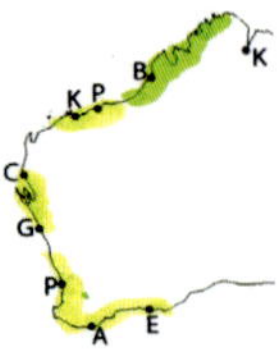

Breeds in several separate areas in the Arctic from Canada through to parts of Russia and the far eastern tip of Siberia. Unlike on the east coast, the number of Red Knot in Western Australia could exceed Great Knot. They feed together which makes identification easier. A medium sized thickset wader with short legs. Thick bill, in length less than head depth. Scalloping on flanks bolder than on Great Knot. There are 5 subspecies in the world and it is still not totally clear which are the main subspecies that migrate to these shores but *rogersi* and *canutus* are considered the primary migrants. **Similar species:** Refer to notes on Great Knot. **Status:** Uncommon to moderately common. **Where to find:** Roebuck Bay. Parts of

80 mile. Pilbara mudflats. In the Gascoyne Estuary, Mangrove Bay and parts of the Shark Bay region including Dirk Hartog Island. Mudflats and estuaries from Murchison to Bunbury and then uncommon from Wilson Inlet to Esperance. In Perth region mainly Alfred Cove and Peel Inlet.

Photo: Martin Withers

Sanderling *Calidris alba* 20-21 cm

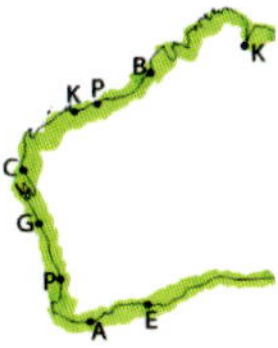

Breeds in various locations as close to the Arctic Circle as possible, just 800 km from the North Pole, having the briefest period in which to nest and raise young before it makes its massive journey to the Southern Hemisphere. It migrates south to coastal beaches in more countries than almost any other wader, including North America, Central America, South America, Africa, the Middle East, the Indian subcontinent and parts of southeast Asia and Australasia. The numbers visiting Australia are not enormous, estimated to be around 10,000 birds. A bird similar and slightly smaller, the Red-necked Stint, arrives on the these shores in numbers around the 250,000 mark. Sanderling are dumpy little birds, smaller than a Curlew Sandpiper and slightly bigger than a Red-necked Stint. Birdwatchers can overlook them in non-breeding plumage, thinking they are Red-necked Stint. Sanderling prefer sandy

Photo: Simon Nevill

beaches but will visit salt lakes such as on Rottnest Island. **Similar species:** Red-necked Stint. Sanderling are not only larger but much thicker set in body. Upperbody has pale grey plumage not grey-brown. The Sanderling has a distinct black shoulder patch, head looks paler. It is one of the most active waders, typically running backwards and forwards as the tide ebbs, exposing crustaceans and other food. Broad-billed Sandpiper is slightly smaller, with a distinct, longer bill with a down turned end and distinct supercilium eye stripe. **Status:** Moderately common. **Where to find:** Roebuck Bay. 80 mile. Pilbara beaches. Lake MacLeod. Patchily distributed on select beaches. Every year can be found on Rottnest beaches and salt lakes.

Red-necked Stint *Calidris ruficollis* 13-16 cm

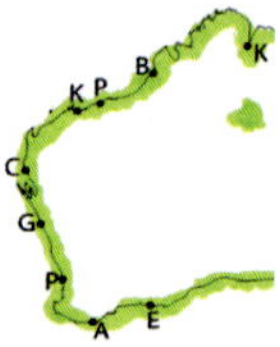

One of the most common waders to arrive on the shores of Australia, breeding directly north of Australia in the far north of Siberia; also a much smaller population in Alaska that flies down to the west coast of North America. Besides the very rare Little Stint and the scarce Long-toed Stint, the Red-necked Stint is our smallest wader to migrate to these shores and smaller than any of our endemic shorebirds. In non-breeding plumage, the Red-necked Stint has grey-brown back, with mottled markings on the neck collar. Slight flecking on the crown, which will turn red-brown with darker brown streaks in full breeding plumage. It has short legs and small straight bill. Does not develop the rich rust red on the face and upper breast until in breeding plumage. Has a white supercilium. In flight on the upperbody it has a white wing bar, not pronounced. What is clearer on the upperbody, which all stints fortunately have, is a white tail with a central black tail coverts showing as a long black centre stripe. Can be found in freshwater as well as saline waters but primarily is a bird of the coastal regions, estuaries, islands and saltworks ponds. Almost always in groups, from a few birds to several hundreds in the south west and in the north west into their thousands. **Similar species:** Long-toed Stint, which is slightly bigger, and could be described as a miniature Sharp-tailed Sandpiper with its black streaking on a chestnut crown and browner back. Longer legs that are yellowish-green. Has a slightly longer neck and will stretch neck to feed or when alarmed which Red-necked Stints do not do as much. Little Stint in non-breeding not easy to split but Little has slightly decurved bill where as Red-necked has a straight bill also the coverts on the upper body have a browner tinge rather than a grey-brown appearance. In breeding

Right: In eclipse plumage

Photo: Simon Nevill

its easier the Little lacks the reddish colouring to the neck and the coverts are richer brown with more rufous edges. **Status:** Very common in the north and common in many parts of the south west. **Where to find:** Roebuck Bay. 80 mile. Most of the larger mudflats on the Karratha coast. Lake McLeod all the way down to Hamelin Bay including Dirk Hartog and Useless Loop. Murchison down to Busselton and Augusta to Cape Arid. Common on islands particularly Rottnest.

Little Stint *Calidris minuta* 12-14 cm Vagrant

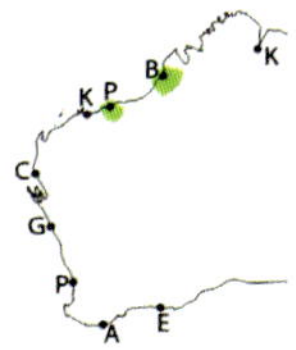

A rare vagrant to Australia but may be more common than we may think, for two possible reasons. Both Little and Red-necked Stint overlap in their breeding range in the Russian Tundra but Little Stint mostly migrate to Africa, the Middle East and the Indian subcontinent, both species overlapping on the Indian coast. Red-necked migrate south to the eastern coast of India and south-east Asia and Australia, however some of the Little Stint, particularly juveniles that leave the most easterly range in the Tundra, may easily take a flight path down through far-east Asia, particularly if weather conditions effect their normal instinctive flight path. The second reason is more straight forward. Little and Red-necked are almost impossible to identify out of breeding plumage and it takes a very skilled observer to split them in the field. It's so difficult that only in the hand, measuring wing and tarsus length, looking at moult and other features can they be successfully split. **Similar species:** Red-necked Stint. Non-breeding very difficult and the differences are marginal but Little Stint has shorter legs, slightly smaller body size, longer wings, making flight profile slightly different, bill tip thicker, head more rounded, more upright posture. In flight both ventral and dorsal view basically impossible to split. In full breeding plumage difference clear, Little Stint has pinkish-salmon facial disc on ear covets that stops above neck. Red-necked Stint has same pinkish-salmon colour on face but extends down to collar and forms bib on upper breast. **Status:** Rare vagrant. **Where to find:** Possible on any north-west beache with Red-necked Stint.

Photo: Adrian Boyle

Long-toed Stint *Calidris subminuta* 13-16 cm

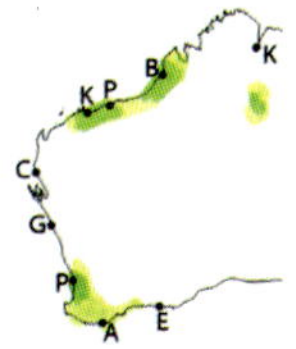

Breeds in central eastern Siberia and parts of Mongolia. One of the few rare waders that have a higher population in Western Australian than in eastern Australia. It is estimated that Australia receives about 1000 birds each year. However I have noticed a mark decline in the species in the Perth area. Prefers shallow freshwater and brackish wetlands, particularly with emergent grasses. A more solitary wader found normally in no more than groups of 5-10 birds although larger numbers have been seen. **Similar species:** Has a distinct shape, being slightly bigger than Red-necked Stint, with small head and longer neck, which it will stretch, to feed or when alarmed standing upright. Dark streaking over chestnut crown. Darker back and shorter rear body. Legs are longer with distinct yellow-green tinge that shows well in good light. Flies at a steep incline when alarmed, with a distinct call. The author often thinks of it as looking like a miniature Sharp-tailed Sandpiper with its slightly rufous crown with faint streaking and heavier markings on the scapulars. **Status:** Scarce throughout the state. **Where to find:** Found on freshwater grass and mud fringes. Lake Campion. Lake Eda. Roebuck Plains. Taylors Lagoon. Parrys Lagoon. Lake Argyle. Samphire Flats floodplain. Most birds are seen on the Swan Coastal Plain with Jandabup, Lake Thompson and Lake Kogalup being the favoured wetlands when they have sufficient water. Lake MacClarty is a particularly good site. During the last 30 plus years, the author has seen several small grassy wetlands that were favoured by Long-toed Stints, filled in for new housing developments particularly in the Jandakot and Banjup suburbs.

Photo: Alan Collins

Pectoral Sandpiper *Calidris melanotus* 19-24 cm

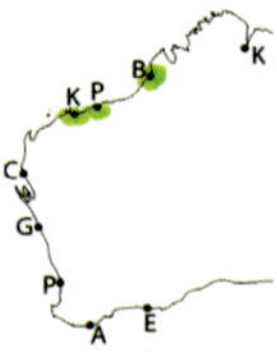

Breeds in the far north of central and eastern Siberia and northern Alaska and Canada. Most birds fly down to South America but a few migrate to the Australian shores each year, however it is generally scarce in Australia with a few coming to the South West. A medium to small wader with small head on longish neck. Body tapers at rear. Well streaked brown crown, nape and upperbody. Bill often slightly down curved with yellowish tinge to the wide

Right: Showing clearly the pectoral band Photo:: Alan Collins

Photo: Alan Collins

part of the bill. Has a distinct break from the streaked throat and breast to a pure white belly. Shortish yellowish legs. Found on freshwater wetlands but will feed on the remaining exposed mud on large lakes. **Similar species:** Sharp-tailed Sandpiper. Lets compare them; the Pectoral Sandpiper is best seen with Sharp-tailed Sandpipers as you can compare the differences quicker. One of the main diagnostic features of the Pectoral Sandpiper and from which it derives its common name is the clear cut division of the heavy brown streaking and beige background on the neck and upper breast to the pure white on the lower breast and underparts. This remains a constant wether its male or female, breeding plumage or non-breeding plumage. If you are lucky to get a male in eclipse or even better breeding plumage, the markings between the upper breast and the white lower breast are one of the strongest contrasts of any wader. The Pectoral Sandpiper has a slightly longer neck and which when you get used to them stands out from the Sharp-tail's neck length. Mention is always made of the longer and slightly down curved bill of the Pectoral Sandpiper but I find this not a great feature as beaks vary slightly in length and they are very close in length and anyway the Sharp-tailed has a slight down curve anyway not as pronounced but still there. The Pectoral Sandpiper shares the same feeding habitat that of emergent grasses in freshwater. Has a more upright stance than the Sharp-tailed Sandpiper.. Similar markings in flight make it difficult to split. **Status:** Uncommon throughout the state. **Where to find:** Roebuck Bay. Karratha mudflats. Lake MacLeod. On the Swan Coastal Plain but rare to scarce on Lake Thompson. Can be seen on any freshwater wetland in the southwest with shallow well grassed margins. Seen at Lake Warden, Esperance and best at Lake McClarty.

Sharp-tailed Sandpiper *Calidris acuminata* 17-22 cm

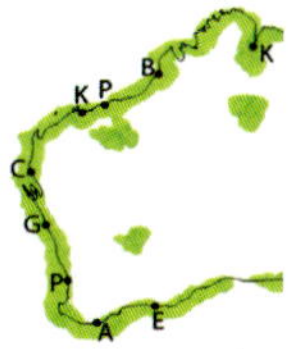

It is an interesting fact that the Sharp-tailed Sandpiper breeds in the same remote Siberian region as the Pectoral Sandpiper, both in large numbers, but one flies almost entirely down through the Americas to South America and the other chooses to come to Australia. Both look similar and both require similar feeding habitats. The Sharp-tailed Sandpiper is another common visiting wader with an estimated population of 140,000 birds. A reasonable number come down the west cost to the South West region. Sharp-tailed Sandpipers are a small-medium sized wader with rufous streaked crown, white supercilium eye stripe, a longish slightly down curved bill, short yellow-green legs, long rear pointed body. The main feature is not so much size but the overall colour that is a richer brown compared with many of the waders that are greyish looking. The Sharp-tailed Sandpiper has warmer colours

Photo: Simon Nevill

particularly the rufous streaked crown. Has a habitat preference for feeding on muddy edges of freshwater and brackish wetlands. Feeds in water that has emergent grasses and sedges but will feed on open mud when the need arises. **Similar species:** Refer to Pectorial Sandpiper. **Status:** Common on selected wetlands throughout the range. **Where to find:** Found from Cape Levique down to Hamelin Bay occasionally on the coast but mostly in flooded samphire or the many bores with overflow pools that create wet grassy areas. Found throughout many wetlands on the Swan Coastal Plain. In Perth can be found on a few lakes with wet grassed margins and receding waters like Thompson, Forrestdale, Jandabup and Joondalup Lakes. Masses can gather on the edge of Peel Inlet in the samphire marshes also Lake MacClarty near by. Also Vasse Estuary. Harvey Estuary and Busselton Wetlands mainly Broadwater. Less common on south coast until the Esperance region where there can be larger numbers. Common locally in wet grasslands in the Kimberley and seen way out in the desert regions as in Lake Gregory.

Curlew Sandpiper *Calidris ferruginea* 18-23 cm

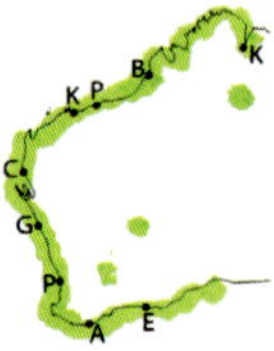

Breeds in the far north of central Russia migrating to Africa, the Middle East, Indian sub-continent and East Asia and Australasia. One of Australia's most common waders. One of the easiest of the small waders to identify with its long down curved bill from which it gets its name. It has longish legs for its size, slender neck and small head. An important diagnostic feature is that most sandpipers and stints have black tail coverts and central tail feathers running down the length of the tail. It contrasts well with the white outer tail feathers. The Curlew Sandpiper has an all white rump and upper tail coverts. Has distinct white wing bar. Found in sheltered bays and estuaries, preferring softer mud conditions. Will feed in freshwater lakes and will go inland although not as much as those that visit Africa. **Similar species:** Only the White-rumped Sandpiper which is extremely rare, has a shorter and blunter bill and is a smaller and slimmer wader. **Status:** Common throughout much of the state. **Where to find:** Roebuck Bay. Karratha mudflats. Port Hedland. Lake McLeod, Mangrove Bay all the way down to Hamelin Bay and then much of the Shark Bay Region. Also will visit bore overflows and ephemeral wetlands near coast. Common on the Swan Coastal Plain particularly large drying lakes like Thompson and Forrestdale. Peel Inlet can get several thousand birds in a good season. Is less common along the southern coast to Esperance.

Eclipse plumage

Photos: Simon Nevill

Broad-billed Sandpiper *Limicola falcinellus* subspecies *sibirica*

16-18 cm

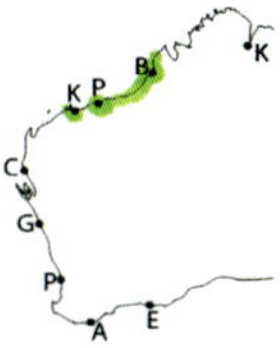

One of the uncommon waders in the world found breeding in a few fragmented areas from northern Europe and Russia. Most birds remain in the north west of WA. Massive numbers for such an uncommon bird have been recorded in the Leslie Saltworks, Port Hedland, with over 3000 birds sighted. A few come down the coast to the Swan Coastal Plain and rarely to the southern coastal belt. It is a small wader with a distinct bill, some people presume it is called Broad-billed because its bill is relatively broad at the base, which it is, but when looking from above the bill, the tip slightly broadens, which is an unusual feature in waders. The Broad-billed Sandpiper is a small wader not much bigger than a Red-necked Stint and smaller than a Curlew Sandpiper. The head is distinctly striped, seen well even in non-breeding plumage. The bill is not only broad at the base but also relatively long, nearly as long as a Curlew Sandpiper in relation to head size, also bent downwards near the tip. Legs short. In the South West not seen in large groups but mostly solitary with Curlew Sandpipers and Red-necked Stints. There are two subspecies and only one *siberica* which breeds in Siberia, comes to Australia. **Similar species:** No other wader visiting Australian shores. **Status:** Uncommon Roebuck Bay. Scarce elsewhere except Dampier Saltworks (no access). **Where to find:** Roebuck Bay. Lake McLeod. Gascoyne Estuary and Mangrove Bay. Found south to Hamelin in the more muddy beaches and estuaries. More likely to fly down to Swan Coastal Plain and Peel Inlet. Best place in Perth is Alfred Cove.

Above: Showing lateral stripes on crown.

Photos: Simon Nevill

Ruff (and Reeve-female) *Philomachus pugnax* 28-32 cm Vagrant

Breeding range western Europe across to eastern Siberia. The males perform the most extraordinary display ritual of any wader, showing off the most elaborate head plumes while they display and perform mock fights at their breeding lecks. Ruff mostly over-winter in Africa, often roosting in huge numbers, however small numbers regularly migrant to Australia. A wader with a distinct jizz, having a very deep body often described as having a 'pot-belly' and 'hunched back', longish neck and legs, small head and smallish decurved bill. In non-breeding plumage male and female very alike, grey-brown back, faint grey-mottled markings on neck and on breast, neck and under-body colouring is a cream, legs pinkish-orange, walks with deliberate slow motion with head extended forward, in flight shows white ovals either side of a dark central shaft which terminates at tip of the tail in a dark band. Males are bigger but the neck of the female is longer. Male Ruff are noted for their amazing variability, not only in plumage colouration but also in sexual dimorphism. Not that you will see it in Australia but when males are displaying at their leks with other males, one male can have a white head plumes and the other totally different having a dark-brown. What you may see in Australia is different shapes, some will have longer necks or smaller body than another. Regardless of the variations in size, Ruffs are still relatively easy to identify, as their unusual body shape, combined with smallish head, give them a unique jizz unlike any other wader in Australia. **Similar species:** None. **Status:** scarce vagrant. **Where to find:** Often found near freshwater ponds with exposed mud or well grasses margins, fairly nervous when approached, not so common on beaches, often seen with Sharp-tailed Sandpipers which are smaller birds. Has been sighted in many areas in WA including wetlands at Albany, Lake McLarty (nearly every year), Perth, Carnarvon, Roebuck Bay, Lake Eda and the Ord River floodplain.

Reeve Photo: Simon Nevill

Red-necked Phalarope *Phalaropus lobatus* 18-19 cm Vagrant

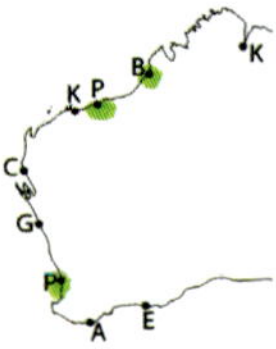

There are 3 species of phalarope in the world and all have been sighted in Australia but regarding WA, we will concentrate on the one regular species that is sighted in this state namely the Red-necked Phalarope. The Red-necked Phalarope is a scarce vagrant but has been sighted regularly in WA although less so in recent years. The other two, Grey Phalarope and Wilson's Phalarope are very rare vagrants. Phalaropes are fascinating birds. Unlike most waders, phalarope spend much of their non-breeding life swimming rather than wading to gain food; this is unusual in the wader world. To do so, they have developed interesting morphological adaptations. Like ducks they have developed water resistant plumage combined with a dense layer of down to aid insulation and also give them buoyancy, hence they can be seen way out at sea although most of their preferred feeding range is within calm tropical waters. They gain their food by either swimming in concentric circles or more often moving from side to side agitating the water surface. The important aspect for optimum feeding habitat is where waters are still and the water surface tension can hold up small aquatic invertebrates. That's why the saline lakes on Rottnest are favoured feeding location for Red-necked Phalarope that have migrated too far south. We know these birds, which have over shot their normal non-breeding range as the sightings are often of lone birds and non-breeding birds are known to gather on staging lakes in Russia in their thousands, so when they do find a prime feeding locality outside their normal over wintering range, like the salt works at Port Hedland they will come together and congregate in small groups to feed. Their main over-wintering near Australia are the calm tropics from east Borneo through Wallacea to New Britain. Most of their feeding is done out at sea in the calm tropical but mostly close to land, they also frequent inland lakes particularly if inclement weather occurs at sea. At sea, rafts of 5000 + birds have been recorded between the islands of Pantar and Alor in Indonesia (R.E.Johnstone). Red-necked Phalarope in full breeding plumage are unmistakable. The female has more distinct markings, the head is slate-grey, throat white surrounded by a fine black margin, rich chestnut collar that passes from behind the eye and broadens on the breast dark grey upper-body with fawn speckling on scapulars, white under-body with grey to upper breast and grey flecking on flanks. Non-breeding plumage, totally different. Gone is the bright chestnut band and strong black colouring. Forehead, neck and under-body white, upper-body, scapular and coverts grey with white margins; phalarope diagnostic feature is black eye stripe curving down behind eye, crown dark-grey both these colours contrast strongly against white easily seen at distance. A small dainty wader, a little smaller than a Sharp-tailed Sandpiper. Feeding technique unique when seen in water, rarely feeds on mud. **Similar species:** In breeding very different to the other two phalarope but in non-breeding Grey Phalarope slightly larger, lacks strong white margins on mantle and scapulars as in Red-necked (perhaps main difference to look for), head looks, and is, larger. Wilson's Phalarope larger, with longer, thinner bill, in flight all white rump, importantly no white wing bar. **Status:** scarce vagrant. **Where to find:** Any inland water both saline (mostly) and fresh water. Mostly not too far from the coast but after cyclonic weather can be found well inland.

Photo: Rohan Clarke

Oriental Pratincole *Glareola malaivarum* 23-24 cm

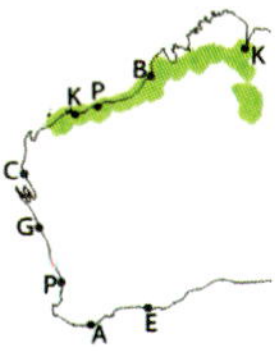

Oriental Pratincole breed in southern Asia and migrate down to Australia in the wet season in massive flocks, often following storms where insects are swept up in the turbulence. Found on open plains where they will run and pick insects or more often fly with incredibly fast manoeuvres hawking for insects. When in full feeding frenzy, it is like watching jet fighters flying in all directions as they fly over the tree-tops taking advantage of swarms of insects in the wet season. Their movements are unpredictable but they can gather in their thousands, such as on Eighty Mile Beach. Often roost in 100's to thousands on some grasslands. Flies fast on elegant long wings with shallow wing beats. Will frequent wetlands margins. **Similar species:** Australian Pratincole, similar in shape only. Australian Pratincole has a brown fawn-

Photo: Simon Nevill

coloured body, the Oriental Pratincole has a grey-brown duller body with a distinct throat pattern seen best in full breeding plumage. Australian Pratincole has a distinct dark chestnut band on the flanks and belly. **Status:** Common in the northern wet season. Rare in the south west, it has been seen in Perth on Jandabup Lake but is extremely rare this far south. **Where to find:** Mainly north of Karratha and common in the Kimberley in summer.

Australian Pratincole *Stiltia Isabella* 19-24 cm

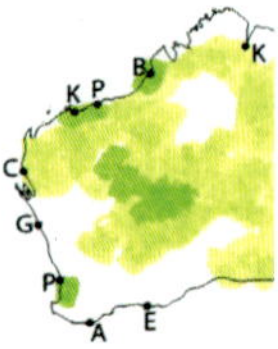

There are 8 pratincole in the world and Australia has two of them, the Oriental Pratincole which migrates here in the summer months and the Australian Pratincole, which is a resident endemic bird and the sole member of its genus. A long slim bird with longer legs than other pratincoles. Has a rust-coloured band on the flanks and belly, which in flight contrasts well with the white vent, has black along the under wing coverts. Bill is slightly down-curved on the tip, and in breeding plumage is bright red from the base to half way along. Has very pointed wings shown clearly in flight and is a graceful flyer. A bird of open stony plains and grasslands of the inland arid regions. Does not hawk for insects as much as Oriental Pratincole but rather runs and picks at insects mainly on exposed stony or very short grassed plains. Even though found on remote open plains, they still need to be near bodies of drinking water. **Nest:** No nest, only a scrape made on bare ground and located where the bird has good all-round vision, often on a rise. Clutch size 2. **Similar species:** Oriental Pratincole which has grey-brown upperbody and not pale chestnut. Legs almost half the length of the Australian Pratincole. In flight easy to split as Oriental lacks chestnut band on the belly but does have pale chestnut axillaries and median coverts. **Status:** Locally common in the Pilbara and north, normally in groups. Scarce in the South West. **Where to find:** Any open stony plain or short grasslands through the mid-west, Pilbara and the Kimberley. Locations such as Camballin floodplains. Roebuck Plains. Eastern Murchison. Rare vagrants have been reported in the Perth region.

Photo: Simon Nevill

Comb-crested Jacana *Irediparra gallinacea* 20-26 cm

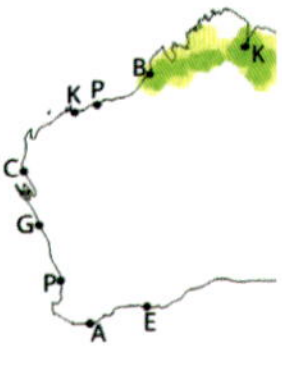

There are two subspecies, the nominate gallinacea is found through out parts of Indonesia, southern Philippines and subspecies *novaehollandiae* is found only in Australia. A bird that requires tropical freshwater water wetlands with floating aquatic plants, particularly water lilies where it uses its extremely long wide toes to distribute its weight, typically running fast between more unstable vegetation to get to a firm footing. It feeds on aquatic insects but also seeds. An unmistakable bird, the adult having a Pinkish-red wattle on forehead, white face with black nape and neck forming wide black breast band with yellow orange tinge to band margins, extremely long legs and toes, upper body dark grey-brown with white vent. Immature birds lack the red wattle on crown, crown brown not black and no black breast. **Similar species:** No other Australian species however there have been at least two sightings of Pheasant-tailed Jacana *Hydrophasianus* in WA but this is a very rare vagrant with totally different head and wing and no wattle on head, a resident markings. Found in the Indian subcontinent to southern China some movement south to over winter in Sumatra and Java hence a few birds flying on to Australia. **Status:** Locally common in the Kununurra area, uncommon else where. **Where to find:** On the lower Ord River above the dam, Parry Lagoon and many parts of Lake Argyle. Patchily distributed around the rest of the Kimberley such as Lake Eda and Taylors Lagoon near Broome, Camballin wetlands, Munkanyarra wetland south of Derby and the odd remote wetlands like Munja Lagoon on the Walcott.

Photo: Simon Nevill

Australian Painted Snipe *Rostratula australis* 23-30 cm

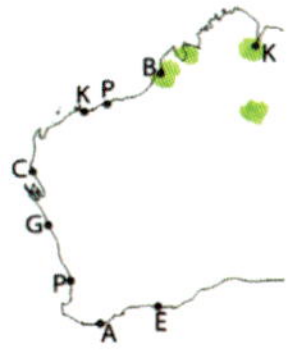

A totally different genus to the genus *Gallinago*. A strongly patterned attractive snipe that was only until recently lumped with the nominate species *benghalersis*, the Greater Painted Snipe found in Africa, the Indian subcontinent and south east Asia. The Australian species is more prevalent in the eastern parts of Australia, requiring freshwater marshes with exposed mud on which to feed. Is semi-nomadic, moving to wetlands where the conditions are suitable. Unlike *Gallinago* snipe, the Painted Snipe has a slightly down-curved bill and their flight is slower with slower wing beats. Unmistakable markings like no other snipe with distinct white band from upper-body to the white under-body. Male has darker head colouring, both sexes have white eye patch, wings heavily mottled with cream spots. The female is slightly larger. **Similar species:** None. *Gallinago* snipe have longer, straight bills and generally have mottled brown colouring, have white supercilium eye-stripe. **Status:** Very uncommon in the

Photo: Rohan Clarke

Kimberley, in the northwest highly nomadic. **Where to find:** Can be found wetlands around Derby. Found occasionally in the wetlands on Roebuck Plains and Lake Eda. Kingston Rest. Ord irrigation fields. Parry Lagoon. Has been sighted throughout the state including the Pilbara, Carnarvon and wetlands north of Perth. Historically breed in the Perth region, particularly those wetlands west of Moora and around Gin Gin.

SNIPE: There are 15 species of the genus *Gallinago*, with 3 species that come to Australia. Latham's, Swinhoe's and Pin-tailed Snipe. Latham's is the most common visitor and the only one considered to be a regular migrant to Australia. The other two species are considered accidental vagrants but this has not yet fully been established, that they are in fact irregular migrants, as Swinhoe's is sighted cited fairly regularly in the Kimberley. Splitting snipe species in the field is perhaps one the most challenging exercises one could wish for in bird identification. Identifying Latham's is possibly the easiest if one is experienced in seeing snipe regularly and I certainly am not. Latham's is slightly larger than the other two species. Even though slightly bigger, Latham's appears to have a more pointed body shape at the rear having longer wings and tail. If one is lucky to get a reasonable sighting in flight, you will see the feet do not pass the end of the closed tail feathers. Pin-tailed is smaller and dumpier body, shortest tail of all the 3 species. In flight, legs project well past tail. Swinhoe's legs also protrude past tail. Regarding Pin-tailed and Swinhoe's, it is almost impossible to separate without identifying structure and number of tail feathers in the hand. To give you a better concept of size, the average weight for a Latham's Snipe is 200 g, for Swinhoe's its 120 g and for Pin-tailed Snipe it's 115 g, so the latter two are very close but Latham's is a distinctly heavier species. Quantity of rectrices (tail feathers): Latham's 18. Swinhoe's 20-22 (normal qty) Pin-tailed 25-27 (normal qty). Pin-tailed wing tips almost reach the end of the tail feathers when closed. Latham's we know passes on its over-wintering path down the east coast but Swinhoe's and Pin-tailed are seen and have both been collected in the north and north-west of Australia but often identification has been based on locality but one must be very careful on basing sighting on this premise as Latham's have been netted in WA as far south as Perth. I have seen this species at Lake Seppings Albany years back. Swinhoe's has been netted and collected more so in the Kimberley than Pin-tailed but again one cannot presume most birds seen in the Kimberley are Swinhoe's in the Kimberley rather than the rarer Pin-tailed.

Latham's Snipe *Gallinago hardwickii* 29-33 cm

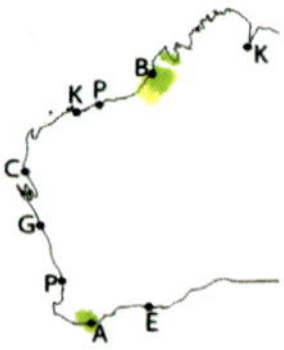

Breeds in Japan and passage migrant to eastern Australia. Largest of the *Gallinago* snipe. Stout long bill, mottled brown upper body, cream stripes on the crown and supercilium stripe contrasting with brown face, short stocky legs. Often seen probing fast into mud or damp grasses in sewing machine fashion just like dowitcher's feed. Often freezes when sighted holds head high while observing, generally shy and prefers longer grass in which to hide, flies fast when alarmed. **Similar species:** As mentioned above, Pin-tailed Snipe and Swinhoe's Snipe. **Status:** Very uncommon in WA. World Population estimated at 36,000 with a large proportion of the population coming to eastern Australia. **Where to find:** Possible on any freshwater wetland with extensive exposed mud and ideally well grassed. Sighted at Lake Kogalup and Lake McLarty (per J. Graff).

Photo: Rohan Clarke

Swinhoe's Snipe *Gallinago megala* 27-29 cm

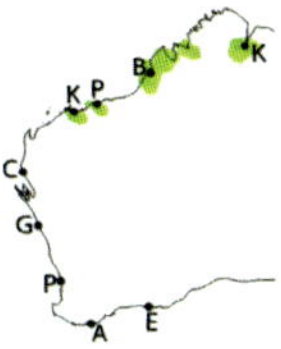

Breeds in central eastern Russia and south into Mongolia and also far eastern Russia north of Korea. None-breeding range mainly eastern Indian peninsula, northern Far East, Indonesia, with small numbers coming as far as Australia mostly in the northern region. General features and habit preference similar to Latham's. **Similar species:** As mentioned above refer Latham's Snipe. **Status:** Uncommon but regularly sighted in the Kimberley of WA. **Where to find:** Possible on any freshwater wetland with exposed mud and ideally well grassed. Many times I have sighted snipe at the Derby sewerage overflow and believe they were Swinhoe's from flight and jizz. The species has been netted regularly and also collected throughout parts of the Kimberley and is certainly not rare. Has been found at Lake Eda and Parry Lagoon. Look not only on wetlands but also along river systems where there are well-grassed overflow areas.

Photo: Rohan Clarke

Pin-tailed Snipe *Gallinago sturnura* 25-27 cm

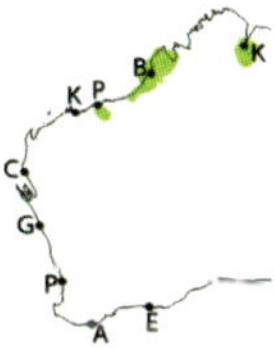

Breeds in central eastern Russia and south into Mongolia and also far eastern Russia north of Korea. None-breeding range mainly eastern Indian peninsula, northern Far East, Indonesia, with small numbers coming as far as Australia, mostly into the northern region. General features and habit preference similar to Latham's. **Similar species:** As mentioned above refer Latham's Snipe. **Status:** Very uncommon but has been recorded in the Kimberley of WA and several other locations, particularly in the Pilbara. I sighted one on an overflow dyke in Perth, thanks to Geoff Shannon. **Where to find:** Possible on any freshwater wetland throughout the state.

Photo: Adrian Boyle

Pacific Gull *Larus pacificus* subspecies *georgi* 50-67 cm

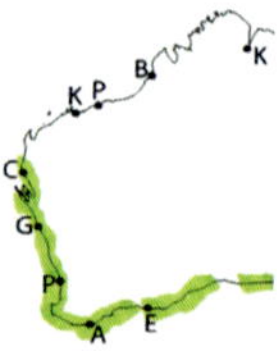

Endemic to the shores of southern Australia, reaching the highest latitude in Australia on the west coast as far as Carnarvon. Scarce north of Carnarvon. The largest gull in Australia with a heavy bill, dark back, wings at rest extending past the tail. Feeds mainly on fish and molluscs but is also a predator, taking the young of other seabirds during the nesting period. Will hunt at night as well as day over some seabird colonies. It is a solitary bird and rarely feeds in groups. Pacific Gulls do not reach full adult plumage until the fourth year and that makes identification of the first, second and third year birds more difficult to split with Kelp Gull. There are two subspecies in Australia with one, *georgi* in the South West. **Nest:** No nest is built but a relatively deep scrape is made and lined with seaweed if near a beach but on islands the nest hollow is lined with any soft material adjacent to nest site. Clutch size 2 to 3. **Similar species:** The Kelp Gull is a smaller less-stocky bird. Bill is not as massive. The nostrils are an interesting feature of the Pacific Gull, insomuch as that they are round, except in the subspecies *georgii* in the Shark Bay region, which has slightly narrower nostrils. The Kelp Gull nostrils are the narrow long slit typical of most gulls. Two important differences involve colour in the adult birds. The bill on the Pacific Gull has a red-orange tip on both the upper and lower mandible; on the Kelp's bill, it is only on the lower mandible. The tail of the Pacific Gull has a distinct broad black band whereas the Kelp Gull has no band but an all-white tail. Juveniles of the Pacific Gull still have the massive bill and bigger more rounded head but are a darker, more-even brown all over, getting less brown into adult plumage by third year, but do not have any yellow on bill until becoming an adult bird in the fourth year. **Status:** Moderately common in the Shark Bay region. Uncommon on the Swan Coastal Plain. Common on the South Coast. **Where to find:** Carnarvon, Denham and most of the Shark bay region. Kalbarri. The Abrolhos Islands. Most islands off the Swan Coastal Plain. Numerous beaches on the South Coast particularly on rocky coastlines. Common in the Albany region and Esperance region.

Immature

Photos: Simon Nevill

Kelp Gull *Larus dominicanus dominicanus* 49-62 cm

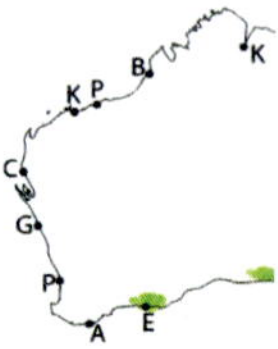

Unlike the Pacific Gull, which is endemic to Australia, the Kelp Gull can be found on the coastlines of the Antarctic, the sub-Antarctic islands, southern Africa and southern South America as far north as Peru. Over the last 60 years the Australian population of Kelp Gull has been expanding on the east coast and its range is still increasing, but in WA it is still very uncommon, with rare sightings on the west coast and several sightings on the south coast. There are two subspecies in the world, *vetula* in South Africa and *dominicanus* in Australasia. **Nest:** Nest site and clutch size is the same as that for Pacific Gull.

Similar species: So close in general appearance to Pacific Gull that for years it was overlooked as different species, which is surprising as there are distinct differences. The Main feature is that Kelp have an all white tail whereas Pacific have black terminal band on the tail. Kelp Gull is a slightly smaller with a smaller head. The bill is not as massive as the Pacific. Red-orange on lower bill. Small white mirror on outer primary. Kelp Gull is more of a flocking gull than Pacific. They also breeds in colonies and their call is totally different. **Status:** Scarce vagrant in WA but moderately common on the east coast with population expanding, very common in Tasmania. **Where to find:** Possible anywhere on the south coast and lower south west. Best chances on south coast in the Esperance region and around the Archipelago of the Recherche.

Photo: Simon Nevill

Black-headed Gull *Chroicocephalus ridibundus* 36 cm VAGRANT

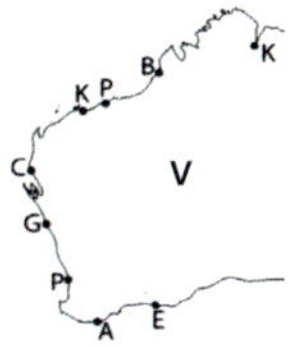

A rare vagrant that has been sighted a few times in Australia and in WA it has been sighted in Broome. Breeds in the northern Palearctic. Some of the European breeding birds remain sedentary others that have bred in colder regions migrate south to the Mediterranean and the northern African coast. Those species that breed in Siberian and northern China partially migrate south to the Japanese, Chinese and far eastern coastline. Always possible to over pass to Australia. Can be found not only on coastal waters but also on inland freshwater areas. **Similar species:** None when in breeding plumage. In non-breeding plumage could be confused with the New Zealand species, the Black-billed Gull that has similar wing pattern. Black-billed is a larger, more thick-set gull. Lacks small black facial mark that Black-headed has. **Status:** Rare. **Where to find:** Any northern or central coastline in WA. Both Franklin's Gull *Leucophaeus pipixcan* and Laughing Gull *Leucophaeus atricilla* although rare in Australia are however sighted more often than the above species but I have not included them in this book.

Photo: Simon Nevill

Silver Gull *Chroicephalus novaehollandiae* 36-44 cm

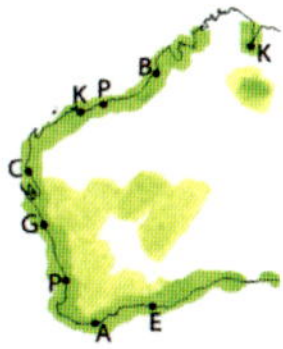

The ubiquitous scavenger we know in our coastal cities throughout much of Australia. It is however an attractive gull, particularly the adult bird in breeding plumage with its bright red legs and bill. Found only in Australasia, it has adapted well to European settlement and most probably increased in numbers. Found along the entire Australian coastline including all the northern coastline, and throughout New Zealand (subspecies *scopulinus*). An opportunistic scavenger that now relies heavily on human organic waste found at fishing harbours, waste dumps, and ornamental parks and gardens. Can be found in the hinterland where ephemeral lakes occur. On the coast fish is still an important part of its diet. There are three subspecies in the world with only one, *novaehollandiae,* in Australia. Breeds mostly on islands that dot the whole South West coast. The Perth region has the largest populations of Silver Gull, nesting on Rottnest, Carnac, Flat Rocks, Gull Rock and Penguin Island. Silver Gulls are basically sedentary in the Perth region and tend not to travel further than 50 kilometres from their breeding grounds. **Nest:** A scrape on the ground lined with any soft material available. Nests mostly on off-shore islands but also on islands on inland lakes. Clutch size 2 to 3.

Photo: Simon Nevill

Similar species: None in WA. In New Zealand there can be slight confusion between immature Silver Gulls and the endemic Black-billed Gull. **Status:** Very common. **Where to find:** Throughout the WA coastline. Least of all the far north Kimberley.

Gull-billed Tern *Gelochelidon nilotica* subspecies *macrotarsa*

35-38 cm

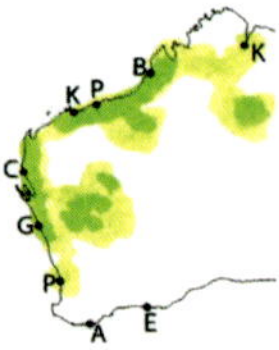

Gull-billed Terns are patchily distributed throughout all the continents of the world. The Australian subspecies macrotarsa occurs both on the coast and inland WA. The subspecies *affinis* that breeds in China comes down to northern Australia and would be extremely rare in the South West. Gull-billed Terns are medium sized terns with a black crown, short gull-type bill which is completely black both in and out of breeding plumage. In non-breeding plumage there is no black crown, only a black patch remaining around the eye and on the ear coverts. They are gregarious birds and will normally be

Above: In non-breeding plumage

Photos: Simon Nevill

accompanied by other Gull-billed Terns. Has a slow languid flight. Even though it can be found on beaches and mudflats in the South West, has a preference for ephemeral freshwater or brackish lakes. Highly nomadic and will disperse to inland lakes like Yarra Yarra Lake, Lake Austin or Lake Anneen. There are 6 subspecies in the world and one in Australia. **Nest:** No nest is made, only a scrape in sand or earth. Located mostly on islands off shore. It does however, nest on islands on inland lakes more than any other tern. Clutch size 2 to 3. **Similar species:** None. **Status:** Generally uncommon but locally common in the north and north east. Uncommon on the Swan Coastal Plain. Scarce in the southern region. **Where to find:** On the coastal belt from Broome to Shark Bay. Inland waters like Lake Gregory. Lake Argyle. Lake MacLeod. Seen at Mc Neil Claypan. Hutt Lagoon north of Geraldton. Yenyening Lakes. Camel Lake north of the Stirling Range. Any inland lake after heavy rains, particularly Anneen, Austin, Wooleen. Yarra Yarra and Hinds.

Caspian Tern *Hydroprogne caspia* 47-54 cm

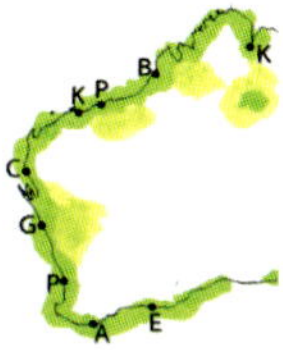

A very large tern found in many parts of the world. Largest of all the Australian terns. Has large orange-red bill, slightly down curved with a black tip, black of the crown passing under the eye to the base of the nape, short black legs. In non-breeding plumage the crown is mottled with some white flecking in the black. Juveniles have black edging to the scapulars on the back. Has short tail with slight wedge, in flight shows black tips to the wings. Often calls with typical rasping harsh note similar to a White-necked Heron's call. Feeds entirely on fish. Found on sheltered inlets and bays, coastal lakes mainly brackish, harbours and reservoirs. **Nest:** No nest is made but places soft material in a scrape. Nests mostly on islands either on sand above the high water mark or in the centre of islands. Can be a solitary nester or form small colonies. **Similar species:** None. **Status:** Moderately common along the WA coastline. **Where to find:** Wyndham to Broome and Roebuck Bay. Eighty Mile Beach. Common throughout the Karratha and Port Hedland coast. Carnarvon region including Lake McLeod. Mangrove Bay down to Hamelin Pool. Throughout the Denham-Useless Loop region in small numbers. From Murchison all the way to Bunbury. In Perth often at Alfred Cove and also on Rottnest Island, where it nests. The Rockingham area and Penguin Island. Peel Inlet. On the south coast from Augusta to Esperance, particularly in the large inlets like Nornalup Inlet and in Oyster Harbour easily seen at Emu Point.

Photo: Simon Nevill

Greater Crested Tern *Thalasseus bergii* subspecies *cristata*

40-50 cm

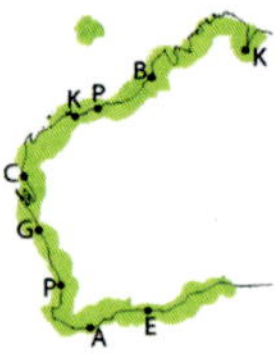

Found in countries abutting the Indian Ocean, west Pacific Ocean and Australasia. A large tern with a black crown and crest, narrow white gap between bill and crest on the forehead in breeding plumage, pale long yellow bill, long forked tail, black legs. Feeds almost entirely on fish, hovering in true tern-like fashion just a few metres above the water to look for fish and then dive and catch. Is a colonial breeder and will also roost colonially but tends to feed individually. Found throughout the coastal beaches of WA, on open beaches, estuaries, and brackish waters, islands and mangrove-lined shores. Breeds on many islands throughout range sometimes in large numbers. On Pelsaert Island in the Abrolhos there can be over 1000 birds in one colony. When breeding, nests very close to other birds mostly just out of pecking range. The crest remains extended in bristle fashion when breeding or threatened, other times crest is dropped. There are four subspecies in the world with only one in Australia *cristate* being one of the smallest sub species. **Nest:** No nest made just a hollow scrape in sand. Nests in colonies, often with other species of tern almost always on islands. Clutch size normally one egg only but occasionally 2. **Similar species:** Lesser Crested Tern, although scarce in this region, is a smaller, slimmer bird. Has a finer orange to orange-yellow bill not pale lemon-yellow. Important feature with adult birds is that the black cap touches the base of the bill on the Lesser Crested Tern. The Crested has a small gap of white between the bill and the black crown. In immature birds it's not a distinctive feature, best to look at size, shape and bill colour. **Status:** Common throughout the WA coastline. **Where to find:** Many beaches from all the way from Cape Leveque south to south coast. Nests throughout the Abrolhos region on selected islands and from the Murchison all the way down to Busselton. Augusta to Cape Arid. In the Perth region nests on Rottnest. Lancelin. Carnac and Seal Islands.

Photo: Simon Nevill

Lesser Crested Tern *Thalasseus benggaensis* 38-43 cm

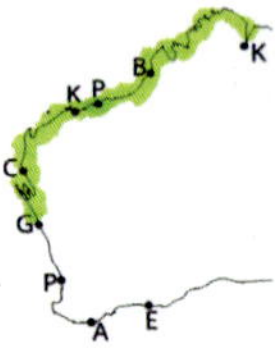

Two subspecies occur *bengalensis* and *torresii*. Australia has the subspecies *torresii*. Lesser Crested Tern is closely related to Crested Tern but prefers tropical waters. Size is the main difference and if the birds are together the difference is clear as Lesser appears less bulky. Luckily its not uncommon for the two species to roost together particularly in Roebuck Bay. Frequents sandy beaches particularly coral cays, preferring to feed in shallow waters and not a pelagic wanderer. **Nest:** Coral cays, sandy beaches sometimes on coral pebbles. Slight depression often with coral debris. Clutch normally 1. **Similar species:** Crested Tern, which is larger. Main feature is that the bill is orange coloured on the Lesser not dull yellow. In breeding plumage Lesser-crested black crown comes close to base of bill where as Crested always has a white break between bill and black crest. Both species often seen together. **Status:** Moderately common in the north west as far south as Shark Bay. **Where to find:** Any sandy beach from Shark Bay to Kalumburu. Common along much of the Pilbarra coast. Often on Roebuck Bay.

Photo: Ian Montgomery

Roseate Tern *Sterna dougallii* subspecies *gracilis* 31-38 cm

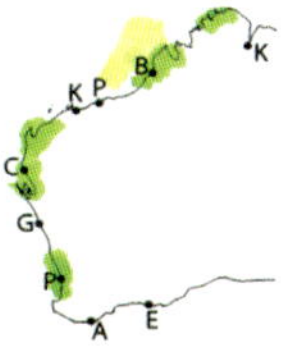

Found throughout many coastal regions in the world including parts of America (on the east coast), Europe, Africa, Asia and Australasia. A beautiful elegant tern with a slick black crown that touches the bill and passes down the back of the nape in the breeding season. Out of the breeding season the forehead and lores become white and streaking into the top of the crown. Has red legs except when a juvenile. Long down curved bill which is black in the immature and non-breeding adults, but during the pre-nuptial and incubation period is bright red then darkens when chicks hatch. Has a long tail. Has a beautiful pinkish flush on the

Photo: Simon Nevill

underbody in the breeding season. Feeds on fish and occasionally molluscs. Prefers to feed in what is known as 'blue waters', that is waters that are clear for feeding. Not found on the south coast. There are at least four subspecies in the world with one subspecies, *gracilis* in Australia. **Nest:** No nest is made just a scrape in the sand. Nests in colonies on islands. Clutch size normally 2. **Similar species:** Common Tern subspecies *longipennis* is very similar but is a bigger and thicker-set bird, the upperbody is darker grey, the bill is shorter and thicker, the tail not as long or as forked, neck shorter, legs are black and slightly shorter and it has a grey flush to flanks. The rare nominate race *hirundo* has a red bill with black tip and red legs closer in colour to a Roseate Tern. **Status:** Moderately common in the north of the region. Common on the Abrolhos. Uncommon south of Murchison. Vagrant on the south coast. **Where to find:** Nests on Ashmore Reef. Lacepedes. Nests on at least another 20 islands including Pelsaert Island in the Abrolhos. Cervantes Island. Off Dirk Hartog Island and in Denham region. Has nested on Meade Island near Dirk Hartog Island. Bird Island. Peron Peninsula. Shark Bay. Cape Leveque. Has nested in the Peel/Mandurah region in past years. Also Bathurst Point on Rottnest Island.

Common Tern *Sturna hirundo* 32-37 cm

Vagrant

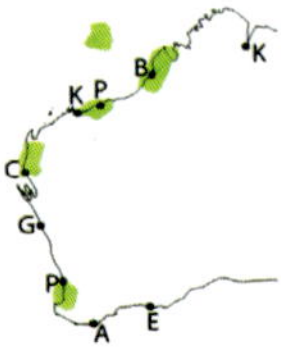

One of the 'commic' terns and is certainly not common in WA. It is named 'common' as the nominate race *hirundo* is very common in parts of Europe, where it breeds. Seen in greater numbers on the east coast of Australia. There are 3 subspecies the nominate *hirundo* breeds in North America. Europe. Middle East and as far east as western Siberia. The subspecies *longipennis* breeds in eastern Siberia and China and the subspecies *tibetana* breeds in Kashmir to western Mongolia *longipennis* over winters south of its breeding range and much of its migration route passes through the far east but many continue down the Australian east coast sometimes gathering in large flocks. This subspecies is uncommon in WA maybe because the flight path is blocked by many islands and is easier to pass eastern New Guinea. A medium sized tern found mostly off shore between 10 and 50 kms. Even though it is uncommon, it is still a regular visitor to WA, visiting between October and March. It was often overlooked until the 1940's and now with more skilled observers this bird is being identified far more frequently. **Similar species:** Not an easy task. Arctic Tern in breeding has white line below black cap that is visible from light grey body. Less black on wing tips. When resting, tail feathers same length as wings on the Common. Arctic is slightly smaller and slimmer (27-35 cm) and head appears smaller. Legs distinctly shorter. **Status:** Uncommon in

Note position of tail

Photo: Simon Nevill

the north-west. Scarce south of Port Hedland. **Where to find:** West Best from Roebuck Bay to Cape Leveque. Can be found anywhere on west coast. This species has been sighted many times in the Mandurah region. Note that the European subspecies *hirundo* has come to these shores and always has red legs where as the more frequent visitor subspecies *longepennis* has black legs which is an important feature. Arctic Tern is a rare vagrant to WA waters so less chance of confusion.

Artic Tern *Sterna paradisaea ridibundus* 32 cm

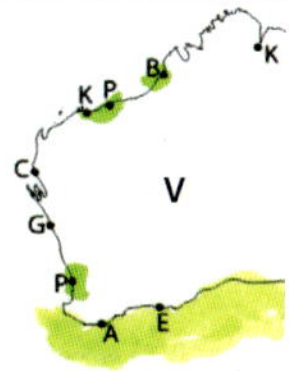

What a journey this delicate tern undertakes, migrating from the far northern hemisphere, all the way south to the coldest and wildest seas of the Antarctic, sometimes making a journey in excess of 20,000 km on their southern journey. They breed in Russia. Europe and North America and several Islands in the Artic Sea. When they over-winter in the Antarctic, they mainly feed in the pack ice region being circumpolar in their movements. When seen in WA waters, it is normally well off the shelf in deeper waters as they move south or north from their over-wintering grounds or northern hemisphere breeding grounds. **Similar species:** Can be confused with Common Tern, which is slightly larger and not as delicate in its jizz. Arctic Tern has a shorter neck, narrower wings and the tail streamers are longer passing the wing tips when at rest. **Status:** Common in the southern Antarctic seas, but rarely seen as they normally fly well south of the WA coastline. **Where to find:** Anywhere in the Southern Ocean.

Note position of tail

Photo: Simon Nevill

Little Tern *Sternula albifrom* 20-28 cm

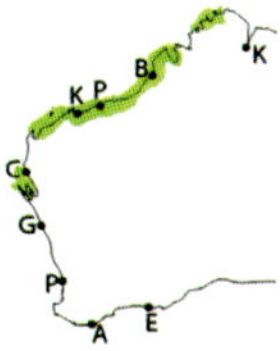

Both Little and Fairy Tern can be found in Western Australia. There is considerable overlap in their distribution range. Little has been recorded as far south as Shark Bay and Fairy Tern has been recorded as far north as Point Sampson an over lap distance of 1000 km. Little Tern by name is the smallest of all the terns found in Australia and breeds in many regions of the World, including Europe. China. Japan. New Guinea and Australia. Little Tern is primarily found along the eastern and northern coasts and the Fairy generally along the western and southern coasts. It is a fast flying delicate bird, that flaps wings rapidly. All white with black cap white forehead narrower band in breeding than Fairy. Black eye stripe terminates on bill. Black on the tip of a yellow bill. Most of the worlds population is threatened as unfortunately, they choose to nest not only on the high water mark but also prefer open sandy beaches that unfortunately are the haunt of sun loving homo sapiens, combine this with the prevalence of beach buggies and you have a disastrous combination. This applies to both species. We need what Europe and North America have done and that is to fence off known sites before we loose more of the breeding populous. Both Little and Fairy Tern males have the lovely habit of flying above females and then landing and presenting a small fish, similar to other unrelated species in the World. **Nest:** Prefers sheltered locations, typically sand spits away from actual beach front. No nest is made, just a small scrape on sand. Nests both on islands and narrow sandy spits or headlands. Clutch size 2-3. **Similar species:** The Fairy Tern is slightly bigger and slightly larger but too small to use as positive identification. In breeding the Fairy Terns bill is more orange-yellow with no black tip and the black eye stripe only just passes the eye and does touch the base of the bill. The black crest stops higher on the crown and the white forehead is distinctly broader. In non-breeding plumage, the Little Terns bill is all black whereas Fairy always has slight yellowish-orange tinge. Also Fairy Tern retains a full black crown except immature birds. The Little Tern crown diminishes to mostly white from the breeding black crown. **Status:** Moderately common locally. Population in WA in less decline than say NSW. **Where to find:** Found from Port Warrender in the north of the Kimberley south to Shark Bay with some birds getting to Carnarvon. Good place to see is in Roebuck Bay Broome.

Photo: Simon Nevill

Fairy Tern *Sternula neresis neresis* 22-27 cm

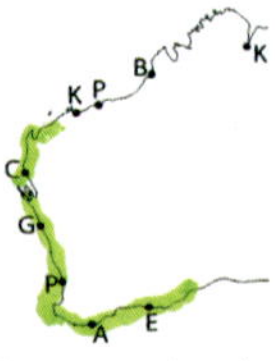

Unlike the Little Tern, which is found throughout much of the coastal waters of the world, the Fairy Tern is restricted to Australasia. Sadly the population of Fairy Terns has been diminishing over the years since European settlement. The Western Australian population is estimated to be around 1400 pairs (Hill et al. 1988). Much of the disturbance and loss of nesting colonies results from the consistent expansion of towns such as Mandurah, and disturbance by recreational vehicles near breeding colonies. The Fairy Tern is a delicate small tern, found mostly in the southern part of Australia including most of the coastline in the South West. Feeds almost entirely on small fish and, like many terns, the male will present to prospective mates an offering of a small fish to try to instigate the breeding process. In breeding plumage has black crown, which drops down to the front of the eye and stops before the bill, showing white between the bill and the black eye stripe. The primaries are an even grey colour. The bill colour is bright yellow-orange. In non-breeding, the bill is still yellow-orange but has some black at the base and a small black band just before the tip. Feet yellow. There are 3 subspecies in the Australasian region with one subspecies *neresis* in Australia. **Nest:** No nest is made, just a small scrape on sand. Nests both on islands and narrow sandy spits or headlands. Clutch size normally 2. **Similar species:** The Little Tern is slightly smaller and slightly slimmer. The bill is yellow colour rather than yellow-orange, with a black tip. The black crest comes closer to the forehead and forms a white eyebrow rather than just coming across the head and straight down to the eye as in the Fairy, and the back is greyer. **Status:** Uncommon throughout the west coast and its islands and the southern coast, except the Abrolhos Islands where it is common. Moderately common locally. Population may be on the decline. The Little Tern does come down to Carnarvon occasionally overlooked. **Where to find:** Good chances in the Denham region as it has nested on Peron Peninsula and also on Dirk Hartog Island. Can be seen on Rottnest Island. Garden Island. Alfred Cove. Woodman's Point. Halls Head. Penguin Island. Mandurah Harbour. Esperance lakes particularly Lake Warden.

Photo: Simon Nevill

Bridled Tern *Onychoprion anaethetus anaethetus* 30-32 cm

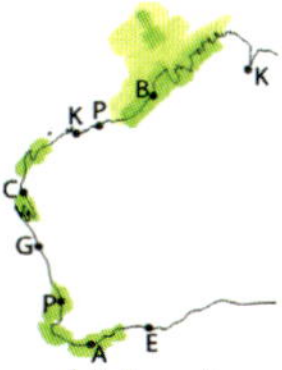

The nominate race is found throughout much of South East Asia and most of Australasia. A medium sized tropical tern very similar to the Sooty Tern with which it shares much of the same distribution, although the Bridled Tern nests in smaller numbers than the Sooty Tern. Bridled Terns nest on many of the South West islands, including islands just off the east side of Dirk Hartog and in Freycinet Harbour including White and Mary Anne Islands. Nests on many of the Abrolhos Islands in small colonies, often amongst the more populous Common Noddy. The estimated population is in excess of 7000 birds, breeding on over 60 islands. Further north out of this region there are larger nesting colonies, particularly on Ashmore Reef. Barrow Island and in the Montebello Islands. Other islands closer to Perth include Cervantes, Lancelin, Rottenest, Carnac. Little (near Whitfords Beach), Garden and Penguin Islands. Bridled Terns have expanded their breeding range over the last fifty years, moving further south. There are 4 subspecies in the world with one, *anaethetus* in Australia. The Western Australian birds are slightly smaller than the east coast birds. **Nest:** No nest is made, just a scrape in the sand. Nests in colonies on offshore islands. Clutch size one egg. **Similar species:** Sooty Tern. The Bridled Tern is slightly smaller and slimmer than the Sooty. Has a dark black-brown upperbody becoming grey-brown below the nape. Shortish neck. The important feature is the white supercilium eye stripe running from the forehead past the eye as a thin eyebrow. Long tail that passes the wing tips at rest. The Sooty Tern is slightly larger and thicker bodied. The main differences from Bridled Tern is that the Sooty Tern's upperbody is black not dark brown. The forehead is white stopping just above the eye as a

Photo: Simon Nevill

broad patch not a narrow long eye stripe passing the eye to a point as in the Bridled Tern. The black of the crown runs down the nape and continues as a solid black back. There is not a break at the base of the nape that changes to grey as in the Bridled. **Status:** Common off the Dampier Peninsula. Common on most islands in the mid-west and off the Perth coast. Scarce on the southern coast. **Where to find:** All the islands listed above. There are over 30 islands off the WA coast used by this species to breed. Easiest to see near Perth on Penguin Island and Rottnest Island.

Sooty Tern *Onychoprion fuscata* subspecies *serrata* 33-36 cm

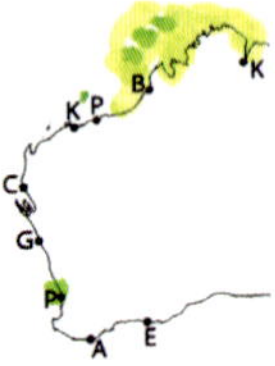

A bird that is found throughout the southern hemisphere oceans. Some of the colonies in the Pacific have as many as 300,000 birds. Colonies in Australia are far smaller but, in terms of many other seabirds, they are still in large numbers and the largest colony is here in Western Australia on the Abrolhos Islands where the numbers are in excess of 25,000 birds. As described under Bridled Tern, the Sooty Terns have a jet-black crown, the black passing down the nape to the upperbody. The forehead is white, stopping just above the eye as a broad patch not a narrow long eye stripe passing the eye to a point as in the Bridled Tern. There is no break at the base of the nape that changes to grey as in the Bridled. They arrive in the breeding season in October and November and depart February and March. Possibly 7 subspecies in the world with 3 in Australia, *serrata* on the Abrolhos Islands, *oahuensis* on Christmas Island and *nubilosa* on the Cocos-Keeling Islands. **Nest:** No nest is made, just a scrape in the sand. Nests in large colonies on off-shore islands. Clutch size one egg. **Similar species:** Refer to Bridled Tern. **Status:** Very common on the Abrolhos. South of the Abrolhos it is accidental and only seen or found dead on beaches after severe storms. Common off the Dampier coast. **Where to find:** Ashmore Reef. Lacepede Islands. The Abrolhos Islands.

Photo: Simon Nevill

Whiskered Tern *Chlidonias hybrida* subspecies *javanicus* 23-25 cm

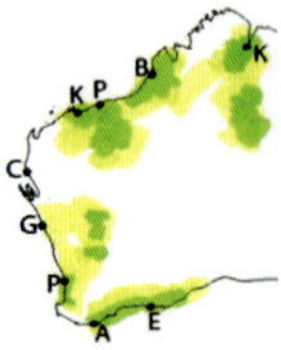

One of the 'marsh terns' found in Africa, Europe, Indian subcontinent, South East Asia, China and Australasia. A bird of freshwater lakes, swamps, reservoirs and flooded samphire. Will nest on inland or coastal wetlands, often with preference for recently flooded lowlands with emergent grasses. The adult breeding birds have a black cap, the black extending from the base of the bill below the eye to the nape. The bill is red and slightly down curved with a white collar below the black. Has a dark grey belly that lightens towards light grey at the top of the breast. Wings and back are light grey. Feet bright red. In non-breeding plumage the coloration changes dramatically. The bill and feet lose their bright red colour, the intense black cap diminishes to streaked black on part of the crown, passing down the rear and below the eye as a black spot. There are three subspecies in the world with one in Australia *javanicus*. The Australian birds are paler in breeding plumage than the nominate race *hybrida*. **Nest:** A small raft of aquatic plants laid against reeds or grasses on inland lakes. Clutch size 2 to 3. **Similar species:** In breeding plumage there is no similar species but in non-breeding plumage the White-winged Black Tern is very similar. The streaked crown on the White-winged Black Tern forms a slightly larger black spot passing below the rear of the eye. Also the streaked black crown stops at the back of the head giving a complete white collar. However the Whiskered Terns' streaking passes down the back of the head almost to the back. Luckily when you see these birds either together or on their own, there is often one adult coming out or going into breeding plumage that helps with the identification. In eclipse plumage the cap can be completely black as in breeding plumage. **Status:** Generally uncommon to locally common. Often in small groups. **Where to find:** Samphire Flats off 80 mile after good rains. MacNeill claypan. Lake Wooleen. Lake Anneen. Lake Austin. Lake Monger. On the Swan Coastal Plain any freshwater wetland like Lake McClarty, normally on the coastal plain in autumn.

Photo: Simon Nevill

White-winged Black Tern *Chlidonias leucopteris* 20-23 cm

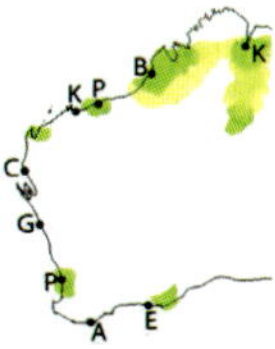

Two freshwater frequenting terns come to these shores, the Whiskered and the White-winged Black Tern. Both can be found throughout many regions of the world, often feeding together on the same wetlands. The White-winged feeds in coastal waters and can be seen in the company of Fairy Terns and Little Terns. The Whiskered Tern nests in Australia but

White-winged Black Tern in breeding plumage

Photo: Graeme Chapman

Photo: Simon Nevill

the White-winged does not, breeding in the northern hemisphere. In breeding plumage, the White-winged Tern is a stunning bird but alas we do not get to see it that often in Australia, except just before they set off from these shores at the end of March to May. In full breeding plumage, the White-winged Black Tern is completely black on the head, under-body and upper-body breaking at the rump to grey. The vent is white and the wings contrast with the black head and body being pale grey. The tips of the primaries are dark grey. The legs are bright red. Adult non-breeding plumage totally different. Refer notes on Whiskered Tern. **Similar species:** Refer to Whiskered Tern. **Status:** Relatively common in the far north and north west in our summer months. Scarce in the south west but is an irruptive species that will come down when cyclones are present. Flies with Whiskered, which is far more common in the South West. **Where to find:** Roebuck Bay. Lake Campion. Lake Eda. Samphire Flats when flooded. Lake MacLeod. Lake Gregory when water in it. Chance sightings any major lake especially Lake McClarty and over Peel Inlet.

Brown Noddy (Common Noddy) *Anous stolidus* subspecies *pileatus* 40-45 cm

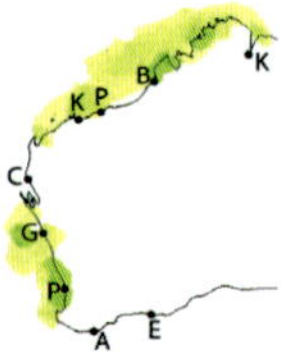

Found throughout all the southern oceans of the world and as far north as the Caribbean. The largest of all the Noddies. A dark brown tern that in West Australian waters leaves their breeding sites in April to May returning in late August through to October. Plumage is dark chocolate brown on all of the body with prominent white cap. The important feature is that the white cap stops abruptly above the eye and above the lores and touches the bill. The white fades down the nape into the dark brown back, level with the eye. There are 4 subspecies in the world with one *pileatus* in Australia. The populations in Western Australia are slightly smaller and paler than those found on the east coast. **Nest:** Consists of a large pile of seaweed and other plants on top of or below prostrate plants on offshore islands. Clutch size one. **Similar species:** Lesser Noddy is slightly smaller and certainly a slimmer bird. Plumage is darker chocolate brown almost black, although not as dark as Black Noddy, which has not been seen in these waters as yet. White cap passes below eye and at base of bill. Importantly, the white cap grades into the dark brown more gradually not reaching its darkest brown until the upper back and belly. So in simple terms, the head of the Lesser Noddy is overall lighter in colour than the Common. In flight when seen together, which is common in the Abrolhos, the Lesser Noddy has faster wing beats. The generic name 'Anous' means silly or foolish in Greek and the early mariners found them easy birds to take as they were so accepting of humans. To walk alongside (not through) colonies of Common and Lesser Noddy is an experience never to forget as they allow such a close approach without fear, a rare privilege in these times. **Status:** Very common in the Abrolhos during the breeding season. Accidental in the south west often found dead after cyclonic storms. They did breed on Lancelin Island in 1989 and 1992. **Where to find:** Common off Dampier Peninsula. Nests Ashmore Reef. Scott Reef. Lacepede Islands. The Abrolhos Islands. Pelsaert Island can have, in some years in excess of 130,000 nests.

Photo: Simon Nevill

Lesser Noddy *Anous tenuirostris* subspecies *melanops* 29-34 cm

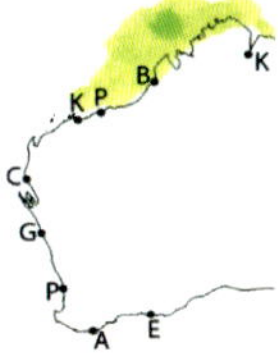

There are 2 main breeding regions in the world for Lesser Noddy, both restricted to remote islands. The nominate *tenuirostris* occurs on a scattered group of islands centred on the Seychelles and Maldives Islands in the Indian Ocean. The other subspecies, *melanops*, occurs in the south west on the Abrolhos Islands. On the Abrolhos they are found on only a few islands, the reason being is that they require mangrove trees as a nesting habitat. On Pelsaert Island, they share the island with a larger colony of Common Noddy but both require different nesting locations. To see a Lesser Noddy nest in mangrove trees is like looking at a Jewish candelabra with various nests branching off the main trunk, all incredibly close together. They have faster wing beats than the Common Noddy. The noise from the whole colony is continuous as parents locate their young. **Nest:** The nest is a small platform consisting of branches and mangrove leaves cemented together with the bird's excreta and placed on the fork of a mangrove branch. Clutch size 1. **Similar species:** Refer to Common Noddy for plumage details. **Status:** Locally very common on a few islands in the Abrolhos. As for Common Noddy, accidental in the South West, often found dead after cyclonic storms.

Photo: Simon Nevill

Black Noddy *Anous minutus* 33-36 cm

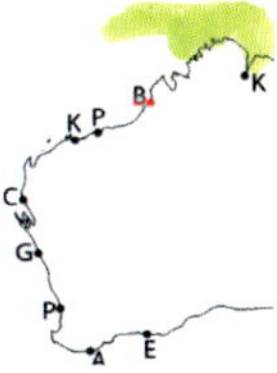

There are 7 identified subspecies of Black Noddy in the world. The Ashmore reef subspecies may in fact well be *warcesteri* as there are breeding populations in Indonesia and not the subspecies *minutes*, which is the east coast subspecies particularly knowing that Black Noddy are basically sedentary, only feeding in waters close to the breeding colonies. Black and Lesser Noddy form a super species group and may be conspecific. Black Noddy have a large world wide range mostly in the mid Pacific, mid Atlantic and Gulf of Mexico, whereas the Lesser has a more restrictive range and certainly more restrictive breeding range. They Black and Lesser Noddy also share the preference for nesting in trees, normally mangroves, whereas Common will nest on low shrubs or directly on the ground. Black Noddy is definitely the darkest of the 3 noddies found in Australian waters. Body all black-brown, white crown that forms loral patch between base of bill and eye and runs level with top of bill to eye. **Similar species:** Lesser Noddy, almost identical in shape but slightly smaller size, more extensive grey on the head passing from the crown to the neck. Common Noddy bigger and more robust, both Black and Lesser have faster wing beats and more buoyant, this can be seen clearly with Lesser and Common on the Abrolhos Islands. **Status:** Common around Ashmore Reef and Scott Reef. **Where to find:** Ashmore and Scott Reef and around Timor Sea. Remember both Black and Common Noddy can be seen on Ashmore and Scott Reef.

Little Penguin (Formerly Fairy Penguin) *Eudyptula minor*
subspecies *novaehollandiae* 40-45 cm.

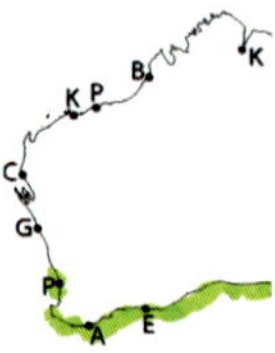

There are several species of penguin that may be driven onto Western Australia's shores after storms or when needing to moult, but the Little Penguin is the only resident penguin. The adult birds do not travel the great distances from their nesting colony that other penguins do; however the immature birds do not return to the colony for at least 2 years, mostly 3 years. Feed on small shoaling fish but will take crustaceans. Most colonies in Western Australia are on islands but there are a few on mainland sites. There are six subspecies They roost during the day on land leaving the breeding site one hour before dawn and start to return at dusk and for at least two hours after dark. Both adults will incubate the egg and rear young. There are six subspecies, only one *novaehollandiae* in Western Australia. **Nest:** Nest is located at the end of small tunnel about one to one and half meters long. The nest chamber has a small quantity of leaves or grasses. Normal clutch size is just 2. **Similar species:** None. The nearest small penguins are the Adelie and Chinstrap which are more than twice the weight. **Status:** Common throughout the south west from Carnac Island all the way around the South West coast to the far eastern colony at Twilight Cove. **Where to find:** Best and easiest location, Penguin Island. Carnac Island. Nest on hundreds of islands along the south west coast most in the Recherche Archipelago.

Photo: Simon Nevill

Red-tailed Tropicbird *Phaethon rubricauda* 95-104 cm (includes tail rectrices 35 cm long)

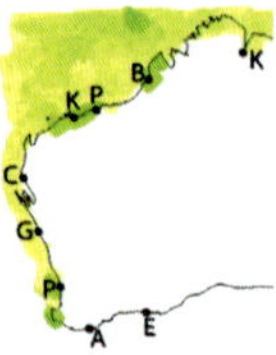

There are 3 tropicbirds in the world and 2 found in Australian waters, the Red-tailed and the White-tailed. The Red-billed-Tropicbird has been sighted in Australian waters but extremely rare and does not nest in Australia. Tropicbirds are so closely related that it is classed as a 'super species'. Seeing the Red-tailed Tropicbird close, one appreciates what a heavy bodied bird it is and to see a breeding pair fly and perform their courtship flight is like looking at two ballerinas in the sky, the pair flying up vertically then doing backward loops with their beautiful red central tail streamers curving behind. A lone species, that wanders the tropical and sub-tropical seas, rarely seen except when at the breeding site. Feeds on fish by diving from a good height (normally not as high as gannets) and staying under for several seconds until securing a fish. **Nest:** Two types of nest location are utilized in Western

Photo: Simon Nevill

Australia: either on rock ledges with no nest material used, or on sandy beaches above the high water mark. Normal clutch size just one egg. **Similar species:** White-tailed Tropicbird, which is a slightly smaller bird with bill and central tail feathers white. White-tailed has distinct black band on the upper wing, unlike the Red-tailed which has no band. **Status:** Scarce in south west. Mid west uncommon. North west uncommon near shore. Common on northern Islands below. **Where to find:** Sugarloaf Rock and the Abrolhos Islands (not always on either). I have seen the species on Rottnest Island, a rare occurrence. and Pelsaert Island. Nest in larger colonies on Ashmore Reef. Christmas Island and Bedwell Island.

White-tailed Tropicbird *Phaethon lepturus* 65-80 cm

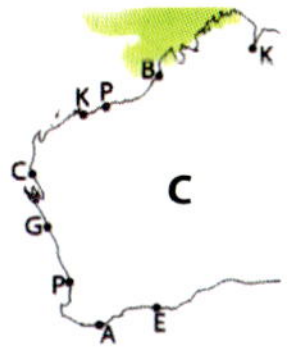

The White-tailed Tropicbird, like all tropicbirds, plies the tropical and sub-tropical seas. It rarely is seen at sea with its partner or other tropicbirds unless close to the breeding site as it is basically a lone pelagic feeder. Breeds on several islands throughout the tropics of the world such including the Ascension Islands. Seychelles and Bermuda but in Australian waters, it nests only on Christmas Island, Cocos-Keeling Islands and on some small islets around Ashmore Reef. Feeds by diving for fish and can swim briefly under water to gain fish. Basically silent at sea but when displaying or at nesting site make strong kek-kek-kek calls reminiscent of a Peregrine Falcon. **Nest:** No nest is made except slight scrape in sand. Sites vary considerably from soft sand to laying on hard rock. Only one egg is laid. **Similar species:** White-tailed Tropicbird can be easily distinguished from Red-tailed Tropicbird by the distinct bold black markings on upper-wing and wing tips. Obviously if close, one can see the white bill and tail feathers. Seen together as at Christmas Island one can see the difference in size. The White-tailed is the smallest of all three tropicbirds and slimmer. **Where to find:** Rarely seen off the north west coast but plentiful on its breeding sites on Christmas Island. Cocos-Keeling Islands and around Ashmore Reef. The golden-yellow tinged morph of the White-tailed Tropicbird found on Christmas Island, nicknamed the 'Golden Bosunbird' illustrated here is a beautiful bird with its golden tinge on its breast, crown and tail feathers. Tropicbirds were originally called 'bos'nbird' by mariners as the central tail feathers reminded them of the marlin-spike, a long steel tool with a spiked end which was used by the bosun on a ship to assist in untying knots etc.

Golden Bosunbird

Photo: Simon Nevill

Lesser Frigatebird (Least Frigatebird) *Fregata ariel* 75-80 cm

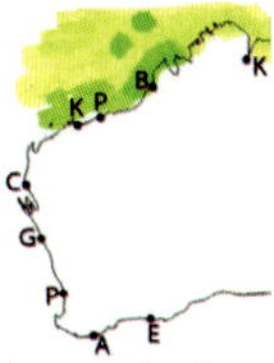

There are 5 species of frigatebird in the World, 3 are found in waters around Australia. Great. Lesser and Christmas Frigatebird. Lesser Frigatebird is the smallest of the frigatebirds. Mostly found in the Indian, far-eastern and west Pacific waters with a small population off the Brazilian coast. It is the most likely species of frigatebird to be seen off the north-west coast. Frigatebirds are the master flyers of the air, often harassing other sea birds to make them disgorge their food, mostly terns, tropicbirds but even boobies returning to feed their young. They often fly high in the thermals, waiting for birds to return to their colony, then dive and harass them. They also take fish from the surface, particularly flying fish, they also patrol beaches for surface fish. Frigatebirds do not like to land on the ground but prefer to perch on branches or any off-ground object, to aid take-off. I have seen them skim over freshwater to drink, even drinking from swimming pools by flying with lower mandible hanging down to catch the water. Frigatebirds are monogamous and most probably form new partnerships with each breeding season. In WA this species breeds on Bedout Island. Adele Island. West Lecepede. Cocos-Keeling and Ashmore Reef. Biggest numbers being recorded on Adele Island with over 5000 birds but the population frequently crashes, due to poor fishing seasons. **Nest:** Loose arrangement of twigs, seaweed etc placed on ground. **Similar Species:** When identifying species of frigatebird it is important when in flight to see the underwing markings. Males: Lesser has an all black under-body but faint white fingers leading from belly across the auxiliaries. Great has a complete black under-body. Christmas Frigatebird has distinct white lower belly to vent. Females: Lesser Frigatebird has a white chest patch with white fingers leading across auxiliaries and complete black head, including throat. Great Frigatebird has a white chest but white leads up under throat and no white in auxiliaries. Christmas Frigatebird is very similar to Lesser Frigatebird but white on the under-belly goes all the way to the vent whereas Lesser Frigatebirds lower belly is completely black. Remember, Lesser Frigatebird is a smaller, slimmer bird if seeing them with other frigatebirds. **Status:** Relatively common. **Where to find:** Often sighted around Broome. Roebuck Bay and along west Dampier Peninsula. Christmas Island, the Lacepede Islands. Ashmore Reef.

Immature

Photo: Simon Nevill

Great Frigatebird *Fregata minor* 80-100 cm

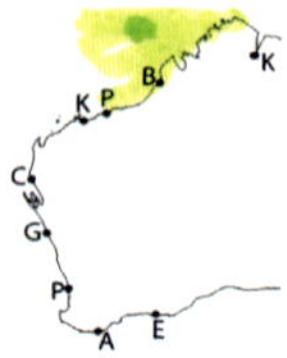

Not a common frigatebird off the north west coast. Larger more robust than Lesser Frigatebird. In WA breeds on Christmas Island. Cocos-Keeling Islands and Ashmore Reef. More likely to be seen along Queensland coast as nesting sites are closer to the mainland. **Nest:** Both in mangroves and low trees but also on the ground. **Similar Species:** Refer Lesser Frigatebird. Remember, Lesser is a smaller slimmer bird if seeing them with other frigatebirds. **Status:** Relatively common but uncommon along WA coast. **Where to find:** In WA best on Christmas Island.

Male

Female

Photo: Simon Nevill

Christmas Frigatebird *Fregata andrewsi* 85-89 cm Endemic

Photo: Simon Nevill

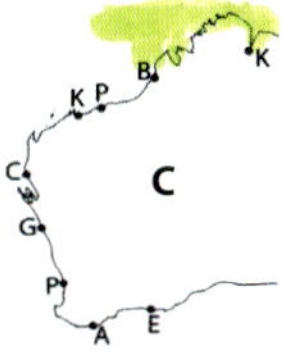

In world terms this is an extremely rare bird, particularly knowing some seabird colonies are vulnerable to cyclonic weather or crashes in fish stock. Its rare as Christmas Island is the only nesting location of this endemic species. In the 1960s the population crashed, mainly due to poaching by Malay's and land reclamation for mining but since strong conservation measures have taken place, the population has remained stable. The numbers fluctuate from year to year but there were 1600 pairs in 1988 (Stokes), others have recorded higher numbers. The population may also be influenced by hybridisation with Great Frigatebirds. **Nest:** Nest in the canopy of tall *Terminalia catappa* on the leeward side of the island or where available cliff ledges occur. **Similar Species:** Refer Lesser Frigatebird. **Status:** Locally common but rare in terms of World population. **Where to find:** Christmas Island. Has been sighted as far south as the Pilbara.

Australasian Gannet *Morus serrator* 85-90 cm

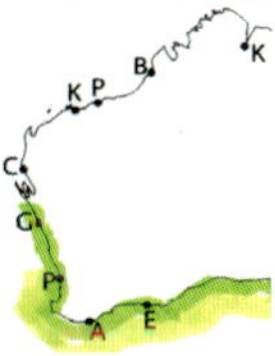

A spectacular diving bird, that plummets from great heights to gain fish. They are relatively strong flyers, handling stormy seas fairly well. There are no breeding colonies in W.A. but they do nest on islands off the Victorian coast, the Bass Strait, islands off Tasmania and New Zealand. Many of the birds that come to the west coast are from New Zealand, visiting mainly in the winter through to late spring. They do come up around the Capes to the west coast but mostly feed off the southern coast. Most birds would never survive the impact of hitting the water from such great heights but gannets and booby's have built-in soft pads above the cranium which cushion the impact and allow them to dive several metres under water. Range as far as Kalbarri but rarely further north. **Nest:** Nests in colonies. The nest consists of a combination of earth, grasses and the birds own excreta moulded into a small dome nest with hollowed-out the top lined with grasses or seaweed. One egg is laid. **Similar species:** None except other species of gannet.

Inmarute

Photo: Simon Nevill

The Northern Gannet is not expected here as it plies the North Atlantic waters but is easy to tell from Australasian, as there is no black in the tail. However the Cape Gannet which has come to these shores and is very difficult to distinguish but if seen together the Cape has a black tip to the tail, whereas the Australasian has just four central black tail feathers with white outer tail feathers. **Status:** Uncommon on west coast although always seen off Perth in winter. Common on the South Coast. **Where to find:** Off the Fremantle North Mole or Woodmans Point. Cape Vlamingh on Rottnest Island, mostly in winter and best when south-westerlies are blowing.

Brown Booby *Sula leucogaster* subspecies *plotus* 70-75 cm

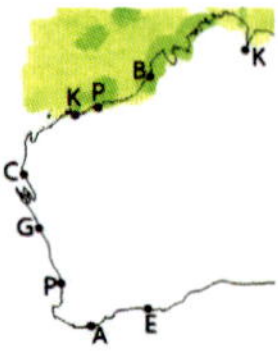

The most common booby found throughout tropical waters of the world. The sub-species *plotus*, is the most common booby on the north west coast, normally not venturing more than 30 km off the north west coast.. A rare vagrant to the coastal shores off the Pilbara coast. An unmistakable booby with a completely brown head and neck and upper body. Nests on the Lacepede Islands and Ashmore Reef. **Similar species:** No adult species. Juveniles Masked Booby's have a mottled brownish head and neck but never as dark or clearly marked as Brown Booby. They also have a yellowish-cream bill whereas Brown have a dark grey-blue bill. **Nest:** Nests in colonies on a bed of sand by gathering sea weed debris or twigs. **Status:** Common in northern tropical waters. **Where to find:** Easy to see around Broome and Roebuck Bay or flying along west coast of the Dampier Peninsula. Nests on the Lacepedes, easy to see there. Christmas Island. Ashmore Reef.

Photo: Simon Nevill

Masked Booby *Sula dactyatra* 75-85 cm

Photo: Rohan Clarke

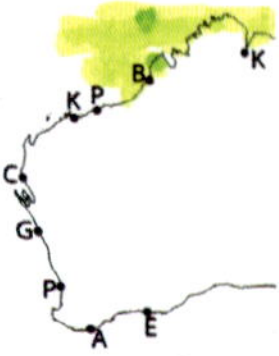

Found throughout all oceans of the world and northern tropical and sub-tropical waters around Australia but not common close to the Australian coastline as Brown Booby, except around breeding islands such as Cocos-Keeling Islands. Adele Island. Ashmore Reef. Bedout Island (40 km off the De Grey River mouth). **Similar species:** Similar in appearance to Australasian Gannet which do not feed in the tropical waters Masked Booby do. Masked Booby lacks the cream coloured crown and has more black in the tail, which is also longer. Abbott's Booby has a complete black upper-wing and in flight, its very long wings are very noticeable. **Nest:** Nests in colonies often mixed with Brown Booby. No nest materials are used, eggs laid on bare sand. Clutch size 1-3 normally 2. **Status:** Locally common around breeding sites but disperses after breeding where the preferred food is flying fish—this applies to most booby's. **Where to find:** Off the coast of the Kimberley but easiest on Cocos-Keeling I. Nest in huge numbers on Bedout Island 40 km north of the De Grey estuary. Ashmore Reef. Adele I.

Red-footed Booby *Sula sula* 70-80 cm

Photo: Simon Nevill

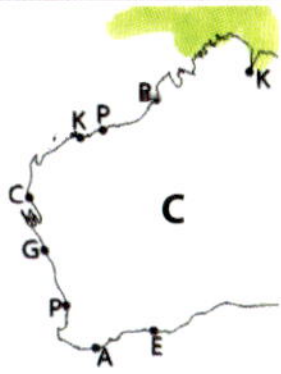

Of the 6 species of booby in the world, this is the smallest and slightest. Rarely seen off the WA coast being more of a pelagic feeder than other booby. Seen more so off the Queensland coast as their breeding sites are closer to the coast. There are various colour morphs of this species but the predominant is the white coloured morph in north-west waters. **Similar species:** Slight similarities to Masked Booby but Mask Booby but does not have bright red legs. Does not have the bright orbital eye mask or bluish beak. **Nest:** Nests in colonies and builds nests in trees on the outer canopy. Clutch size 1. **Status:** Common throughout its world range and the habit of nesting in trees has aided its survival from predation of domestic rats. Locally common around breeding sites but disperses after breeding where the preferred food is flying fish. **Where to find:** Easiest on the Cocos-Keeling Islands where there is one of the largest populations in the world, with an estimated population of 30,000 +. Also Christmas Island, with a smaller population.

Abbott's Booby *Sula abbotti* 80 cm

Endemic

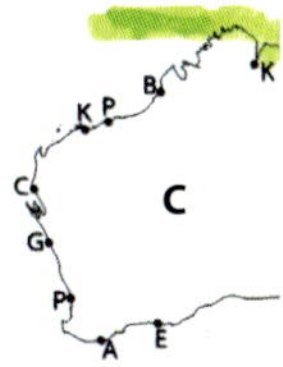

This unique booby was named in honour of W.L. Abbott who first collected them in 1892 from islands north of Madagascar but they have long been extinct from those islands due to deforestation. Luckily they still remain on their one breeding locality, namely Christmas Island. Seen in flight they are unmistakable, having long narrow wings. They require tall open-canopied trees where the young can manage to take flight. An unusually broad beak with the forehead running parallel with the top of the beak. It is presumed that their feeding range is within the region of Christmas Island and Java but this is not clear as the adult birds leave the island for at least 4-5 months. **Similar species:**. No other booby has

Photo: Simon Nevill

an all-black upper wing. Different jizz in flight to all other booby's and far longer wings. **Nest:** Nests in a fragmented colony in the tallest trees mainly in the central parts of the Christmas Island plateau. Clutch size 1. **Status:** In world terms a rare species. Estimated population fluctuates between 1800 and 2000 birds. Locally common around breeding sites but disperses after breeding. **Where to find:** Christmas Island.

ALBATROSS: The taxonomy classification of albatross species has changed many times over the years and is still a contentious field of study, with authorities differing on what species deserve full species status. Differing birding authorities and also differing countries have varied the taxonomic grouping of genera and species. It has been agreed by most authorities to divide albatross into 4 genera; *Diomedea* the great albatross, *Phoebastria* the north Pacific Albatrosses, *Thalassarche* the mollymawks and *Phoebetria* the sooty albatrosses. Establishing genera is one thing but establishing the species is another. The number has varied between 13 to 24 species, but currently there is a general acceptance of 21 species. The seabird species listed over the following pages represent the majority of the more common species that occur in Western Australian waters. There are many rare species that have not been included and also a few more common species but as mentioned it covers the majority and is a general guide only for those starting birding.

Wandering Albatross *Diomedea exulans* 120 cm

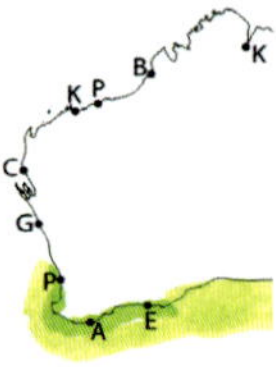

The longest winged seabird in the World and dwarfs the medium sized albatrosses known as mollymawks. Albatross have long narrow wings that have evolved to adapt to gliding rather than for flying with rapid wing beats. They fly with ease in the most turbulent weather. Normally fly within 10-15 m of the surface to gain updrafts from waves. Wandering are long-lived birds and there are several development stages with plumage making identification very challenging. Albatross take at least 9 years to develop full adult breeding colour and go through various stages of plumage colouration. The adult ventral view is all white with black wing tips and thin black margin to rear of underwing. A massive pink bill and with pink legs. Dorsal view in adult birds either all white with black wing tips or black and white mottling on the upper wings merging into the black wing tips. First year birds not common to these shores but they have an all brown body with a white face. The second year Juveniles have brown mottling on the crown and part of the upperwing with a white face, brown patches on vent and pre-auxiliary notch. Back not white but heavily mottled brown and white. Breeds on subantarctic islands, often on high plateaux's. Sadly the populations of the females have been greatly effected, as they feed in more northern subantarctic waters where long line fishing occurs. Need cleared areas to take off and run well for a heavy pelagic bird. Cover vast distance to feed during the non-breeding season. **Similar species:** Royal Albatross which is rare in WA waters. Close view of Royal shows clearly a dark line on cutting edge of upper mandible. **Where to find:** Seen mostly off the southern coast of WA. **Status:** Common in southern WA particularly in winter.

Immature

Photo: Simon Nevill

Black-browed Albatross *Thalassarche melanophrys* 80-95 cm

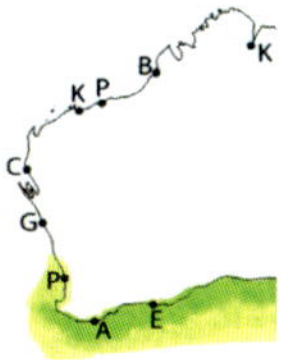

A medium sized albatross with a neat black eyebrow. Feeds primarily in subantarctic and subtropical waters and goes closer to the equator than many other albatross except the Yellow-nosed. Adult birds have bright orange-yellow bill with orange tip. Dorsal view all black wings, black back, white rump, black tail, white head. Ventral view for nominate race, black margin to leading edge of wing with black wing tips and fine black edge to trailing wing edge. Has as complete white head and distinct black eyebrow. Juvenile birds, the ventral view has black edges and black interior of wing with white mottled bars, bill is dull grey with greyish collar narrowing to throat from nape. **Status:** Very common in southern WA waters in the winter but most prolific in the southern Atlantic Ocean. **Similar species:** Campbell Albatross which has a clearly identifiable white iris, not black-brown. **Where to find:** Seen mostly off the southern coast of WA but also of the south-west coast. The nominate species *melanophrys* breeds on the islands of Tierra del Fuego and is the dominant species to be found in WA waters.

Photo: Simon Nevill

Campbell Albatross *Thalassarche impavida* 85-90 cm

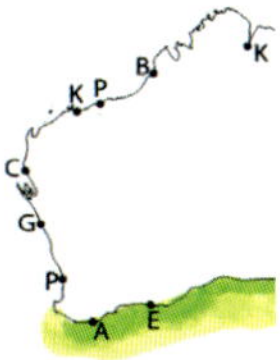

It was only in recent years that this species was split from the far more common Black-browed Albatross *Thalassarche melanophrys* to become a full species, although this is still is in abeyance with some authorities but it is included here as a full species. The primary difference between the two is that Black-browed has black iris where as Campbell has pale-honey coloured iris. This is clearly seen at some distance when one knows what to look for. Also

Note light eye unlike black-browed

Photo: Simon Nevill

the eyebrow is longer and the blackish shade in front of the eye is more extensive. The black underwing margin is broader on Campbell. The species nest on just one remote island and an adjacent islet south of South Island New Zealand. The population crashed in the 70's and 89's due to heavy tolls from long line fishing but the population is climbing back with a decease in the long line fishing of the 80's but still approx. 10% of the population is killed by long line fishing. Adult Campbell mainly ply the seas off South Island and the Ross Sea but non breeding birds and juveniles wander far wider passing the south of Western Australia. **Status:** Uncommon in WA waters. **Similar species:** Black-browed Albatross as mentioned above. **Where to find:** Off the southern coast of WA rarely flying around the Capes.

Indian Yellow-nosed Albatross *Thalassarche carteri* 75 cm

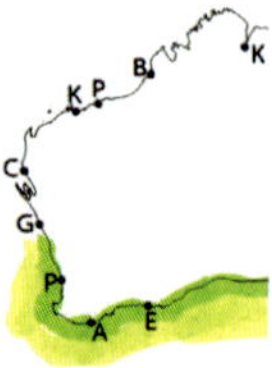

A small mollymawk the smallest all the albatross. The most common albatross to be found in WA waters. It is less pelagic than most albatross frequenting warmer seas over the continental shelf. Its normal range extends as far north as Shark Bay, although there are sighting recorded as far north as the Cocas-Keeling Islands. Adult Yellow-nosed Albatross have black bill with a distinct yellow stripe running down the ridge of the bill to the tip. All white body with pale grey mantle. Slight eyebrow. Immature birds have plain black bill with no yellow on top ridge. **Status:** Very common off the south-west and southern coast. **Similar species:** Buller's Albatross has yellow on both bottom and top of bill and grey neck and face, Yellow-nosed does not. Buller's has grey back not black. Black-browed Albatross immature separated from immature Yellow-nosed by having grey collar. Also Buller's bill at close range is much paler. **Where to find:** Seen off the southern coast of WA and also of the south-west coast as far north as Shark Bay.

Photo: Simon Nevill

Shy Albatross *Thalassarche cauta* 90-100 cm

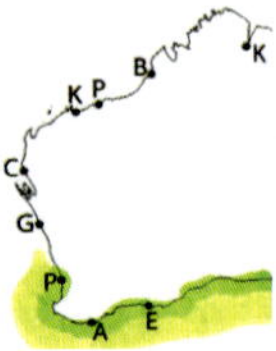

Shy by name but not by nature, comes to boats readily. The largest of the mollymawks and the only albatross in its genus to breed in Australian waters. It flies with the same humpback posture as Wandering does, often with drooped wings. There are 3 similar species Shy Albatross *Thalassarche cauta*, Salvin's Albatross *T. salvini* and Chatham Island *T. eremita*. Shy Albatross breeds on some Tasmania Islands and also on Auckland Island, 450 km south of NZ. Salvin's Albatross breeds on Snares and Bounty Islands NZ and Iles Crozet 1000 km south-east of South Africa. Chatham Albatross breeds on Pyramid stack in the Chatham Island NZ but has never been recorded in WA waters. There are two subspecies of Shy Albatross cauta and White-capped Albatross *steadi* both occur in WA waters but are extremely difficult to tell apart. Some authorities have classed *Thalassarche steadi* a full species, known as White-capped Albatross. So it is a complex arena, that the novice sea-birder finds himself in, when trying to sort out Shy Albatross from those species very similar. Shy Albatross distribution is still not well known but generally circumpolar but records are in the southern Atlantic are few. All species have the small black notch on the leading edge of the wing alongside the body regardless wether adult or immature. All have narrow black margins on wings with black wing tips (less on *cauta's* wing tips). Problems arise from trying to identify juveniles and immature at sea of the 3 species, which are very similar. The differences between the 3 species of juvenile birds require greater information and one needs to refer to more in-depth books on albatross. However adults require less detail. Adult birds are slightly easier to define. Adult *cauta* has a grey bill with pale yellow tip. Light grey tinge on face and neck, with white crown. *salvini* has grey bill with black on end of lower mandible contrasting with yellow tip. Grey on face and neck, stopping at chest. *eremita* bill is pale yellow with black patch on lower mandible, dark grey from face to chest, with grey crown. All races have dark grey backs. **Status:** *cauta* moderately common off the south-west coast. Common off southern coast. **Similar species:** There are two very similar species and they were both subspecies of the Shy Albatross. The characteristic small black notch at the leading edge on wing where wing meets body is the same on adults, juveniles and immature birds of all 3 species and other albatross do not have this feature. It's a small feature but can be seen even at a reasonable distance. Salvin's Albatross has much greyer head than Shy Albatross. Chatham Albatross are a bigger than Shy and the yellow on the top of the bill is brighter. White-capped is also larger than Shy Albatross but yellow on the top of the bill is brighter in Shy. Shy Albatross is smaller than Royal and Wandering but bigger than all other albatross in Australian waters. **Where to find:** The main species to come to WA waters is presumed to be *Thalassarche cauta* from the south-west as far north as Carnarvon and all southern coast. However White-capped Albatross *steadi* is also found in these waters. One should really refer to more in depth references than this guidebook for more details.

Adult

Juveniles

Photo: Simon Nevill

Note in the photograph how the two juveniles of the Shy differ in amount of grey on the head and collar. Also note the very dark head of the immature bird resting at sea. It is very similar to Salvin's and it takes a keen sea-birder to split these species.

Sooty Albatross *Phoebetria fusca* 85-89 cm

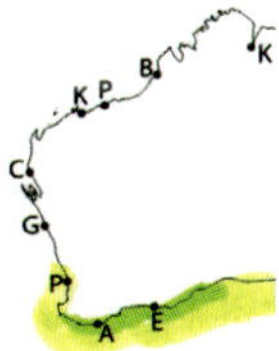

Breeds on islands in the southern Indian and Atlantic Ocean. In the southern Indian ocean islands the species nests on the Crozet group, along with the Light-mantled Albatross being of the same genus. Light-mantled is more common of the south-eastern waters of Australia but the Sooty is more common off the Western Australian waters. An all dark-brown albatross with fine yellowish line on the side of the black bill (the sulcus) whereas Light-mantled Albatross has a fine bluish line on the sulcas. In flight when tail closed, appears as a long pointed tail. Also flaps wings rarely, almost an effortless flight with tail rarely spread. **Status:** Uncommon. **Similar species:** Light-mantled Albatross which has as the

Photo: Simon Nevill

name implies, a lighter grey-brown mantle which contrasts well with a dark-brown head. **Where to find:** Off the south coast. Mostly south of the continental shelf and occasionally venturing around the lower south-west waters.

Southern Giant-Petrel *Macronectes giganteus* 88-100 cm

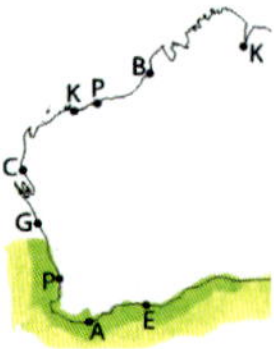

The largest of all petrels. A bulky thick-set bird with a massive tube nose. Flies with the appearance of a hunched back, almost looking like an immature Albatross but unlike Albatross, has an ungainly, laboured flight. There are various coloured morphs in adult birds and also immature birds have differing colouring. Adult birds all dark brown body with white head leading to mottled specks at base of neck. Adult white morph (10% of population) completely white but covered in black-brown flecks to rear of head and upper body. Immature birds are totally brown-black but start to develop white flecking around head. Breeds in subantarctic waters including Macquarie I. Most birds sighted in Australian waters are immature birds, adults tend to roan the Antarctic and subantarctic waters closer to the breeding colonies. **Status:** Common. Less common in the south-west waters than Northern Giant Petrel. **Similar species:** Dark morph and immature very similar to Northern Giant Petrel *M. halli*, which has dark red tip whereas Southern Giant-Petrel greenish tip giving it an even coloured appearance. Adult Northern have whitish facial and throat markings where as Southern have all white head. **Where to find:** Most found off shore from southern WA to the mid west coast of WA.

Photo: Simon Nevill

Northern Giant-Petrel *Macronectes halli* 80-95 cm

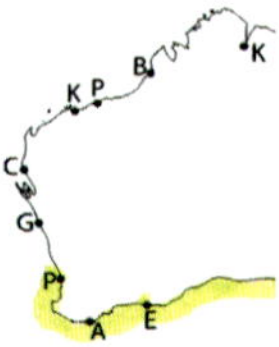

Similar to Great Southern Petrel. The adult birds have an overall dark brown body but face and chin white with brown freckling. Immatures very similar but darker freckling with less white on face. Juveniles have an all dark brown body. There is no light phase morph as there is with the Southern Giant-Petrel. Massive bill with tubes two-thirds the length of the bill, all pale pink with dark pink-red tip. Breeds on several islands in the subantarctic zone with large colonies on many islands south of New Zealand. **Status:** Uncommon in the south-west. **Similar species:** Immature Northern Giant Petrel can be confused with immature Southern Giant-Petrel. Bill tip on Southern Giant-Petrel is dull grey whereas Northern Giant-Petrel is rose-pink.. **Where to find:** Most found off shore in the southern ocean but also off the south west coast. More common than Great Southern. Most birds juveniles and immature birds.

Photos: Simon Nevill

South Polar Skua *Stercorarius maccormicki* 53 cm

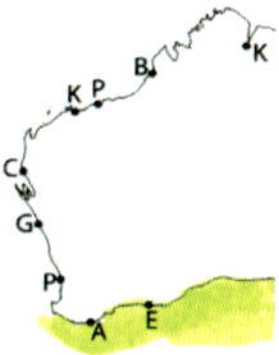

At a distance for first time viewers, this species looks like an over weight, thick set immature Pacific Gull, but the white irregular bulls eyes in the wings make it easily recognizable. Also pale morph adult birds are easy to identify from Brown Skua having a much paler all grey-fawn body and dark head. However, the dark morph form of the South Polar Skua is not so easy but the under wing on the Brown Skua, has more contrast between the dark brown leading edge of the underwing (coverts) and the grey-brown trailing edge (the secondaries). Fortunately the dark morph of the Polar Skua is not common. All plumages show the white patches in the upper and lower wing on the bases of the primary feathers. Breed in the Antarctic on off shore islets but also on the Antarctic mainland. Circumpolar also venturing far north into the northern Atlantic, northern Pacific and northern Indian Ocean. Like all Skuas they are a scavenging species often hassling other birds as in the photograph. They take chicks and eggs at the breeding colonies of other birds and attend rubbish sites at Antarctic research bases. For all their often unpleasant feeding habits, they remain loyal to their partner having a long-term monogamous relationship, reuniting each year at the same breeding site after travelling often thousands of kilometres. **Similar species:** Great Skua as mentioned above. **Where to find:** Found off the south coast mostly several kilometres south of the shelf being a passage migrant. **Status:** Scarce passage migrant.

Photo: Simon Nevill

Brown Skua *Stercorarius antarticus* 63-66 cm

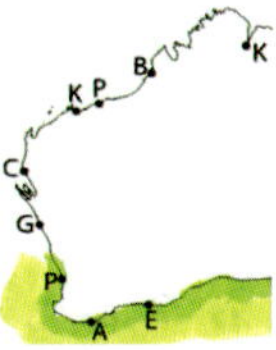

An all brown, thickset bird, looking similar to an immature Pacific gull at first glance. Breeds on islands in the subantarctic and Antarctic leaving, their breeding colonies to feed in waters further north. Basically a scavenging and predatory bird either chasing breeding birds near breeding colonies but more often roaming through colonies to eat eggs or chicks. They are not as skilled at harassing birds in flight as Jaegers or Frigatebirds as they can't manoeuvre so well in flight. However, they take many seabirds by diving from a great height and knocking them down, and then with their hooked beak they can inflict severe damage and normally kill their prey but this is not a common practice as mostly they rely on scavenging. In Australian waters they do take fish. In flight shortish tail with slight wedge appearance. Faint white markings on primaries gives bulls-eye appearance. Thick hooked bill. **Similar species:** South Polar Skua. Adult pale morph far lighter with cream coloured body and contrasting brown wings. Dark morph very difficult to split. There are anatomical differences but most birds in flight will be hard to pick. The main thing is if you are used to seeing the more common Great Skua its slightly bigger and brown colouring slightly lighter, which may assist. **Where to find:** In Perth region often seen around Fremantle and off Woodman's Point in autumn to spring. Seen along the south-west coast as far as Shark Bay, normally as individuals. **Status:** Moderately common.

Photo: Alan Collines

Parasitic Jaeger (Arctic Jaeger) *Stercorarius parasiticus* 40-46 cm

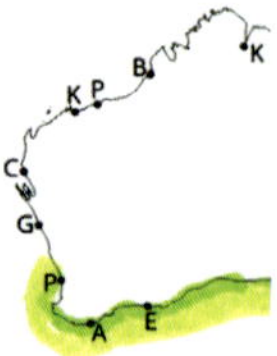

Masters of flight, Jaegers harass other seabirds, particularly gulls and terns to force them to disgorge food. They follow at great speed turning with every move of the pursued bird. This predatory behaviour is however far more prevalent around breeding colonies in the Northern Hemisphere and in Australian waters they actually take more food from the surface of waters rather than harass other species. Jaegers are cosmopolitan in distribution travelling from their breeding grounds all the way to the Southern Hemisphere, mostly hugging the coastlines rather than venturing through the open seas except to migrate. They arrive in Australian waters between October and November moving south to Antarctic waters and return March and April on their long journey north. There are 3 species of Jaeger all of them come to Australian waters. Identification of the different species of Jaeger is difficult compounded by the fact that there is not only colour changes with age but there are different morphs. Arctic Jaeger juvenile birds are over all brown with slightly darker wings. Fine brown barring on rump. They develop barring on rear of body and underwing but still overall brown. This changes in breeding plumage to white under body, dark wings with barring, dark brown barred rear body and tail, dark collar. Adult birds pale morph in breeding have grey-black cap, white body with buff flanks, dark brown collar not meeting at centre of chest, dark-brown wings and importantly rump and vent is completely brown, in none

Photo: Rohan Clarke

breeding the rump becomes barred. Adult dark morph has black cap all over brown but wings and back darker brown. **Similar species:** Pomarine Jaeger the reader should refer to more in depth characteristic differences from more authoritative works but suffice to say Pomarine Jaegers are bigger, bulkier and thicker necked birds. They have a heavier bill and when you get familiar with Jaeger's you will notice the base of the wing is broader near the body. Long-tailed Jaeger, again, not easy although in breeding plumage the long tail shows well. Has greyer under body. It's a slimmer bird with longer wings that make it almost tern like in flight. **Where to find:** Moderately common around Cockburn Sound from April to May and then again in October to November. Small number remain through our winter. **Status:** Common in the northern Hemisphere more dispersive in the Southern Hemisphere. More likely to be seen in WA waters than Pomarine Jaeger. Locally common in Cockburn Sound generally uncommon elsewhere although can be seen all along the WA coast to the Timor Sea.

Pomarine Jaeger *Stercorarius pomarinus* 65-78 cm

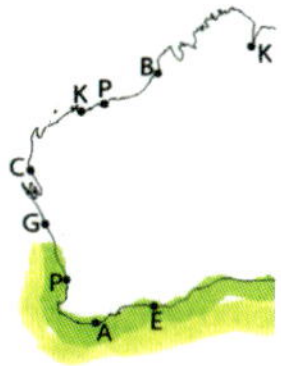

A bird with World wide distribution very similar to Parasitic Jaeger (Arctic Jaeger). Refer notes on Parasitic Jaeger but basically a bigger more robust bird with broader wings. Bill larger than both Arctic and Long-tailed Jaeger. Breast band wider than both other two species. Flight more laboured and heavy. Breeds further north than Parasitic Jaeger in the Russian Tundra. Birds reported from Tundra fly overland to the Indian Ocean. Two morphs occur light and dark as for Arctic. Uncommon in WA from March to April and then again in October to November. Arctic out number Pomarine Jaeger in WA waters. Also Pomarine is more prevalent on the south-east coast of Australia. Tends to cover more pelagic areas than Parasitic, as Parasitic tends to hug the continental shelves when possible. **Status:** Uncommon in the northern Hemisphere more dispersive in the Southern Hemisphere. Scarce in WA waters more common on lower south east coast but seen all along WA coast to Timor Sea. **Where to find:** Moderately common around Cockburn Sound from March to April and then again in October to November but sightings occur throughout the year. Those interested in splitting Jaegers should refer to more definitive books for information, as plumage difference with the 3 Jaegers is complex due to colour change with age, various morphs and wether in breeding plumage or not.

Photo: Rohan Clarke

Long-tailed Jaeger *Stercorarius longicaudus* 35-58 cm (not including 18 cm for tail streamers)

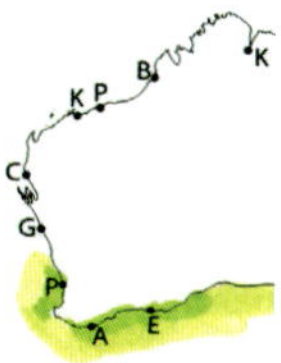

Smallest and slimmest Jaeger, with a more buoyant flight than others. Uncommon close to shore lines more pelagic than Parasitic. Breeds in the northern arctic of Canada and Russia. Master flier as it pirates other seabirds for food but is the least likely to harass birds of the 3 jaegers. Take more food from sea surface. On a side note of interest, they take Lemmings in their breeding area such is the diverse range of its diet for this predatory species. **Where to find:** Scarce in WA waters. Uncommon south-east coast of Australia. Tends to cover more

Photo: Simon Nevill

pelagic waters. **Status:** Uncommon in the northern Hemisphere more dispersive in the Southern Hemisphere. Scarce in WA waters more common on lower south-east coast particularly southern Tasmanian waters. Those interested in splitting Jaegers should refer to more definitive books for information, as plumage difference with the 3 Jaegers is complex due to colour change with age, various morphs and wether in breeding plumage or not.

Antarctic Prion *Patchyptida desolata* 27-28 cm

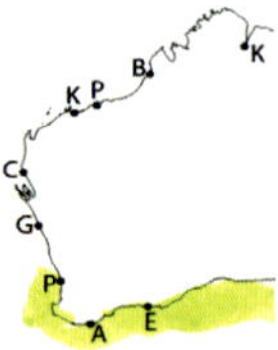

One of the more abundant of the prions to be found in Australian waters but still not common to see close to the shelf, found more often as a beach derelict after severe storms on beaches and occasionally way into the Wheatbelt, such is the power of storms to push birds far and wide trying to avoid the epicentre of storms. All prions are very small, very similar in size, colouration and jizz. In fact they test the most ardent sea birders, always a challenge to split. Identification is compounded by the fact that many of the same prion species breed on different islands often many thousands of kilometres apart and this understandably leads to subtle variations in either colour patterning and sometimes body size. It's range is almost circumpolar although knowledge in the mid southern ocean not well known. It nests in the French territories of the Crozet Islands. Also South Orkneys, Kerguelen Island in the Indian Ocean. On Macquarie Island in Australian waters and Auckland Islands. It feeds off the southern tips of South America, South Africa and Australian and New Zealand waters as well as closer to the south pole. Like all prions, it has the characteristic pale grey-blue upper body and the distinct dark 'm' running from wing tip to wing tip. Black terminal band on tail. Underbody white. Bill slate-grey. Feet and legs pale blue. Crown dark grey-blue bleeding into lighter grey-blue collar. Flies fast and manoeuvres only in flight when feeding. **Status:** Uncommon close to shelf Southern Ocean. Very uncommon but not rare to south west coast. Slender-billed is the most commonly seen prion off the southwest coast followed by Salvin's and then Antarctic. **Similar species:** Very similar to Salvin's Prion and almost impossible to tell at sea if not close. If seen close, the slightly wider bill on Salvin's is clear and Broad-billed Prion is far wider than Antarctic. Slender-billed similar plumage, with subtle differences, slightly smaller and slimmer, wings lay central of front and rear body and are straighter. Base of bill not as wide as Antarctic. Slender-billed neck remains more extended, whereas Antarctic seems more compact and tucked into the body. Broad-billed Prion is far easier to split as it is a larger prion with a very wide bill, larger head with steep forehead, also flies slower with less wing beats. Bill is black rather than bluish-black. Slender-billed Prion bill is longer and slimmer with broader supercilium. Fairy Prion has a smaller head, shorter bill, a much paler grey-blue head and tail marginal band is broader. **Where to find:** Almost circumpolar, found as far north as Geraldton but mainly seen off the southern shelf of the Southern Ocean.

Photo: Rohan Clarke

Salvin's Prion *Patchyptila salvini* 27-28 cm

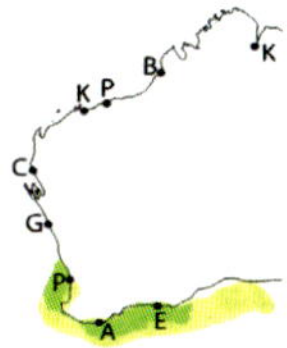

There are two subspecies. The nominate subspecies salvini nests on the Crozet and Prince Edward Islands the other subspecies *Salvin's macgillivrayi* nests on Amsterdam and St. Paul Islands. Salvin's Prion has a closer morphology to the Broad-billed Prion but is more closely related to the Antarctic Prion *Patchyptila desolata*. At sea it is almost impossible to separate from Antarctic Prion both in coloration and flight. The bill on Salvin's is slightly broader and slightly longer. Head closer to body giving a thicker neck profile. It has a lamellae within the bill that filters seawater to gain food. Its bill is only slightly narrower than the Broad-billed Prion *Patchyptila vittate*, hence one of its early common names was Lesser Broad-billed Prion. Often found washed up on beaches after severe storms off the south west coast. **Similar species:** As mentioned above similar to Antarctic and also Broad-billed Prion. The more common Slender-billed Prion *Patchyptila belcheri* is slightly easier to separate as it has a slimmer body, slightly smaller and bill slightly longer and narrower. All prions for those starting pelagic bird watching are a real challenge and it takes many visits out to sea to slowly come to terms with the subtle differences. **Status:** Uncommon throughout the south west and southern coast. The Slender-billed is more common off the southwest but the species has not been illustrated in this book. **Where to find:** Off the Perth the southwest coast and southern coast.

Photo: Rohan Clarke

Cape Petrel *Daption capense* 36-41 cm

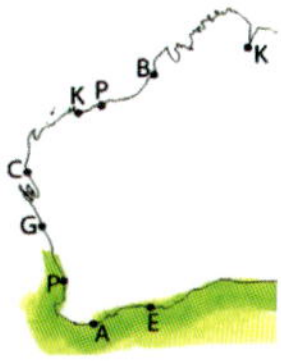

A stocky unmistakable petrel relatively common in south-west WA waters. Black-brown head. White on belly merging from chin to vent. White underwing with black margins. No strong demarcation of patterns but mottled edges and mottled auxiliaries. Dorsal black head, black wing margins, white patches to all coverts and primaries within black margin. Rump black blotches on white, tail band black both top and bottom of tail. Breeds on many islands within circumpolar region as far north as the New Zealand offshore islands. **Status:** Common off the south coast less so the southwest coast. **Similar species:** None really, only species that could confuse is Antarctic Petrel, which has no mottled markings and all margins where black and white meet are clean cut. Rump on Antarctic is white not mottled however a rare species in WA waters. **Where to find:** Mostly found

Photo: Simon Nevill

off south west coast to just north of Perth more prolific in southern seas. Generally less numbers than compared with sightings on the east coast but still one of the more common pelagic birds in WA.

Hutton's Shearwater *Puffinus huttoni* 37 cm

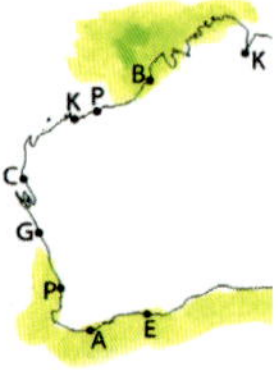

A shearwater with an all black-brown dorsal. The ventral region is white extending from throat to vent and on wings the white extends to the base of the primaries. On the ventral side, the wing margins are dark brown and covering most of the primaries. The auxiliaries are brown leading into the wings as an underwing band at the front of the wing but not touching leading edge. Brown colour on crown and neck runs down to throat forming a dark collar useful diagnostic feature when splitting Manx. This species is almost identical to Fluttering Shearwater and has an interesting history. It was first described in 1912 and for years was considered a subspecies of Fluttering. It was not until as recent as 1965 that the breeding location of this elusive species was found over 4000 m up in the Seaward Kaikoura Range on South Island New Zealand. **Status:** Uncommon but certainly not scarce off the south coast. **Similar species:** As mentioned, this species closely resembles Fluttering Shearwater the only subtle differences are that the collar band on Fluttering is narrower and does not reach as far as the central throat area. The auxiliaries are darker in Hutton's but this is not easy to pick up. **Where to find:** Mostly along the south-west coast as it leaves its breeding grounds in NZ to over winter in subtropical and tropical waters. Can be very common on passage from late July to early September with passage rates up to 3,500 per hour recorded near Cape Naturalists (per John Graff).

Photo: Rohan Clarke

Great-winged Petrel *Pterodroma macroptera* 39-43 cm

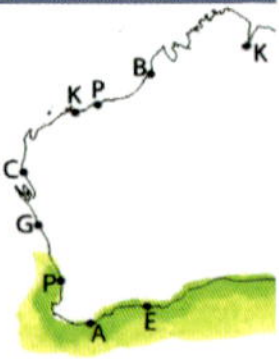

A large long winged petrel with a short thick neck. Uniformly dark brown. Has a pale cream patch at base of bill on face not always evident. Bill is black. Flies well, turning and manoeuvring with ease. There are 2 subspecies. Nominate *macroptera* has the greatest breeding range nesting on many islands in the subantarctic belt in the southern Atlantic and southern Indian Ocean. Subspecies *gauldi* has a more restrictive breeding range on North Island NZ. White facial patch more extensive covering more of the forehead and chin. **Status:** Common in south-west waters. **Similar species:** Providence Petrel mainly confusion with the subspecies *gouldi* which breeds on Lord Howe Island and can be found in the same waters as the Great-winged Petrel, but paler overall brown with main difference being white patches on the underwing on the primaries but Providence Petrel has not been recorded in WA waters. **Where to find:** Most of the south-west and southern coast and as far north as Geraldton. Nests on several islands off the south coast of WA, mostly in the Recherche Archipelago.

Photo: Simon Nevill

White-headed Petrel *Pterodroma lessonii* 40-46 cm

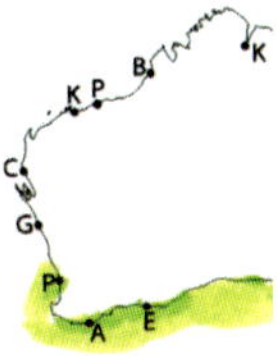

A large, attractive petrel with long narrow wing, held forward. Underwing all dark grey. Back and wings dark-grey, lighter grey on mantle and crown. A black eye patch, being the important diagnostic feature on this bird. Flies with wedge tail except when breaking flight to feed. Mostly a solitary bird at sea, its range being circumpolar in the subantarctic. Nests on few islands in the southern Indian Ocean, also the Macquarie Islands south of New Zealand. **Similar species:** Similar to light phase of the Soft-plumaged Petrel, which has no black eye patch and has dark crown. Also has grey collar. Kermadec Petrel has 'bullseye' type white patches on underwing and no black eye patch. Kermadec not found on the west coast. **Where to find:** Anywhere in the southern seas or south west to approx Perth. Has been sighted in the Pilbara but that is way out of its normal range. **Status:** Uncommon winter visitor mainly sighted in Southern Ocean.

Photo: Rohan Clarke

Soft-plumaged Petrel *Pterodroma molis* 32-36 cm

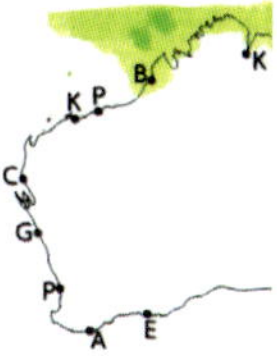

One several attractive species in genus *Pterodroma*. Soft-plumaged Petrel's are regularly seen on pelagic trips off the south-west coast and would be the most common of the Pterodroma complex. It could be that the birds we do see are in fact immature birds like so many pelagic species that pass the WA coastline, as the adult birds are often found around the waters not far from their breeding grounds. The under body clearly marked with a dark 'M' shaped black band contrasting against light grey wings. Grey crown flows into grey collar narrowing on throat. Grey-blue upper body with dark grey M band running from rump to wing tips. Light morph shows white in primaries giving bulls-eye appearance. Grey on crown passes below eye. Black bill. Almost circumpolar in distribution but appears not to occur in the subantarctic South Pacific which is surprising as they nest on islands south of New Zealand. Nests in subantarctic islands in the Indian Ocean. South Atlantic. Islands south of NZ. **Status:** Fairly common away from coastline lower south-west and southern coastline. **Similar species:** White-headed Petrel has no grey collar, white crown not grey, has distinct black eye patch clearly seen at distance. **Where to find:** Off Perth coastline and Southern Ocean. Common on west coast and southern pelagics. Common as passage migrant on southern coast.

Photo: Simon Nevill

Tahiti Petrel *Pseudobulweria rostrata* 38-40 cm

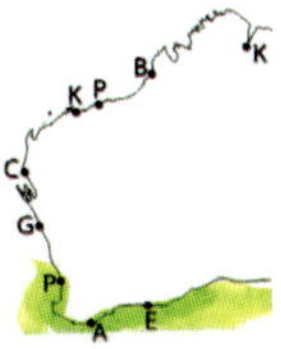

An attractive cleanly marked 'gadfly' petrel found in tropical and subtropical waters. Head all black to the top of chest. All black upper body with faint white mottling on rump margins. Tail generally closed and pointed in flight. On underwing, lesser coverts and primaries black with light brown median band forming margin between rest of

Photo: Rohan Clarke

dark brown underwing. Wings in flight rather strait in appearance. Breeds on volcanic islands in the central Pacific. **Status:** Although the species breeds on several islands in the eastern Pacific, the individual colonies are small and with the introduction rats, populations are being threatened although the World population is estimated to be approximately 10,000 pairs. Not threatened yet but vulnerable. Maybe more common in north-west waters than we think. **Similar species:** Very similar to Phoenix Petrel but this species is rare in Australian waters and has never been recorded in WA waters. Tahiti Petrel has whitish medial band on underwing, Phoenix does not. Phoenix has white band on underwing on front leading edge, Tahiti does not. Tahiti has slight white mottling on outer edge of rump, Phoenix does not. **Where to find:** Now that more serious birding boat trips are being under taken off the north west coast combined with increased pelagic knowledge this species is being identified more often in the north west waters and around Christmas Island and Ashmore Reef.

Streaked Shearwater *Calonectris leucomelas* 46-48 cm

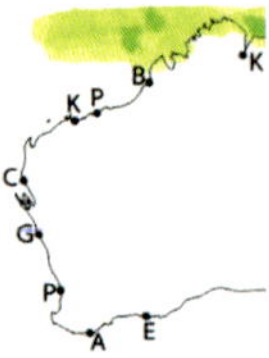

Breeds in the northern Pacific on islands off the Korean, China and Japanese coasts and over winters around New Guinea and as far south as northern Australia. There seems to be an increase in the sightings of this species off the north west coast and also recently, sightings of this bird off the Perth coastline. Its unusual, as this is certainly not a bird of the cold southern seas but with variations in sea temperatures and food availability birds will adapt and move out of their normal feeding range. Also with the increased journeys by skilled observers to locations like Scott Reef, Ashmore Reef and Christmas Island, Streaked Shearwaters are being found fairly regularly. Streaked Shearwaters have a dark brown upper body and crown. Underbody white with primaries dark brown. Tail covets white covering dark brown tail forming distinct V dark-brown border. Tail is wedged in flight and fans when banking. Has slim, long neck which is very clear when close to other shearwaters of the *Puffinus* genus, freckled forehead and face, white around eye. **Status:** Uncommon to moderately common in northern waters particularly in summer. **Similar species:** Only Buller's Shearwater which apparently has not been recorded in WA waters. **Where to find:** Northern waters off the Broome coast to Ashmore Reef and across to northern Queensland.

Photo: Rohan Clarke

Wedge-tailed Shearwater *Ardenna pacifica* 38-46 cm

Photo: Rohan Clarke

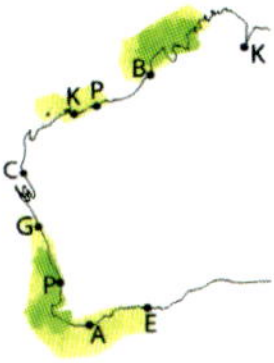

A shearwater of tropical and subtropical seas, that tolerates greater fluctuations in sea temperatures than many southern subantarctic shearwaters. All dark-brown colouring. Longish wedge shaped tail held well closed when in normal flight. Wings held forward and bent not straight. In calm weather flight is buoyant and unhurried often close to water. Nests in many locations throughout the Indian and Pacific Oceans. **Status:** Very common off the southern coast and south west coast. Breeds on many Islands off the Pilbara coast particularly in the Montebello Islands also the Dampier Archipelago. Found as far north as Scott Reef and Ashmore Reef. **Similar species:** Read notes on Fleshy-footed for similarities to that species. **Where to find:** Anywhere along the west coast of WA. Ventures further off the continental shelf than many shearwaters.

Short-tailed Shearwater *Ardenna tenuirostris* 40-43 cm

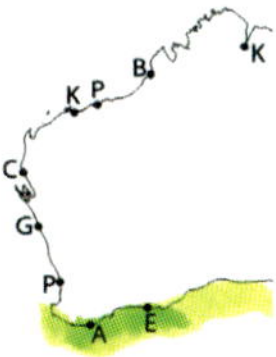

A medium sized shearwater with a slender bill. Tail is short and toes just extend past the tail. Dorsal is all dark brown but under body shows pale central band along wing, more pronounced in light morph than dark morph. This may well be the most prolific shearwater in Australia but in WA waters it is not so common but not common on the west coast. Forms in very large groups close to breeding the colonies and will rest as huge rafts. In the Tasman Sea the species can be seen in their thousands. Nests in far greater numbers on the south-east coast islands of Victoria and Tasmania. The WA nesting population is centred on the Recherché Archipelago on at least 6 islands with an estimated total population of 50,000 individuals. The total Australian population however is estimated to be around 23 million. **Status:** Uncommon far more common on the east coast. **Similar species:** The Sooty Shearwater is the most similar looking species but not common in WA waters. It is a slightly larger bird and the pale brown patches on the underwing more distinct but this is not an easy split. **Where to find:** Anywhere along the southern coast as far as the Australian Bight and across to the south-east coast. It is found as far north as southern Queensland a higher latitude than the species are found in WA.

Photo: Simon Nevill

Flesh-footed Shearwater *Ardenna cameipes* 40-45cm

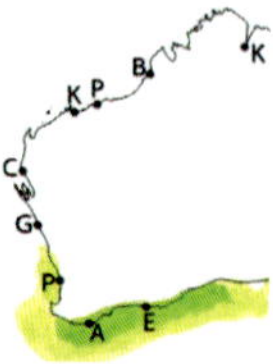

A large shearwater with a robust head. Entirely dark brown plumage. The flight has a distinct quality more languid than most shearwaters. The main features to split from other shearwaters is the larger light coloured bill with dark tip. Seen mostly on the southern coast and to a lesser extent on the lower west coast basically a reversal of the Wedge-tailed Shearwater. This is because Fleshy-footed breed on the south coast inlands from Cape Leeuwin to the Australian Bight. It is very common breeder in the Recherché Archipelago whereas Wedge-tailed Shearwaters nest on many west coast islands from Carnac Island in the south as far north as Ashmore Reef. Away from the breeding colonies seen mostly alone at sea but will occasionally be seen with large groups up to 100 in good feeding areas. Found mostly over continental shelf rather than over deep oceans. **Status:** Very Common in Southern Ocean and moderately common on the west coast as far as Perth. More records in winter. **Similar species:** Split from other Puffinus shearwaters by its larger paler bill. Wedge-tailed Shearwater has longer tail and feet further away from tail tip. In flight wings Wedge-tailed holds wings further forward where as Fleshy-footed wings are straighter. Short-tailed is a smaller bird with a fine bill. **Where to find:** Anywhere in the southern seas.

Photo: Simon Nevill

Little Shearwater *Puffinus assimilis* 25-30 cm

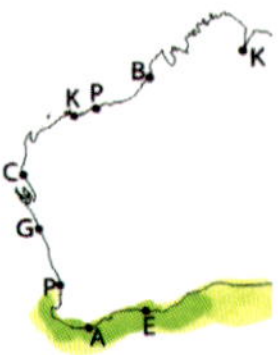

This is the smallest of shearwaters to be found in Australian waters. The species has a world wide distribution with 7 subspecies occurring in the World but their distribution is very fragmented and research shows that their feeding range away from the nesting location is often limited, suggesting they maybe more sedentary. So the subspecies status for some may change to full species status with time. The subspecies *assimilis* stat breeds on many islands off the WA coast, including on the Houtman Abrolhos group. Islands just west of Albany and on several islands in the Archipelago of the Recherche. Flies normally only a few metres above the surface of the sea, the flight is direct, fast, with rapid wing beats interspersed by short gliding periods. Easily identified from most shearwaters with its distinct markings. Dorsal is all black. Under body is all white. Wings have thin clean-cut black margin on trailing edge and wing tips. An important feature is that the white on the head passes over the eye giving it a white eye patch appearance, making it identifiable from other similar shearwaters.

Status: Fairly common. **Similar species:** Audubon's Shearwater has similar markings but rare in Australian waters and even rarer in WA waters. Little is one third smaller with shorter tail. Audubon's black on back and crown passes below eye, so does not have white-eye patch appearance as Little does. Hutton's Shearwater that is common in WA waters is slightly larger and heavier bodied. Longer winged. But the main features that stand out are that the underwing patterns are not cleanly defined, so lacks the black white contrast as in Little. Hutton's brown markings on the auxiliaries and dark brown head with no white-eye patch. **Where to find:** Highest numbers off the Abrolhos and the south coast.

Photo: Adrian Boyle

Wilson's Storm-Petrel *Oceanites oceanicus* 15-19 cm

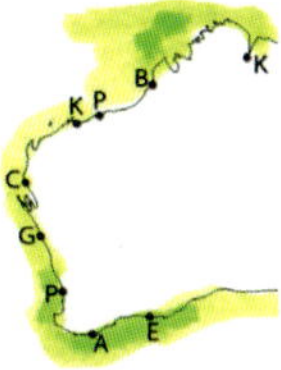

Storm-Petrels are small tubenosed seabirds, found throughout the oceans of the World. They are the smallest of the Worlds pelagic birds. They are split into two subfamilies. Oceanitidae there are 7 species with 5 found in Australian waters. They are generally of the southern hemisphere, are long legged, short winged most with square tails. The other subfamily Hydrobatidae has 14 species but only two have been sighted in Australian waters. They are generally from the northern hemisphere. Shorter legged. Longer wings. Mostly forked or wedge shaped tails. The key to identifying the different species is to identify tail shape, body colour patterns and wing patterns. Length of legs. In the hand, shape and length of nasal tube. Wilson's Storm-Petrel is one of the most common storm-petrels to be seen off the WA coast. Over winters throughout all the oceans of the World mostly to the Indian and Atlantic less so the Pacific. Breeds in the Subantarctic and Antarctic. An all brown storm-petrel with a distinct broad white rump. Square tail. Black medial band on tail. Grey wing covets. Legs protrude past tail. Wings shorter than many storm-petrels. Follows ships. Seen dancing on water when feeding. A storm-petrel that tends to be seen closer to continental shelves more than most and hence it is seen more than most other species from the coastline. **Similar species:** The other storm-petrel that is seen in Australian north-west waters is Matsudaira's Storm Petrel *Oceanites matsurarae* and it is easily identified having no white rump, tail is forked (not good ID feature). The other similar dark storm-petrels is rare in the north west waters is Swinhoe's Storm-Petrel. It is very similar and has now been recorded in the north-west of WA. It has a faster flight, has white on leading edge of shoulder. Tail is forked deeper. It breeds off the coast of Japan, Korea and eastern China.

Photo: Simon Nevill

Over winters in Indian Ocean, so it may be over looked in Australian waters. **Status:** A very common storm-petrel. **Where to find:** All WA coastline and found throughout the Australian coastline except in the Gulf of Carpentaria and from Cape York to where the Barrier Reef terminates.

Matsudaira's Storm-Petrel *Hydrobates matsudairae* 24-25 cm

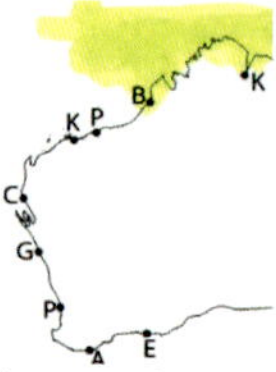

Slightly larger than most Storm–Petrels. An all dark storm-petrel with forked tail. Buffish colouring on across dorsal. One important feature is that the base of the primary shafts feathers are white and this shows clearly on most birds, showing as a white patch. It flies with fast wing beats. Found in tropical and sub-tropical waters. The life cycle of this little known storm-petrel is interesting. We know it breeds on the island of Volcano and Bonin Island off the Japanese coast but is absent out of the breeding season from June to December. Instead of concentrating most of its feeding route across the Pacific, which is adjacent to the Japanese coast, much of it migratory route takes it as far as the eastern Indian Ocean and off the east coast of Africa, above Madagascar and around the Seychelles. Researchers are not sure if its route takes it through the Straits of Malacca to the Indian Ocean or through the Molucca Passage to the Arafura Sea, which is most probably the case as the bird has been sighted regularly in north west waters around Christmas Island and Ashmore Reef. **Similar species:** The only common storm-petrel that is seen in Australian north-west waters is Wilson's and it is easily identified having a white rump, tail less forked (not good ID feature) and no white patch on primaries. Other similar dark storm-petrels are rare in north west waters but Swinhoe's Storm-Petrel is very similar and has been recorded recently in the north-west waters of WA. Swinhoe's is smaller at 19-21 cm than Matsudaira's, faster flight and no white on leading edge of shoulder. Fork less deeply forked. Swinhoe's also breeds in Japan, but also Korea and eastern China. Over winters in Indian Ocean. **Status:** Little is know about the population of Matsudaira's Storm-Petrel but said to out number Wilson's 20-1 in the Timor Sea (Johnstone and Storr). **Where to find:** Timor Sea and off north west coast. Around Christmas Island and Ashmore Reef. Has been sighted off Pilbara and Kimberley coast.

Photo: Rohan Clarke

White-faced Storm-Petrel *Pelagodroma marina* 18-21 cm

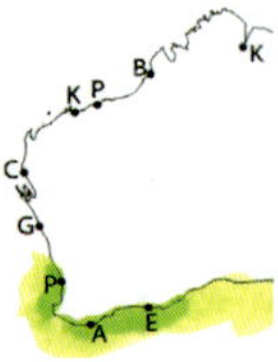

Breeds in many locations in the World. In WA, it nests on the Abrolhos Islands. Islands off Cape Leeuwin. Islands west of Albany and in the Recherche Archipelago. Birds that breed in WA over winter in the Indian Ocean with the Abrolhos being there southern wintering limit most birds are recorded around the Seychelles and East African coast. A well marked storm-petrel having dark crown and collar, white face and under parts, dark stripe below eye, light grey rump, square tail, feet extend past square tail. Rarely treads water like Wilson's but touches with both feet in a hoping action. Feeds while swimming. Unlike Wilson's rarely seen near land, very much a pelagic bird away from the breeding site. **Status:** Fairly common on south-west coast. **Similar species:** No other storm-petrel has a broken facial head pattern of white and blue-grey. The markings looking like a prion. **Where to find:** Mid-west waters with high concentrations around the Abrolhos October–February.

Photo: Simon Nevill

Eastern Osprey *Pandion cristatus* subspecies *leucocephalus*

50-65 span 160 cm

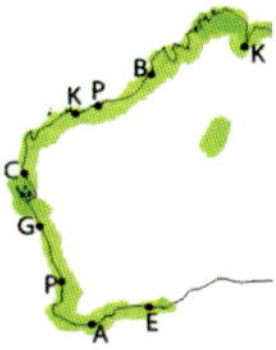

A large 'fishing eagle' found throughout the world. Feeds almost entirely on fish. Adult birds remain in their breeding territory all their lives if the nesting site and feeding area are secure and food supply adequate. Can be harassed by the larger raptor, the White-bellied Sea Eagle on some of the island sites. Found along all of the South West coast line except due east of Cape Le Grande where it becomes scarce. In flight looks generally white underneath, with bands of dark brown on wings and tail and open black wing tips. When in stoop flight, tail is closed. Often seen patrolling the coastline. Female has brown flecking on breast. Male is pure white with no brown flecking. There are four subspecies in the world and just the one subspecies *leucocephalus* in Australia. **Nest:** Their nests can be huge structures, particularly those on the ground such as on the Abrolhos Islands, sometimes well over a metre and a half high. They nest mostly on the ground on islands but also in dead

Photo: Simon Nevill

trees over water on inland waters. Clutch size: Mostly 2, sometimes 3 to 4. **Similar species:** The immature White-bellied Sea Eagle could be confused with an Osprey but is far larger and the tail more wedged-shaped. **Status:** Uncommon to common from Esperance north to the Kimberley coast. Common on the Abrolhos Islands. Islands in the Dampier Archipelago. Throughout the Kimberley coast particularly on off-shore islands. Uncommon from Augusta to Cape Arid. **Where to find:** In Perth found in Alfred Cove and the Swan foreshore opposite Royal Perth Golf Course. Also on Rottenest Island. North of Perth: Green Island. Wedge Island. Lancelin. South of Perth: Peel Inlet. Hamelin Bay. Torbay. Lower King River. Quoin Head (Fitzgerald River NP). Abrolhos Islands. Shark Bay. Carnarvon. Common along most of the Pilbara coast. Broome to Cape Leveque.

Black-shouldered Kite *Elanus axillaris* 36 cm span 90 cm

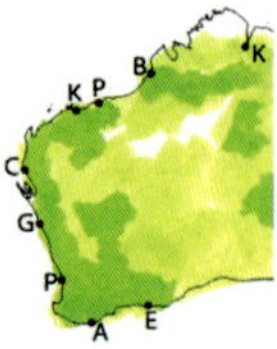

An elegant pale hawk, often seen hovering with rapid wing beats, looking for small rodents or reptiles. White body parts seen in flight, with black carpals which lead to dark grey primaries. Adults have a completely white under body. Juveniles have an orange brown tinge to upper-chest feathers, seen clearly in flight. An irruptive species seen more often when mice plagues occur but remains in areas where the food supply is constant. Prefers open country rather than forested areas, rarely seen in the Darling Range **Nest:** placed near the top of a tree, often a solitary tree or small group of trees normally with cleared surroundings. The nest is a small twig nest, lined with fresh leaves. Clutch size: 3 to 4. **Similar species:** Letter-winged Kite, which has distinct long black band running from the axillaries near the body to the primary coverts. The band follows the line of the underwing medium covets. The black stands out clearly as a band, as the shoulders (marginal coverts) are pure white and also the rest of the underwing covets and primaries are white. With the Black-shouldered Kite, the black is just seen on the underwing secondary covets from the body to the centre of the leading edge wing. The Black-shouldered Kite also has a distinct black eyebrow that is not only on the lores as in the Letter-wing but also passes the eye making it look like an extended eyebrow. **Status:** Moderately common on the Swan Coastal Plain and throughout the Wheatbelt Region and the southern coastal belt except the deep South West corner. Moderately common from Perth all the way north to Carnarvon. Many of the birds coming to the South West are non-breeding visitors and can come at any time of year. As the species is irruptive, it can be

Black-shouldered Kite

Letter-winged Kite

Photos: Simon Nevill

seen in station country when conditions favour rodent explosions. Uncommon to common throughout the Pilbara. Common in the southern parts of the Kimberley, less so in the north. **Where to find:** The dry grass areas of the Swan Coastal Plain below the Darling Range. All of the inner Wheatbelt, particularly from Bindoon all the way to the Murchison. Over Kwongan heath such as those in the Fitzgerald River NP. Any grass plains or floodplains in the north, particularly around Lake Gregory where Letterwing also occur.

Letter-winged Kite *Elanus scriptus* 33-38 cm span 90 cm Vagrant

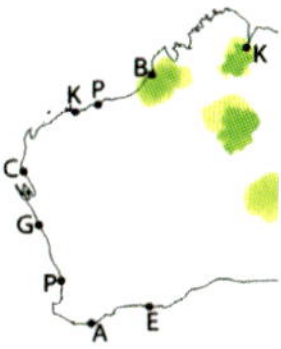

Rare in the South West. An interesting species that has evolved and adjusted to take advantage of the higher numbers of marsupials that are active at night. May not have the nocturnal vision abilities of owls but has developed a large retina to assist with the low light levels in which it hunts. Letter-wing Kites will often roost communally in small numbers particularly along tree-lined watercourses. They leave to hunt just after sunset or sometimes later and return at the latest one-hour before sunrise. The primary source of prey is the Long Haired Rat (*Rattus vilosissimus*). It is interesting that the distribution of Long-haired Rat stretches from northern South Australia to where the Strzelecki Desert starts through the Simpson Desert and all the way across the southern deserts of the Northern Territory, then into the Tanami Desert in Western Australia and then narrows to a small band of grasslands that terminate almost at the coast near Derby and Broome. Hence the plains around Roebuck Bay often get an influx of Letter-wings. The birds breed profusely where the rats are in large numbers but when the population of the rats declines, the Letter-wings disperse far and wide and that is often when the South West will receive the odd visiting bird. Letter-wings have bred in the South West but only a few pairs have been recorded. The Birdsville and Strzelecki Deserts region is the core centre of their breeding population. **Nest:** Loose twigs and branches with fresh leaves in the centre, placed normally above 7 mostly in eucalypts along

Photo: Simon Nevill

watercourses. Clutch size: 3 to 4. **Similar species:** Black-shouldered Kite. Refer notes on Black-shouldered Kite. Note the photograph of a Letter-wing Kite (taken on Roebuck Plains): in flight the differences between the two species is easily visible but there are other subtle differences, as is illustrated in the two head photographs. The crown is more rounded in the Letter-wing Kite and there is no eye frown, also the eyes look larger. Note the eyebrow and the slightly conical crown of the Black-shouldered Kite. **Status:** A rare vagrant to the south west. Occasionally seen in the Pilbara but more so in the lower Kimberley after eruptions of rodents on the floodplains, particularly at Roebuck Bay. **Where to find:** Best chances in Western Australia are on the grasslands of the Kimberly and particularly around Lake Gregory in favourable years. Letter-winged Kites have been recorded in few locations in the South West Region but are rare sightings. Higher chance at Lake Gregory. I have only seen communal roosting in South Australia and in Southern Queensland. In Western Australia, he has seen one in broad daylight perched in a dead tree adjacent to Lake Gregory, also on the Roebuck Plains (the bird illustrated) and Parry's Lagoon.

Square-tailed Kite *Lophoictinia isura* 55 cm span 130 cm

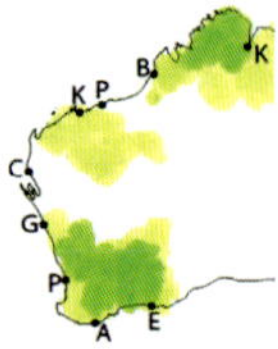

A large, long and broad-winged Kite with open slots on the primaries. It has two favoured hunting habitats in the South West, one over trees in open or dense woodland and the other over heathland. In both habitats, it will circle often only 10-15 metres above the trees or heath looking for live prey. Most of its prey consists of small birds, although it can tackle bigger birds. When hunting, it takes a high percentage of fledglings such as the young of Yellow-plumed Honeyeaters. It will take reptiles and small mammals if the chance arises but does not like carrion. When it hunts for prey above the tree canopy, it flies in tight concentric circles, tighter than a Little Eagle but will alternate the direction of the circle more than a Little Eagle. It will return to the same nesting site or area year after year. It migrates down from northern Australia in spring, arriving in mid to late August and nesting from September to December. Birds still are present in late January in the far south of the region such as the Torbay and Denmark region. In the drier areas like Dryandra Woodland Reserve, birds will often have left by mid-December. **Nest:** placed mostly on top of a large fork of branches either horizontal or sometimes vertical, normally 8-18 metres high and consists of large bowl structure, lined with fresh leaves. Has a preference for Wandoo trees or lower branches in Karri trees in the South West. Clutch size:

Photo: Simon Nevill

2 to 3. **Similar species:** Possible confusion with Little Eagle, which has a shorter and less-broad tail. Does not twist its tail as much as Square-tailed Kite and wings are not as long and certainly not as long 'fingers' on the primaries. The Little Eagle's body is stocky and thick set. Its wings are not as long and fingers on the primaries far shorter. Both have rufous colouring on parts of underwing but the Little Eagle's rufous plumage is confined to the leading edge of the wings. The Little Eagle's under body is pale in the light morph, the most common morph in the South West. The Square-tail has an all rufous under-body and has light patches on the middle of the primaries looking almost like faint bull's-eyes, though nowhere near as prominent as a Black-breasted Buzzard's. Little Eagle has stronger contrast between underwing patterns. **Status:** Uncommon in the South West and southern coastal regions. Scarce north of Perth. The Square-tailed Kite would have been in greater numbers in the south west before European settlement as one of its favoured hunting grounds is Wandoo (Eucalyptus wandoo) woodlands, which extended from the edge of the Jarrah Forest almost to the Vermin Fence before it was cleared for farming. **Where to find:** In the Perth Region: On the scarp of the Darling Range. In the South West Region: Dryandra Woodland NOR. Stirling Range NAP. Torbay. Fitzgerald River NP. In the Transitional Woodland anywhere from Vermin Fence to Norseman. Rarely seen on its migration north but always present in the Kimberley in the winter months.

Black-breasted Buzzard *Hamrostra melanosternon*

56-60 cm 145-155 cm

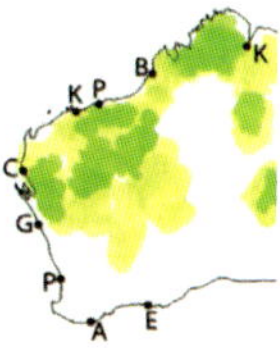

An unmistakable raptor found in the north and north east of the South West. Their diet is more varied than some raptors, possibly taking a higher percentage of reptiles than most species except the Pacific Baza. Takes many ground nesting birds. Will eat carrion as per photograph

Photo: Simon Nevill

taken near Gascoyne Junction. Favoured hunting areas are open country often close to their nesting ground, which is normally in River Gums along inland river systems. Easily identified in flight by its jizz i.e. very short tail, stocky body, distinct dihedral rocking flight more angular than Wedge-tailed Eagle's. White bull's-eyes in wings, both in the immature bird and adult bird. It is believed that some of Australia's raptors are of an ancient lineage, not easily placed in raptor systematics. At present, it is placed in its own genus. Superficially in flight it reminds one of the African Bateleur, with its rocking motion from side to side when

Similar species: In adult plumage there is no similar species. As immature birds, they can look similar to Wedge-tailed Eagles but the White-bellied Sea Eagle has a distinct white patch on the primaries and also the tail is rounded, not diamond-shaped as in the Wedge-tail. The Osprey has similar colouring but is far smaller and tail is slimmer. **Status:** Moderately common on islands to the west off the Geraldton coast and the Dirk Hartog region. Scarce on the Swan Coastal Plain. Uncommon on southern coast. Locally common from Shark Bay to Cape Keraudren and Broome all around the coast to Wyndham. **Where to find:** Abrolhos Islands. Dirk Hartog Island. Peel Inlet. Vasse Estuary. Broke Inlet. Wilson Inlet. Oyster Harbour. Fitzgerald coastline. Esperance Region including the Esperance freshwater lakes. Shark Bay. Dampier Peninsula. Common on the Pilbara Islands and moderately common from Onslow to Port Hedland. Broome. Moderately common along north-west Kimberley coastline and islands.

Little Eagle *Hieraaetus morphnoides* 45-55 cm span 110-135 cm

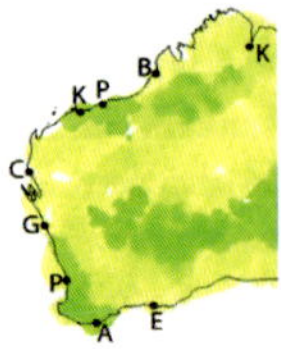

A small compact eagle with varying colour morphs throughout Australia, although in the South West, it is almost entirely the light phase. Has leg feathers on top of the feet like all true eagles. It is an attractive eagle with a small crest on the nape. Seen typically soaring over wetlands or open woodlands not far above the tree canopy at a similar height to the Square-tailed Kite. Flies in tight concentric circles more consistent and tighter than the Whistling Kite. Its habitat varies from open woodland to desert sand dunes. It has declined since the reduction of rabbit numbers in the South West and possibly never common before European settlement. Little Eagle takes more live prey than e.g. Whistling Kite, although they will eat carrion. One of 6 species in the world of the genus *Hieraaetus*. There are two subspecies in Australasia, *weiskei* in New Guinea which is smaller, and the nominate race *morphnoides* in Australia. **Nest:** A large structure of branches and twigs. In desert regions can be quite low but in the South West mostly above 12 metres, rarely in solitary trees, mostly in groups of trees. Clutch size: 1 to 2. **Similar species:** Some similarity to Whistling Kite which however has less contrasting colours on the underwing and body. Little Eagle has a richer orange-brown on the underwing covets from the body to the carpal joint of the wing and its tail is shorter and more square cut. Again jizz is important, even in flight, as the Little Eagle looks stocky and robust where as the Whistling Kite is slimmer and longer-tailed. **Status:** Moderately common throughout the much of WA except in the deep South West corner. Less common than the Whistling Kite. Prefers to be near wooded country with cleared country adjacent. Found more in the arid regions. The further north one travels the greater the chace of seeing Little Eagle. One of the few raptors to be seen in the remotest desert areas. **Where to find:** Throughout the whole of the state. Less so the Nullarbor.

Photo: Simon Nevill

Whistling Kite *Haliastur sphenurus* 50-60 cm span 120-140 cm

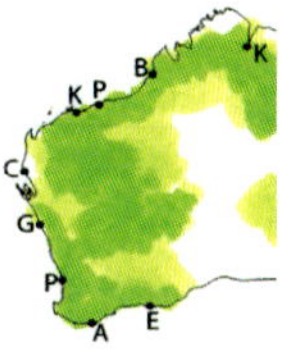

Found throughout the state. Has a preference for wooded country adjacent to wetlands in the South West. Scavenges more than Little Eagle, eating more carrion but does take live prey, particularly in the breeding season bringing fresh kills to the nest. Ducks and waterbirds are constantly harassed by the Whistling Kite. Does not flock to the same degree as Black Kites, although will mass with Black Kites in the northern regions after fires. **Nest:** A bulky twig nest often placed in a grove of trees near water, rarely in a forested region. Clutch size: 2 to 3. **Similar species:** Has similarities to the light morph of the Little Eagle, the main form found in the South West. Refer Little Eagle for details. Flies more erratically than the Little Eagle making more turns and manoeuvres while looking for prey. **Status:** Common on the Swan Coastal Plain, less so on the south coast. Uncommon throughout the Wheatbelt. More common the further north one travels. Common in the mid-west and Pilbara. Very common in the Kimberley. Uncommon in the sandy deserts. **Where to find:** In Perth it is quite common around the major wetlands and throughout the Swan Coastal Plain. Easily seen throughout most of the state.

Photo: Simon Nevill

Black Kite *Milvus migrans* subspecies *affinis* 45-55 cm span 120-140 cm

Immature

Photo: Simon Nevill

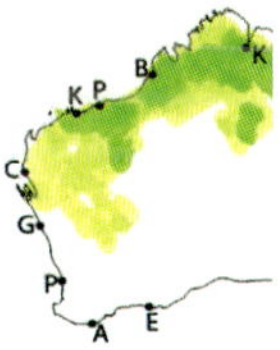

Found throughout Europe, Africa and Asia, being one of the most numerous raptors in the world. It is a common bird in the north of Australia. Not found in the south west, the nearest areas being in the mid west station country but Whistling Kite is still the dominant raptor there. As one travels to the Pilbara and then onto the Kimberley, the numbers increase. In the Kimberley they can congregate in hundreds over abattoirs, rubbish dumps or stockyards, as they are basically a scavenger and opportunist feeder. Will often be found in country that is still burning, its fork tail constantly twisting as it manoeuvres in flight to take insects and reptiles in front of the fire. During the breeding period, a higher percentage of live prey may be taken but at other times it relies heavily on carrion. There are 6 subspecies in the world with the Australian subspecies *affinis* being the smallest. **Nest:** A large stick structure located in groups of trees sometimes along watercourses. Nest is lined with fur or wool or other soft materials. Clutch size: 2 to 3. **Similar species:** Should not really be confused with any other raptor, as the tail is so distinct. The Whistling Kite is overall a lighter bird, noticeably so when perched; the Black Kite has an even, darker brown plumage, whereas the Whistling Kite has a light fawn head and under-body and upper-body. **Status:** Uncommon in the mid west. Locally common in the Pilbara, very common in the Kimberley, very uncommon in the desert regions. **Where to find:** Anywhere in the north, scarce down to Geraldton and Mullewa.

Brahminy Kite *Haliastur indus subspecies girrenera* 45-50 cm span 110-120 cm

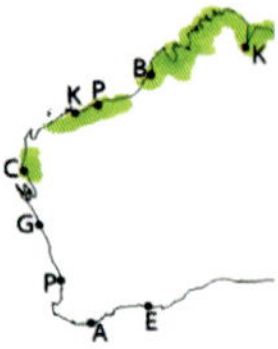

A bird of coastal waters, with a preference for mangrove-lined shores where it also nests. An attractive bird that feeds mostly on carrion. It does not have the powerful talons of the White-bellied Sea Eagle but can still take small live fish. Seen mostly flying low over mangroves or along beaches being an opportunistic hunter. There are four subspecies in the world and only one subspecies *girrenera* in Australia. **Nest:** A bulky structure lined with grasses, seaweed or leaves. Mostly nests in mangroves often only 4-5 metres above ground. Clutch size: 1 to 2. **Similar species:** None really, although the juvenile bird may be confused with Whistling Kite and Little Eagle; however, it is rare to see Little Eagles over mangroves and beaches though Whistling Kites are certainly commonly seen over beaches. At first glance the under-body of the Little Eagle has very similar markings and colour to the juvenile Brahminy Kite. The juvenile Little Eagle's markings are clearly defined rust on the shoulder breaks, with the greater covets pure white and tail clearly barred and squarer. The under-body on the Whistling Kite has plain brown markings no heavy contrast, brown to white. The main thing is the jizz, the Whistling Kite slim-bodied with longer slimmer tail and flies with drooped wings, the Brahminy has a stocky body and shorter tail. **Status:** uncommon around Carnarvon. Moderately common on the coastal mangroves of the Pilbara, more common around the mangroves of the Kimberley. **Where to find:** Around Carnarvon between Mangrove Point and Miaboolia Beach. Onslow. Dampier. Point Sampson. Port Hedland Harbour. Broome to Cape Leveque. All of the Kimberley coast where mangroves are numerous.

Photo: Simon Nevill

Collared Sparrowhawk *Accipiter cirrocephalus*

30-40 cm span 55-80 cm

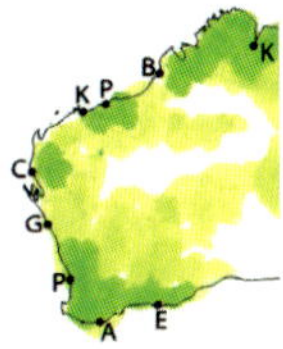

A slim, sleek small accipiter, extremely fast, having the ability to manoeuvre at high speeds between branches and vegetation in its pursuit to catch small birds. It can, however, tackle prey nearly twice its weight. It has a preference for well-wooded areas where it will perch and wait for prey to move. It may start hunting before the sun rises as do so many of the falcons. It knows it needs prey to survive, so it maximizes the day, taking birds that are just adjusting to the first morning light and may not be as alert as during the day. Similarly it will hunt at dusk when birds are moving to roosting sites and similarly may not be as alert as in full sunlight. The female is nearly twice the weight of the male bird, the male weighing on average 140 g and the female 235 g. If you are close to the female, you can see that it is bulkier in the chest than the male Sparrowhawk, however, it's harder to split from similar-sized male Brown Goshawk. Sparrowhawks in the breeding season can often be heard using their rapid high -pitched contact call, which will sometimes lead you to a dependent juvenile. There are 4 subspecies in Australasia and New Guinea with 2 occurring in Western Australia. The nominate species in southern Australia *cirrocephalus* is larger than the subspecies *quaesitandus* in northern Australia. **Nest:** A small flat twig structure lined with fresh leaves often well above 15 metres in fork near top of tree. Clutch size: 2 to 3. **Similar species:** Brown Goshawk. Jizz is so important here: the Sparrowhawk is certainly a slimmer bird, including the female, than the male and female Goshawk. The much discussed frown on the Goshawk is certainly evident but one has to have it well in the binoculars to see it clearly. You can pick up more quickly, the more wide-eyed look of the Sparrowhawk. The tail is square on the tip in the Sparrowhawk, more rounded in the Goshawk but when the Goshawk is in moult, it's not an easy thing to recognise. However in flight, even at speed the Sparrowhawk shows what appears to be a longer tail due to the fact that it is narrow and square cut in both sexes. Mention is often made of the collar on the Sparrowhawk as being more defined but the author finds this not a good diagnostic feature. What he does find, which is not often mentioned, is when perched the Sparrowhawks legs look decidedly thin and appear longer in proportion to the body than the Goshawk. The Sparrowhawk is less often seen soaring in flight than Brown Goshawk, which is one of the differences in hunting technique as the Goshawk is willing to tackle bigger prey in open country, whereas the Sparrowhawk normally relies on surprise in denser vegetation, often chasing smaller birds into thick foliage. **Status:** Moderately common throughout the state but as mentioned prefers to hunt in well-wooded areas, in the arid zone has a preference for river systems where water is present and birds have to come to drink. **Where to find:** Throughout the region being a chance encounter in any habitat zone even mangroves. Preference for riverine habitat and dense woodland.

Photo: Simon Nevill

Brown Goshawk *Accipiter fasciatus* 40-55 cm span 70-85 cm

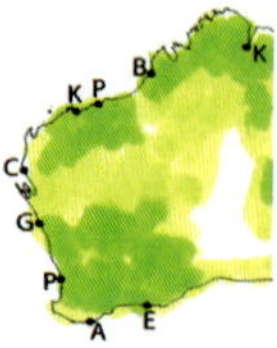

Accipiter is a large genus, there being 50 species throughout the world. Australia has 4, the Brown Goshawk, Collared Sparrowhawk, Grey Goshawk and Variable Goshawk. The rare Red Goshawk is a different genus. The Brown Goshawk has no hesitation in tackling prey close to its own weight such as Ring-necked Parrot. In fact the parrots are a favoured group of prey. If one approaches water holes particularly in the station country in the summer, there is often a resident Goshawk waiting for birds to drink, although Sparrowhawks do the same but the Goshawk will soon chase off the Sparrowhawk. When taking international birding groups around Australia, the author found that even after 16-20 days, many people could still not differentiate between the two species, which is understandable. So if one has trouble with these two species, take heart, you are not alone. There are at least 12 subspecies in Australasia, parts of Indonesia, New Guinea and some of the south eastern Pacific Islands. There are 2 in Australia all found within Western Australia, the nominate subspecies *fasciatus* in the South West, *didimus* in northern Australia. The subspecies *natalis* on Christmas Island still remains as a subspecies . **Nest:** Similar to Collared Sparrowhawk but slightly larger nest, clutch size 2 to 4. **Similar species:** Collared Sparrow Hawk. When the Brown Goshawk is soaring which it often does, the tail is often fanned out, and it is then you can clearly see the rounded end to the tail. When fanned, the Sparrowhawk shows clearly a clean cut square tail almost like a miniature Square-tailed Kite's tail. For more details refer to the Collared Sparrowhawk text for similarities and differences between the two species. **Status:** Relatively common throughout the whole state. **Where to find:** No specific locations. It can be seen in any area, often soaring much higher than Collared Sparrowhawk will do.

Photos: Simon Nevill

Christmas Island Goshawk (Variable Goshawk)

Accipiter natalis 35-37cm

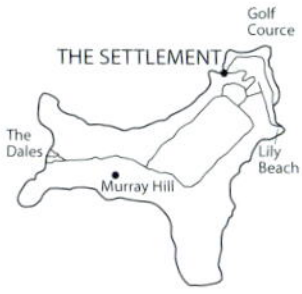

There is much debate regarding the taxonomic status of this species. This species is lumped with Brown Goshawk (*A.fasciatus*). More recent conjecture is that it is closely related to Varied Goshawk (*A. hiogaster griseogularis*) and I certainly agree with this assumption. In some ways I am surprised that BARC and the IOC have not recognized this species as distinct from Brown Goshawk. Based on my observations and knowing its morphology I have placed this in this publication as a separate species as per Johnstone 2004. I must state however that am certainly not qualified to make this assumption but simply concur with Johnstone's taxonomic classification at this stage until further studies prove conclusive. The Christmas Island Goshawk is a robust hawk with a rich rufous body forming a rufous collar with barring on lower body, not always well-defined. Head and wings are a dark blue-grey. Legs and feet are yellow with contrasting black talons. Orange iris. Black bill with contrasting yellow cere. Immatures are heavily barred as per photograph. Typical of so many island species, it is not afraid of humans and on the contrary, is highly inquisitive, which is amazing for a raptor. **Nest:** Nest large platform of twigs lined with leaves placed on fork, mostly above 10 m. Clutch presumed 2-3. **Similar species:** None on island. **Status:** Moderately common but there has been extensive clearing due to mining activities which has reduced their habitat. **Where to find:** Anywhere on the island as it has adapted to all habitats but best seen in the central rainforests. Very confiding.

Immature

Photo: Simon Nevill

Grey Goshawk *Accipiter novaehollandiae* 40-54 cm

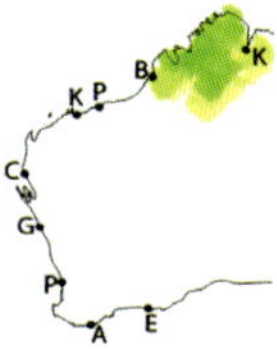

Most powerful species in the accipiter genus. Like all *accipiters,* the female is much larger and bulkier. Two morphs occur, grey and white. Both are beautiful in their own right but the white morph is so unusual in the raptor world that it often takes you by surprise when first seen. Both the grey and white morph have the jizz of a Brown Goshawk but are slightly bulkier birds. Grey morph has pale grey crown and back with darker grey wings and black wing tips. Yellow-orange legs and cere. The under-body is white with faint grey barring on the chest. The white morph is white all over, only the yellow legs and yellow cere and dark red eye remain as a contrast to the white body, truly a lovely raptor when seen well. Like many accipiters, females hunt more often on the edge of forests where the habitat is more open but the males, being lighter and more manoeuvrable in flight, will perch and hunt more within the forest proper. In WA the majority of Grey Goshawks are found in areas where rainforest gullies merge into the tall mangroves areas. Apparently the grey morph is more prevalent on the northwest region of the Kimberley and the white phase in the east Kimberley (Johnstone 2004). I have only seen the white morph on the Mitchell Plateau near Port Warrender and along the Prince Regent River and near Sale River, however I don't sail the off-shores anywhere near as much as Broome birdos so its not a common bird for me in the Kimberley. I have seen the species many, man-y times in Tasmania, central NSW and QLD giving the reader an indication of its scarcity in WA. **Nest:** Nest large platform of twigs lined with leaves placed on fork, mostly above 10 m. Can return to same nest site year after year. Clutch presumed 2-4. **Similar species:** Not really. The lighter-backed form of the Brown Goshawk subspecies *didimus*, found in northern Australia may possibly be confused from a rear view but dark head, chestnut-barred chest and chestnut collar should make it clear when seen properly. **Status:** Scarce in mangroves, monsoonal rainforest and heavy woodland. **Where to find:** Remote north west coast. Best to do boat trips then walk gullies, ideally with Kimberley bird guides. White morph has been seen around Kununurra area.

Grey form Immature

White form

Photos: Simon Nevill

Red Goshawk *Erythrotriorchis radiatus* 48-60 cm

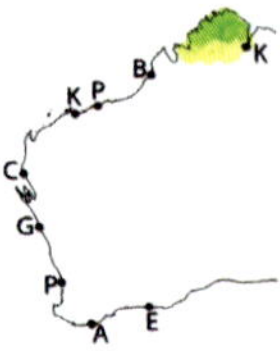

A powerful bird, it thinks nothing of taking a Red-tailed Black Cockatoo in flight, even Barking Owl, such is the power of this raptor. One of only two species in this genus, the other the Chestnut-shouldered Goshawk found in the mountains of PNG. People think of this bird as an accipiter as it has similar morphological features, but it's a far more bulky bird than all the Australian accipiters. The largest accipiter the Grey Goshawk (the female) weighs 680 g but the Red Goshawk weighs on average 1100 g, so it a heavily built bird weighing even more than Black and Peregrine Falcons who both like the Red Goshawk, are efficient hunters of birds. Seen from below, the adult male bird on the under body has rich rufous on the forewing, chest and vent, then dark barring on the outer primaries and tail with a broader sub-terminal band and moderately long fingers, pale cream legs, head pale with dark-brown flecking. Female is similar but only pale rufous on chest throat and vent white with dark-brown streaks. When perched, upright stance like accipiters. Powerful long legs with heavily long feathered trousers a distinct feature of this species. **Nest:** Nest large platform of twigs with finer twigs in nest cavity lined with leaves, placed mostly in a high position above 15 m. Starts breeding August young leaving Nov-Dec. Clutch 1-2. **Similar species:** Not really. In flight some similarities to Square-tailed Kite but Square-tailed has longer wings, more open fingers but smaller body. Flight slower and flies mostly in tighter concentric circles over canopy. Has whitish diffused bulls-eyes like Black-breasted Buzzard. **Status:** Rare, said to be approx. 700 mature pairs in Australia. **Where to find:** North of a line from Derby, Fitzroy Crossing to Cockburn Range. I have only seen this bird in the Kimberley on the Drysdale River. Ellenbrae Station (on tour). Kalumburu Road (on tour) and a few times over the years at Bachstein Gorge. The Kalumburu Road and Ellenbrae birds were just luck but the others I knew their habitat preference and localities but it normally took several hours to find them. The rivers to walk for better chances are the Hann. Charnley. Gibb. Drysdale. Calder and King Edward. A favoured habitat in the Kimberley is where the very big cadjeput occur along river systems. They require habitat with pools on rivers where bird life is prolific.

Photo: Adrian Boyle

Swamp Harrier *Circus approximans* 50-60 cm span 120-145 cm

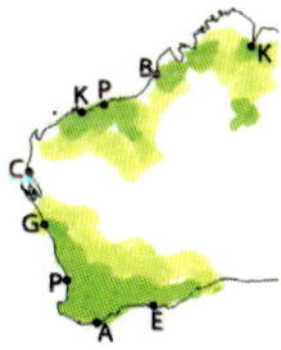

A long and broad-winged raptor having a relatively lightweight body in proportion to the wings which allows it to glide effortlessly over reed beds often only a few metres above the water. The facial disc on the Swamp Harrier is large and flattened and is thought to assist in its hearing capacity, particularly as it flies so low, so that anything that moves may be heard. Both Australian harriers have this feature. The Tasmanian Swamp Harriers hunt over forests, grasslands, lakes and estuary margins but the western birds are almost entirely confined to swamp habitats and lake margins although can be seen over fields but never too far from wetlands. Whether there is a migration from the southwest population to the northern region as in Tasmania is not well known. What is known is that birds say at Herdsman Lake can remain all year but locations like Lake Gregory however will get Swamp Harriers in good seasons but disperse when not, so they certainly can move great distances when conditions are favourable. **Nest:** A trampled reed bed platform located in dense extensive reed beds. Occasionally in tall grasses close to water. Clutch size 3 to 4. **Similar species:** Spotted Harrier, however they do not share the same territory. The grey back and barred tail stand out on the Spotted Harrier and it is a far more colourful raptor than the Swamp Harrier. The colour of the Swamp Harrier is an even brown colour on the whole body with its diagnostic white rump seen well in adult birds. The juvenile is darker, becoming slightly lighter with age. **Status:** Locally common on some wetlands on the Swan Coastal Plain. The southern coastal birds can be seen over heathland and fields adjacent to swamps. **Where to find:** In the

Perth area: Herdsman Lake. Bibra Lake. Lake Joondalup. Lake Kogalup. Forrestdale Lake. Thomsons Lake. On the Swan Coastal Plain and South West Region: Benger Swamp. Vasse Estuary. Torbay-Elleker. Two Peoples Bay. Cheyne Beach region. Esperance lakes. Cape Le Grand. Any wetland north that has sufficient reed beds but not so common as on the south coast.

Photo: Simon Nevill

Spotted Harrier *Circus assimilis* 50-62 cm span 120-150 cm

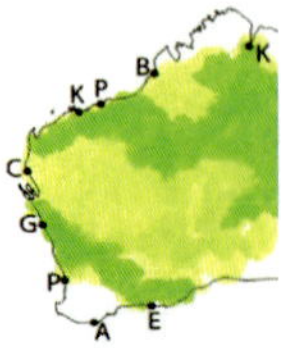

This has to be one of Australia's most beautiful raptors. Of the 13 harriers in the world, this is the only harrier in its genus that nests in trees. Its favoured habitat is wide open grassed plains, open bluebush plains and low spinifex plains, both in rocky country and sand dune country. However, since European settlement, it has adapted to feeding over cereal crops, the closest thing to native grassland habitat. It hunts just a few metres above the crops or grasses looking and listening for mice, small marsupials and small birds such as quail, songlarks or button-quail. It will take reptiles but rarely carrion. After

Photos: Simon Nevill

breeding there is most probably after breeding, some movement from the South West to the northern parts of Australia as is the case in the eastern states. Sightings are certainly fewer from late summer to mid winter. **Nest:** The nest is located mostly in a group of trees rarely a solitary tree and mostly above 10 metres on a horizontal fork. A flat saucer structure lined with fresh leaves. Clutch size mostly 3 but varies from 2 to 4. **Similar species:** None really. Mention has been made of the juvenile Swamp Harrier but the chances of overlap in Swamp Harrier's territory would be uncommon in the south west. **Status:** Uncommon in the south west, particularly the deeper forested region of the south West. Uncommon but seen frequently in the Wheatbelt. Locally common in the grasslands of the Pilbara and desert areas. **Where to find:** In the south west, the best opportunities are in the Wheatbelt region from August through to December. Rare on the lower Swan Coastal Plain. Uncommon to moderately common from Moora to Geraldton. Then uncommon throughout the rest of the region. Note: Uncommon in the drier regions does not mean you will not see them, it's just that they are few and far between as they require big territories of open grasslands and spinifex covered hills to feed from. Best chances are on the grass plains of the Pilbara and also throughout the desert areas where grassland and spinifex plains occur.

Brown Falcon *Falco berigora berigora* 40-50 cm span 90-120 cm

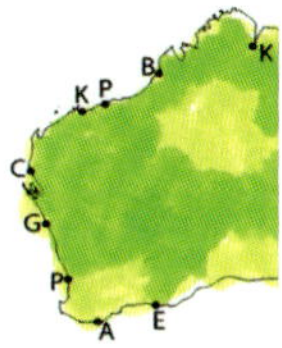

An extremely common raptor, found throughout the state with a range of colour morphs from very light brown birds to blackish-brown, almost as dark as a Black Falcon; most however are just a fawn-brown. It is an unusual falcon, not related to any of the 60 world falcons except possibly the New Zealand Falcon, *Falco novaeseelandiae.* Most falcons have trouble walking on the ground as the legs and talons are designed to take prey in the air. The Brown Falcon however has long legs and short talons making it easier to walk and also grasp smaller prey such as large insects and reptiles, particularly snakes. It is one of the raptors that will sit for long periods waiting for prey to move, although most of the time it will glide and hover, pouncing on small prey. It does not develop the fast speeds in flight or the power of other Australian falcons and lacks the robust talons to take large prey, so has become more opportunistic in its hunting, taking anything small that moves, even taking the opportunity of feeding on grasshoppers when they are in plague proportions. It can, however, take parrots, pigeons and other medium-sized birds but this is not a common occurrence as it lacks the speed. There are 2 subspecies in Australasia one only in Australia *berigora.* **Nest:** Normally another raptors or an old corvid's nest. Found at varying heights but rarely placed in solitary trees in wide open spaces as does the Black Falcon's. Will re-line the nest with leaves. Clutch size: 2 to 3. **Similar species:** The lighter tanned morph of the Brown Falcon can sometimes be confused with the Kestrel but the Kestrel is smaller, has a black band on the tail in both the female and male, and when flying, the Kestrel's wings are sharper at the end and not as broad as the Brown Falcon's. There are also similarities with the

Photo: Simon Nevill

more powerful Black Falcon which, when perched, has an unmistakable jizz once you have become familiar with it. The first thing that you become aware of is its powerful shoulders with a relatively smaller head in proportion to the body than the Brown Falcon. Another feature is that when at rest, the Black Falcon will perch with an upright vertical posture, but in a hunting mode when perched it will be the opposite of a Brown Falcon standing more horizontally and the longer wings and tail are clearly visible. In flight the jizz again shows a more powerful flyer with longer pointed wings and all black-brown body. **Status:** Very common throughout the state. **Where to find:** Throughout the state.

Black Falcon *Falco subniger* 45-55 cm span 95-115 cm

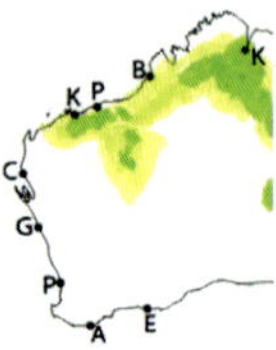

An impressive falcon and always a joy to watch in flight. One of the few raptors that can cruise in flight fairly slowly but quickly gain height by soaring and then if it sees prey, can stoop and fly faster than most raptors. In horizontal flight when moving at high speed, its wings droop only slightly with occasional rapid wing beats. Most of the time it soars high in the sky looking for prey, being an opportunistic feeder. When fires are sighted it will move in from afar to take advantage of mammals or reptiles leaving the fire, as will the Brown Falcon. The female is the largest of all the falcons in Australia. When perched seen front on, it is almost unmistakable with its powerful square shoulders and the head looking smaller than Brown Falcon in proportion to the body. One thing to notice is the tarsi (legs), are mostly covered by feathers whereas the Brown Falcon shows slightly longer exposed tarsi. **Nest:** Like the Brown Falcon it will not build a new nest but uses other old raptor or corvid nests. It will return to the same nest in subsequent years. Unlike the Brown Falcon, it will nest in solitary trees, often on wide open stony plains or occasionally along watercourses. Clutch size normally 3 but can lay 2-4. **Similar species:** Refer to Brown Falcon. **Status:** Extremely rare in the south west region and most recorded sightings are misidentified birds, mostly Brown Falcons. Scarce in the Pilbara, Southern Kimberley, Northern Canning Stock Route to Sturt Creek. The furthest South West I have seen this bird, was at Lake Anneen south of Meekatharra. Other locations seen by myself in WA were all in the Kimberley namely: Camballin Station, Kingston Rest, Roebuck Plains, Parry Lagoon and Lake Gregory. I have seen far more in the Queensland and South Australian deserts and ironically this is not a rare bird in Victoria and southern NSW coming as far south as Werribee, something I have

Photo: Simon Nevill

never grasped why there and not more prevalent in our deserts let agricultural habit as in NSW. **Where to find:** Scarce in WA, mostly over the grass plains of the Kimberley, Pilbara floodplains and northern deserts. Also surprisingly one of our rarest raptors the Grey Falcon is seen more often than Black Falcon in the Pilbara.

Nankeen Kestrel *Falco cenchroides cenchroides* 30-35 cm span 60-80 cm

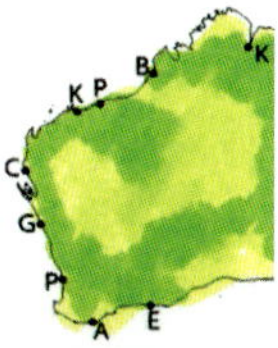

Perhaps more common in the south west than the Brown Falcon, the Nankeen Kestrel is the smallest Australian falcon and seen in many different habitats. It has adapted well to the environmental pressures of European settlement. Seen close to the Perth city centre and throughout the entire farming region. With the clearing of land and the influx of the European mouse, it has never been short of prey. Seen hovering over beach dunes, roadside verges, fields and native grasslands. Has a diverse diet, including mice, small birds and large insects. 2 subspecies occur in Australasia, one in Indonesia and one, *cenchroides* in Australia. Nests: Nests in a variety of locations, including trees, rock ledges as on Rottnest Island, also in hollows in trees. Will also use old corvid nests. Clutch size 3-4. **Similar species:** Refer Brown Falcon. Status. Very common throughout the state. **Where to find:** Throughout the state.

Photo: Simon Nevill

Australian Hobby *Falco longipennis longipennis* 30-36 span 67-90

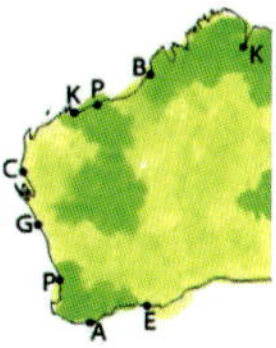

Even though the Australian Hobby is less than half the weight of a Peregrine Falcon, it still can attain extremely fast speeds when chasing prey. However, unlike the peregrine which will take on prey its own weight and more without hesitation, the hobby mostly chases much smaller birds and although it has taken birds up to a third more its weight, the vast majority are less than half its weight. It regularly takes large insects in flight as other hobbies do throughout the world. The author has always been intrigued with which bird species will fear, which type of raptor. Watching a Pied Butcherbird in the same dead tree whilst singing its beautiful early morning song with an Australian Hobby preening no more than 3 metres away, makes one realise that every species, knows what birds to fear; it is doubtful whether the butcherbird would do the same alongside a peregrine. Much of the hobby's hunting is around trees, catching prey unawares, and less likely to soar and dive like the peregrine. Prefers open wooded country like Wandoo, Salmon Gum, York Gum and mulga country, and River Red Gums along inland river systems. Less likely to be in heavily forested areas of the deep South West. There are 3 subspecies in the world with two in Australia, *longipennis* in Tasmania and coastal Australia and *murchisonianus* in inland Australia. **Nest:** Does not build a nest but uses old raptor or corvid nests placed high in a tree. Clutch size mostly 3 but varies between 2 to 3. **Similar species:** Only the Peregrine Falcon. When perched, it is far easier to differentiate than when in high soaring flight. The peregrine is not only bigger, nearly 3 times the weight, but its jizz is a bulkier and a thicker set bird. The peregrine has a white collar passing from the white throat to the centre of the neck. The hobby's collar not only goes to the centre of the neck but also curves around to almost the height of the eye. Hobbies always have a rust-coloured chest whether adult or immature. The adult male peregrine does have a slight rusty tinge on the under-body but not as strongly and the female peregrine has no rust tingeing at all. The cere is clearly yellow on the peregrine and white on the hobby. When soaring, it's not always easy to tell as height makes size deceptive but generally the hobby has slimmer wings and a slimmer body. The peregrine shows its white and dark brown barring much more in flight. In general, the hobby has sharper wing tips and it beats its wings faster. The tail is squarer cut. **Status:** Uncommon to common throughout the state. **Where to find:** Anywhere on Swan Coastal Plain particularly woodland adjacent to wetlands. Wandoo woodland particularly near the Stirling Range. Throughout the Wheatbelt and station country particularly along river courses in arid country. Found throughout the Kimberley but not common.

Photo: Simon Nevill

Peregrine Falcon *Falco peregrinus subspecies macropus* 35-50 cm span 80-105 cm

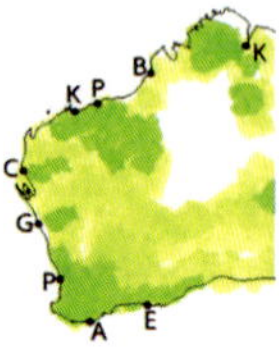

So much has been written of this master of flight. They can be found in the cold Tundra perhaps hunting for Dunlin, Curlew Sandpiper or Ruff, in North America maybe a Ring-necked Duck, maybe Torrent Ducks in the high Andes, Teal in Europe, Sandgrouse in Africa, White-bellied Pigeon in Japan. The Peregrine Falcon is truly a cosmopolitan hunter throughout the world. Here in Western Australia, it has in the past been listed as endangered, but in fact, many people miss or do not see peregrines regularly as they are such efficient hunters. They have often finished taking their prey in the first half hour of light. I have seen them at pre-dawn a peregrine leave its roosting site on a cliff face and come back regularly with a Spinifex Pigeon before the sun has even risen. It will then slowly eat the prey at leisure and may not fly all day until hungry again. So they are often around perched on a cliff face somewhere but may not always be seen. Certainly, they are not common but there are countless locations in the south west and throughout Western Australia where peregrines have their eyries, particularly in the goldfields. The peregrine's diet is almost totally restricted to birds. They will take small mammals but birds are their preferred kill. The majority of birds are taken in flight. There are many ways it will do this, either waiting on a perch and then chasing prey but more often stooping from great heights and taking birds on impact. The most spectacular kill I ever witnessed was at Trephina Gorge in the Northern Territory while looking at a flock of Galahs. There suddenly appeared high in the sky something falling at high speed. It was a peregrine in a full stoop flight at an incredible speed. The wings were closed and it had gained so much momentum it passed through the centre of the flock and was able to turn and just take a Galah trying desperately to gain height and fly off. To watch the technique and timing was absolutely astonishing. There are between 16 and 22 recognized subspecies in the world, varying with different authorities. The peregrines of the south west were once recognised as a separate subspecies *submelanogenys* but they have now been grouped back with *macropus*. **Nest:** The majority of south west peregrines nest in trees, not the chosen location as for most inland and northern birds. They will also nest in trees in the Wheatbelt and in other regions if food is plentiful but cliff sites are not readily available. Clutch size 2 to 4 but mostly 3. Established pairs will remain and nest in the same territory year after year. **Similar species:** Australian Hobby. Refer to the section similar species under Peregrine Falcon. **Status:** Uncommon. **Where to find:** Fitzgerald River NP. Stirling Range. NP (several pairs there). Porongurup NP. Kondinin. Peak Charles. There are many more locations north of Perth. Prefer not to be too specific as egg collecting still prevails with some of our rarer raptors, particularly falcons. Found on the rock face of many open cut mines. Many high rock faces throughout the state may have Peregrine nests where their young are secure from predation.

Photo: Simon Nevill

Grey Falcon *Falco hypoleucos* 30-45 cm span 85-95 cm

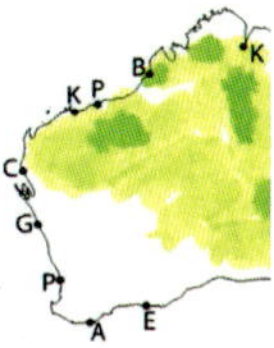

The rarest falcon in Australia and compares with the Red Goshawk in terms of being one of Australia's rarest raptors. For years I just could not find this bird. He had seen Red Goshawk several times before I saw this magnificent falcon. The first sighting was in its stronghold territory of the Birdsville and Strzelecki Tracks and on several repeated visits there but it was several years before he managed to find them in WA on the northern Canning Stock Route, in fact three of them performing aerial displays, then two days later another single bird and since that year in several other localities but never in the south west, although it has been seen by others and even sound recorded on the Murchison River and it certainly is possible to see it in the Gascoyne region and catchment area of the Murchison River. The bird is smaller than a Peregrine but not much smaller. Its flight is certainly similar. It will soar high and then glide over open country taking its prey on the wing. Even against bright sunlight, the pale body is still recognisable with slightly longer and broader wings than the Peregrine, which is surprising as it is over one third less the weight of the peregrine. Its feet are bright yellow with long feathered socks covering the legs. Has a bright yellow cere. Mostly feeds on birds. **Nest:** Normally placed in an old raptor or corvid nest located along an inland watercourse. Clutch size is 2 to 4. **Similar species:** None. **Status:** Rare in Western Australia. **Where to find:** Mostly seen in the Kimberley. Pilbara and desert regions in Western Australia. Roebuck Plains. On the road to Willare Bridge from Broome. The Fortecue floodplains. Marble Bar region. The northern Canning Stock Route and Breaden Hills. Sturt Creek. Lake Gregory. Carrowine Gorge. Perhaps seen more often in the eastern Pilbara than any area of WA as they breed here.

Photo: Simon Nevill

Pacific Baza *Aviceda subcristata* 35-45 cm

Photo: Simon Nevill

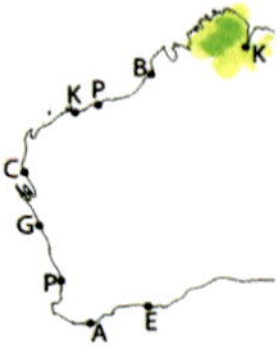

An attractive raptor and the only species of this genus in Australia. Unmistakable plumage. Grey head and neck. Dark grey erect crest on rear of crown, very diagnostic. Darker grey face. Brown back with grey wings on upper-body. Heavily barred brown and white chest. Rufous vent. In flight underwing unmistakable, all white with heavy black barring on primaries. Long open fingers. Chestnut coverts. Barred chest shows well against chestnut vent. Tail white with black barring and wider black terminal band. Feed mostly on large insects, frogs and reptiles including snakes, particularly tree snakes. They will take small birds when they can but do not have the manoeuvrability in flight as accipiters do. Their main technique is to fly in tight circles only a few metres above the canopy and hold wings back diving into the canopy to secure small prey. In WA, this is a bird of the Kimberley only. I have travelled to the Kimberley many, many times but I have admit I have only seen this raptor a few times in this region, whereas when I would travel to the Atherton Tablelands at least twice a year, we normally would see at least one bird on a trip and I would say the Kimberley population density is lower, like the NT, compared with the wetter regions of QLD. In the Kimberley it's most often encountered along river systems with tall trees, like the Hann, Durack or the Drysdale River. **Nest:** Neat smallish platform of small stick located in tall trees, nest lined with fresh leaves. Breeds before the onset of the wet so chicks develop just before the wet season commences.. Clutch said to be 2-3. **Similar species:** None. **Status:** Uncommon to scarce in the Kimberley but widespread range north of the Great Northern Highway. **Where to find:** You come across this species mostly in flight as they often remain hidden under the canopy, perched motionless high up in trees. Best is when you see Baza's flying in tight circles above the tree tops with its strong contrasting black bars on the white wing tips. Also barring on the chest and heavy black terminal band on the tail. Be careful not to confuse in flight with the rare Red Goshawk which shares similar habitat. The Red Goshawk is far larger, different jizz in flight but barring on wings is far less noticeable and has the unmistakable rufous under-body and underwing coverts. Brown Goshawk barring far less contrasting. Funny enough I have found both these species on the same river the Drysdale. River systems to look along Isdell. Charnley. Carson Durack and Chamberlain Rivers.

Rufous Owl *Ninox Rufa* subspecies *rufa* 46-57 cm

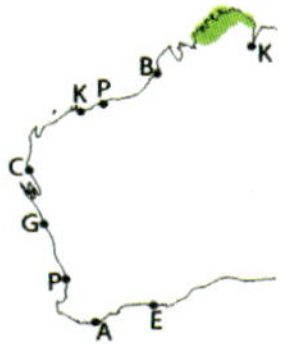

A large rufous coloured owl. In wetter times this beautiful owl's distribution range would have covered all of the Kimberley region but now the extensive rainforests are gone and it remains in the remnant rainforests and deciduous vine thickets that are found on the north west part of the Kimberley where the rainfall is highest. There are 3 subspecies in Australia with *rufa* found in the Kimberley and NT. Rufous Owl has a rounded head with dark brown eye spectacles, bright yellow eyes with grey orbital ring, cream coloured under-body with heavily marked, narrow horizontal rufous bands, dark brown back, short tail with wider grey-brown bars, powerful pale cream legs. Pairs alike with plumage calls differ. Rufous Owl pair for life and hold onto territory strongly, if you find one bird the other may not be too far away. It will start hunting at dusk as normally under closed forest woodland. The *Ninox* genus lacks the high audio sensitivity that species like the Masked Owl genus *Tyro* have, so they are very reliant on sight although obviously still hear well. Their prey is varied but there is no shortage in the north-west Kimberley, as the marsupial species have not been so ravaged by introduced predators as much as southern marsupial have. The remote sandstone boulder country with its deep dissected gorges is home to small marsupials like the Narbelek, Monjon and the Golden-backed Tree Rat, There are few fruit bats and flying foxes such as Least Blossom Bat, Black Flying-Fox and Little Red Flying-Fox so

Photo: Simon Nevill

there is no shortage of prey. Rufous Owl can tackle large prey as well noted by David Hollands and the intrepid John Young whilst filming in north QLD they both witnessed a male killing a roosting Jungle Fowl. However some studies have shown that birds do not make up a large percentage of the kills b*ut* mostly its arboreal mammals and then to a lesser degree, terrestrial mammals, then bats and birds last. Diet however changes with the seasons and breeding needs also location must influence diet. For birds close to vine forests near mangroves, bats may be the prime prey. Most of the hunting is said to be not under the canopy but in the adjacent surrounding open areas. Its call is similar in tone to Powerful Owl but softer a repeated woo –hoo. **Nest:** Hollows in large gums that line creek. Clutch size large normally 2 rarely 1. **Similar species:** Could be confused with Barking Owl subsp. *peninsularis* that does not have horizontal fine bands but heavy dark chocolate brown vertical streaks, white spectacles with grey margins not black-brown spectacles and is a smaller owl. **Status:** Scarce although maybe not as scarce as we think as there are in that remote north-west, many rainforest pockets with a healthy nocturnal life but not easily accessed. **Where to find:** Best taking a boat trip with experienced bird guides who know the west Kimberley coast region and can take you to areas like Calder River, Sale River and Prince Regent River, these are typical locations and walk the gullies where the rainforest occurs. If not, then you may have luck on the Mitchell Plateau. I have found them in lush vine forests above the Calder River.

Barking Owl *Ninox connivens* subspecies *connivens* 39-44 cm

Photo: Simon Nevill

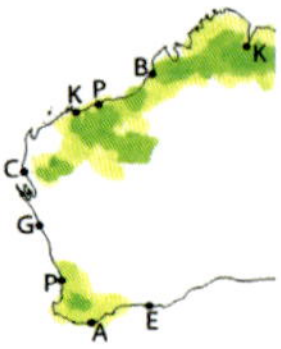

There are 4 subspecies of Barking Owl in the world with two being found in Australia, both occur WA. The south-west birds (subspecies connivens) are found in the deep south-west region and are very scarce. The Subspecies peninsularis links with the Kimberley birds. Barking Owl is a bigger bird than the common Boobook Owl and can tackle much larger prey. Not a lot is known about the south west birds, as their numbers are few. We know a major part of their diet is small marsupials so need to live where marsupials are fairly abundant and that restricts their location in the south west. Their call is very distinct reminiscent of a barking dog, a rapid 'woof woof.....woof woof......woof woof', hence the name Barking Owl. If you get a duelling pair calling, it can last for at least 15 minutes, rising in pitch and speed as they call. It is by far the most vocal of the owls in the Kimberley. They also will start hunting earlier than most owls, often just after sunset when reasonable light still available. **Similar species:** Boobook Owl, which is similar in shape but smaller and the head looks larger in proportion to the body. The Boobook subspecies *ocellata* is lighter brown and the Barking is more grey-brown. The Boobook has dark patches around the eye like eye patches whereas the Barking has a light whitish ring around eye. The barring is heavier and bolder on the under body of the Barking. It is not easy identifying owls by colour alone at night and the best is the jizz of the bird that gives them away. **Nest:** in hollow from 5-20 metres high. Clutch size 2 to 3. **Status:** Common in parts the Kimberley and to a lesser degree the Pilbara. Scarce to rare in south west. **Where to find:** Has been seen in the Busselton region, Augusta region and Esperance region. Also in the Katanning to Cranbrook area. Common in the Pilbara. Found throughout the Kimberley, mainly in woodlands adjacent to wetlands and river and creek systems, particularly along the Fitzroy River. Interestingly, talking to Ron Johnstone of the WA museum he mentioned that they occur in the paperbark woodlands behind 80 mile beach at the back of Samphire Flats a long way from Pilbara birds and Kimberley birds.

Southern Boobook *Ninox novaeseelandiae (Ninox boobook. IOC.)*

subspecies *ocellata* 30-35 cm

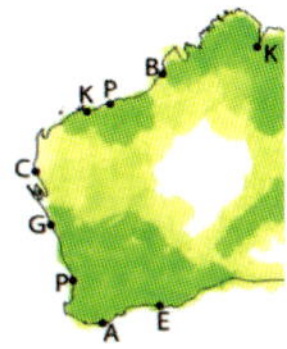

A very variable owl in terms of plumage colour. It is Australia's smallest owl and by far the most common, covering most regions in this state and throughout the south west. Found in thick forests, open woodland, river systems and heath with woodland. In the south west mainly sedentary but some movement north. Feeds on reptiles and invertebrates but mostly small mammals and birds. It is a stocky owl with a large head for its size and rounded wings and short tail. Birds in the deep south west tend to be darker than those in the drier regions. The well known 'boobook' call is its main contact and territory call, which it uses mostly in the first hour of darkness. It has a variety of other calls one of which is an extended 'brrrrrrrrrr' call, which it uses in defence of its territory. Will take advantage of floodlights and remain in that area, then swoop on anything that passes on the ground. **Nest:** In hollows from 5-15 metres high. Clutch size 2 to 3. **Similar species:** Barking Owl which is larger and with a smaller head in proportion to body; refer Barking Owl. **Status:** Common throughout most of the south west. Resident in Perth and many other towns. **Where to find:** Any extensive woodland in the Wheatbelt. The Jarrah Forest and the Karri and Tingle Forests of the South West. Found in the Mulga, Pilbara, Deserts and Kimberley regions of WA.

Photo: Simon Nevill

Australian Masked Owl *Tyto novaehollandiae* subspecies *novaehollandiae* 33-41 cm

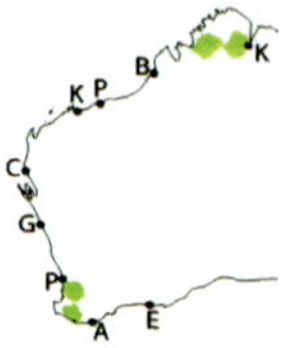

There are 8 subspecies in Australasia with 4 on the mainland of Australia. The south western subspecies *novaehollandiae* may one day become a separate species from the eastern states birds as it is larger bird and also has longer wings and thicker tarsus. It is the largest owl found in the south west, restricted to the thicker humid forests of the South West Region, particularly in the Pemberton and Manjimup area and along the Murray River in the Lane Poole area. The subspecies *kimberli* is found in the remoter parts of the north west Kimberley and is often overlooked as people miss-identify thinking they are Barn Owls. They need marsupials and birds to feed on. In the south west one of their favoured prey is the Ring-tailed Possum and their heavier tarsus and feet are well able to tackle bigger prey than their close relative the Barn Owl. The declining population of many of the south west mammals may have greatly affected the Masked Owl's population. In Tasmania where mammals are far more prolific the population of Masked Owl is greater and more stable. They roost mostly in large hollowed trees, coming to the entry while there is still light. The owl's call or one should say screech is stronger than the Barn Owl's with a richer quality and more extended. The Barn Owl is a more rapid, sharper and a shorter screech. **Similar species:** Barn Owl, which is smaller, sits more upright. Facial disc is more heart-shape, coming to a point at base. Tarsus much thinner and slightly longer in proportion and feet smaller. Crown has lighter coloured flecking not contrasting so much with white facial disc. Flecking on under body, smaller dots and not so dark. Female Barn Owl is darker on the back than the male, so is closer in colour to the Masked Owl. The Masked Owl has a distinct hunched rounded shoulder look as if leaning forward. The facial disc is more circular even though it comes to a point at the base. The back is darker even in the lighter phased birds. The crown has darker flecking that makes the white facial disc stand out. The legs are decidedly thicker, you can imagine this bird tackling Ring-tailed Possums whereas the Barn Owl's tarsus only looks fit for what it does most, tackling small rodents. **Nest:** In large Marri, Jarrah and Karri trees. Clutch size 2 to 3. **Status:** Uncommon in the lower South West Jarrah Forests. Masked owls are never common and with the demise of many of our larger marsupials, particularly the Ring-tailed Possum, its population may be on the decline. **Where to find:** Pemberton, Manjimup area and Lane Poole area. Rarely sighted seen in the Kimberley and may be well be overlooked but still a very scarce bird there.

Photo: Simon Nevill

Eastern Grass Owl *Tyto longimembris* 30-38 cm

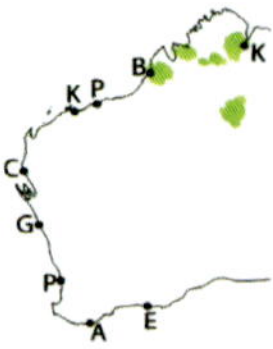

The body length of Grass Owl falls between the size of a Barn Owl (smaller) and Masked Owl (bigger) but it has the longest legs of all the large masked owls. Rarely seen due to the specialized habitat requirements of long grassed areas (often rank grass) with high rodent numbers. Grass Owls are nomadic less so in the wet tropics where higher density of prey occur and rainfall generally is more consistent. In the Kimberley conditions tend to vary more and when rodent populations crash, Grass Owls are then forced to seek new feeding areas. Prime habitat is dry and rank tall grass adjacent to wetlands. They will call at dusk and early morning but there are reports of feeding into mid morning, presumably when prey is scarce. Out of the breeding they are known to occasionally feed with other Grass Owls and on the east coast up to 11 birds have been sighted over one grassed area. Judging from the paucity of records in the Kimberley the populations density is obviously low, but numbers will vary as they will fly long distances when conditions are favourable, hence the many sightings in the Channel Country and Lake Eyre Basin where desert rodents populations spike into large numbers after good rains in those arid areas. When nesting the female stays with the dependent young for at least 4 weeks and the male brings the food to a cleared feeding area and then the female will take food through a short tunnel between the grasses to feed the young (per Hollands/Young). Unfortunately Grass Owls are often overlooked being mistaken for Barn Owl's. **Nest:** Interestingly and contrary to early presumptions, Hollands and Young found that Grass Owls did not flatten grass to form a nesting platform but actually cut stems and gathered them together forming the nest platform. Clutch size 3-8 in good seasons. **Similar species:** Barn Owl. Eastern Grass Owl has darker rich brown back with darker brown speckling with small white spotting on back and wings . Facial disc longer (slightly), more heart shaped than oval, importantly, crown is a very dark brown and also longer legs, under-body similar so important to check back and facial disc and contrast of dark crown with white face. **Status:** Generally scarce in the Kimberley but small populations do occur when conditions are favourable but would presume population less than NT and definitely far less than the east coast. Masked Owl also has darker back and crown but refer under **Where to find:** Sightings have been made on the Roebuck Plains, Kingston Rest and the Ord River Flood plain.

Photo: Simon Nevill

Eastern Barn Owl *Tyto alba* subspecies *delicatula* 29-38 cm

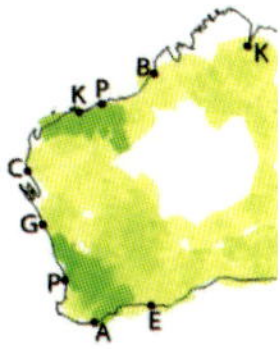

An owl that is found in every continent of the world, except the Antarctic, there being about 31 subspecies in total. After the Boobook Owl, the Barn Owl would be the second most common owl in the WA. Smaller than the scarce Masked Owl, it feeds almost entirely on rodents particularly the introduced European House Mouse (*Mus musculus*). With land clearing and the introduction of vermin like the House Mouse, the population of Barn Owls in the Wheatbelt Region has increased. Some small reserves like Dongolocking Nature Reserve have high numbers of Barn Owl and there would be more if it were not for the lack of suitable nesting sites. In the Pilbara and Kimberley found in eucalypt woodland close to open grasslands. The Barn Owl is a tall slim owl, rather pale particularly the male which is always lighter. Its preferred habitat is open grasslands but it requires woodland to both nest and roost in. In the drier arid regions outside the Wheatbelt it nests along creek and river systems, particularly in Red River Gums, although has been recorded nesting in caves. It is a silent bird when hunting, just gliding over crops or grasslands when it hears movement it hovers and drops to take prey. Calls just after sunset for a good 20 minutes in the breeding season.

Photo: Simon Nevill

Its hissing screech shorter and more rapid than the bigger Masked Owl. **Nest:** Is often located in Wandoo trees but also most large eucalypts throughout the south west. Found in River Red Gums. Clutch size 3 to 4. **Similar species:** Masked Owl has distinct dark orbital ring around face, refer also to notes on Masked Owl. **Status:** Moderately common throughout most of the south west. Scarce in thick Jarrah Forest. Dependent on mouse populations will move from areas when mouse populations collapse. **Where to find:** Darling Range on the edge of farming country. Most Wheatbelt reserves like Dryandra Woodland NR. Any eucalypt woodland alongside large rivers that run through grassland areas and Spinifex areas in the mid-west, the Pilbara and the Kimberley.

Christmas Boobook *Ninox natalis* 26-29 cm — Endemic

A small hawk-owl confined to Christmas Island. Orange-brown body, under-body cream with close, fine horizontal barring, short wider barred tail, bright yellow eyes, sexes similar, female slightly larger, call similar to Boobook's boo-book call with the rapid but not far carrying por-por-por-por uttered in breeding season. Formerly range covered the entire island but with development it mainly occurs in primary forest however it is seen in the main settlement but not common there. Most of its prey is large insects but some small invertebrates taken, hunts both in the open and under the canopy. **Nest:** Hollows in large *Syzgium nervosum*. Clutch size not known; Normally would be max 2, typical of *Ninox*. **Similar species:** None on Island. **Status:** Population stable but classed as vulnerable (Garnett 1993). Estimated population said to be 562 +or – 105 pairs (Hill & Lill 1998). **Where to find:** Throughout Island but the golf course is a good place to try, don't worry about the giant robber crabs, just walk around them. I dipped in my brief stay on Christmas Island looking for this bird as the winds at night were far too strong, but they are not rare so don't give up.

Photo: Ian Montgomery

Tawny Frogmouth *Podargus strigoides* 35-40 cm

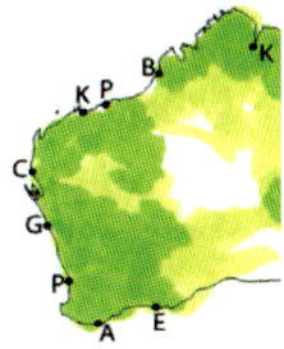

There are 4 subspecies in Australia with two in WA, namely *brachypterus* in the south and mid west and *phalaenoides* in the Kimberley. Unlike the other 3, the one found in the south west, *brachypterus*, is the only subspecies that is always a grey morph; the other subspecies are variable from grey right through to chestnut-brown plumage. The south west birds are larger than the eastern states birds. Unlike many bird species, the male is a larger heavier bird with a wider bill. Frogmouths have exceptionally wide bills adapted to catching moths and other insects. They often roost in sheoak woodland that has dense covering during the day. When approached both day and night, they will develop the vertical stick posture with bill held high, looking like a broken off trunk. If you walk around them slowly you will notice the head moving with you almost rotating 180 degrees. Unlike the owls and Owlet-nightjar, the frogmouths build a stick nest. The plumage, although overall fawn-grey, is exquisitely marked with mottled coverts to the wings and back and black streaking on the under body. If you find a bird during the day; in the breeding season, there is a good chance it is a female as the male is the one that mainly incubates the eggs during the day, at night both sexes take turns, as that obviously is their feeding period. Tawny Frogmouths are found throughout the state in every habitat. They not only feed on large moths and other insects but also will take small mammals, pounding them to a pulp before digesting. Their call is a distinct deep resonant call as if coming from the depths of the throat, slow 'ummm….ummm….. ummm', repeated rapidly for at least a minute then repeated again after a short break. **Nest:** A loose collection of short twigs occasionally with a few fresh leaves on the top, placed on a horizontal branch on a vertical fork often in sheoak trees. Clutch size 2 to 3. **Similar species:** The two other frogmouth species but not found in this region. **Status:** Common throughout the region and in some areas, densities possibly higher than many people realise. **Where to find:** Found in a variety of habitat but if sheoak woodland is nearby, your chances will increase in finding them. Spotlighting in vehicles will normally find them in the right type of woodland like Wandoo, either seen gliding or perched. Common in all Wheatbelt reserves. Very common throughout the Kimberley.

Photo: Simon Nevill

Spotted Nightjar *Eurostopodus argus* 27-35 cm

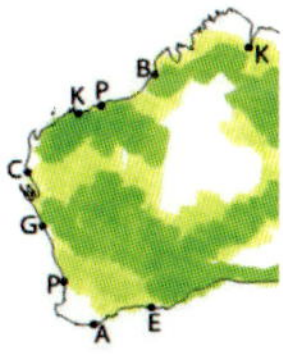

Nightjars are found throughout the world and have adapted well to taking insects in flight. In Australia we have 3 nightjars with only one occurring in WA. I have always been intrigued why Large-tailed is not found in the Kimberley as there is habitat that would support them. Two of the nightjars have the classic 'bullseye' on the wings except the White-throated Nightjar found on the east coast. The Spotted Nightjar's white bullseye's are clearly visible at night when seen in headlights or spotlights. The wings are very long and broad, its flight very manoeuvrable; it catches insects particularly moths on the wing with its wide mouth. Often seen in the middle of gravel roads waiting for large flying insects to move. The birds are well camouflaged with their mottled markings, I have walked only two metres away from a Spotted Nightjar on the side of a gravel track and not seen it, only to turn back a few metres past it and it has finally taken flight, such is their ability to freeze and blend in with foliage. You will always know when they are near you at night as their characteristic laughing chuckle call will give them away, hence one of its common names is 'laughing owl'. It has 3 distinct calls; the main contact and territorial call starts as a slow laughing 'whoo – whoo-whoo-whoo, whoo,whoo' ending in a fast gobbling set of notes, hard to describe but very distinct. It calls before the sun has set, mainly in breeding season; it will however call during the day particularly on overcast days but the call is very brief. On the southern coast, it has a preference for open heath lands with gravely soils where it sallies for insects on the wing, flying only a few metres above the vegetation tops or just sitting on gravel ridges and roads waiting for moving insects. It can also be found in thick mallee, which abuts open farming country. In the Mulga and Gascoyne region, has a preference for stony ridges in breakaway country. **Nest:** No nest is made, just a scrape on the ground and the one egg laid directly on leaf litter or between stones. Clutch size just 2 to 3. **Status:** Moderately common in the southern coastal heaths from Waychinicup to Cape Arid. Uncommon in the inner Wheatbelt. Common in most large northern Wheatbelt reserves. Common in the Mulga and stony ridges. **Where to find:** Locally common in the Fitzgerald River NP. All the national parks containing heath habitat with gravel, from Hopetoun to Cape Arid. All stony ridges throughout the Mulga Region. They are in fact a common night bird in some areas as. An example, I camped in at least 4 large separate northern Wheatbelt reserves, I noted that at every camp spot there were Spotted Nightjars calling. They are common in many stony areas of the deserts, Pilbara and Kimberley.

Photo: Simon Nevill

Australian Owlet-nightjar *Aegothleles cristatus* 19-25 cm

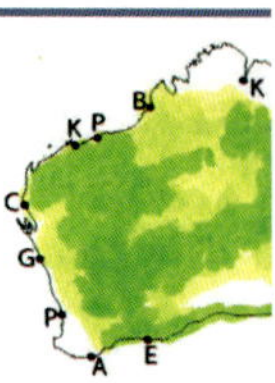

Smallest of all the Australian nocturnal birds, looks like a combination of a small owl and a nightjar. Feeds on much smaller insects than do nightjars or frogmouths. Uses the same technique as Nightjars of hunting on the wing for insects but fly less than a nightjar and pounces more like frogmouths, taking small insects from the ground. It has 3 distinct calls but the territorial call from a nest or roosting hollow is the most common; it has a funny laughing quality totally different to the nightjar's. A three-noted bubbling call that starts on an extended note followed quickly by 2 short notes. It will give this call 4 or 5 times during the day either from the entrance of the nest or roosting hollow or within the nesting hollow.

While taking bird groups I would often hear the call and look for the appropriate hollow, and then get people to look at the hollow with binoculars, while the author scratched the side of the trunk with a branch to imitate a crawling monitor lizard climbing up the trunk. Often, out would pop its head, checking who was there, much to the amazement of the group, nothing special, just knowing the call certainly helps. Owlet Nightjars pair for life and will use the same hollow year after year. They are not always easily seen as they are well camouflaged with their dark black-brown stripes that pass through the eye over and over the top of the crown. **Nest:** Nest in a great variety of trees in small hollows. Clutch size 2 to 3. **Similar species:** None. **Status:** Scarce in deep South West. Moderately common throughout the state with adequate small nest and roosting hollows. **Where to find:** Locally common in mature Wandoo Woodland around the Stirling Range. Most Wheatbelt reserves with adequate woodland. Salmon Gum woodland. Any large river system in the arid region. I have seen them on the Canning Stock Route in small white gums way out in the desert.

Photo: Simon Nevill

Rock Dove (Feral Pigeon) *Columba livia* 31-34 cm Introduced

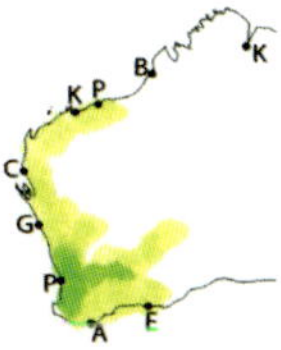

The pigeon that is so common in many cities of the world is now widespread in most large southern cities of Australia being introduced in the mid-1800s. Its origins are difficult to define, as there is so much interbreeding between domestic pigeons and wild pigeons throughout the many countries in which it occurs. They most probably originated in northern Africa and the Middle East where wild populations still occur, in fact there are still wild populations in the north of Britain. Even though introduced to Australia, they still instinctively go back to their roots, nesting in rock ledges as can be seen on Penguin Island where they nest in the small limestone caves and crevasses on the island. The wild populations are often augmented by stray racing pigeons. There are approximately 14 subspecies spread throughout the world. **Similar species:** None. Status. Common to uncommon in large towns. **Where to find:** Found in most of the larger south-west cities in Western Australia from Esperance all the way to Geraldton.

Photo: Simon Nevill

Laughing Dove *Streptopelia senegalensis senegalensis* 25-27 cm

Introduced

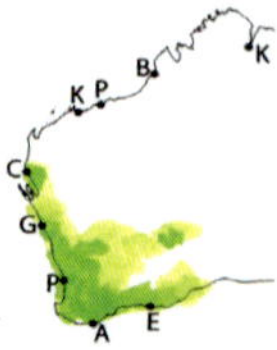

Laughing Turtle Dove is a small pigeon with general pinkish tinge to the upper body and breast. Grey on the wings with distinct black and pink mottling on the throat, its main diagnostic feature. Introduced to Perth in 1898 its origins are not clear, however there are 6 subspecies in the world, found on the African continent, part of the Middle East and the Indian subcontinent. One of its common names is Senegal Dove from where the nominate race occurs. Laughing Turtle-Dove in Australia only occurs in the South West. The dove has spread to most of the major towns in the South West. In one of his maps, Sedgwick showed that the Laughing Dove had reached Geraldton, Kalgoorlie, Esperance, Albany and Bunbury by 1958. It is uncommon north of Geraldton. Like all pigeons, its staple diet is grain and consequently it has followed the railways lines that fan out from Perth. Luckily, Laughing Doves do not inhabit native woodlands but have a preference for city parks, roadside verges in cities and railway lines wherever seed is abundant. There are 7 subspecies in the world and it is presumed the nominate *senegalensis* is the one that was introduced into Perth. **Nest:** The nest is a flimsy structure of twigs placed in a variety of trees or bushes mostly exotic. Clutch size 2 to 3. **Similar species:** Spotted Turtle Dove, which is a larger pigeon with a distinct black collar with white spots forming a collar, that passes from the centre of the neck around to the back forming a half collar . Has a grey head, its tail is longer and dark grey brown not pink. Laughing Dove's tail is blue-grey. Spotted Dove's eye is yellow and Laughing Dove's eye is black. **Status:** Locally common. **Where to find:** Throughout the major towns in the South West as far north as Carnarvon south to Esperance and east to Kalgoorlie.

Photo: Simon Nevill

Spotted Dove *Spilopelia chinensis* 30-33 cm

Introduced

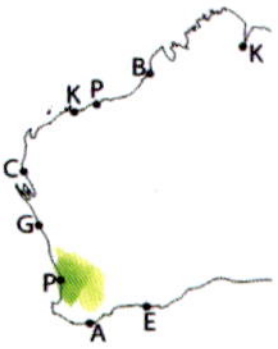

An introduced bird from the Indian subcontinent and South East Asia. The first birds released into the wild were those from Perth Zoo between 1898 and 1899. Other birds were released over the years in all the eastern states unlike the introduced Laughing Turtle-Dove, which was only introduced to Perth. There are 8 subspecies in the world, 2 subspecies, *tigrina* and *chinensis*, have been introduced to Australia and interbreeding has certainly occurred between them. The Spotted-Turtle Dove has spread not quite as far as the Laughing Turtle-Dove in the South West. Has expanded down to Augusta but has not passed Dongara in the north. Was also released in Kalgoorlie but is not found in the outer Wheatbelt or most of the Southern Coastal Region. **Status:** Locally very common around Perth. Uncommon to scarce south and north of Perth. **Nest:** Clutch size and nest structure similar to Spotted Turtle-Dove. **Similar species:** Refer to Laughing Turtle-Dove. **Where to find:** Most coastal major towns including Perth, Rockingham, Mandurah, Bunbury, Also many of the Darling Range suburbs on the outskirts of Perth.

Photo: Simon Nevill

Common Bronzewing *Phaps Chalcoptera* 28-36 cm

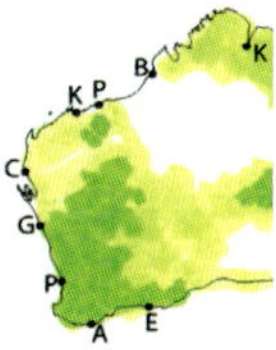

Found throughout the state, but rarely in the sand dune desert regions. A bird of open woodlands, open fields with adjacent cover, roadside verges, Mulga Woodland, Mallee Woodland and rocky terrain. It is a generally a solitary feeder but when grains is spilt and food abundant small groups can congregate to feed. Feeds on seeds and some insects, however cannot bite so must swallow whole. Also must come to water to drink. Often found by hearing its low booming call of 'ooom..ooom…ooom' repeated slowly. A nervous bird that will often leave its nest if on eggs, when one is not really close to the nest. There are small differences between the male and the female. The male has a pale yellow-cream patch on the forehead, female none. Male has red-brown crown, female plain brown. Male has pink flush to all the underbody, female grey-brown. Bronzing on wing coverts more iridescent in the male. **Nest:** A loose collection of twigs in a saucer shape placed from 3-15 metres on a fork or horizontal limb. Clutch size normally 2.

Male

Female

Photos: Simon Nevill

Similar species: Brush Bronzewing which is smaller in size with shorter tail, bronzing on wing restricted to two central bars on the median and greater coverts. Forehead on male has more buff-cream and goes to just above front of eye. Dark chocolate brown above white eye stripe. Blue-grey under body and throat. Most important feature is a distinct deep red-brown throat patch on the male. Nape and shoulder on male rich chestnut-brown passing to a point on the upper breast. Female lighter copper-brown nape and neck contrasting with the blue-grey neck and under body. **Status:** Common to uncommon throughout the range. Moderately Common in parts of the acacia woodlands of the Pilbara. Uncommon in the Kimberley. Scarce in the deserts. **Where to find:** Darling Range near Perth. All of the Wheatbelt particularly where there is remnant woodland and dams close by. Common in all the inner Wheatbelt reserves. Found throughout the Mulga Woodland.

Brush Bronzewing *Phaps elegans* subspecies *occidentalis* 25-33 cm

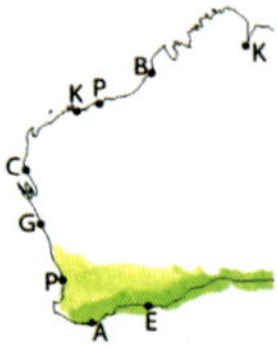

There are two subspecies in Australia, elegans in eastern Australia and *occidentalis* in Western Australia. There are subtle differences in the Western Australian subspecies. It has a buff-cream forehead not orange-brown, back and nape darker chestnut-brown and lighter grey under-body than nominate race *elegans* in the eastern states. A shy bird found close to the coast in coastal thick scrub, heath and wet gullies from just south of Dongara all the way to north of Perth. Most of its former habitat on the Swan Coastal Plain is now cleared so does not become locally common until the Busselton region and then found all along the coastal belt and just inland from Cape Naturaliste to Cape Leeuwin. Becomes more common from Augusta to Cape Arid, where its favoured habitat of lusher heaths and well-watered gullies occurs. The Jarrah Forest is not one of its preferred habitats, although it certainly is for the Common Bronzewing. However when the Wandoo Woodland commences after the Jarrah Forest in the inner Wheatbelt, the Brush Bronzewing can be found, particularly if there is an understorey of heath and plants of the poison pea group (*Gastrolobium*). Does not congregate in a group as does the Common Bronzewing. The Brush Bronzewing has the bronzing on its wing restricted to two central bars on the median and greater wing coverts. Forehead on male more buff-cream, which goes to just above front of the eye. Dark chocolate brown above white eye stripe. Blue-grey under body and throat. Important feature is a distinct deep red-brown throat patch on both sexes. Nape and shoulder on male is a rich dark chestnut that passes to a point on the upper breast. Female lighter copper-brown nape and neck contrasting with the blue-grey neck and under body. **Nest:** Similar to Common Bronze but much lower and often placed in thicker foliage. Clutch size 2. **Similar species:** Refer to Common Bronzewing. **Status:** Uncommon in selected habitat. **Where to find:** Coastal heaths between Dongara and

Photo: Simon Nevill

Yanchep. Between the two capes in the deep South West. All along the south coast particularly in heath, including the following parks: Stirling Range NP Two Peoples Bay NP Waychinicup NP Fitzgerald River NP Stokes Inlet NP Cape Le Grand NP Cape Arid NP Also many of the Wheatbelt reserves including Boyagin Rock NR Tutanning NR Dryandra Woodland NR Dongalocking NR Tarrin Rock NR. Unlike the Common Bronzewing, you will not find this bird in the dry Mulga Woodland and beyond. Unusual remnant population on the Abrolhos Islands. Certainly not found around Geraldton so could be remnant populations remaining when the coastal belt had a higher rainfall and lusher vegetation.

Flock Bronzewing *Phaps histrionica* 28-31 cm

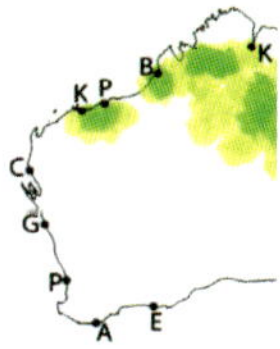

A gregarious species that is centred in the Barkly Tableland in the Northern Territory and the southern channel country in Queensland and the northern South Australian deserts. Uncommon in the north of WA. Found in open grassy plains mainly in northern grass plains of the Pilbara, but mostly in the Kimberley and Great Sandy Desert particularly near Lake Gregory. Main feature to look for is a white mask to front of head contrasting with a black band that covers the head and throat, broken with a white circle surrounding the lores and with patch on upper breast. Both features occur in the male and female but crown and lores are brown instead of black on the female. Fawn upper body in both sexes, blue-grey chest and under body in male contrasting with white throat. Female grey-blue only on the under body. If you see a large flock of pigeons flying faster than any other species of pigeon you have ever seen before, then the chances are you have just seen Flock Bronzewing as no pigeon flocks in the numbers that Flock Bronzewing does. If you see them land normally away from water, try to get close and well hidden from the dam or pool you think they are coming to. They will eventually one by one walk through the grass checking all is clear then group and come down to the water and drink fast, and then all will fly off together at the same time, circling first and then flying off. In flight the white facial patch on the male stands out from a chestnut-orange back and wings as well as distinct white patch on the leading edge of the shoulders (inner primaries and outer secondaries) which is clearly visible in flight but not shown in many books. **Nest:** No nest is made, only a slight hollow if soil is soft enough to scrape. Mostly placed between high grasses on black soil plains. Clutch size 2. **Similar species:** None. **Status:** Scarce in the north but locally common in some areas like the Camballin Floodplain. Lake Gregory. **Where to find:** Remote chance in grassed areas on the Gascoyne floodplain. Grasslands on the lower Maitland and Fortescue Rivers. Sometimes near Lake Eda. Black soil plains on Meda and Kimberley Downs Station, Camballin, Parry's Lagoon, Lake Gregory. Less chance in the Pilbara although chances on the grass plains between the Jones, Sherlock and Yule Rivers in the north west or on the Fortescue Floodplain.

Photo: Simon Nevill

Crested Pigeon *Ocyphaps lophotes* 31-36 cm

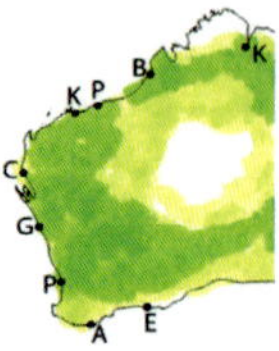

A bird that is found throughout most of the state except the lower South West forests and the Salmon Gum Woodland region east of Hyden. Also rare in sand dune country and the Nullarbor. It was a bird of drier regions being the dominant Bronzewing in the arid areas and still is, but since land clearing in the early nineteen hundreds the Crested Pigeon has slowly encroached into the South West Region eventually coming to Perth. You can imagine the Swan Coastal Plain where Perth stands today, was either thick with coastal heath or thick Jarrah Forests or swamps with heath land, not the habitat preferred by Crested Pigeons but that of Brush Bronzewing. Now with crop fields, it is understandable that this bird that requires dry grains finds food in abundance in the South West with permanent water readily available on dams, which it must have. Even though the Crested Pigeon has moved into the South West it is basically a sedentary bird and will remain in the territory that it feeds in. It is a gregarious pigeon and will often be seen with other birds, particularly coming to drink. It is an unmistakable pigeon with its high crest. There is no marked difference between the male and the female, the main feature is that the female has a duller lustre to the bronzing on the wing and the body a little duller grey. In the South West there are 2 subspecies, the nominate *lophotes* found in the southern part of Western Australia and *whitlocki* found in central and northern Western Australia, and which has a narrower white tail-band. The central and northern birds are slightly smaller. **Nest:** Is the same as Common Bronzewing. Clutch size 2. **Similar species:** None. **Status:** Common throughout much of the state. **Where to find:** On the outskirts of Perth and throughout the Wheatbelt and in the mulga coming to dams. Throughout the Pilbara and most parts of the Kimberley, less so the north Kimberley. The subspecies *whitlocki* is centred in the Pilbara but can come as low as the Gascoyne area.

Photo: Simon Nevill

Spinifex Pigeon *Geophaps plumifera* and subspecies *ferruginea*

20-24 cm

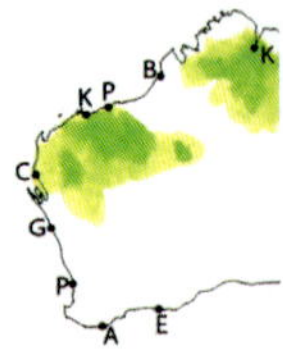

The Spinifex Pigeon subspecies *ferruginea* comes down to the Gascoyne region although the main populations are in the Pilbara, The nominate species *plumifera* is found in the southern and eastern Kimberley. Subspecies ferruginea has an entirely rust-red under body rather than a white under body as in *plumifera*. It also is slightly smaller. It's a bird of rocky terrain with spinifex being the principal plant from whose seed it feeds. The Spinifex Pigeon can remind one of a clockwork toy as it scurries fast with its short feet and body hardly moving as it runs

Photo: Simon Nevill

over stony ground and rocks. Mostly seen coming down to water points in small groups as a safety measure. Its crest is a very distinct feature and it has an all orange-red body. If you see one bird there are likely to be others hidden very close. **Nest:** They nest on the ground next to spinifex clumps, often between small boulders. Nest consists of a few leaves or grasses either placed alongside spinifex plant or on a rock ledge. Clutch size 2. **Similar species:** None. **Status:** Common in the Pilbara and Kimberley. **Where to find:** Occasionally in the Dairy Creek region on the Gascoyne River. Throughout the Hamersley Range. Purnululu. Wyndham. Cockburn Range. Even way out in the desert, where rocks occur near Purnumu west of the Canning Stock Route.

Diamond Dove *Geopelia cuneata* 19-24 cm

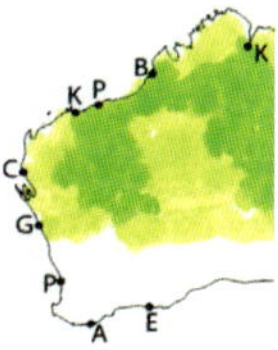

The smallest Australian Dove, with an overall grey body and darker grey wings covered in white spots on the scapulars and coverts. Has a prominent red orbital eye ring that stands out from its blue-grey head and pink legs. Found in the semi-arid regions. More common above the Murchison River. Found throughout the north. Often found feeding in small group up to 10 birds. Prefers even drier country than the Peaceful Dove, found throughout the mulga, although it overlaps with the Peaceful Dove. **Nest:** A small saucer shape of twigs placed in any position in a tree or bush, normally between 3-10 metres at the most. Clutch size 2. **Similar species:** Peaceful Dove, which is slightly larger and which has a light blue orbital eye-ring rather than a bright red eye ring. Has black barring on the nape, running down to the neck and chest, forming a band. Dark edges to the wing coverts and scapulars. Barring on back, rump and tail. **Status:** Common throughout the northern regions. **Where to find:** Kimberley. Pilbara. Much of the northern desert regions. Yalgoo region. Murchison Settlement. Wooleen Station. Austin Downs.

Photo: Simon Nevill

Peaceful Dove *Geopelia placida* subspecies *clelandi* 20-24 cm

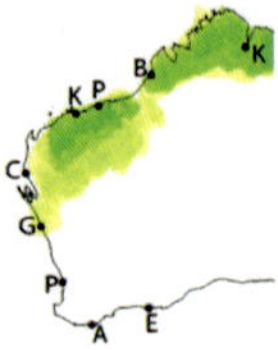

Most common in the north of Western Australia, preferring less-arid parts of Australia. Here in Western Australia it is found throughout the Kimberley, being the subspecies *placida*. Then the Great Sandy Desert breaks the distribution until the Pilbara and Gascoyne River region where the subspecies *clelandi* occurs, which is overall paler and has fawn tinge to the grey upper body and tail compared to the Kimberley birds. Like the Diamond Dove they both have attractive markings. The Peaceful Dove has a light blue orbital eye-ring and has black barring on the nape, running down to the neck and chest, forming a band. Dark edges to the wing coverts and scapulars. Barring on back, rump and tail. Their pleasant 'cooing' call is always a reminder that you are in the north where the warmer climates occur. **Nest:** Clutch size is the same as for the Diamond Dove. There are 6 subspecies in the world with 2 in Australia. **Similar species:** Refer to Diamond Dove. **Status:** Common in the Gascoyne catchment and Murchison Settlement all the way north to the Kimberley. **Where to find:** Carnarvon and along the lower Murchison where pools and river gums (*E. camaldulensis*) occur. On the Butchers Track on the Murchison River where the Wooleen road crosses. Along the Gascoyne River to the start of the Lyons River. Throughout parts of the Pilbara and Kimberley. Not common in the desert areas.

Photo: Simon Nevill

Bar-shouldered Dove *Geopelia humereralis* 30 cm

Photo: Simon Nevill

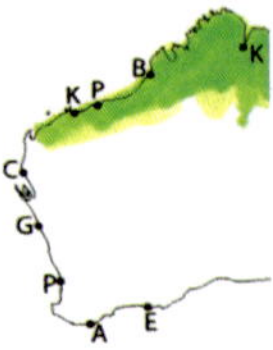

A dove of the north and north-west of WA. Found in various habitats but highest concentrations are within 50 km of the coast. The Bar-shouldered Doves of the Pilbara coastline belong to the subspecies *helendi* . They have a duller coppery hind neck and nape and an overall greyer upper-body. Bar-shouldered Dove is common in mangrove woodland and woodland close to river systems. **Nest:** Nest often located on the fork of a branch with loose twigs as base. Will often nest in pandanus frons. Clutch size 2. **Status:** Very common on the Kimberley coast as far south to the Ashburton River. **Similar species:** Possibly Peaceful Dove, which is smaller and has a complete mottled black and white neck collar. Has light blue eye ring not red. **Where to find:** Throughout the Kimberley south to the Pilbara coastline. Common around Broome.

Rose-crowned Fruit-Dove *Ptilinopus regina* 23-24 cm

Photo: Simon Nevill

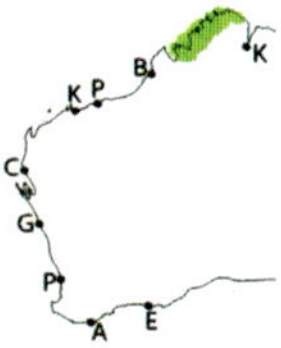

An attractive fruit-dove, found in semi-deciduous vine forests on the north-west coast and islands of the Kimberley. Feeds both on the ground and in trees gathering fruit. Often its location is given away by its 'woop' 'woop' call that starts slowly and finishes very rapidly. Moves into towns like Derby in the wet season, taking advantage of gardens, with fallen fruit. **Nest:** Nest often located in horizontal fork or in bunched twigs. Clutch size 1-2. **Status:** Uncommon on the Kimberley coast as far south to Broome. **Similar species:** None. **Where to find:** Monsoonal vine forests and mangroves on islands and not far from coast. Found in vine thickets at Cape Leveque. Extends range during the wet.

Common Emerald Dove *Chalcophaps indica* 24 cm

This species can be found as far north as southern China to West Papua in the east, to the Indian sub-continent in the west, south to the Sundas, Molucccas Islands and Christmas Island. On Christmas Island the species occurs throughout the island. This species is noticeably smaller than the Pacific Emerald Dove on the mainland of Australia. The male Common Emerald Dove has a pale white eye stripe and grey crown which the Pacific does not have. **Nest:** Same as per Pacific Emerald Dove. Clutch size 2. **Status:** Common through much of the island. **Similar species:** None on the island. **Where to find:** Found throughout the Island but common in the 'Dales' area and the track leading to Winifred Beach.

Photo: Simon Nevill

Pacific Emerald Dove *Chalcophaps longirostris* 26 cm

Photo: Simon Nevill

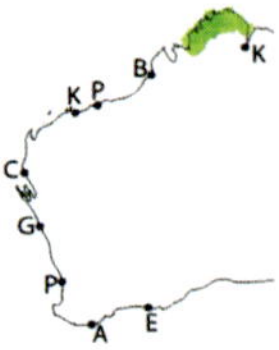

There where 8 subspecies of Emerald Dove previously classified in the World and 3 where found in the Australian territories. Now the IOC has named the Australian mainland bird Pacific Emerald Dove *Chalcophaps longirostris* and the Christmas Island bird is now named Common Emerald Dove *Chalcophaps indica*. The Emerald Dove in WA is found in the Kimberley. Found mainly in larger monsoonal rainforest pockets in the far north west of the Kimberley, also in the denser mangrove fringes adjacent to rainforest areas. A ground feeder taking fruits and seeds. **Nest:** A loose collection of twigs placed normally at the junction of a horizontal branch, typically from 2-5 m high. Clutch size 2. **Status:** Locally common. **Similar species:** None. **Where to find:** In larger sub-humid forests mainly north of the Walcott Inlet. Mitchell Plateau. Carson River escarpment. On the Ord River north of Lake Argyle.

Christmas Island Imperial-Pigeon *Ducula whartoni* 43-48 cm

Endemic

Photo: Simon Nevill

A large fruit pigeon found only on Christmas Island. Not a common pigeon on the Island and mostly single birds are seen. On my 7 days visit to the island, I did not find this species super abundant, although it may be overlooked as it prefers the outer canopy of the taller trees in the central plateau. In 1975 the population was said to be around 50 – 100 breeding pairs (van Tets). Latest count in 2010 puts the population between 1000 -10,000 so large discrepancies in various surveys. I would have certainly question the later figure. Feeds mostly on fruits on the canopy. **Nest:** Nests high in the canopy, normally over 5 m. Clutch size 1. **Status:** Uncommon to moderately common. **Similar species:** None. **Where to find:** Throughout island from above the settlement into the hinterland. Driving on the main roads from the settlement to either South Point or Murray Hill should provide this species.

Torresian Imperial-Pigeon *Ducula spilorrhoa* 39-44 cm

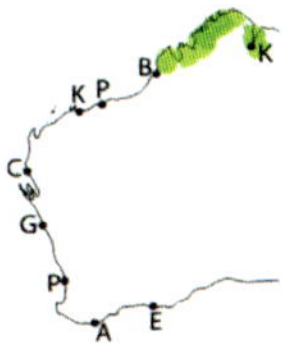

Found through most of the north-west region of the Kimberley, mostly no more than 20 km inland. Found in similar habitat to Rose-crowned Fruit-Dove but in larger numbers and will spread out from semi-deciduous vine forests to adjacent fruiting trees, particularly to off shore islands, so has a much broader range of distribution. Disperses locally with the onset of the wet but unlike the Queensland population, the Kimberley populations is fairly sedentary. Flock sizes far smaller than east coast birds. **Nest:** Nests normally over 3 m high but will nest in mangroves. Makes basic platform with minimal twigs. Clutch size 1. **Status:** Locally common on north-west coast and islands from Collier Bay around the northern coast to Cape Lambert. **Similar species:** None. **Where to find:** Mitchell Plateau. Calder River. Sale River. Lower Drysdale River. Most Islands from Augustus Island around to Cape Londonderry.

Photo: Simon Nevill

Partridge Pigeon (Kimberley yellow-faced race)

Geophaps smithii area subspecies *blaauwi*

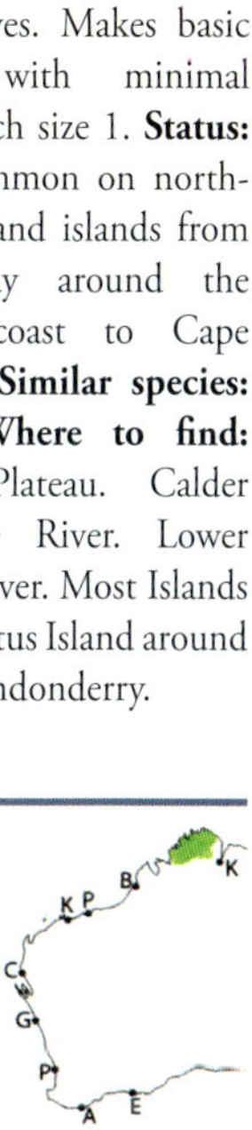

A pigeon of the north-west Kimberley and Northern Territory. The subspecies *blaauwi* found in the Kimberley has a bright yellow facial patch unlike that of subspecies *smithii*, of the NT, which has a red facial patch. The two populations are over 400 kms apart and generally sedentary, so there is no interbreeding between the two races. They have a preference for grassed country on boulder-strewn slopes leading to alluvial soils in the valley floor. It's preferred feeding territory is now being compromised by continual fire regimes that take place in the winter. I have noticed a drop in the population since first coming to the Kimberley in the early 1980's. Often flushed from grasses. I have found them both in open stony country hill-tops and also in tall grasses, adjacent to creek lines, also in country recently burnt. **Nest:** Nests on the ground, normally at base of a clump of grass, often with

Photo: Simon Nevill

over hanging grasses. Clutch size 2. **Status:** Uncommon. **Similar species:** Possibly Common Bronzewing which has different head patterning and tend to frequent more flat open country mostly further inland in the Kimberley. Yellow facial disc stands out on Partridge Pigeon, plus it has a white belly. **Where to find:** North-west region of Kimberley where the sandstone country occurs. North of the Charnley River to Mitchell Plateaux. Most visits to the Mitchell Plateau should reward you with sightingsgive you this species.

White-quilled Rock-pigeon *Petrophassa albipennis*

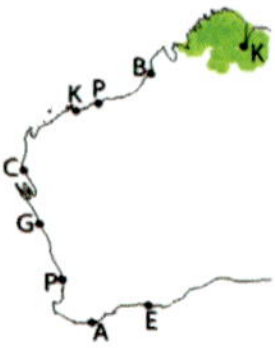

A rock pigeon of the escarpment country and stony terrain. Range more extensive than Partridge Pigeon. Often fond on the top plateau escarpments or making their way down rocky hillsides to drink, as they have to drink regularly through the day being seed eaters. Their flapping wing beats often give them away as they drop down from higher rock country to feed on lower grasses. Their white bulls-eyes in the wings are clearly visible in flight. Will sit in the heat of the mid day under rock ledges. **Nest:** Nests on rock ledges always with an overhang above. Base of nest has a scant gathering of twigs, almost no nest is made. Clutch size 2. **Status:** Common in stony, rocky country. **Similar species:** None. **Where to find:** From the Fitzroy River north to Kalumburu and from Kimbolton in the east through to Kununurra. Close to Kununurra at Hidden Valley. The Grotto south of Wyndham and throughout the King Leopold stone country of the north-west.

Photo: Simon Nevill

Red-tailed Black-Cockatoo *Calyptorhynchus banksii* subspecies *samueli* 55-60 cm

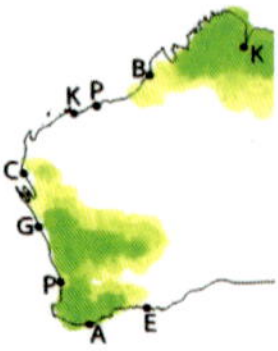

Female

Photo: Simon Nevill

One of the largest cockatoos in Australia. In Western Australia there are 3 subspecies: *macrorhynchus* found in the Kimberley and the rest of northern and eastern Australia, subspecies *samueli* in the Pilbara down to the northern Wheatbelt and also central Australia and subspecies naso found solely in the South West. Of the two subspecies in the south west region, *macrorhynchus* and *naso,* there is clear evidence from historical records that the mid west birds (*samueli*) are slowly moving south into the south west. They have in fact increased in numbers in the northern Wheatbelt where they rarely occurred before. With the advent of land clearing and permanent water on dams the Red-tailed Black-Cockatoos have found abundant new sources of food, with seeds from weeds and grains. Some of the largest flocks in the state, often

Male Photo: Simon Nevill

over 200 birds can be found in the northern and north eastern Wheatbelt adjacent to the transitional eucalypt woodland. Red-tailed Black Cockatoos will begin nesting in the northern Wheatbelt in the months of April and May. The South West bird, the subspecies *naso*, may be on the decline due to loss of habitat. They have adapted to feeding almost entirely on the South West eucalypts, mainly Jarrah, Marri and Black Butt. The male's plumage is entirely glossy black except for the broad bright red panels half way along the tail feathers, seen best in flight or when alighting. The female's entire body is a duller black without the lustre of the male but is covered with pale cream-yellow spotting, which occurs on the head and wings as flecking. The underbody is slightly greyer black with cream-yellow barring. The tail, seen best in flight, has orange-yellow barring that goes closer to the end of the tail shafts but also bleeds into a brighter orange colour near the outer tail feathers. The Red-tailed Black-Cockatoo's call has similarities to the White-tailed Black-Cockatoo but is much softer, more mournful and less harsh. **Similar species:** The Short-billed and Long-billed Black Cockatoo's are similar large black cockatoos but both species have white in the tail, and have different calls from the Red-tailed. Crest on the Red-tailed is longer and extends higher when alarmed or making contact calls than the crest on the White-tailed black cockatoo. **Status:** Subspecies *macrorhynchus* is common throughout most of the Kimberley. subspecies *samueli* is less common in the Pilbara and locally common in the Transitional Woodland. Subspecies naso is uncommon in the Jarrah Forest found in small groups of 8-14 birds. **Where to find:** subspecies *naso*, in the Hills suburbs like Roleystone, Kalamunda. Bungendore Reserve. Throughout the Jarrah Forrest. Seen on the Brookton and Albany Highways particularly around North Bannister. Margaret River area. Subspecies *samueli*: seen best in the northern and north eastern Wheatbelt Region and the Transitional Woodland, particularly in the area around Eurady Station in the York Gum woodland. Wandana NR. Along the Warriedar Road east of Perenjori. The Ninghan area. On the road from Beacon to Bimbijy. In the Salmon Gum woodland in reserves like Korrelocking NR. near Wyalkatchem. subspecies: *macrorhynchus* can be seen through certain parts of the Kimberley, often in areas that have been recently burnt.

Carnaby's Black-Cockatoo (Short-billed Black-Cockatoo)

Calyptorhynchus latirostris 54-56 cm **Endemic**

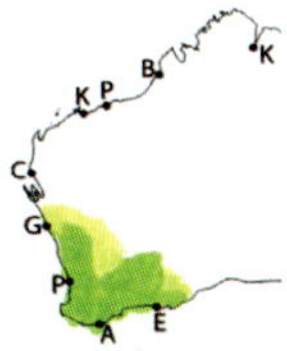

There are two very similar white-tailed cockatoos in the South West both endemic to WA, the Short-billed and the Long-billed Black-Cockatoo. The Short-billed Black Cockatoo (also known as Carnaby's Black-Cockatoo) favours a different habitat to Long-billed Cockatoos (also known as Baudin's Black-Cockatoo). However, they do share similar habitats at times particularly in the Perth region and nesting Short-bills have been found in the Jarrah Forest close to the nesting sites of Long-bills. The main fact is that the majority of the Short-billed Black-Cockatoos, particularly the southern and northern populations, prefer heathland habitats. Sadly much of their preferred habitat has been cleared and there has been a movement of the northern populations into the deeper South West. The northern populations can be found in their breeding and feeding areas of Lesueur National Park, Eneabba, Watheroo, Coorow, New Norcia and Rica Erickson Reserve. Also the armed force's prohibited site north of Julimar State Forest and the north west section of Julimar Forest. They require habitat that combines their nest sites, primarily *Eucalyptus wandoo*, with rich flora habitat to feed from which includes plants mainly of the Proteaceae family, namely dryandras (soon to be grouped as banksias), hakeas and banksias found on the heaths of the northern sand plain. With the loss of much of their territory, many now move after the breeding period finishes from January through to April into areas where there are pine plantations, often massing in numbers of over one hundred birds. Mostly though they break up into flocks 10-40 birds. One can see them in the summer feeding in suburban gardens. They can strip fruiting trees of all their seed and are catholic in their choice of food, from native plants as diverse as bottlebrush (*Calothamnus*) to the eucalypt Illyarrie (*E. erythrocorys*). The southern populations have the same problem as they too have lost much of their favoured tree, the Wandoo to nest in, as well as countless hundreds of square kilometres of Kwongan heath that they would have previously fed from. They are found throughout much of the southern Wheatbelt Region, including the Stirling Range NP. Fitzgerald River NP. and woodlands in the eastern Wheatbelt east of Hyden where there are extensive heaths. On the author's previous property adjacent to the Stirling Range, there was a breeding group of 14 birds with 4 nesting pairs, the remainder being dependent young. They selected nests sites in the third week of September, two pairs returning to their previous nest hollow from the season before. Sometimes, the female would improve the entry to the nest hollow, while the male would give encouragement to the female, calling constantly with a repetitive rasping soft call, totally different from his harsh contact call. The small group would call just after dawn, making contact with each pair from the group. They would gather at a central meeting point, normally a large dead Wandoo tree and then leave together for their feeding grounds on the heaths on the eastern side of the Stirling Range. The female becomes quite aggressive to the mature dependent young, while she and the male are feeding the newborn chicks and possibly this constant harassment, combined with the harshest screeching calls that they ever make, possibly the trigger for the dependent young to finally leave the breeding area. This practice is common among so many species whose dependent young remain with the parents for extended periods. Clutch size is 2. **Similar species:** The Long-billed Black-Cockatoo is almost identical and certainly gives even the most knowledgeable eastern states birdwatchers some initial difficulty. When you can get close views with binoculars, you can soon see, as in the photograph, that the Long-billed has a long fine bill and if you manage to view it front on, you will notice it is a slim bill.

Photo: Simon Nevill

The Short-billed has a broader bill and shorter top mandible particularly near the tip of the bill. The call is slightly harsher than the Long-billed and has slightly longer intervals between their repeat calls but you need to be accustomed to listening to the two species regularly to hear this difference. **Status:** Overall uncommon but can be locally common. There is concern for the Short-billed Cockatoo's status whose population has indeed decreased since European settlement but in general it has stabilized. But there are a few far more vulnerable species like the Crested Shriketit subspecies leucogaster, also the Australian Bittern and, most crucial of all, the western subspecies of the Ground Parrot. **Where to find:** Mt Lesueur area. Near Manmanning NR. Rica Erickson NR. North Julimar Forest. Durokoppin Reserve. Between Carnamah and Perenjori. Dryandra Woodland NR.(but does not breed there). Stirling Range NP both east and northern boundaries. 70 km east of Hyden. Esperance region to 50 km north. Throughout north and central Swan Coastal Plain including Yanchep and Gnangara Pine Plantation between January and May. Moves south through Fremantle to the southern suburbs February to April.

Baudin's Black-Cockatoo (Long-billed Black-Cockatoo)

Calyptorhynchus baudinii 52-57 cm **Endemic**

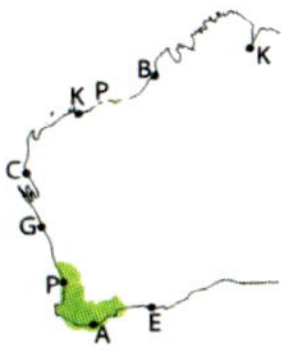

Found in the South West Region, principally in the Jarrah Forest although Jarrah (*Eucalyptus marginata*) is not the primary eucalypt from which it feeds, that being Marri (*Corymbia calophylla*). Found as far north as Muchea and east between Wandering and Pumphreys Bridge and south to the Porongurups and Albany, all the way across to Augusta. The Long-billed Cockatoo as the name implies, has a longer bill than the Short-billed and sometimes, looked at from the side, its like the upper mandible tip appears so fine that it would simply break with the hard marri nuts it feeds on, however the long bill is used to dislodge the seed from the internal base of the nut in just the same manner as the Red-capped Parrot while the heavier cracking of nuts is done at the back of the two mandibles. Long-bills also have the unusual habit of ripping bark from the trunks of Jarrah and Marri trees, looking for larvae (see photograph). The male has the distinct red orbital eye-ring, a common diagnostic feature in many male parrots whereas the female has a dull grey eye-ring. Nests: Selected hollows in jarrah and marri trees. Clutch size is 2. **Similar species:** Refer to Short-billed Black-Cockatoo. **Status:** Some reduction in population since European settlement but relatively stable

Baudin's

Carnaby's

Photos: Simon Nevill

population now restricted to the Jarrah Forest. Is still shot by orchardists, as the cockatoos will eat cultivated fruit and take the seeds. **Where to find:** Throughout the Jarrah Forest: including Bungendore Reserve. Jarrahdale. Dwellingup. West of Quindanning. Darkan. Albany. Porogurups. Denmark. Walpole and all the deeper South West Region, which has the highest concentrations of this species.

Author's note: It is interesting to note the IOC has recently named these two species the same as my previous publication. It may upset some that it is not Carnaby's or Baudin's and I understand that very well but it certainly makes it a clear statement describing the species for layperson .

Galah *Eolophus roseicapillus roseicapillus* 35-37 cm

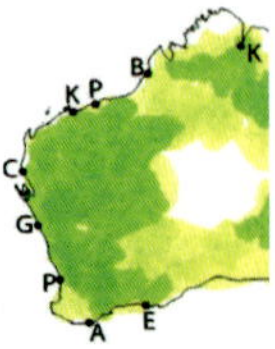

Since European settlement this species has expanded its range, with the onset of land clearing, and availability of abundant crops producing seed and permanent water from farm dams. It has also moved into the Swan Coastal Plain, a region it never was found in before. It has adapted to urban environments and is seen in most areas of Perth. It typically masses in large flocks throughout its range not only in the farming country but also in the Mulga Region. The habitat it least prefers, is the dense forested areas of the lower south west. There are three subspecies in Australia, two of them in WA: *kuhli* in the Kimberley and the rest of Northern Australia and *roseicapilla* found in the Pilbara, the south west and central Australia. In the context of this book the differences are so marginal but suffice to say that the subspecies found in the south west *roseicapilla* and has a larger and more extensive crest extending partly down the nape, The 'play' behaviour of Galahs are not entirely understood but they certainly play, turning somersaults on wires, hanging on windmills, sliding down corrugated roofs similar behaviour to the Keas in New Zealand, often accompanied by screeching. **Nest:** In a variety of eucalypts throughout its range. Clutch size 3-5. **Similar species:** None. **Status:** Common throughout the state, locally common, sometimes in flocks in excess of 200 birds. Uncommon on much of the south coast. **Where to find:** Throughout the state but least of all the sand dune deserts.

Photo: Simon Nevill

Long-billed Corella *Cactua tenuirostris* 40 cm Introduced

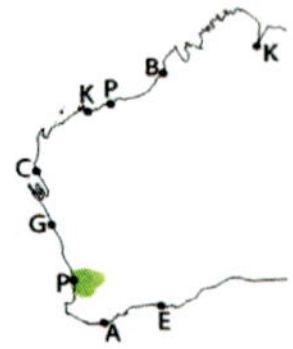

An introduced bird from Victoria, its natural population occurring in an area centred in the region of the Grampian Range. Now introduced to Western Australia in Perth, where captive birds were released. Often seen with Little Corellas and occasionally with Western Corellas. Has the longest top mandible of all corellas and has extensive bright scarlet-pink around lores and lower forehead above the base of the bill: scarlet passes over the eye and finishes at the rear of the eye. The only corella with the upper chest feathers coloured scarlet. Has distinct scarlet colouring to the base feathers on the top of the chest. When the moves neck more scarlet can be seen on the base of the feathers on the neck. **Similar species:** The most similar species is the Western Corella which also has a long top

Photo: Simon Nevill

mandible and scarlet to the lores but the colour is not as intense or as extensive, being restricted to between the eyes and the base of the bill and does not pass over or behind the eye as in the Long-billed. Does not have the red on the upper chest feathers. The crest of the Western Corella is taller and more erect than Long-billed when extended. Also the Western Corella is larger than the Long-billed. The Little Corella is the least similar with a much shorter bill and only a faint pink patch between the eye and the base of the bill, and it is slightly smaller than the other corellas. **Status:** Only a small population in Perth. **Where to find:** Lake Joondalup. Herdsman Lake. Lake Monger. Bold Park. Manning Lake.

Western Corella *Cacatua pastinator* subspecies *derbyi* 40-45 cm

Endemic

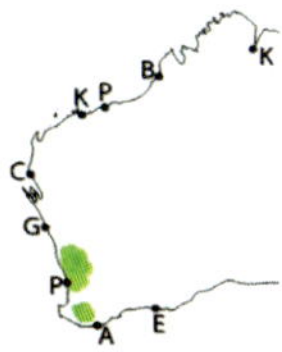

There are two subspecies in Western Australia, derbyi occurs between Dongara. Moora and Three Springs in the north, mainly in the following areas: Hill River in the west, Reagans Ford, New Norcia, Goomalling in the south and Wubin, Latham, Kalanie in the east. The other subspecies

Western Corella

Long-billed Corella

Little Corella

Photos: Simon Nevill

pastinastor is centred in the Lake Muir, Rocky Gully, and Frankland areas. The Western Corella has a high crest that, when displaying or agitated raises the crest forward. Has a very long pointed bill and slightly more yellow on the underwing than the Little Corella. The Western Corella can be found digging up roots in crop stubble in the process of which the underbody gets covered in red ochre dust so the birds tend to have a red ochre coloured underbody. Normally in large flocks from 30-60 birds. For such a restricted species it is very common in the central areas of its range like Wongan Hills, Moora and Wubin. The other main area for the northern subspecies *derbyi* is around the Wongan Hills, Mt Lesueur and Hill River area. **Similar species:** Refer to Long-billed Corella. The only species that overlaps the northern population of Western Corella is the Little Corella. The Long-billed and Western Corella only overlap in the Perth region where they both occur. The differences between Little and Western Corella are far easier to define. The long bill on the Western is the obvious feature and can be noticed even in flight such is the bills shape and size. It is also is a longer and larger looking bird. The pink between the base of the bill and eye is always weak on the Little Corella but even so not a good feature from which to identify the two species. **Nests:** Located in tree hollows of Salmon Gum, Wandoo and sometimes York Gum. Clutch size 2 to 3. **Status:** Locally common. **Where to find:** subspecies *derbyi* can be found around Moora, where they also nest in the few remaining Salmon Gums north of the town on the road from Moora to Three Springs. Perhaps the closest site to Perth is just south of New Norcia and Moora where there are resident flocks. Another general region is between Wongan Hills and Wubin and east to the first vermin fence; however, Wubin can get Little Corella so be careful. In the Hill River and Mt Lesueur area is another population. The subspecies *pastinator* has a more restricted range centred around Boyup Brook in the north and south around the Lake Muir area.

Little Corella *Cacatua sanguinea* subspecies *westralensis* 35-40 cm

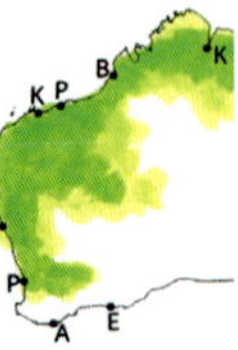

Little Corellas are found in much of the arid and semi-arid regions of Australia including Tasmania. The only areas where they are not found are in some of the central deserts, the Nullarbor and obviously rainforest. In the Pilbara and part of the south west only the subspecies *westralensis* occurs. It is slightly larger, has more red on the lores and longer wings than the eastern states subspecies *gymnopis*. There are more than 1000 kilometres between these two subspecies. In the northern Wheatbelt Little Corellas are often found near wheat bins or out in farming paddocks digging for seed and can be found in very large numbers in excess of 200 birds. They overlap with Western Corella particularly near Three Springs. There are four subspecies in Australia with two in this state *sanguinea* in the

Photo: Simon Nevill

Kimberley and *westralensis* in the Pilbara and south west. **Nest:** Nests in a variety of eucalypts. Clutch size mostly 3. **Similar species:** Refer to Western Corella. **Status:** Very common locally and uncommon throughout the north of the region. **Where to find:** Throughout the Pilbara and mid west. Occasionally as far south as Dumbleyung. Throughout the Kimberley down to Lake Gregory. A population now established in Perth, presumably from escapees.

Pink Cockatoo (Major Mitchell's Cockatoo) *Cacatua leadbeateri*

subspecies *mollis* 39 cm

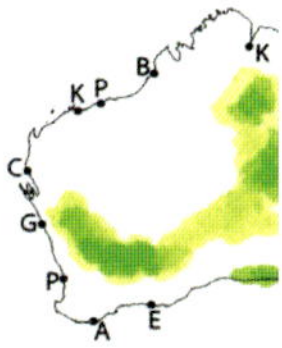

The distribution of the Pink Cockatoo in Australia is intriguing, as there are distinct populations in specific areas. Patchily distributed in WA. Found on the northern Canning Stock Route. On the Gary Hwy. East of Warburton in the Blackstone Ranges. Docker River. Northern and Southern Nullarbor. Often a daily visitor to drink at Eyre Bird Observatory. In the south west, they are found in two areas: along the Murchison River where it is not common and in the northern and eastern Wheatbelt, through the Transitional Woodland to Lake Barlee, the largest flocks are centred north of Bonnie Rock and east to Chiddarcooping NR. A beautiful white cockatoo, with pink plumage on most of the head, chest and underbody, as well as pink to the edges of the tail feathers and white on the wings and back. When the crest is raised, it shows a strong scarlet-orange band with white tips to the crest feathers. Its distinct contact call, often given in flight, leads one to their location. At rest when perched, the crest is normally laid down but raised when making contact calls or alarmed. In flight, shows pink in underwing. In the Bonnie Rock and Beacon area, they have adapted to feeding on grain in open paddocks. Tends to hug the Transitional Woodland line where York Gum (*E. loxophleba* subsp. *loxophleba*) grows and sometimes shares the same habitat with Red-tailed Black-Cockatoos. There are 2 subspecies in Australia, the nominate *leadbeateri* in central eastern Australia and *mollis* in west and central Australia. Western birds have no yellow banding in the crest. **Nest:** Nests in variety of eucalypts mainly Salmon Gum, Red River Gum, York Gum, Ghost Gum and occasionally large Callitris Pines. Clutch size 2 to 4. **Similar species:** None really. The Galah has a grey back and tail, stronger red-pink underbody and a small white crest with no red. **Status:** Overall uncommon but locally can be in large numbers. **Where to find:** Between Mereberie and Milly Milly but scarce. Around Karara area, north east of Perenjori. Southern end of Lake Monger off the Great Northern Highway down to Kalannie. 40 km south of Bimbijy on the Beacon Road. North east of Mukinbudin and Bonnie Rock (some of the largest flocks in the state here). Desert and Nullarbor locations discussed above. Northern Canning Stock Route south of Breaden Hills (not common). Docker and Warburton area and the Blackstone Ranges. Mallee areas south of the Nullarbor scarp.

Photo: Simon Nevill

Sulphur-crested Cockatoo *Cacatua galerita* 50-55cm

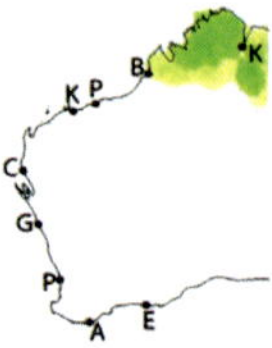

In the eastern states flocks of over 1000 birds can be seen of this species. Here the Little Corrella is the more prolific bird in the Kimberley, it certainly is not a common bird. To be honest, its not a favourite species of mine purely for selfish reasons, the times I have been walking very quietly in rainforest towards an unusual species when all of a sudden the loudest screech echoes through forest. At least it's a good sentry for other birds. A very large parrot, with all white body and large bright lemon-yellow crest which it extends when either agitated or displaying to other birds. There are four subspecies with two in Australia, is found in south eastern Australia and Fitzroy in the tropics of Australia with the Derby region being its furthest western range. There is a remnant population of the subspecies galerita in the Perth region that has now largely been eradicated. A few captive birds were released in 1935 and reached numbers between 300 and 400 in the late 70s. In the Kimberley it can either be found in thicker forested areas, particularly coastal near vine forests and mangroves but mostly small groups and more often single or paired birds. It can also be found in larger groups mostly in the Fitzroy flood plain or the Ord River irrigation area where it feeds on bulbous roots and seeds. **Nest:** In hollow normally above 5 m in a variety of eucalypts throughout its range. Clutch size 2-3. **Similar species:** Possibly Little Corella at distance which is smaller and less bulky lacks the yellow crest, wings less rounded. **Status:** Uncommon throughout the Kimberley region, locally common. **Where to find:** Throughout the Kimberley region but very uncommon west of Willare bridge on the Fitzroy River.

Photo:Simon Nevill

Cockatiel *Nymphicus hollandicus* 29-32 cm

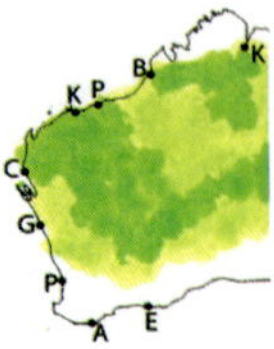

Even though distribution maps show cockatiels located in the south west region, they are casual to the south west. Found principally north and north east of the Eucalypt-Mulga line, north to the southern Kimberley. Least in the Nullarbor. In flight when they call, their old nickname of 'weero' is very apt as they give their contact call, often while flying extremely fast. Cockatiels are birds of semi-arid regions but not so much the arid deserts. They must come to water and the mulga country is ideal for them where they feed on open grasslands within the mulga habitat and along watercourses. Most large inland river systems have their share of cockatiels where they also nest. A gregarious parrot normally in small groups of 10 to 16 birds but bigger groups can occur in the summer months. They are nomadic and will move to areas with the best feeding conditions. The female has a greyer head, less white on the face and the orange spot on the lores is duller. **Nest:** Nests in a variety of eucalypts,

Photo: Simon Nevill

mostly in dead hollows of Red River Gums. Clutch size 4-5. **Similar species:** None. **Status:** Common in the upper Gascoyne catchment. Increasing in numbers to Minilya, Ashburton and Fortescue Rivers. Common parts of Pilbara. **Where to find:** Mainly in the Mulga station country. Lower parts of the Pilbara. Southern Kimberley Fitzroy floodplain.

Rainbow Lorikeet *Trichoglossus moluccanus* 28-30 cm

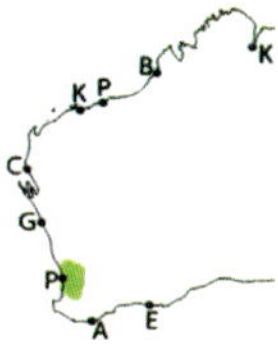

Rainbow Lorikeet distribution range extends from Bali in the west to New Caledonia in the east. The species was split into no less than 16 subspecies. With the establishment of new species nomclature there are now two full species to consider in Australia. Rainbow Lorikeet *Trichoglossu moluccanus* and newly established Red-collared Lorikeet *Trichoglossus rubritorquis*. The Rainbow Lorikeet range extends from western Cape York south to Tasmania and west to Kangaroo with the disjunct introduced population in WA in the Perth Region. The eastern Australian birds introduced to Perth subspecies *malaccans* has a yellow collar on the rear of the neck, below the nape whereas *rubritorquis* has an orange collar. Also *malaccans* has a completely green back and yellow on the side of the breast whereas *rubritorquis* had blue purple patch below the orange collar as well as a complete orange breast. Rainbow Lorikeets are pugnacious parrots and, even though considerably smaller than the endemic Ring-neck Parrot, they will challenge and displace them from nesting in hollows that the Rainbow Lorikeet wishes to use. **Nest:** Nests in a variety of eucalypts particularly

Photo: Simon Nevill

introduced eucalypts. Clutch size 2. **Similar species:** None. **Status:** Subspecies *moluccanus* is increasing in numbers and should have been eradicated years ago when the population was small. Locally common in central Perth and moving to the outer suburbs including the hills suburbs. **Where to find:** Kings Park. Perth Domestic Airport. Cottesloe, adjacent to the beach in the Norfolk Pines. Perry Lakes. Throughout much of the metropolitan region.

Red-collared Lorikeet *Trichoglossus rubritorquis* 28-30 cm

Photo: Simon Nevill

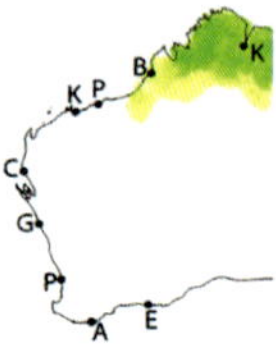

Red-collared Lorikeet distribution range from the west Kimberley to the Gulf of Carpentaria. Found throughout most of the Kimberley to the edge of the Great Sandy Desert along the Fitzroy floodplain. More common in the central Kimberley and the Kununurra area. A gregarious parrot forming small to large groups as they move to flowering areas. **Nest:** Nests in a variety of eucalypts particularly normally of the 'white' barked varieties. Clutch size 2. **Similar species:** Varied Lorikeet ref under that species. **Status:** Relatively common. **Where to find:** Mainly near the river gum woodlands and also wherever there are blossoming eucalypts throughout much of the central Kimberley.

Varied Lorikeet *Psitteuteles versicolor* 19-20 cm

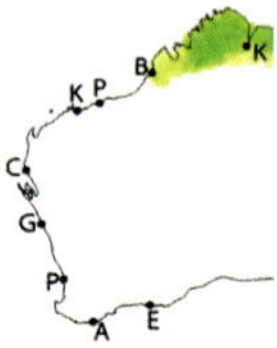

A monotypic species only found in Australia. A lorikeet not always easily seen until one hears their soft chattering call. Their calls are less abrasive than Rainbow Lorikeets. Found throughout the Kimberley region but not a common species and can be easily be missed when visiting the Kimberley as they are nomadic and move to areas whereever there are flowering grevilleas or eucalypts, particularly the bloodwoods. They are brightly coloured, have a vivid red crown, white eye ring that contrasts well with dark eye, pink bill, lime green body with yellow flecking on back and under-body, faint rose coloured breast streaked with yellow, shortish tail. In flight shows no red on the under wing, a very fast direct flyer. **Nest:** Normally in a horizontal hollow on an overhead thick branch. Nest in a variety of eucalypts throughout its range. Clutch size 1-2. **Similar species:** In region, only Red-collared Lorikeet, which has totally different head colouring being violet blue with orange collar, much longer tail, in flight on the underwing, Rainbow Lorikeet has bright red on the forewing. **Status:** Uncommon throughout the Kimberley region, but large groups found when in a feeding flock. **Where to find:** Throughout the Kimberley from Broome across into NT and south to the Fitzroy. Driving the Gibb River road and the Kalumburu road gives you the best chance as you have to get them feeding and they can be anywhere so by driving the length of these roads you pass through prime feeding habitat.

Photo:Simon Nevill

Purple-crowned Lorikeet *Glossopsitta porphyrocephala* 17-18.5 cm

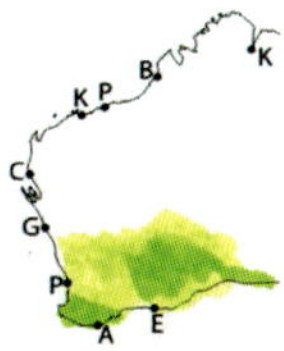

Western Australia's smallest parrot, feeding almost entirely from the pollen and nectar of flowering eucalypts. It is highly nomadic in its search for flowering trees and will move into areas when certain eucalypts are in full bloom. Often seen flying at very fast speeds over the tree-tops, while the group call continuously with their high pitched 'zit zit' to keep the group together. They can be difficult to locate when feeding high up in the tall flowering eucalypts but when a good view is had, the multi coloured head makes it well worth finding. Its range in the South West is linked to the occurrence of stands of flowering eucalypts. In the deep South West it feeds off the Karri (*E. diversicolor*), Marri (*Corymbia calophylla*) and Jarrah (*E. marginata*), in the Wheatbelt Wandoo (*E. wandoo*). Powder-bark Wandoo (*E. accedens*), Wheatbelt Wandoo (*E. capillosa*) and Salmon Gum (E. *salmonophloia*). In the mallee region they feed on many mallees particularly Sand Mallee (*E. eremophila*) and Capped mallee (*E. pileata*). A major region with high populations of Purple-crowned Lorikeets is the Transitional Woodlands of the south western Goldfields from the Vermin Fence 50 km east of Hyden to Norseman and beyond. There are well over 100 species of eucalypt in this region and there are always one or two eucalypt species in bloom at any given time of the year. **Nest:** Nests in a variety of eucalypts including Karri, Wandoo, Swamp Yate, Goldfields Blackbutt, Black Morrel and Salmon Gum. Clutch size 2 to 3. **Similar species:** None. **Status:** Locally very common. **Where to find:** The deep South West where Karri grows, from Pemberton all the way along the southern coast to well past Esperance, including: the Walpole National Park, Porongurups, Stirling Range and Fitzgerald River National Park. On the Hyden to Norseman and the Lake King to Norseman roads. Occasionally in Dryandra Woodland NR depending on how the Karri Forest is blooming in the deep South West Region. Can be found on the southern edge of the Nullarbor in the mallee. It is a monotypic so Eastern States birds are the same nominate species.

Photo: Simon Nevill

Regent Parrot *Polytelis anthopeplus* subspecies *anthopeplus*

40-42 cm

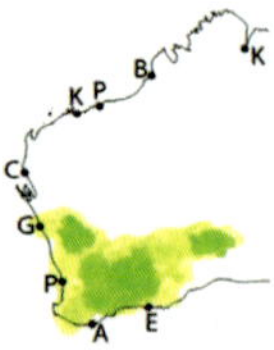

There are 2 subspecies of Regent Parrot in Australia; in the south west is the subspecies *anthopeplus*. There are very slight differences in the two subspecies. The eastern Australian birds are a duller yellow, the Western Australian birds have slightly darker lime green body and other minor differences. All the Polytelis group are extremely fast flying parrots, normally identified in flight with their contact calls, a plaintive soft screech. The call is similar when perched but softer to the ear. Has an extremely long tail although not quite as long as Princess Parrots. The male is a brighter lime green with brighter red on the wings, bill brighter red, contrasting

Photo: Simon Nevill

vividly with the lime green head. Has a long dark blue tail. Female has similar colouring but body is duller and darker lime green. Bill is a less intense scarlet. Main difference is the tail, which is dull green not dark blue. They favour certain eucalypts from which to feed on and also nest in. On the Swan Coastal Plain it is the Tuart (*E. gomphocephala*). They do not favour Marri or Jarrah hence they are rarely found in the deep south west. In the Wheatbelt they favour Wandoo and Yate Woodland in which they also nest. In the Salmon Gum woodland east of the Vermin Fence between Hyden and Norseman they nest and feed in a variety of large eucalypts. In the north-eastern Transitional Woodland inland from Wubin, they feed and nest in York Gum (*E. loxophleba* subspecies *loxophleba*). *E. oldfieldii* and Salmon Gum (*E. salmonophloia*). **Nest:** Normally above 15 metres. Clutch size 3 to 4. **Similar species:** None in this region. **Status:** Overall uncommon, moderately common locally. The Regent Parrots population in the South West expanded with the land clearing and the growing of grain crops and by the mid 1940s it was one of the most common parrots of the Wheatbelt and in fact was declared vermin; but then numbers drastically declined. It has slowly reduced since the 1950s particularly in the northern Wheatbelt. Only in the last 20 years has their population stabilized but they do not occur in the numbers as previously years. **Where to find:** There are distinct areas where Regents may be seen and this is related to where they have nesting sites, although they will move in summer returning back to favoured nesting locations. The Wandoo Woodlands in the South West on the western and northern sides of the Stirling Range. The Woodanilling—Kojonup area. Narrogin to Dryandra Woodland NR. On the Swan Coastal Plain on the Mandurah to Bunbury road and in the Pinjarra area. On the Hyden to Norseman road and the Lake King to Norseman road. 60 km north of Beacon on the Bimbijy road and between 40 and 60 km north east of Wubin.

Princess Parrot *Polytelis alexandrae* 42-45m

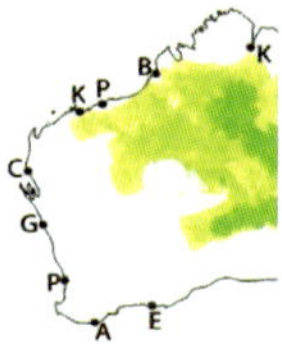

What a bird. Excuse the emotive statement but when I think of this bird, I think of my first sightings way back in the early 1980s when few birdwatchers had ever seen this bird. it was not so much that they lacked observational skills, no it was more about the remoteness of this bird's preferred habitat, mainly centred in Great Sandy Desert in those days. I have always had a love of the deserts and never feared venturing into remote areas as I still do, and often alone not a wise move, but I do go well prepared. I first saw the Princess Parrots not on the Canning Stock Route but on the Gary Highway north of Windy Corner. In subsequent years when running tours I started to go south from the Breaden Hills down the Canning always going as far as Lake Tobin. I have taken the journey 5 or 6 times but only once did I dip on this species. I found it fairly approachable. On one of my tours, I'd seen them fly fast over a ridge and we managed to sight where they had landed. After a careful approach with the group, the birds just stayed under the canopy of a tree, we did not even need binoculars and it was a lovely feeling for us all to just walk away and leave these stunning parrots well alone. Princess

Photos: Simon Nevill

Parrots are highly nomadic. Besides the Canning Stock Route and the Gary Highway I have seen them east of Neale Junction but only once ,even though I have travelled it a few times. I started to keep a record of the various sightings of mine plus other records including Peter Kelly and partner Janet Morris also recording the notes of Mike Carter and others plus a few Atlas records. The distribution range extends from Laverton in the west to east of the Macdonnell Ranges and from Well 47 in the north to the Anne Beadell Highway. There are sightings out of this range but I would surmise this is the normal breeding range. Having travelled throughout all the WA deserts and down the Tanami a few times, most of my sightings have been between Well 38 and Well 45 of northern Canning. Also between Gary Junction and just east of Jupiter Well, however that was mostly in the 90s The prime habitat in these areas is typified by high parallel dunes but the important aspect are the swales in between the dunes. I have seen them feeding on *Aleuta* sp (genus previously Thryptomene) both on the ground and on the plant itself. It's these plants, plus the fruiting bloodwoods that could be their prime source of food. Being highly nomadic Princess Parrots will certainly move from their central core ranges, hence recent irruptions to the north east of Alice, however irruptions to the Alice region and the Olgas region are not new but many decades may pass before they are sighted in these areas suggesting that there may be two core areas, one the Sand Dune Bloodwood (*E. chippendalei*) region of the Great Sandy and south in the Victoria Desert region as many sightings have been made down there particularly between 2006 and 2012. In talking to Peter Kelly (one of the best and most experienced desert tour operators I know) he mentioned that on 10 trips made along the Anne Beadell Highway from the mid 90s to 2008, roughly 80% of them, they managed to see Princess Parrot. **Nest:** Not well known but tree hollows in River Red Gums have been recorded as a favoured site but that is because those sites are around the Alice Region where recent nests were observed. It still begs the question if the Lake Tobin–Jupiter Well area has been a possible core centre, where are they nesting as there are no river gums, that I know in that region. Its possible that the Desert Bloodwood (*E. terminalis*) or Sand-dune bloodwood (*E. chippendalei*) are possible nesting sites but in general do not produce large tree hollows and Desert Sheoak (*Allocasuarina decaisneana*) does not produce hollows. In the Victoria Desert there is certainly no shortage there as Marble Gum (*E. gongylocarpa*) certainly has large nest hollows. Princess Parrot male has light blue crown, pink throat, pale white-cream under-body, lime-green shoulders, olive-green back, pale violet rump, pink vent, an exceptionally long tail, longest of all the *polytelis* genus. Female colouring very similar but overall duller with shorter tail. Flight like all *polytelis*, extremely fast and very direct, makes contact calls while in flight to keep contact just the same as Regent Parrots do. Clutch size 4-6. **Similar species:** None in the region. WA Regent Parrots do range as far east as Zanthus but still not in the range of Princess. **Status:** Locally common, perhaps generally uncommon but not scarce as it just occurs where few people travel. **Where to find:** The Great Sandy Desert. Victoria Desert and western Tanami Desert. Spasmodic irruptions to central Australia.

Western Rosella *Platycercus icterotis* subspecies *icterotis* 25-30 cm

Endemic

The Western Rosella is the smallest of all the rosellas in Australia. There are two subspecies in the South West: *icterotis*, which occurs in the wetter forests of the deeper South West and *xanthogenys* in the Wheatbelt and eastern and north eastern Transitional Woodland. There is a hybrid zone, roughly running from north of Perth down past the Stirling Range to Bremer Bay. The Wheatbelt *xanthogenys* male birds' main contrasting feature, particularly in areas like Hyden and Bruce Rock, is a red to the back (mantle) rather than green. They also have more blue in the wing coverts and a dark blue tail rather than a dark blue-green tail. There are other minor differences in rump and cheek colour. Also the outer Wheatbelt birds are slightly smaller and slimmer than the deep South West Region birds. The female of the subspecies *icterotis* has duller colouring, lacking a bright red head and under body, and a red patch on the forehead and mottled green crown and upper body. Cheek patch is duller cream-yellow as compared to bright yellow as in the male. The female subspecies *xanthogenys* is far redder on the under body and redder on the head like the male. The South West birds are particularly prevalent in the lower strata of the Karri Forest feeding off acacia and other understorey plants. Both in the wetter zone and the drier zone they prefer to feed off casuarinas and allocasuarinas. Often when walking through groves of Rock Sheoak (*Allocasuarina huegeliana*) in the outer eastern Wheatbelt, suddenly you can come across these beautiful small parrots that seem so out of place in drier environments compared with the other subspecies that occur in the damp Karri forests in the deep South West. All rosellas throughout Australia have as part of their calls a bell-like note and this will often lead you to where the birds are located. Nests: The South West birds nest in Karri. Marri. Jarrah and Wandoo the eastern Wheatbelt birds nest in Wheatbelt Wandoo and Black Morrel as well as other Transitional Woodland eucalypts. Clutch size 3 to 5. **Similar species:** None. **Status:** The subspecies *icterotis* is common in the wetter forest of the deep south west. Uncommon on the Swan Coastal Plain. Moderately common in the Darling Range. Moderately common in the marginal zone of the Wandoo Woodland. *xanthogenys* is uncommon in the outer Wheatbelt areas and the Transitional Eucalypt Woodland. **Where to find:** The subspecies *icterotis*, throughout the Karri Forest. Also from Dunsborough all the way along the southern coast to Bremer Bay. Most prolific between Dunsborough and Albany and the inner Wheatbelt from Jurien Bay across to Tuttaning NR. down through Dryandra Woodland NR. Dongolocking NR and the Stirling Range NP. subspecies *xanthogenys* is found in the outer Wheatbelt and well past Norseman into the huge Dundas NR. as far as Newman Rocks, also north to Southern Cross.

Photo: Simon Nevill

Northern Rosella *Platycercus venustus* 29-32cm

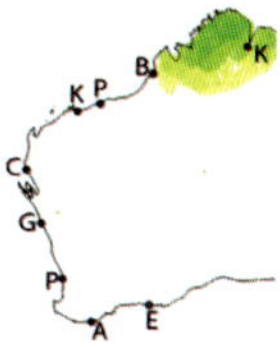

Not a common parrot. Mostly found in open woodland or close to creek lines. A medium sized rosella. I have come across them feeding on acacia seeds only a few metres high, making no noise at all, not even chattering to the few that are nearby. Normally solitary or in pairs or small groups, never in large groups. Colouring not bright, body pale yellow, wings darkish blue not iridescent, red vent, black cap, longish tail black tail, under-tail blue, typical rosella facial disc in this case white surrounded by blue margin wider in the Kimberley subspecies *hilli* than the NT birds subspecies *venustus*. Typical rosella undulating flight with a rosella bell-like call whilst in flight, call softer than other rosellas. **Nest:** Tree hollows. Clutch size large 2-3. **Similar species:** None in region. **Status:** Uncommon throughout the

Photo: Simon Nevill

Kimberley. **Where to find:** If entering the Kimberley from the west they are normally encountered from the Napier Range into the Kimberley proper, with most sightings in the hilly parts of the Kimberley. Always possible off the Gibb River Road or Kulumburu Road all the way to the Mitchell Plateau. Found in the Kununurra area towards the WA border. Often seen in the Bungle Bungle range particularly around the Bell camping area.

Australian Ringneck *Barnardius zonarius* subspecies *semitorquatus* 32-44 cm

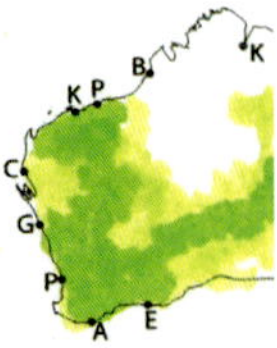

One of the most prolific birds to be found in the whole of the south west. In Australia, authorities have split the Ringnecks into 3 subspecies groups (some argue for four). Here in Western Australia we have 2, *zonarius* and *semitorquatus*, with a hybrid zone between them. *Zonarius* is a bird of the dry country, its southern boundary running roughly from Jurien in the far west through to Wyalkatchem, Hyden to Israelite Bay. It stretches north to the Great Sandy Desert and right across central Australia excluding the Nullarbor. In the northern parts *zonarious* is a markedly smaller and slimmer bird. As you move through the range of the Australian Ringneck, one will find that there are subtle differences in call. The main plumage variation is that there is no red dot above the bill on the forehead. There is a distinct contrasting break on the under body from a green chest to a bright yellow belly. All the green plumage is paler, particularly on the back, *semitorquatus* has a distinct red patch just above the bill on the forehead, an all green under body, darker on the breast, lighter on the under body. The bird is much thicker set and longer bodied. When pairing or making contact with a group, will give a chattering call while shacking their tails from side to side; also will give a high pitched alarm call when disturbed, and the characteristic 'twenty eight –twenty eight' is the normal contact call. *semitorquatus* has certainly increased in numbers since European settlement and has become a major problem for orchardists. **Nest:** Nests in a variety of

Photo: Simon Nevill

eucalypts throughout its range. Clutch size 3 to 5. **Similar species:** None. Status. Common in the South West. Very common in the Wheatbelt. Moderately common in the drier regions. **Where to find:** Throughout the state north to latitude 20. Not common in the Karri Forest and heath lands of the Esperance region. Both hybrids and the distinct subspecies can occur together in localities like Dryandra Woodland NR which is in their transitional zone.

Red-capped Parrot *Purpureicephalus spurius* 34-38 cm Endemic

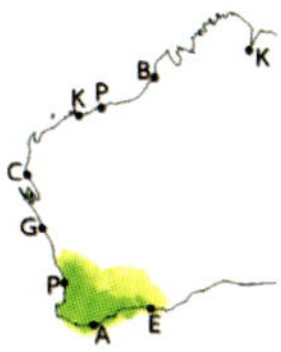

The male bird is one of Australia's most stunning parrots with a distinct red crown, lime green cheeks, purple under body, red vent, green back and wings, yellow rump and long green tail. Many birdwatchers presume the duller juveniles are the female of the species but in fact both the juvenile male and female have a duller and more subdued plumage, having a dull green not red crown, pale green cheeks, vent mottled lime green and red, under body pale grey tinged purple with no bright purple under body. The difference in male and female juveniles is very difficult to see and too subtle to warrant inclusion here. They retain the duller plumage through the summer until the following breeding season. The female adult is very similar to the male but the crown is slightly duller, more dark maroon than red, the back is duller green and the vent has more yellow flecking in the red but all in all they are difficult to split, the main difference is that the male's colours are more intense, particularly the red crown and lime green cheek patch. The Red-capped Parrots long down curved bill is designed expressly to remove the seed from eucalypt fruit capsules, particularly the fruit from the Marri tree. It grasps the large fruit capsule with one foot just like the larger Cockatoos but, unlike the Cockatoos bill that has the ability to severe a hard nut in two, the Red-capped Parrot delicately probes with the long bill cutting free the internal seeds. They feed quietly, the only sign of their presence often being the sound of falling eucalypt fruits from high up in the canopy. They will feed with other parrots but generally are in very small groups of 2-6 birds or on their own. After the breeding season through the summer months, they can congregate in small groups of 10-15 birds. Besides the fruits of Marri, Jarrah and Tuart, they will also eat the fruits of Dryandras and Hakeas. Their range in the north almost follows the limit of the Marri tree from Hill River, then south east to Dumbleyung and across to Esperance where it is very common. **Nest:** Nests in Marri, Jarrah, Flood Gum and Wandoo. Clutch size 3 to 5. **Similar species:** None. **Status:** Moderately common in the wetter forests of the south west and along the southern coast. Moderately common in the inner Wandoo belt, uncommon in the outer Wheatbelt. **Where to find:** Throughout the Swan Coastal Plain and particularly the Darling Range to the south coast. Highest numbers occur from Margaret River down to Pemberton and then along the coast to Esperance. Particularly abundant in the Walpole to Albany region and the Esperance Region.

Photo: Simon Nevill

Mulga Parrot *Psephotus varius* 27-32 cm

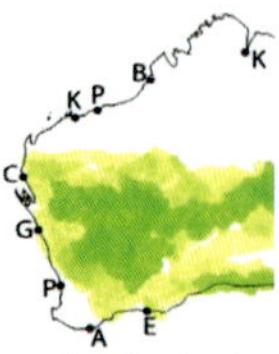

Well named, as its range follows almost entirely the extent of the Mulga (Acacia aneura), roughly as far north as Mount Augusta but they can be found north of here. It does come into the Wheatbelt region, but in the main its primary habitat is north and northeast of the Transitional Woodland line. The male is brightly coloured with a turquoise head, under body and mantle and wings. On the very top of the crown is a small patch of red. On the forehead is a small patch of yellow. The shoulder is bright yellow and stands out from the turquoise wing. The belly is bright red bleeding into a yellow vent. In flight, the blue in the wings becomes prominent and the red rump shows well. The female is much duller. The main body and wings are shades of olive-green. The shoulder is a dull red unlike the bright red of the male. The crown and the nape have a red patch more extensive than the male but a dull red-brown, also on the forehead there is a small yellow patch above the bill. It feeds on seeds on the ground, mostly on native grasses, hence in the Wheatbelt, it is invariably found along roadside verges. In the semi-arid zone it starts feeding soon after sunrise and then rests for much of the day. Comes to cattle troughs and waterholes in the morning and late in the evening particularly in the summer months. Often missed midday as it sits quietly in mulga foliage like Scarlet-chested Parrots and Bourke Parrots. Has a pleasant soft whistle but does not call much out of the breeding season except in flight to keep contact. **Nest:** Nests in a variety of eucalypts as well as mulga trees. Clutch size 4 to 6. **Similar species:** Red-rumped Parrot, not found in this state. **Status:** Uncommon in the central and northern Wheatbelt. Scarce in the Transitional Woodland. Common in the Mulga Woodland. **Where to find:** Throughout the sheep station country in the Mulga Region.

Male

Female

Photos: Simon Nevill

Bluebonnet *Northiella haematogaster* subspecies *narethae*

30-35cm

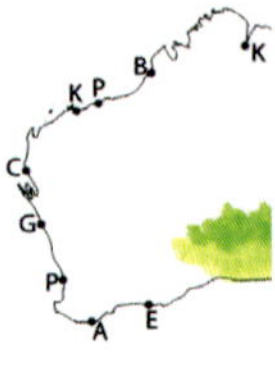

There are 4 subspecies of Blue Bonnets in Australia, most of them but not all, are found in dry habitats. The WA subspecies *narethae* certainly occurs in a very dry part of this state. It was known as the 'Naretha' Blue Bonnet named after the small railway siding on the Trans Continental

Line, which was thought to be the centre of its range. Until recently, I thought that Naretha and Rawlinna were the epeicentre of this disjunct population cut off from the other subspecies, whose distribution did not overlap but since some birdwatchers have travelled along the Tjuntjuntjarra Track (note - Aboriginal private track restricted access only) the range of Blue Bonnets has been extended and in fact the Naretha subspecies flows into the SA subspecies population of *pallescens*. Often seen in small groups from 5 to 10 birds. They are particularly common where the acacias Myall (*A. papyrocarpa*) and Miljee (*A.oswoldii*) grow which are two species of tree that they will nest in. They feed on a variety of seeds but particularly the chenopod seeds that grow on the Bluebush (*Maireana sedifolia*), Grey Bush (*Cratystylis conocephala*). Blue Bonnet's are grey on the upper-body and head, breast a lighter grey, face blue, pale lemon-yellow on shoulders, reddish tinge on lesser coverts, blue on primary coverts, and bluish-black primaries. The Nullarbor subspecies *narethae* is the only subspecies to have no red on the centre of the belly, it only has red on the vent but has a two tone blue face with light blue on the forehead splitting at the eye to darker blue on the lores. *narethae* is also slightly smaller than the nominate species. Flight is relatively fast and undulating. **Nest:** Tree hollows but often in the cracks between the splitting trunks of Myall, as hollows are few. Clutch size large 3-8 average 5. **Similar species:** None in region, although co habits with Mulga Parrots in the northern parts of its range which are smaller, slimmer birds, male colouring can not confuse but at a distance females could confuse but no blue face. **Status:** Locally common. **Where to find:** Along the Trans Continental Line from east of Kitchener. The northern part of the track that leads down from Rawlinna to the Eyre Highway (let people know if you are using this track) should produce this species and where there are water troughs, the parrots can be seen to come in early morning or late evening. They can be found on the southern part of the Connie Sue Track where the Myall and saltbush still occurs.

Photo: Simon Nevill

Budgerigar *Melopsittacus undulates* 18 cm

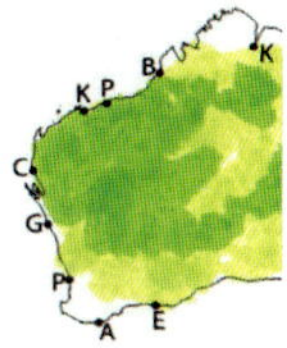

One of the quintessential Australian birds, found throughout the arid regions of Australia. Highly nomadic, its movements are related to rainfall which governs the growth of native and introduced grasses. It occasionally comes into the south west when the arid regions are experiencing drought and is likely to occur in the Gascoyne River region, however the Pilbara is far more reliable having permanent water and profuse grasses. Budgerigars can be occasionally found in any part of the northern, central and eastern part of the Wheatbelt in summer but never in the numbers that can be found in the Pilbara and the Mulga Region and the Gascoyne catchment. When conditions are good, river

Photo: Simon Nevill

systems with Red River Gum (*E. camaldulensis*) can be alive with birds all chattering and quite approachable near their nest sites, with every tree and nest hollow being taken up with resident pairs. To witness great flocks circling over waterholes is a wondrous sight, with the swirling mass of 3-4 thousand birds making turns in total unison; the sound of the swirling wing feathers is also unforgettable as the flock twists and turns and then breaks up and falls like confetti above the water holes to drink. **Nest:** Nests in a variety of eucalypts, particularly River Red Gum and Coolibah. Clutch size 4 to 6. **Similar species:** None. **Status:** Scarce to Wheatbelt (never in large numbers). In the mulga can be abundant or scarce depends on the season. Common in the Pilbara, deserts and throughout the southern Kimberley. **Where to find:** Gascoyne, Fortescue, Lyons and Ashburton Rivers. Seen mainly in the lower Mulga Region when the Pilbara and deserts are drier. Seen best in the Ashburton and Pilbara regions. Throughout all WA deserts. Fitzroy River system and south.

Bourke's Parrot *Neopsephotus bourkii* 20-23 cm

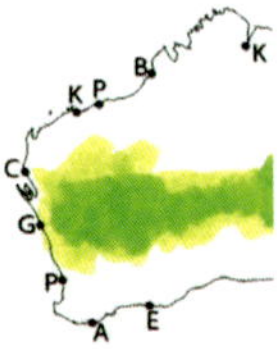

The Bourke's Parrot can be found in a similar range to the Mulga Parrot but ironically is far more dependent on mulga woodland than the Mulga Parrot whose range spreads further than the Mulga. Unlike the nomadic Budgerigar, the Bourke's Parrot is basically sedentary and unless there is severe drought in the area, will remain in the same region providing permanent water is available. They come to drink just before first light and after sunset, perhaps as a precaution against raptors such as Collared Sparrowhawk and Brown Goshawk, making soft twittering calls as they come to drink in the semi-darkness, one group after another slowly walking to the edge of the water. The female differs slightly from the male, mainly in having a lighter grey-brown back with stronger pink colouring on vent. The main distinct difference, and one that requires a good view of the bird, is that the male has a fine light blue supercilium stripe that passes around the forehead. The female has no such stripe. Like all the Neophema group, Bourke's Parrots sit for hours resting at midday, particularly when hot, so can

Photo: Simon Nevill

easily be missed. Most active early morning. **Nest:** Nests in a variety of eucalypts and also in Callitris pines. Clutch size 4 to 5. **Similar species:** None. **Status:** Uncommon to common in the Mulga Woodland but moderately common from Muggon to Laverton and Mt Magnet in the south to Kumarina in the north. Moderately common in parts of the Pilbara where extensive stands of mulga occur. **Where to find:** Wooleen Station. Austin Downs. Nallan Station. Erong Springs. Doolgunna and Mooloogool Stations. Laverton Downs Station.

Elegant Parrot *Neophema elegans* subspecies *carteri* 22-25 cm

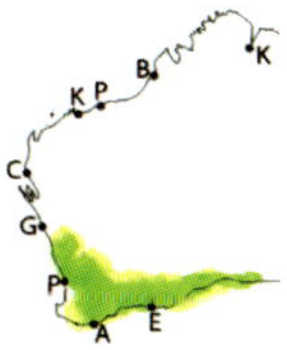

There are two subspecies of Elegant Parrot in Australia, over 1000 kilometres apart. The eastern states subspecies is the nominate race *elegans*, in Western Australia the subspecies is *carteri*. There are no differences in plumage, only size, *carteri* being smaller with a shorter tail. Elegant parrots inhabit open woodlands with grasslands. Since European settlement, it has extended its range into the south east. Although not common, it can now be found as far south as the Albany region in fact only reaching Albany and the Swan Coastal Plain only in the late 1960s. Its main northern limit is approximately to Morawa, although has be been seen further north. It avoids the coast until just south of Perth. Found inland to Wubin and across to Merredin and then past Salmon Gums down to Condingup. Elegant parrots feed almost entirely on the ground on seed, small fruits and occasionally insects. Often found on roadside verges in the inner Wheatbelt or amongst native grasses in the Wandoo Woodland. The female is very similar to the male except the back is duller it has a blue and yellow band to forehead and is thinner and duller. The male has more blue in the tail. The female looks overall duller, almost the colours and hue of a Rock Parrot. **Nest:** Mostly in Wandoo trees in the south-west. Clutch size 4 to 5. **Similar species:** Rock Parrot, which is a generally duller bird, more so than even the female Elegant and closer to the immature Elegant Parrots colouring. The Elegant adult bird is a brighter green overall. The specific difference is that the Rock Parrot has no yellow on the lores but a two-coloured blue band that goes from the eye to the forehead and is less contrasting in blue tones than the Elegant. The Elegant has a distinct light blue and dark blue band. The Rock Parrot has a dull blue eye ring whereas the Elegant has a faint yellow eye ring, the Elegant has yellow in the outer tail feathers whereas the Rock Parrot has green. The Elegant certainly does not feed on succulent plants on beaches and sand dunes, as does the Rock Parrot. The Rock Parrot nests in rock crevices or under on offshore islands whereas Elegant Parrots nest in tree hollows. **Status:** Uncommon in open woodlands with grassed understorey. Scarce in the outer Wheatbelt. **Where to find.** Boyagin Rock NR. Dryandra Woodland NR. Stirling Range NP. and particularly the surrounding farmland abutting the Stirling Range. Locally common from Tenterden down to Albany.

Elegant Parrot

Rock Parrot

Photos: Simon Nevill

Rock parrot *Neophema petrophila* 22-24 cm

There are two subspecies of Rock Parrot: the nominate race in Western Australia *petrophila* and subspecies *zietzi* in South Australia. A bird of the seashores and islands along the south west coastline, does not venture far from coastal shores. Feeds on succulent plants like pigface as well as other seeding plants like beach spinifex and dune arctotheca. Found along beaches and on sand dunes running from one clump to another; normally one or two birds, or in small groups of up to 6 -10 at most . Will also feed on low bushes that are close to the sea. On Rottnest, comes to the centre of the island to feed on samphire seed adjacent to the salt lakes. **Nest:** Nests under rock overhangs and crevices on off-shore islands, flying to the mainland to feed. Clutch size 3 to 4. **Similar species:** Elegant Parrot. Refer to notes on Elegant Parrot. To confirm sighting of Rock Parrot look at the lores. If you know the colouring of the Elegant, note how dull the overall plumage of the Rock Parrot is, being more olive-green than the brighter lime green of the Elegant Parrot. Status. Scarce in the Shark Bay Region on small islands in Freycinet Reach between Useless Loop and Peron Peninsula. Uncommon from Jurien Bay to Lancelin. Moderately common between Cape Naturaliste and Cape Leeuwin. Moderately common from Walpole to Esperance. **Where to find:** Lancelin. Rottnest (not so easy nowadays but they are there). Carnac Island. Easiest to find on south coast: Torbay Beach and beaches opposite Shelter Island. In Albany walking along Middleton Beach to Emu Point or on grassed lawns adjacent to Oyster Harbour. Two Peoples Bay at Little Beach. Bremer Bay at the inlet at John Cove and Short Beach and Fisheries Beach. In Esperance along coast and very good chances on Woody Island.

Photo: Simon Nevill

Scarlet-chested Parrot *Neophema splendida* 19-21 cm

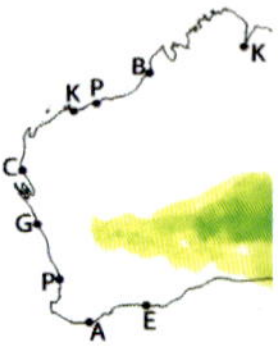

The closest this species comes to the south-west is occasionally the Mt Jackson and Paynes Find area. I have only only seen this bird in the Great Victoria Desert and west of Warburton. Over the last few years I have seen them several times along the Anne Beadell Hwy. Its stronghold seems to be centred on the Western Australian and South Australian border and across into Maralinga and Yellabinna Regional Reserve and Glue Pot in South Australia. It is an irruptive species and will move to better feeding areas out of its normal range and it is often sighted a long way from its central range. The male is a stunning little bird with its bright red chest, blue face and yellow belly. Feed on seeds from native grasses and herbaceous plants. Sit in bushes and trees for long periods through the day so can be easily missed; often flushed from the side of desert tracks by vehicles. Their call is a soft whistle. I camped for a few days on the Anne Beadell Highway where they were feeding in mallee and from native seeds on the ground. By the second day they became used to my presence and allowed a fairly close approach on a few occasions, a real privilege. **Nest:** Nests in eucalypts mainly in the Victoria Desert. Clutch size 3 to 5. **Similar species:** None in their habitat. **Where to find:** In localities like Plumridge Lakes. Northern end of Connie Sue. Warburton. Along the Anne Beadell Highway into South Australia. On a tour all the way from Yeo Lakes to Volkes Hill Corner in South Australia, not one was seen even though a few years previously, the largest flock was sighted with over 200 birds just south of Volkes Hill Corner where we travelled, such is the nature of this elusive bird.

Male

Photo: Simon Nevill

Western Ground Parrot *Pezoporus flaviventris* 30 cm

Endemic

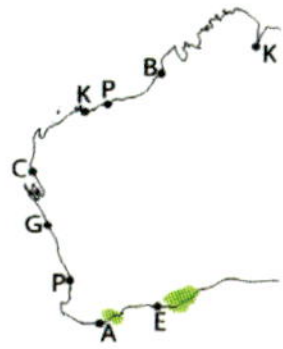

There were two subspecies in Australia: *wallicus* in the eastern states and *flaviventris* in Western Australia and possibly in the future the Tasmanian birds may be split to a full species as they are certainly darker than the mainland birds. The WA birds have now a full species status and one of our rarest birds. The Western Australian birds have lighter yellow-green underparts and fainter barring on belly and are generally lighter green all over. The Ground Parrot is basically an all green plumaged bird, with black and yellow flecking covering much of its body. Has a relatively long tail with a distinct red dot above the bill at the base of the forehead. The Ground Parrot is perhaps Western Australia's most endangered land bird (excluding the Night Parrot). Its population since European settlement has drastically declined, primarily due to the introduction of the fox and even more so the feral cat which has had a devastating impact on all ground dwelling birds. Also in some of their last remaining strongholds, fires have greatly affected their population. The bird feed almost entirely on the ground, bending grasses and sedges down to beak height to eat the seeds. They feed on buds, green seeds and heath plants (*Epacridaceae*). They apparently derive enough moisture from the plants they feed off and do not need to go to water to drink like most parrots. It will climb into low bushes to feed but more commonly walks through the undergrowth selecting seed. It has a plaintive call reminiscent of a Tawny-crowned Honeyeater which also calls in the same region, but it is a stronger, and longer lasting call and once heard is easily remembered. It calls mostly twenty minutes or so after sunset before flying to its roosting site. It calls well before sunrise as much as 90 minutes and will then fly to its feeding area, so chances of seeing flying birds are remote and finding them in the bush even harder also throughout their range Ground Parrots in Australia, require dense low heath which often hard to walk through. **Nest:** Nest a shallow scrape on the ground below dense bushes or clumps of grass, lined with short lengths of grass. Clutch size 2 to 3. **Similar species:** None. **Status:** Extremely rare.

Where to find: Four known populations: Waychinicup (population tenuous over the last few years they have not been heard in the region over the last few years). Fitzgerald River NP. Cape Arid NP. Nuytsland NP. Looking for Ground Parrots is discouraged due to the fragile nature of their population. However, volunteers from time to time are welcomed to assist with surveys but must be reasonably fit. Organised under the guidance of the government agency the DPaW (previously DEC) and mostly co-ordinated through the agency's office in Albany.

Western Ground Parrot young are greyer green than the eastern states birds

Photos: Brent Barrett

Night Parrot *Pezoporus occidentalis* 23 cm

A brief note on the elusive Night Parrot. A species closely related to the Western Ground Parrot. Historically, an uncommon desert dwelling parrot, which occurred in the Spinifex plains of much of central Australia. The population by the early 1900s was almost decimated by a combination of factors, including the introduction of livestock, feral animals like the camel and fox but most of all the feral cat. A recently dead bird was found in Queensland in 1990. Since then there have been numerous reported sighting. The early naturalist G. A. Keartland, found the bird with the aid of indigenous guides near Lake Austin in September 1854. Since then, it has been sighted by two ornithologists in recent years near the Fortescue Marshes in the Pilbara. The most important information I ever gleaned from someone about this species was from Robbie Robinson, an indigenous elder from Yuendemu settlement on the Tanami Track. I was showing him pictures of the Princess Parrot, as I was interested in their distribution in the NT. He commented 'No, we don't get this bird. We get him Balgo way'. Well I knew he knew his facts as he was spot on with the general location. However, what did amaze me was when he was flicking the pages of the guidebook, he pointed to the Night Parrot and said 'We get, this bird but we get him dead, snake get him'. I asked him where would he find it, he replied, 'Tanami'. Refering to the vast spinifex plains before you get to the Tanami Mine on the main track. The uncertainty surrounding the species may now be put to rest as the ever, resourceful John Young of Queensland has reputedly found the bird in central Queensland and at this very moment, he is having discussions with the Western Australian Museum. They have been conducting DNA studies from a few feathers that John collected from a roosting site in the Lake Eyre Basin. Much rests on the development of this finding, as immediate protection should occur as soon as possible, to protect this rare species.

Red-winged Parrot *Aprosmictus erythropterus* 32-33 cm

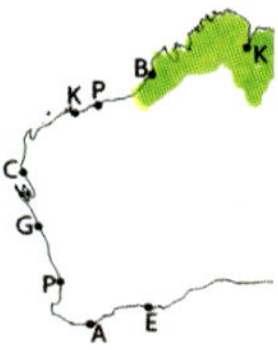

A bird of tropical and subtropical Australia with a range from Cape Missiessy, on Eighty Mile Beach, across the Kimberley into the NT. The male has bright lime green body, black mantle, blue on the centre the back, rump lime-green, tail green with lime-green tip, shoulder (lesser coverts) bright red shows well when at rest, in flight shows two bright red patches on upper-wing contrasting with lime-green rump. Male in flight shows bright red wing patches well. Flight is sluggish, undulating and erratic in movement, not dissimilar to a Galah in flight. Female is different, the greens are duller, only small amount of red on the shoulders, no black on mantle but green, bill paler pink. Juvenile male very similar to adult female, having no black on mantle and has a small amount of red on lower shoulder. **Nest:** Not only in tree hollows but broken off tree stumps. Clutch size large 4-6. **Similar species:** No other species has the colour combination of Red-winged Parrots. **Status:** Fairly common throughout the Kimberley region but normally found as a solitary bird or small groups. **Where to find:** Throughout the Kimberley it's one of the few tropical parrots that extends its range into the arid region, hence when driving north of Sanfire Flats, you will start to see them in the Pindan Woodland as they love acacia seeds which are very prevalent in the Pindan.

Male

Female

Photo: Simon Nevill

Pallid Cuckoo *Cuculus pallidus* 31 cm

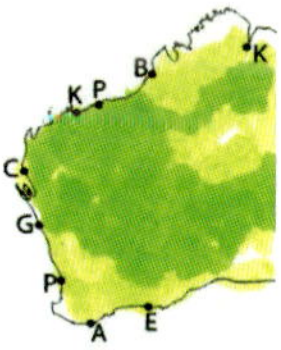

The movements of many cuckoo species are still not well known, mainly due to the fact that some birds stay longer in the south, and some remain in the north and do not migrate south during the spring and early summer breeding season. There is however a marked movement after the breeding season of Pallid Cuckoos moving north in the months of October to January and then arriving again in the south mostly between May to July. Pallid Cuckoos prefer open country that is lightly wooded or open acacia woodland. The Mulga Region is one of their preferred habitats. They also can be found along creek and river systems in dry country and throughout the desert regions. They do however come to northern and central parts of the Wheatbelt Region. The male has an all grey plumaged head with a dark line running through the eye to the shoulder. The female differs in having a brown stripe along the crown to the nape with a white spot on the nape. Generally, the male is greyer with no brown on the crown. Their call is a single whistle that is rapidly repeated in an ascending scale. Most often seen in a dead tree or on an exposed limb either calling or looking for food in open areas. They not only call for long periods during the day but also at night even when there is no moon. They call soon after arriving in areas in the mid west, typically in late winter and early spring. Several species are parasitised by the Pallid Cuckoo, the majority have open nests and must be insect feeding birds. The biggest percentage are honeyeaters, others include flycatchers, robins and wood swallows, all well represented in the semi-arid regions. Pallid Cuckoos lay through much of the year but the high point is August through to September when breeding by other species is at its maximum peak. All cuckoos in Australia will lay only one egg in the host species nest and laying in a number of other nests, allows them to maximize their prodigies' chances of survival. How many they lay throughout the year is still not well known but one would presume it to be between 3 and 5 eggs per season. Like many cuckoos it feeds on caterpillars, removing the guts and toxins by hitting the caterpillar against a branch or rock. It supplements its diet with other insects but caterpillars certainly make up the largest percentage of its food. **Similar species:** Oriental Cuckoo: Has strong black and white barring on under body. Grey head with no black eye stripe and yellow iris not black. **Status:** Scarce in the lower South West. Moderately common in the central and eastern Wheatbelt. Common in the northern Wheatbelt. Very common in the Mulga Region. **Where to find:** Dryandra Woodland NR. Tutanning NR. Wongan Hills. Lake Magenta NR. Found throughout the north all the way to the Kimberley.

Photo: Simon Nevill

Fan-tailed Cuckoo *Cacomantis flabelliformis* 24-28 cm

Photo: Simon Nevill

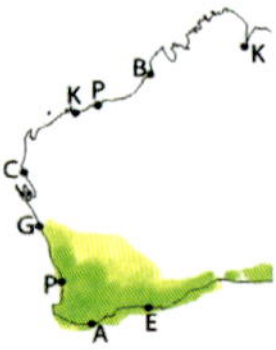

Unlike the Pallid Cuckoo that prefers semi-arid open woodland, the Fan-tailed Cuckoo prefers the taller wetter forests of the South West Region, often in areas with a thicker understorey. Perches at the mid canopy level, rarely in the high canopy. It has a distinct call very different to the Pallid Cuckoo. It has several calls but its primary contact call, is a whistle with descending notes that are rather plaintive and mournful. It is found in a variety of habitats including the tall Karri Forest, Jarrah Forest, and Wandoo Woodland. Thick riparian gullies. In trees adjacent to heath. Its main breeding area is the deep south west and southern coastal belt where it arrives in August to October; after breeding, it stays in the lower south west, leaving the region in April or May and moving northwards to the northern Wheatbelt and beyond through the winter months. In the Wheatbelt, the birds breed earlier, normally in August early September. Most female cuckoos will watch birds building a nest and will remain in the same locality feeding until the host species is ready to lay. They parasitise mostly domed nest such as those of the thornbills, fieldwrens, redthroats, scrubwrens, even fairy-wrens. They will parasitise some open nests but have a preference for domed nests. They feed on caterpillars, beetles, moths and any other form of insect. Their breeding range extends from the Lower South West to Kalbarri in the north, down to Koorda and east to Norseman and down to Cape Arid and all areas south of this line. There are 4 subspecies in the world with one in Australia. **Similar species:** None. **Status:** Moderately common in the Lower South West and uncommon in the Wheatbelt particularly furthest away from the lower south west. **Where to find:** Pemberton region all along the southern coast from Augusta to Esperance. Stirling Range NP. Dryandra Woodland NR. Throughout the Darling Range. Northern limit Kalbarri and eastern range along the southern boundary of the Nullarbor in the mallee.

Black-eared Cuckoo *Chrysococcyx osculans* 19 cm

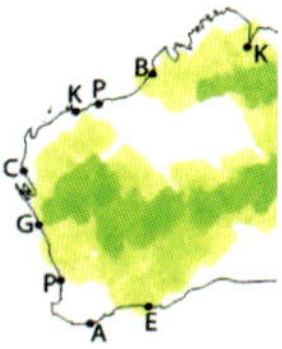

Found in the central semi-arid and arid regions of WA, also in the Kimberley. In the south west, found predominately in the Mulga Region. There is still much to learn about this species; we know they breed in the central arid regions particularly between Billabong in the west and Ashburton Downs in the north to Laverton and beyond in the east and to Bimbidji in the south. They do occur in other areas particularly the Pilbara, the lower deserts where mulga occurs, also in the central and southern Kimberley. They come to the central arid regions in April to July and, after the breeding season, leave to travel north in October to December. Black-eared Cuckoos parasitise mainly domed nesting birds such as Southern Whiteface, Western Gerygone, Redthroat and Slaty-backed Thornbill. Occasionally Grey-crowned Babbler and White-browed Babbler. They have been recorded nesting in open nests such as the Red-capped Robin's. Feed entirely on insects particularly caterpillars. Are mobbed instinctively by birds such as thornbills even if not near the thornbill's nest. Feed from perches low down, pouncing on prey or hoping on the ground in search of insects. Their call is very similar to the Horsfield's Bronze-Cuckoo with whom it often shares the same territory. There are several calls but the most common contact call of both species can be described as follows: the Black-eared Cuckoo's call is stronger than the Horsfield's a longer drawn out 'seeeeuu...seeeeuu…seeeeu' that goes up and then down. The Horsfield's call is a shorter, softer and more rapid 'seeu seeu seeu'. **Similar species:** None, although confusion can occur with Horsfield's Bronze-Cuckoo which has barring on the underbody and a green metallic tinge

Photo: Simon Nevill

to the upperbody, rust-brown in the outer tail feathers, eye stripe faint. Black-eared Cuckoo has plain olive-grey back, strong unmistakable eye stripe, no barring on underbody just plain cream coloured plumage. It also is a slightly larger bird. **Status:** Uncommon throughout range unless in country that has received above average rains and then can be moderately common. Where found: Mainly the mulga areas such as stations like Wooleen (can stay on the property). Thundelarra. Nallan (can stay at property). Gascoyne Junction particularly around Yalbalgo Station. Sandstone region. Between Kalgoorlie and Leinster. Uncommon in northern Wheatbelt but occasionally comes down to a line from Wongan Hills across to Kellerberrin and Southern Cross. Scarce below this line. Found in the Central Ranges and northern deserts.

Horsfield's Bronze-Cuckoo *Chrysococcyx basalis* 16 cm

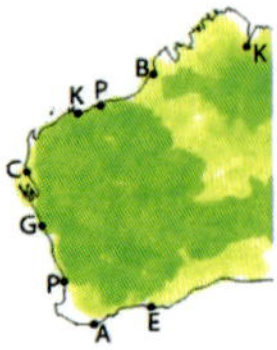

Comes down from the north particularly the Kimberley from March to May, although some do over-winter in the north. Breeds throughout the central regions as far north as the Pilbara and as far south as the edge of the wetter south west, virtually taking the place of the Shining Bronze-Cuckoo which only occurs in the wetter lower south west; however there is an over-lap with some birds sharing the marginal zones. Horsfield's return to the Kimberley and some even go further north to New Guinea and Lesser Sundas Islands between October to January. The Horsfield's Bronze-Cuckoo is more wide-spread in choice of habitat than the Black-eared Cuckoo and also more common and, like the Black-eared Cuckoo, parasitises birds that build domed nests but more so the fairy-wrens being a smaller cuckoo. The range of nests varies from habitat to habitat. In the north grasswrens, emu-wrens, Redthroats and thornbills are often selected. In the south, fieldwrens, fairy-wrens and thornbills are parasitised. In the Fitzgerald River National Park the author watched for a full hour or so, a poor Southern Emu-wren battling to keep pace with the demands of a well developed Horsfield's Bronze-Cuckoo four to five times its own weight. **Similar species:** Shining Bronze-Cuckoo, which has a more lustrous metallic green back. The important difference is that the Shining has no prominent eye stripe, and barring on underbody is stronger and the bars pass right around underbody. The Horsfield's has a duller green-bronzed back and the barring is incomplete in the centre of underbody with bars tapering and fading to points. Has a distinct black line running from the eye, broadening on the ear coverts to the shoulder. Calls are totally different refer

Photos: Simon Nevill

Shining Bronze-Cuckoo. The Horsefield's is a long drawn out 'seeeuu…seeeuu…seeeu' that goes up and then down with each call. **Status:** Uncommon throughout parts of the Wheatbelt and southern coast from Bremer Bay to Cape Arid. Moderately common in the northern Wheatbelt into the Mulga Region. **Where to find:** Best chances north from Wongan Hills to Kellerberrin. East from Kondinin to Norseman. Fitzgerald River NP. to Cape Arid. Most prolific in the lower regions of the Mulga Region.

Shining Bronze-Cuckoo *Chalcites lucidus* subspecies *plagosus*

14-18 cm

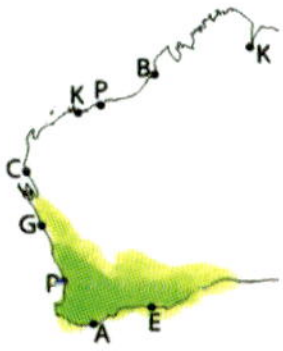

The Shining Bronze-Cuckoo arrives in the deep south west mainly from July to August and leaves after the breeding season between December and March. Their preference is for heavier forested areas but unlike the Horsfield's Bronze-Cuckoo that does not come into the thicker wetter forests and into the Wandoo only rarely, the Shining Bronze-Cuckoo breeds not only in the wetter south west but is found in open woodlands which are semi dry such in the New Norcia region and even further north. It still is the dominant Bronze-Cuckoo nesting in Dryandra Woodland Reserve even though one would not class the area as a wetter south west reserve. Inner Wheatbelt reserves like Dryandra Woodland Reserve are a fascinating transitional zone, as all four cuckoos that enter the south west, namely Fantail Cuckoo, Pallid Cuckoo, Shining Bronze-Cuckoo and Horsfield's Bronze-Cuckoo, can be seen in that reserve, showing how transition zones allow overlap of all known cuckoos that enter the south west. Shining Bronze-Cuckoo is more selective than the Horsfield's Bronze-Cuckoo on what type of nest it will parasitise. They almost exclusively parasitise enclosed domed nests such as those of White-browed Scrubwren, Yellow-rumped Thornbill, Western Gerygone, Southern Emu-wren, Inland Thornbill, Splendid Fairy-wren and Red-winged Fairy-wren. Rarely will they use open nests. The call is unmistakable and is sometimes given the nickname 'walking the dog' call. A repetitive, rapid whistle with each note descending 'seee seee seee' finishing with an extended 'seeeeeuu' which has an undulating whistle-warbling tone. All bronzewing cuckoos share the occasional habit of stretching their wings out while singing their contact/breeding call. There are 4 subspecies of Shining Bronze-Cuckoo in the world, only two migrate and breed in Australia, *plagosus* and *lucidus*, the other two are sedentary, *layardi* on New Caledonia and New Hebrides and *harterti,* on Renell and Bellona Islands.

Photo: Simon Nevill

In the west, plagosus migrates south from either northern Australia or from New Guinea or the Lesser Sundas Islands, most probably the west coast birds come from the Lesser Sundas. Some come south from the Kimberley and Northern Territory. **Similar species:** Horsfield's Bronze Cuckoo. Calls totally different. Horsfield's has barring on underbody broken at centre, strong eye stripe, duller bronzing on upperbody. Shining Bronze's barring on underbody is complete although thinning near centre, broader and bolder on flanks. No eye stripe. Brighter green-bronzed upperbody. **Status:** Moderately common in South West. **Where to find:** Karri Forest in the south west. The Jarrah Forest. The inner Wandoo Woodlands. All along the southern coast.

Little Bronze-Cuckoo *Chrysococcyx minutillus* 14-17 cm

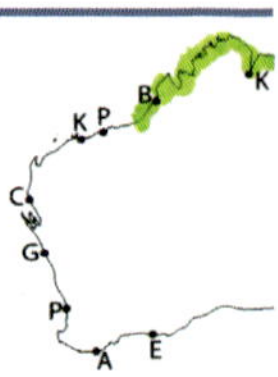

When you put pen to paper to write about the Little Bronze and Gould's Bronze Cuckoo it's like opening up a can of worms. It's hard enough initially for those just starting birding to get their head around the genus *Chrysococcyx*, but it's made doubly difficult for experienced birdwatchers when there are conflicting facts and at times vague hypotheses regarding the two species and it is certainly not yet conclusive but then, that's the joy of birding and establishing species status particularly regarding Little Bronze-Cuckoo *Chrysococcyx minutillus* and Gould's Bronze-Cuckoo *Chrysococcyx russatus*. Clements lists 13 subspecies for Little Bronze of which there are various complexes. What we do know is that hybridisation does take place between the overlapping subspecies, which certainly does not help identification as there will always be colour changes when you have a species ranging from the Malay Peninsula to South East Australia. As Parker himself wrote whilst studying the Little Bronze Cuckoo complex 'differences between Little and Gould's not always striking and difficulties in distinguishing them exacerbated by varying nature of contact zones'. Recent DNA studies by Leo Joseph and others questioned the 'merit of recognising the taxon russatus' within the range of Australia. Luckily the range of this book covers just WA and the authorities identify only one species in this state, namely the nominate subspecies of Little Bronze-Cuckoo *Chrysococcyx minutillus*. Little Bronze Cuckoos range in the north from Port Smith around all the coastline of the Kimberley into NT. There are some Atlas records within the Kimberley proper but I have never seen them well inland only in mangroves and adjacent vine thickets. Little Bronze Cuckoo has black unbroken barring running from face down to vent with wider bars on tail, fine barring on face, emerald green cap and back with metallic sheen, rufous on central tail feathers. Important feature is its red eye and red orbital eye ring. Female has pale bluish eye. Flight is similar to all Bronze-Cuckoos, slow and undulating. **Nest:** In host's domed nest, possible in Dusky, Mangrove, Large-billed or White-throated Gerygone. Clutch size large 1. **Similar species:** None in region. Horsfield's does not share same habitat, has

Photo: Simon Nevill

distinct black eye stripe black-brown forehead not green, breast bars broken, no red eye red orbital ring, call different. **Status:** Moderately common throughout the mangrove and vine thicket region of the Kimberley. **Where to find:** Port Smith. Barred Creek. Willie Creek (all near Broome). Derby Wharf.

Brush Cuckoo *Cacomantis variolosus* subspecies *dumetorum* 22-26 cm

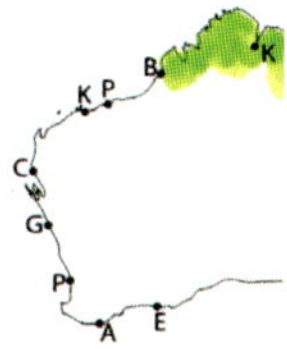

A plain looking cuckoo with no distinct features other than having a very distinctive call, which is far-carrying but has a mournful tone, each note slowly descending. The east coast subspecies *variolosus* migrates to Wallacea and NG but the WA subspecies *dumetorum* is said to be sedentary. Brush Cuckoo male has a grey head and body, buff lower belly and vent, grey-brown wings, black eye with fine pale cream orbital ring, tail grey-brown with white patches on outer tail feather margins. Female similar but rather than grey head and upper-body, the head is grey on crown, pale cream face, pale cream breast. Brush Cuckoo's are often heard before seen, they sit on open branches waiting for movement of insects particularly their favoured food the caterpillar, but they will forage in the canopy. They parasitise the nest of a large range of species but have a preference for flycatchers, honeyeaters and robins. **Nest:** In host's open nest although has been recorded as having laid in a dome nest. As mentioned, have a preference for flycatchers, honeyeaters and robins open nests. Clutch size large 1. **Similar species:** Fantail Cuckoo. Is slightly larger, darker grey head, has distinct fine yellow orbital ring, pale rust breast, dark grey back no hint of brown, heavier barring on tail. Pallid Cuckoo much bigger, Pale off white head with distinct black-brown eye stripe, lighter under-body greyish white, call ascending not descending. **Status:** Moderately common throughout the Kimberley. **Where to find:** Anywhere throughout the Kimberley and (throughout) the Fitzroy River system.

Photo: Simon Nevill

Oriental Cuckoo *Cuculus saturatus* 30-32 cm

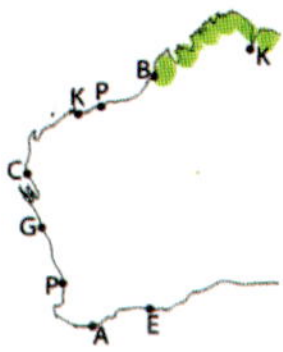

A large cuckoo breeding in East Pakistan, northern China, Russia, Kamchatka, Japan and Sumatra, Lesser Sundas and northern Borneo. Birds in the northern breeding zone migrate but many in S.E. Asia are sedentary. Those few that migrate to Australia most arrive in the tropics and most remain in the north but birds do fly south on the east coast as far as southern NSW. Those birds that I have seen have always arrived in mid November and this coincides with reports that large numbers of the subspecies *optatus* pass through Korea and Japan on their migration south and this most probably is the subspecies that comes to Australia, particularly knowing that many of the south-east Asian subspecies breed in the warmer mountain regions in South East Asia and can drop down to the coastal areas out of breeding season so no need to migrate. Oriental Cuckoo remains fairly quiet and rather shy when in Australia, in fact when I have approached these birds from a distance, I have found them very flighty. Often found in thick paperbark woodland along creek lines often close to the coast and in the early morning on the margins of thicker woodland. Oriental Cuckoo are slightly larger than Pallid Cuckoo with a grey head, clear demarcation on middle breast where heavy horizontal barring runs down the body, bars opening on the vent, back and upper-tail dark grey, white tip on tail, under-tail coverts grey with white tips, yellow-orange feet, bright yellow-orange eye and orbital ring, female almost identical. Rufous morph, scarce in Australia, rich rufous coloured back and crown with dark brown scalloping, throat and under-body white with black barring. **Similar species:** Grey morph similar to Pallid Cuckoo but different jizz when perched, Oriental holds head more upright, Pallid holds head slightly forward giving it a slightly hunched back look, Pallid has distinct black eye stripe, yellow orbital ring not yellow-orange, no barring on under-body, back is paler. **Status:** Scarce throughout the Kimberley. Has been rarely sighted as far south as the Pilbara. **Where to find:** Possibly anywhere in the Kimberley but normally close to coast. Ord River woodlands are accessible and best chance after mid November on, but prime habitat is in the northern Kimberley such as the Mitchell Plateau, which is normally inaccessible in the wet.

Photo: Simon Nevill

Channel-billed Cuckoo *Scythrops novaehollandiae* 56-70 cm

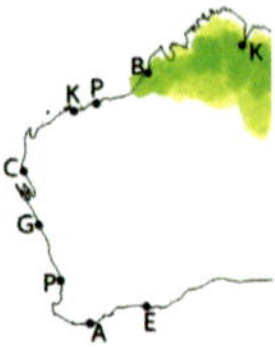

A massive cuckoo and certainly the largest in the world. It mainly parasitises the nests of currawongs, corvids and magpies. Most probably, the birds migrating to the Kimberley use the nests of corvids as there are no currawongs in that region. Like the Koel, they are primarily a fruit-eating bird however they will eat large insects particularly grasshoppers. They migrate to New Guinea, the Molucca Islands and Indonesia in our winter returning to breed in our summer, arriving in late September to October and departing from late January through till early April. When they do arrive they can be seen high up in the sky making their way to areas of fruiting trees; when seen close by, you can see that they fly with bill angled towards the ground looking like a flying giant walking stick. Their call is raucous and loud and can be heard at least a kilometre away and they will gather in small groups before breeding. Light grey body and head, distinct pinkish-red orbital ring with red eye and red lores, faint dark grey barring on lower flanks, dark grey scalloping on back and wings, long broad grey tail with black-grey sub-terminal tail band contrasting with white tail tip, under tail large broader tail bands, males only slightly larger than female, massive light pink bill, so massive that early naturalists thought they were a species of hornbill. Female has shorter thinner bill, paler under-body with darker grey V shaped scalloping, broader under-tail bands

Photo: Simon Nevill

than male. **Nest:** In host's open cup nest. The main host species would be Torresian Crow (Corvus *orru* subspecies *cecilae*). Host bird does all the raising of young. Clutch size large 1-3 **Similar species:** None. **Status:** Uncommon throughout the subhumid zone of the Kimberley, less common than Koel but extends its distribution slightly further into semi-arid zone than Koel. **Where to find:** All Kimberley with most sightings in Kununurra region particularly along the Ord River but less common than Australian Koel. To see this species, you must visit the Kimberley in the early summer months or late summer to see this species.

Pacific Koel (Australian Koel) *Eudynamys orientalis* 41 cm

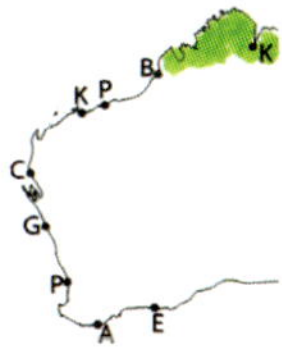

Photo: Simon Nevill

There is debate regarding this species in terms of relationship with Asian Koel *Eudynamys scolopaceus* and it may prove that they are conspecific. We know that Asian Koel is now sighted regularly on Christmas Island and there is no reason why they may not turn up in the Kimberley but there is still further work needed to assist with identifying colour markings of females of the two species. Australian Koel is one of several cuckoo species found in Australia but at a quick glance the male looks more like a Trumpet Manucode than Australia's other larger cuckoos. A summer breeding migrant to Australia, normally arriving just prior to the onset of the 'wet' hence their other common names 'Rainbird' or 'Storm Bird'. Normally arrive from mid October (can be much earlier) and will start departing as early as mid January onward to mid to late March. They over winter on several islands in Wallacea and in New Guinea. They often come back to the same breeding area where fruiting trees occur. (I should know as year after year they would call in a massive fig tree, just outside our rooms at the Pine Creek Motel, it would drive some of my clients up the wall, calling well into the night). It's diet is almost entirely restricted to fruits, both native and introduced. They have a very wide gape to enable it to consume fruits with large seeds. After breeding the males rarely call. Both sexes feeding high in fruiting trees being hard to see. I have seen them feed on fallen over-ripe mangoes, pecking at the surrounding fruit. Being a cuckoo they parasitise other birds. They require an open nest, mostly birds of similar size like Figbird, Spangled Drongo and Oriole which are the favoured species as their diet is similar to the Koels', however nests of Friarbirds and Blue-faced Honeyeater are used as well as other species. Sexes are very different in colouring. The male is an all glossy blue-black bird with a white bill and vivid red eye. Male immature birds are similar but under-body is mottled creamy-white. The female has striking markings, black head with white crescent bar running from base of bill broadening to the collar, all white under-body, breast has yellowish tinge, heavily mottled white spots on brown-black back, vivid red eye, distinct brown-black barring on white tail, ivory bill. In flight female under-body heavily barred on wings, body and tail. Upper-body spotting on wings and back, tail barred. Male in flight all blue-black. **Nest:** In host's open cup nests such as Figbird, Spangled Drongo and Oriole. Clutch size large 1. **Similar species:** Very similar to Asian (Common) Koel. Males almost identical but bill on Australian is pale grey to bone colour whereas Asian is lighter pale green. Female Asian Koel upper-body is plain brown with light brown flecking, tail less contrast between brown and cream barring. Australian Koel female has grey-brown with brighter contrast of white flecking, tail more contrast between white and grey-brown. Male Cicadabird much smaller, dull sooty grey colour not glossy blue-black. **Status:** Moderately common throughout the subhumid zone of the Kimberley, perhaps more common than thought but few people travel in the Kimberley proper in the wet. Uncommon south of the Fitzroy River floodplain. **Where to find:** Mainly from Derby across to NT. Locally common in the Ord River system where there are fruiting trees.

Pheasant Coucal *Centropus phasianinus* subspecies *melanurus* 54-63 cm

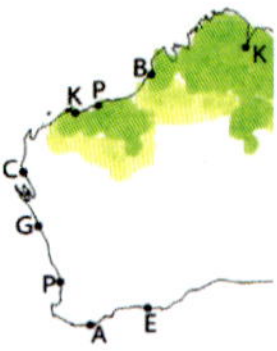

Said to superficially resemble old world pheasants but for many observers they don't even look superficially like them and certainly are unrelated. Most of this family occur in Africa and Asia. Australia has only one species of the genus *Centropus*, with two subspecies, *phasianinus* and *melanurus*. Mainly a ground dwelling bird that lives in long grassed habitat, both rank and dry. Its diet is varied, taking frogs, lizards, insects, and small rodents and nesting birds. Its streamlined profile allows it to manoeuvre through thick grass having a very strong bill. It runs fast and often chases prey. When alarmed either slowly moves off through the grass or if really agitated will alight in trees. Always looks awkward when doing so but again manoeuvres its way through the foliage very well considering its size and length. A cryptic looking bird, both sexes look similar, the female is slightly larger, out of breeding both are brown all over with cream streaking on head to the

Photo: Simon Nevill

mantle and under-body, strong barring on wings and tail, short legs, males do most of parental care. In breeding, both develop a complete black head down to the mantle and under-body, tail darkens but barring still shows, back and wings become richer chocolate-brown, bill darkens to black. You know when the birds are in the area as their far carrying call echoes across the woodlands, an unmistakable woop….woop…woop,.woop…woop…. getting faster at the end **Nest:** A domed nest lined with leaves either placed near ground or up in thick foliage particularly creepers. Clutch size large 3-4. **Similar species:** None. **Status:** Common throughout the Kimberley. Found in suitable grassed habitats way out in the Great Sandy. Patchily distributed throughout the Pilbara as far south as Newman. **Where to find:** Anywhere in the Kimberley where long grasses particularly cane grass occur, often not far from water, very common in the Ord River basin and central to northern Kimberley. I have even seen them in the Breaden Hills way out in the desert region off the northern Canning Stock Route.

Pacific Swift (Fork-tailed Swift) *Aptus pacificus* 18-21 cm

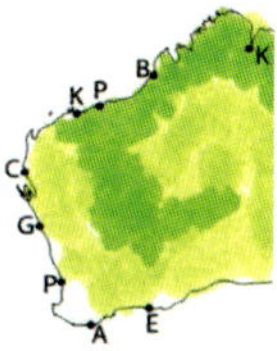

A fascinating bird that breeds from Siberia in the north, to Japan in the east and Thailand in the south. A migrant to Australia. A very large swift with long curved scythe-like wings and a distinct white rump seen well in flight, and a small white throat not so easily seen. If ever a body was designed for living much of its life on the wing, flying at times at great speed, it is the larger swifts. Pacific Swifts spend most of their life on the wing, they feed, drink, rest, sleep and even mate on the wing. The Fork-tailed Swift is a gregarious bird that flies as high as 1000 metres to gain food but mostly between 50 and 200 metres where most of the insect life occurs. They call constantly, keeping the group together even though they zoom in from all angles and directions while feeding, and it's amazing they don't collide head on with the speeds they attain flying often in opposite directions. The call is a single high-pitched metallic note 'zit…zit…zit' made with long gaps between each 'zit'. Records of actually roosting in trees are extremely rare, roosting almost entirely on the wing. This is a scarce bird to the south west; I have only seen these birds a few times in the south west in late summer but many times along Eighty Mile Beach and the Kimberley in summer. They take advantage of the cyclonic turbulent winds, which aid rapid flight without requiring much effort. Often seen at the head of a storm and away from the epicentre, as thousands of insects are whisked up in the massive thermal drafts. The picture shows a storm on the highway near Eighty Mile Beach in December. Driving through the eye of the storm to the north side, I was was confronted by hundreds of swifts in the grey stormy sky hawking for insects. **Similar species:** On the east coast White-throated Needletail, often flies with Pacific Swift and are similar looking birds being larger with a thicker set body, longer but less curved wings and no white on rump but only on the vent. **Status:** Rare in South West. More common in the Pilbara and very common in the summer in the Kimberley. Mainly seen between mid November through to April. **Where to find:** Any where in the north of the state. Not always with storms.

Typical storm where Fork-tail can be sighted

Photos: Simon Nevill

Oriental Dollarbird *Eurystomus orientalis* 28 cm

Photo: Simon Nevill

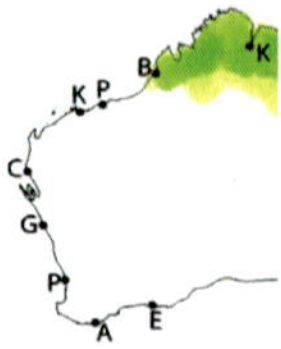

Widespread throughout Asia and the Indian sub-continent, north to Korea and Japan there being 9 subspecies with *pacificus* being the subspecies to visit Australia. One of the 'roller' family. Nests here in our late spring, arriving in mid October and later and leaves to over winter in New Guinea and the Indonesian Islands between March and April. The male being very active when first arriving finding a suitable nesting habitat. Then performs his aerial flights giving its harsh grating 'Kak.. Kak..Kak' call and showing its clear white bulls-eyes in flight to attract the female. It mostly alights on high, dead exposed branches where it also sallies for insects, its primary food. Will feed well into the early evening taking cicadas on the wing. **Similar species:** None. **Status:** Not so common as on the east coast but moderately common throughout parts of the central Kimberley. **Where to find:** Along parts of the Fitzroy basin and around the Kununurra region.

Laughing Kookaburra *Dacelo novaeguineae* 40-48 cm Introduced

Photo: Simon Nevill

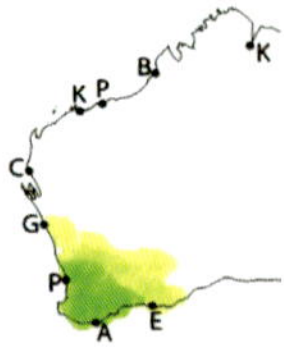

One of the 'Tree' kookaburras of the world and also one of the largest. Many people forget that this bird was an introduced bird to the south west, similar to those introduced into Tasmania. The Western Australian birds were released from the Perth Zoo between the years 1897 and 1912, often in the vain hope of removing snakes from many areas of the city. It did not take long for the bird to become established in Perth and by 1927 they had expanded their range to Denmark on the south coast. They now cover a range as far north as Dongara and then follow a line through Moora and across the southern Wheatbelt to Ravensthorpe. Sadly, many of the native birds have not adapted to the hunting skills of a kookaburra as they can sit motionless save for the occasional movement of the head for long periods of time looking for any sign of movement from a unsuspecting bird or reptile.

Often it's a robin or flycatcher returning to the nest and exposing the chicks as prey. Also the reptilian world has been greatly affected, as a higher percentage of their prey is snakes and other reptiles. They are a bird primarily of the wetter south west, preferring heavily forested areas and open woodland. The call needs no introduction. The female has almost the same colouring as the male but lacks the blue patch on the rump of the male seen well in flight, and tail is not as long. **Nest:** They nest in hollows of both dead and live trees. Clutch size 2 to 4 mostly 3. **Similar species:** Blue-winged Kookaburra, the subspecies *cliftoni* in the Gascoyne. Pilbara and Kimberley region is slightly smaller but has longer bill. Light blue shoulders. Light blue rump on both sexes. Male has strong dark blue tail. Head and crown is white with small black streaking from forehead to nape. No black-brown eye stripe or medial crown band. Call is different. **Status:** Common in much of the South West, particularly the Karri and Jarrah Forests and less so the inner Wandoo Woodland. **Where to find:** Most metropolitan parks with extensive woodland and grassed areas particularly on the base of the Darling Range scarp. Most picnic grounds throughout in the South West.

Blue-winged Kookaburra *Dacelo leachii* subspecies *cliftoni* 39 cm

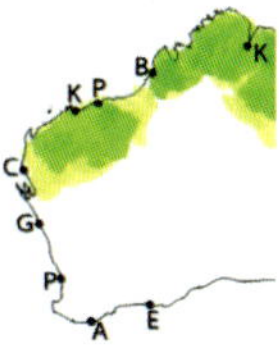

A kookaburra of the northern tropics and dry regions of northern Australia south to the Gascoyne River. Slightly smaller than the Laughing Kookaburra, the subspecies *cliftoni* is restricted to the Gascoyne River system north to the De Grey River system, cut off from the Kimberley and Northern Territory birds by the Great Sandy Desert. The subspecies *cliftoni* differs from its nearest subspecies *leachii* of the Kimberley and northern Australia, by having finer black streaks in the crown making the head look whiter and a buffy under body and tail marginally longer. Its cacophonous call is similar to that of the Laughing Kookaburra but, once heard, one can easily recognize it as a very different call, almost like a series of loud gurgling laughs, very hard to describe but dominating any other birds that call in the area. They behave in the same manner as Laughing Kookaburra having strong family bonding and defending territory strongly. The group may often come together on one branch with maybe 3 to 4 birds all calling with bills held upwards giving a raucous song. This is normally done after sunset and even into the first hour of darkness; it is also one of the first calls heard at first light. **Nest:** Location and clutch size same as Laughing Kookaburra, 2 to 3. **Similar species:** Laughing Kookaburra, refer to notes on Laughing Kookaburra. **Status:** Common through most of its distribution, both in the Pilbara and Kimberley.

Photo: Simon Nevill

Red-backed Kingfisher *Todiramphus pyrrhopygia* 23 cm

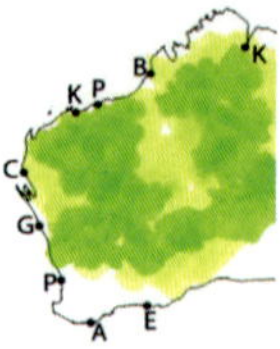

Found from the southern Mulga all the way north to the Kimberley. Mostly in open woodland and along river systems. Also found way out in the desert regions, as they require little water gaining their fluid source from mostly reptiles. Basically a sedentary bird, although some movement in the southern areas. They require a feeding location that is near firm sand or creek bank where they can excavate a nest hollow and have trees nearby from which to hunt for prey. Feeds on insects, grasshoppers and reptiles, by sitting on branches for short periods before moving to new areas to check prey and waiting for movement so they can then pounce. Will whack prey against a branch to make sure it is killed before digesting. Red-backed Kingfishers don't have a red back but have a rust red rump seen well in flight. The crown is white with heavy black streaking (male) with a prominent black eye stripe. Wing coverts are light royal blue; it has a blue-green upper body and blue tail. Female has a duller green back and the crown streaks are dark green, it's not easy to split the sexes in the field. **Nest:** Hollow in earth banks, nest placed at the base of the tunnel, nest chamber being from 50-120 cm into the bank but never close to water levels. Clutch size 3 to 4. **Similar species:** Sacred Kingfisher has green crown with no streaking. Buff-white dot between bill and eye. Collar narrower. Back greener. Flanks buff coloured. No rust red rump. Nest in tree hollows. Calls quite different. **Status:** Common along major rivers of the north. Common in the mulga and desert regions. Uncommon in northern and north eastern Wheatbelt. **Where to find:** Middle and upper Murchison and most of the Gascoyne River and Lyons River. Found on stations where no rivers occur such as Wooleen and Nallan Station. Through out the Pilbara and Kimberley, particularly in the Fitzroy basin. A species that can occur in very dry desert regions.

Photo: Simon Nevill

Sacred Kingfisher *Todiramphus sanctus* 21 cm

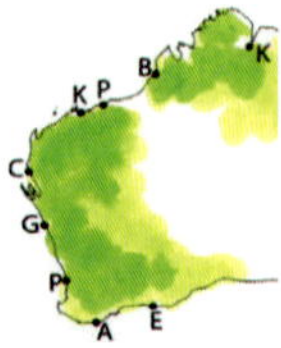

Found throughout much of the south west, the Sacred Kingfisher lives in a variety of habitats, mostly lives in open woodland or along river courses, where there is not too much undergrowth. It is not a bird of the inland arid deserts; the Red-backed Kingfisher. Takes its place there. They do occur in mangroves and this creates problems for those still not familiar with Collared Kingfishers. Sacred Kingfishers migrate north from the South West from early December through until January and do not return till August or September. They feed on a variety of prey from insects through to small rodents, reptiles and even young birds. Sacred Kingfishers have a blue-green crown and buff spot on the base of forehead. Strong black eye stripe that passes around the nape of the neck, green back and blue-green wings, blue rump and blue tail, white under body with buff flanks. **Nest:** They nest in tree hollows often returning to the same location to nest. Clutch size 3 to 5. **Similar species:** Collared Kingfisher larger. Bill massive. Crown brown not turquoise. Under body pure white no buff on flanks. Red-backed Kingfisher, refer notes on that species. **Status:** Moderately common in the South

Photo: Simon Nevill

West during spring and early summer. Common Pilbara and Kimberley. **Where to find:** Any of the larger Wheatbelt reserves including Tutanning. Boyagin. Dryandra Woodland. Bungendore State Forest. Found in parts of the eastern Darling Range and Swan Coastal Plain including Rottnest Island. Common in the Pilbara and the Kimberley.

Azure Kingfisher *Alcedo azureus* subspecies *ruficollaris* 17-18 cm

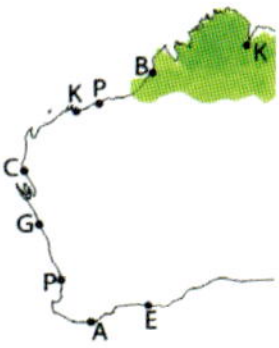

A stunning small kingfisher. 7 subspecies with 4 found in Australia. The Kimberley subspecies has a more rich coppery hue to the orange-rufous coloured under-body and violet-blue to the side of the under-body (flanks), violet-blue head and upper-body including tail, white stripe on side of neck, white throat merging into orange-rufous breast, oval loral spot above base of bill, small size only slightly bigger than Little Kingfisher. Found along the freshwater streams of the greater part of the Kimberley particularly pandanus lined creeks, prefers creeks and ponds that have numerous exposed root systems that line the banks of rivers and creeks from which to perch. They sit patiently bobbing their head occasionally as they look for fish, perhaps to aid viewing to compensate for refraction on water. You often hear them before you see them, as they give a high pitched whistle, often as they fly away as they often see you before you see them as they are generally very wary. They are sedentary birds, keeping a territory of creek or river 200–500 m long. **Nest:** Make long tunnels into soft banks, can be as long a metre. Clutch size: large 5-8. **Similar species:** None and Little Kingfisher not normally found in Kimberley but smaller with white under-body with no rust red coloration. Has white spot on forehead. **Status:** Moderately common but often overlooked. **Where to find:** Throughout the wetter parts of the Kimberley, although found along a few creeks and pools near Halls Creek. Good chance along Ord River, particularly taking a boat trip from Kununurra. Any of the central rivers such as the Charnley, Drysdale, Isdell, Hann and numerous small creeks. Can be found in mangroves but not as frequently as Sacred or Mangrove Kingfisher.

Photo: Simon Nevill

Little Kingfisher *Alcedo pusilla* 12 cm

Vagrant

Photo: Simon Nevill

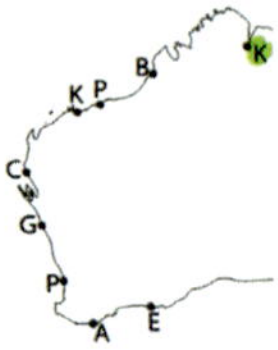

This species has been included as a pictorial reference, as it has been sighted in WA but it will most probably remain a rare bird as it needs undisturbed lush forests with freshwater pools or creeks, normally with a closed canopy over water and numerous root systems from which it can sally for small fish close to its perch. **Similar species:** None, extremely diminutive and only the Azure comes any where near its size and coloration of the Azure Kingfisher is totally different. **Status:** Rare in the Kimberley. **Where to find:** Possible on small backwaters in the Kununurra and Ord River region.

Forest Kingfisher *Todiramphus macleayii* subspecies *macleayii* 19-23 cm

Vagrant

This species is rare in WA which is surprising as there is much favourable habitat. It has been sighted in Kununurra and has also been seen a few times in Keep River National Park NT, which is very close to WA. Forest Kingfisher have a bright blue upper-body, darker wings, dark blue tail, white under-body with faint beige flanks, white collar, darker blue-black eye stripe that blends into dark blue cap, distinct white spot either side of forehead, stout black bill with contrasting cream on base of lower mandible, in flight very noticeable pair of white bulls-eyes on wings. Call in breeding season, a very distinct trill repeated rapidly. Often found near water, perches at length waiting for movement of insects or reptiles before pouncing to take prey. **Nest:** Makes its nest mostly in arboreal termites nests, will often make opening by flying at nest using bill as a drill to gain initial hole and then peck out last part of shallow tunnel to make nesting chamber. Clutch size: large 3-6. **Similar species:** Sacred Kingfisher which has green back; cap greenish blue not dark blue, no distinct white spot on forehead but longer fawn spot, black eye stripe much more distinct than Forest, Main thing to look for is with Forest Kingfisher the blues are bright rich royal blue on back whereas Sacred Kingfisher has greenish, duller back and no distinct pure white spot on each side of forehead. Most importantly, in flight has no white bulls-eyes. **Status:** Scarce in Kimberley **Where to find:** Found occasionally near the WA and the NT border near Kununurra.

Photo: Simon Nevill

Collared Kingfisher *Todiramphus chloris* subspecies *sordidus and pilbara* 23-27 cm

A large kingfisher with a robust bill that requires good stands of large mangroves. There are few species that have been split into so many subspecies around the world. There are at least 50 subspecies and this only covers half the planet's latitudes from its western range on the Ethiopian coast to its eastern range on Tonga and Samoa in the western Pacific. Both subspecies occur in WA. The range of subspecies *pilbara* starts from the Exmouth Gulf in the south to Cape Keraudren in the north and the range of subspecies *sordidus* starts at Port Smith in the south and then around the Kimberley coast and into the NT. Collared Kingfisher subspecies *pilbara* has a pure white under-body and wide pure white neck collar, importantly olive-brown cap and back, wings pale and tail green-blue, the black eye stripe narrows around nape as a fine line, the black band contrasts well with pure white collar, bill is very robust and thick, has two small white spot on lores, female so similar details would confuse in the field. The subspecies *sordidus* found in the Kimberley, has olive-green back and cap not olive-brown as in the *pilbara* subspecies. Collar slightly wider and bill longer and more robust. Harsh far carrying call kik- kik…..kik-kik. A sedentary bird defending and holding territory very aggressively to other Collared Kingfishers. Feeds on crabs and other crustaceans, small fish, mud skippers, insects and retiles, kills its prey not so much by diving into water but by pouncing and killing prey against hard surfaces as mangrove trunks using its strong beak. **Nest:** Makes its nest mostly in arboreal termites' nests or in large mangroves like *Avicenna marina*. Clutch size: Normally 3. **Similar species:** Sacred Kingfisher which often can be found in mangroves but do not normally share same locations as Collared Kingfisher. They are smaller, bill smaller, normally buffy colour on neck colour and flanks where as Collared is pure white. Sacred Kingfisher's upper-body blue-green, cap lighter green-blue. So main things to look for Mangrove is a bigger bird, pure clean white under-body and collar, bill larger, cap darker and importantly, black eye stripe does not contrast with very dark-green cap whereas Sacred black eye stripe contrasts well with brighter green-blue cap, call different. **Status:** subspecies *pilbara* generally uncommon but locally common around Point Sampson area. subspecies *sordidus* moderately common around much of the Kimberley coast but requires good stands of large mangrove. **Where to find:** Locally common near Derby and Wyndham. Port Sampson. Port Hedland.

Photo: Simon Nevill

Rainbow Bee-eater *Merops ornatus* 25 cm

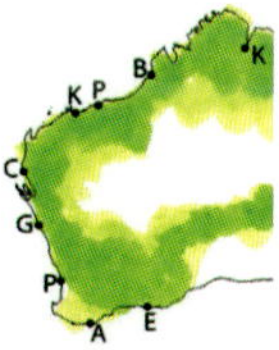

Found throughout the state except the desert regions. A breeding migratory bird to the South West arriving mainly in September to October and departing the South West January to April. They often return to the same area the following year digging out nests in open areas where there is relatively soft but firm sands. The angled nest is either on flat ground or in the side of a sandy bank. They feed on insects particularly large flies, dragon flies and bees, sitting on exposed branches darting out to take prey on the wing, then whacking the insects repeatedly against a branch, before swallowing. Their pleasant trill often leads one to their nesting area. After the breeding season February to March, both adult and immature birds congregate in small groups, normally between 8-10 birds, still feeding in the same locality getting ready for their migration north. When they finally depart, they often meet other groups heading north and the group's size can increase to 30-40 birds as they move north. They over-winter as far south as the Gascoyne but most travel to the Kimberley and many fly on to Indonesia, not to return until September and October. **Nest:** Excavated tunnel is from 50-120 cm long with large chamber at base, the nest lined with soft grasses. Both sexes excavate nest tunnel. Nest location is almost invariably close to trees with exposed dead limbs from which they can hawk for insects. **Similar species:** None. **Status:** Common throughout the state. **Where to find:** South west, any clearing with loose soils in wooded areas or open areas near trees. Common on the eastern side of the Darling Range. On the Swan Coastal Plain including Perth in gardens and playing fields. Throughout the Wheatbelt. Easily seen in the summer months in the Kimberley as well as over wintering immature birds.

Photo: Simon Nevill

Rainbow Pitta *Pitta iris* subspecies *johnstoneiana* 16-18 cm

To quote the early naturalist John Gould who wrote of the Rainbow Pitta in quaint English ' …the boldness and richness of its markings render it a most attractive object in the bush'. Birdwatchers around the world adore seeing pittas and no wonder as nearly all the 30 plus species are multi coloured, often with the most vivid combination of colours. To see them through the dim light of the rainforest and then see them suddenly step into a shaft of sunlight is wonderful sight. Pittas are found in Africa to the Solomon Islands in the east. Australia has 3 species with at least two rare vagrants coming to these shores. In WA the Rainbow Pitta is restricted to the far north west Kimberley coast and its adjacent hinterland. The primary range is from Walcott Inlet in the south to the Mitchell Plateau in the north. They are found in monsoonal vine forests and rainforests and prefer the thickest patches of relic rainforests. They are found on several of the larger islands and also a long way inland in some areas like the Prince Regent area. They can survive in patches of rainforest only a few hectares in size. Rainbow Pitta has an all black under-body, bright red vent, green back and wings with bright blue iridescent shoulder patch that shows well in sunlight, distinct orange-brown eyebrow stripe that forms a crescent around the nape. The differences between the Kimberley subspecies *Johnstoneiana* and the nominate subspecies *iris* are very slight, *Johnstoneiana* subspecies differs in that the orange-brown crescent is broader on the nape and slightly smaller in size. Pittas get their food by continually turning over leaves taking insects and can only survive where there is a reasonable layer of leaf litter. They suffer very badly in poor wet seasons. They have a liking for land snails and will use selected stone anvils from which to crack open the snail shells, an interesting use of a 'tool' to aid in feeding. Generally shy birds and will hop away when approached but will take to high branches if alarmed. Call at dawn and regularly through the day, more frequent in the breeding season, very territorial defending feeding territory strongly. **Nest:** Well woven dome nest placed between buttressed roots of trees. Clutch size: Normally 3-4. **Similar species:** Noisy Pitta not in region however there have been two rare vagrant species sighted in WA. One: the Blue-winged Pitta (has been sighted at least 4 times) and Two: the Fairy Pitta. Both form part of sub group of pittas closely related to the Australian Noisy Pitta and the also the Mangrove Pitta of the Malay Peninsula. All 4 have subtle variations to plumage and reference should be made to more local definitive

guide books to check for eye stripe size, crown colour, supercilium stripe colour in some, under-body colour variation and extent of red vent, bill size (for Mangrove Pitta). Some of these pittas are passage migrants from China and over winter in Sumatra and Java, so it's not surprising that a few would over fly to Australia. **Status:** Uncommon with a restricted range. **Where to find:** Rainforest pockets on several islands and adjacent coastline from Walcott to Mitchell Plateaux, some inland rainforest pockets like at Camp Creek on the Mitchell Plateaux. If you are not concerned about seeing a different subspecies then NT is a far easier place to see this species as there is a higher number in NT. Common in the Darwin area, Buffalo Creek, East Point. Holmes Jungle. Out of Darwin, Howard Springs. Fogg Dam. Gubara in Kakadu.

Noisy Scrub-bird *Atrichornis clamosus* 23 cm   Endemic

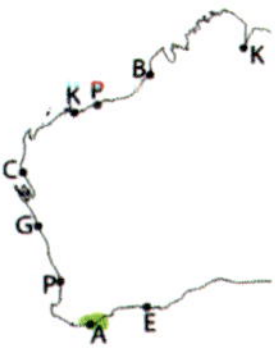

The Noisy Scrub-bird is a fascinating bird, with its evolutionary development going back to connections with species like the lyrebird which have no wishbone, which is the main bone structure that flight muscles are attached to and is required by most flying birds. That's why you will not see a Scrub-bird fly far; it will make a short flight burst and semi-glide as it does so. Most of the time they will run and that they do extremely well, having very strong thighs just like their distant relative the lyrebird, which also does not like flying and only does so to roost in trees or glide a short distance down steep hillsides. The story of the Noisy Scrub-bird involving its rediscovery and recovery in Western Australia is a credit to all the hard working research workers and volunteers who were involved in its recovery program, for it was thought that the Scrub-bird was extinct. In 1961 the amateur birdwatcher Harley Webster of Albany found the bird on Mount Gardner at Two Peoples Bay. By a systematic program of translocating captured birds and releasing them in varying sites like Mount Many Peaks, Cheynes Beach and Bald Island, the population has gone from at the most 100 birds prior to the 1970s to over 2000. With two major fires over the last ten years, I am not sure of the estimated population now but it will always be far greater than the fragile population that occurred before re discovery. Its population has now well surpassed the rare Western Ground Parrot which is found over a greater range but its status is very tenuous and researchers and volunteers are working hard to get to grips with its needs and habitat. Well, we know its needs but what to do to protect it is the big question.

The Noisy Scrub-bird is one of the hardest birds to see, even more so than its east coast relative the Rufous Scrub-bird. Some birdwatchers get lucky at Cheynes Beach and see the bird in the open but generally it is extremely difficult to see. I have had groups of birdwatchers only metres away from where the birds have been calling, only to have the bird run behind us or through us, such is its ability to run through thick undergrowth and still avoid detection. Its call is amazingly powerful for such a small bird and can carry several hundred metres. The call starts slowly and builds up rapidly, finishing in a crescendo burst. It also has a warning call, a sharp grating 'chip' similar to the Western Bristlebird's warning note, using it normally when it sees you first. First year males try to develop the power and coordination of the full adult call but it lacks the full volume and tends to be broken in parts. They will call away from the main occupied gully territories. Males will call in summer on overcast days, even at midday but with nothing like the quantity of calls in the pre-breeding and breeding season. The dominant males tend to select and dominate the lushest, thickest gullies. The feeding path often follows the line of a very small creek or runoff where the riverine vegetation has a taller canopy and the understorey is incredibly dense with thick areas of sedge. Other less dominant males have to occupy territories with less of an over storey but the species always requires thick undergrowth, more so than bristlebird. They feed on small macro-insects in the leaf litter, turning over the soil to get to insects etc. The male is a much bigger bird, has a stronger white throat with strong scalloping on the centre of the throat and upper breast, pale underbody. **Nest:** The nest is an oval dome structure made from rushes with a side entrance often with a platform entry. The base of the nest has a hard pulp consistency made from masticated rushes. The female builds the nest and she will often use the same nest or build another in close proximity to the old one. She lays just one egg. Surprisingly for such a wet cool region it lays in the middle of winter mainly July, just like its ancient ancestor the lyrebird; however, there may be replacement clutches in late spring into October. **Similar species:** Western Bristlebird which is smaller. Has longer tail but body looks and is slimmer. Distinct scalloping on underbody and mantle. Call totally different being more musical and is a shorter four noted call, first two notes extended last two very rapid. **Status:** Overall rare in previous territories but locally common. **Where to find:** Two Peoples Bay. Mount Many Peaks. Waychinicup. Cheynes Beach. Has been released in a few other locations many kilometres away in former territories known in the mid to late 1800s. Note: Please do not use tapes to attract these birds as they have been overplayed in some areas creating disturbance and stress for the birds. At Cheynes Beach or Two Peoples Bay there is a reasonable chance of seeing the bird without the need for tapes.

Noisy Scrub-bird female

Photo:: Graeme Chapman

Rufous Treecreeper *Climacteris rufa* 17 cm

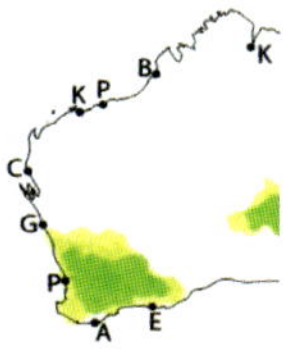

An attractive, all rufous-coloured Treecreeper, found in the South West and in the Great Victoria Desert, with another population in lower south east of South Australia. It is a sedentary bird, requiring much fallen timber and mature large eucalypts with tree hollows. Found in varying eucalypt woodlands in the south west including Wandoo, Karri, York Gum and Salmon Gum woodlands, to a lesser extent Jarrah Woodlands. Some of the highest population densities occur in the Wandoo Woodland. Feeding both on the ground and on tree trunks, their rasping call often warning other birds of your presence. Both sexes similar but the male has fine distinct vertical white flecking with black edges to the margins located on the chest. The female has just white flecking and lacks the strong black edging. In flight they show the prominent cream band on the centre of the wings. **Nest:** Like all treecreepers, nests in tree hollows, normally placed between 2-10 metres up in a tree often in an angled spout. A nest is built at the base of the hollow and a matt of grasses and bark used. Clutch size 2 to 3. **Similar species:** White-browed Treecreeper same family and same shape and behaviour but colorations totally different being an over all dark brown not rust red colour. **Status:** locally common in the larger remnant Wandoo Woodlands and Transitional Woodlands on the east and northern Wheatbelt. Now scarce on the Swan Coastal Plain, where it was once common before European settlement. Locally common south of Neale Junction. **Where to find:** Boyagin Rock. Dryandra NR. Dongolocking. Stirling Range. Bungendore. The eastern boundary of the Jarrah Forest where it blends into the Wandoo Woodland. East of Perenjori and north east of Wubin, in the Transitional Woodland. The area near Neale Junction and further east along the Anne Beadell Hwy.

Photo: Simon Nevill

White-browed Treecreeper *Climacteris affinis* subspecies *superciliosa* 14-16 cm

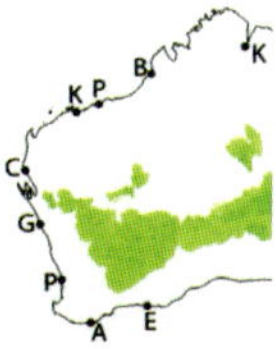

A bird of the semi-arid to arid regions, its distribution is fragmented over its range. In the mid west and southern parts of WA, it is found in the mulga woodland and casuarinas woodland north of the Nullarbor. Has a preference for areas where the native pine (*Callitris glaucophylla*) grows, where it feeds in the deep crevices and also nests in the hollows of those long lasting old dead standing trees, which the Mulga trees do not have. They also feed on the trunks and in the branches of acacias, principally the broad and narrow leaved Mulga. They are locally common in certain locations but never in large groups there normally being only 4-8 birds in each feeding group, and patchily distributed throughout the mulga. Their call often gives them away, a high-pitched rapid trill slowing at the end 'stzz stzz stzz stzz – stizzz –stizzz' often uttered from the top of dead exposed branch. Has an undulating flight as it flies from tree to tree. Feeds on insects from crevices in the bark starting at the base of a tree and working up to the mid level before dropping down to the next tree trunk. There are two subspecies in Australia, the nominate subspecies *affinis* in eastern Australia and subspecies

Photo: Simon Nevill

superciliosa in central and western Australia. The Western Australian birds have a browner back and rump not grey-brown as in *affinis*. There are other small colour variances with the scapulars and throat but very subtle differences. **Nest:** As mentioned, placed in a tree hollow; sometimes the bird enters a hollow in the middle of a trunk and exits through the top of the hollow trunk. Clutch size 2 to 3. **Similar species:** Brown Tree Creeper, not found in WA. **Status:** Generally uncommon but occasionally locally common. **Where to find:** Pindabunna Station (no tourist facilities-working station). Wooleen Station. Nallan Station. Billabalong Station (No tourist facilities—working station). North of the Mt Jackson area. Goongarrie NR. Locally common in the woodlands north of the Nullarbor. Odd populations half way along the Gary Hwy near Neale Junction.

Black-tailed Treecreeper *Climacteris melanura* subspecies *wellsi*

16 -19 cm

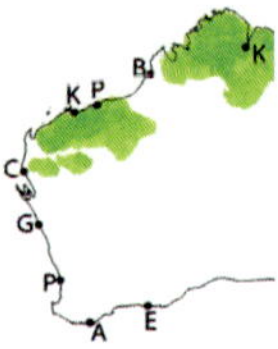

A bird of the Kimberley subspecies *melanura* and the Pilbara subspecies *wellsi*,. Furthest south is a small restricted population along the middle Gascoyne River. It has the same feeding habits as the other two Western Australian treecreepers. On the Gascoyne it frequents the Red River Gums that run in a broad band along the central Gascoyne River between Rocky Pool and Bidgemia Station. The male differs from the female by having a white throat with black vertical streaks and darker rust red underbody. The female has a white throat with no black streaks. The subspecies *wellsi* in this region has smaller wings and tail but body size is the same. Also the underbody is much lighter than the Kimberley birds being rufous rather than black-brown. In flight both sexes show a light cream band when flying that spreads across the centre of the wings. **Nest:** In tree hollows, sometimes in a horizontal hollow branch. Clutch size 2 to 3. **Similar species:** None. **Status:** Uncommon throughout its southern range. More common in the Kimberley. **Where to find:** From Rocky Pool to Bidgemia on the Gascoyne River. In the Pilbara uncommon

Photo: Simon Nevill

found at Duck Creek, on the middle Fortescue River north east of Wittenoom. Northern limit the De Grey River. Parts of the Oakover River. Lots of locations in the Kimberley but patchily distributed but population over all, greater than Pilbara. Found in open woodlands particularly in the central Kimberley as in the track into Bell Gorge or Charnley River Station. Track into the northern tip of Lake Argyle near the WA border.

Splendid Fairy-wren *Malurus splendens* subspecies *splendens*

11-13.5 cm

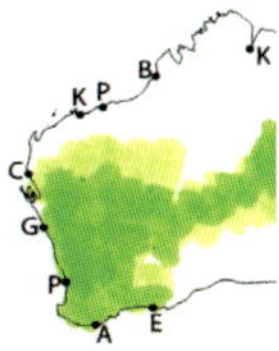

There are 2 subspecies of Splendid Wren within WA. *splendens* occurs throughout most of the South West but above Kalgoorlie and up to Wiluna east of that line, hybrids start to occur between *splendens* and the desert subspecies *callainus* which used to be known as the Turquoise Wren. If you see it in the central deserts set against the rust red sand dunes, it certainly is a stunning subspecies, showing its distinct turquoise back. In the South West, *splendens* occupies relatively thick undergrowth often overlapping the territories of the larger Red-winged Fairy-wren. Both have their food niches. Splendid Fairy-wren is found throughout most of the South West into the Mulga Woodland and into the Pilbara. It can be seen scurrying around the bases of mulga trees with no groundcover unlike South West birds. They feed mostly on the ground or within low bushes but will occasionally feed higher up in low trees. In the Wandoo Woodland it occupies a different location to the Blue-breasted, which is mostly in denser vegetation in the Wheatbelt. The male plumage varies with age; an older dominant bird often holds its breeding plumage longer and the moulting period is often only short. In full breeding it is unmistakable as per the photograph, in non-breeding it retains its blue tail, which is a stronger brighter blue than the females. Also on the wings it retains a bluish tinge, which the females do not have. It also lacks the orange-brown orbital eye ring that the female possesses. The female has a dull blue-green tail. No blue on the wings at any period. The main diagnostic feature is the orange-brown orbital ring that goes around the eye and extends to the bill. **Nest:** They build the typical Fairy-wren dense ball nest low to the ground with topside entrance. In the South West they nest from September through to December but the more inland birds will nest as early as late July and August. They will breed twice and the author has seen nests with young in April after good rains in the Mulga Region. Clutch size 2 to 4 eggs. **Similar species:** None. **Status:** Common in the South West. Uncommon in the Wheatbelt. Common past the Vermin Fence and to the desert regions. **Where to find:**

Photo: Simon Nevill

Prolific in the South West and all along the coast to Cheynes Beach then a distinct drop in numbers as one goes east into the arid zone. Common in parts of the Swan Coastal Plain all the way to Carnarvon but after Carnarvon becomes very uncommon. The range of the subspecies *callainus* commences almost where the Spinifex starts in the Great Victoria Desert and Central Ranges. It occupies acacia and eucalypt woodland in the central ranges but also the flatter Spinifex country of the Victoria Desert.

Purple-backed Fairy-wren (previously Varigated Fairy -Wren)

Malurus assimilis 12-15 cm

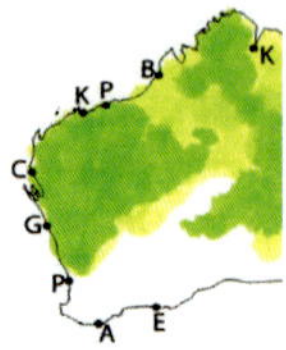

The far eastern sub sq *lamberti* is now known as the Variegated Fairy Wren. This may as one would say be 'drawing a long bow' as the colour variation between *assimilis* and *lamberti* is marginal, the ear tufts on *assimilis* are paler than the colouration on the crown whereas *lambertii* colouration on the crown and ear tufts are the same. Most bird watchers will find it hard to differentiate in the field. Also within Western Australia the bird's colour gets paler the further north they occur. The subspecies *rogersi* occurs throughout the Kimberley and in some areas, shares the same habitat with Black Grasswrens namely rocky hill sides. It has a darker chestnut mantle and small amount of blue on the upper flank. Tail is darker than the southern *assimilis* birds. Where the Red-winged Fairy-wren requires moist gullies of the lower South West, the Variegated survives in far drier habitats. Most common on the coastal belt as far south as Perth. It is one of the common fairy-wrens of thick coastal heath, particularly north of Jurien all the way to Kalbarri. It is often asked how long do these little fairy-wrens live; well its not unusual for a fairy-wren to live for about 6-7 years. In the non-breeding plumage the male differs from the female by having a dark bill. Lacks the orange orbital eye ring, instead has a smaller pale eye ring. Has a light blue tail, female has a dull grey-blue tail and orange bill and orange orbital eye ring. **Nest:** Is similar to the Splendid Fairy-wren but slightly bulkier. **Similar species:** Blue-breasted Fairy-wren, which has black-blue breast seen in sunlight only. An even violet-blue crown and ear coverts although there is a slight change to lighter coverts. With Variegated the contrast is greater and the crown is lighter blue and the coverts even lighter still. The breast is black on the Variegated, even in full sunlight. There are only subtle differences between females. The Blue-breasted female has a brighter orange orbital eye ring than the darker brown-orange of the Variegated. The bill on the female Variegated is bright orange whereas the female Blue-breasted is a brown-orange colour. **Status:** Common on the coastal belt. Uncommon through the arid regions. **Where to find:** Kalbarri National Park particularly near the coast. From Lancelin to Dongara and parts of the northern Wheatbelt, past Moora to the Mulga Woodland. Moderately common in the Shark Bay region, uncommon to locally common in the Pilbara and central deserts. Kimberley birds more common in the rocky bolder country of the north-west.

Subsp. *Rogersi* Photo: Simon Nevill

Blue-breasted Fairy-wren *Malurus pulcherrimus* 12.5-14 cm

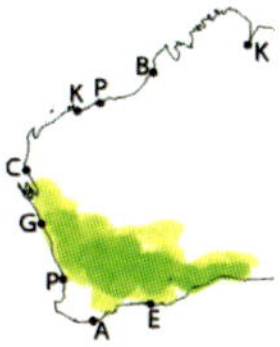

Sadly much of the territory that the Blue-breasted occurred in has been cleared for farming and some of the populations in the Wheatbelt remain tenuous. It has a particular preference for sand plain heaths, known as Kwongan Heath and also for open Wandoo Woodland where the Sand plain Poison (*Gastrolobium microcarpum*) occurs. It does not occur in the lower wetter South West, so there is never any confusion with Red-winged Fairy-wren except at a crossover region around Cheynes Beach and east. The difficulty with identification is between Variegated and the Blue-breasted; on the coast north of Perth and the northern Wheatbelt there is an overlap particularly in the Badgingarra area. Most of their territory starts where the Red-winged Fairy-wren's range finishes and goes through the Wheatbelt and Transitional Woodland to where the mulga starts and the Variegated Fairy-wrens occur. Its distribution also spreads as a narrow band at the base of the Nullarbor Plain and then its population recommences on the Eyre Peninsula in South Australia. The non-breeding Blue-breasted male has similar plumage colours to the female but the bill is black not brown-orange and there is no orange orbital eye ring or lores. **Nest:** Similar to Variegated Fairy-wren but often built very close to the ground often in one of the prostrate dryandra bushes. **Similar species:** Refer to the notes on Variegated Fairy-wren. Red-winged Fairy-wren is slightly larger and slightly heavier bodied. The crown on the male is a lighter silvery blue and the ear coverts even lighter again. The lores on the female are larger and more chestnut in colour. The mantle seen best in flight is a much lighter blue. The chestnut shoulder patch slightly narrower than on the Blue-breasted. **Status:** Locally common in some reserves but overall throughout its range uncommon. **Where to find:** Wongan Hills. Watheroo. Tutanning. Dryandra Woodland. Dongolocking. East Yorkrakine. Magenta. Dragon Rocks and many other reserves. The lower Transitional woodland east of Hyden. Mallee along southern Nullarbor.

Photo: Simon Nevill

Red-winged Fairy-wren *Malurus elegans* 15 cm **Endemic**

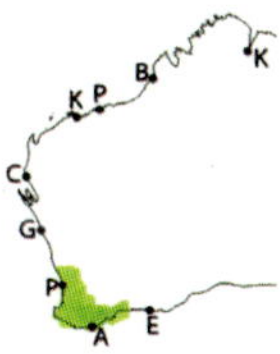

Found in the wetter parts of the South West. It is the largest of the fairy-wrens found in the South West and also the largest of all the Australian red backed fairy-wrens. Its preferred habitat is lush riverine vegetation or thick undergrowth of the heavily forested Jarrah, Karri and Tingle Forests of the lower South West. They range as far north as the remnant population on the Moore River, then a break to the northern part of the Darling Range broadening their range in the South West corner. They extend to Waychinicup National Park and inland to the Porongurups. The male has a pale silvery blue crown and the ear coverts are even lighter.

The mantle is the same light silvery blue, seen best in flight. The lores on the female are larger and more orange-chestnut in colour than Blue-breasted Fairy-wren's. Red-winged Fairy-wrens were found on the Swan Coastal Plain when settlers first arrived but the clearing of land along the river systems has almost removed all populations except on rivers like the Harvey River. **Nest:** Similar to Variegated but placed often higher in a bush. **Similar species:** Refer to the notes on Blue-breasted Fairy-wren. **Status:** Moderately common throughout its range.

Where to find: Most rivers and creeks that run down the scarp of the Darling Range such as Serpentine, Wungong, Dandalup Rivers. Piesse. Bickley and Churchman Brook. Throughout the lower South West, from Dunsborough to Albany. Particularly common from Walpole to Elleker. Common in the gullies of the Porongurup Range.

Photo: Simon Nevill

White-winged Fairy-wren *Malurrus leucopteris* subspecies *leuconotus* 11-13 cm

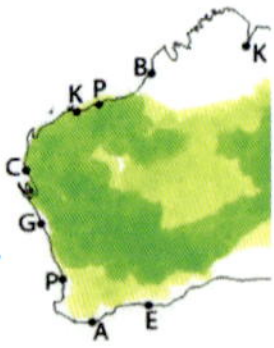

There are 3 subspecies of the White-winged Fairy-wren all occur in Western Australia. It is interesting that the nominate subspecies *leucopteris* was collected first on Dirk Hartog Island where there is an unusual relic population which instead of having the normal cobalt blue body and head is in fact all black, quite a stunning little bird. The other restricted subspecies

Photo: Simon Nevill

edouardi occurs on Barrow Island and is very close to the black and white birds of Dirk Hartog with very subtle differences that this book does not need to cover. The subspecies *leuconotus*. occurs throughout the vast arid zone from the west coast across the central deserts and from the mid Great Sandy Desert south to mid south west. In the South West, it comes as far south as Perth although is losing its coastal territory each year. It can be found in parts of the northern Wheatbelt. Throughout the Mulga into the Pilbara and the deserts. Calls of the other fairy-wrens are very similar and hard to split unless you know the calls exceptionally well. However, the White-winged Fairy-wren is easier to pick, as its call is far louder and instead of being soft short whistles with breaks in between, it is a strong rolling reel that is extended for several seconds 'treeeeeeeeit' then a break 'treeeeeeeeit'. They have a particular preference for chenopod vegetation, samphires and blue bush, also dense coastal low heaths and margins of ephemeral inland lakes that have lignum growth. The female is very similar to non-breeding male and one has to see them well to split them. The non-breeding male has a black bill not pale pink and the central tail feathers are slightly tinged blue not easily seen best to check the bill. Luckily the dominant male normally remains in moult for only a short period. **Nest:** The dome nest is built by the female and situated low only 20-40 cm off the ground. **Similar species:** None, although the females are similar to other female fairy-wrens. **Status:** Subspecies *leuconotus* moderately common on coastal heaths to Dongara. Common north of Geraldton to Kalbarri. Common in Shark Bay region. Uncommon but locally common on preferred habitat in the drier inland regions. **Where to find:** Guilderton to Green Head. Kalbarri coastal heath. Throughout the Shark Bay region, mainly on Peron Peninsula. From Gladstone to Carnarvon. Throughout much of the mulga and sandy desert country. Also the Nullarbor.

Red-backed Fairy-wren *Malurus leucopterus* subspecies *cruentatus* 10-13 cm

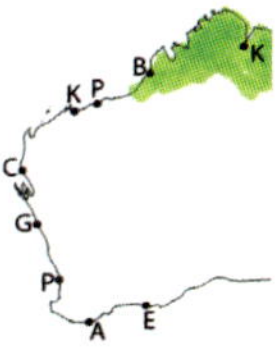

Smallest of the Fairy-wrens, the male having a vivid contrasting plumage of bright scarlet contrasting with jet black, the black has a slight metallic sheen when viewed in sunlight, has all black head and under body black-brown wings, black tail, black bill. The male can take up to 4 years to develop full breeding plumage, in the 3rd year the male develops patches of red on the mantle before moulting the next season. Female pale brown darker on wings off white underbody, dark brown tail, pale beige lores, pinkish bill. The subspecies *cruentatus* of the Kimberley and NT has a pure scarlet mantle whereas the subspecies *melanocephalus* from eastern Australia has an orange-scarlet mantle, shorter tail than other fairy-wrens. Found throughout the Kimberley and along the parts of the coastal belt of 80-mile beach, including country around the salt lake system near Samphire and as far as Cape Keraudren. Found in cane grass country and rank grasslands along water courses, wide spread.

Photo: Simon Nevill

Nest: Dome nest built entirely by the female placed in and around grass stems or with low bushes. Clutch size: 3-4. **Similar species:** Male in nuptial plumage none. Female possibly with female Variegated Fairy-wren but Variegated has bluish-grey mantle and crown, longer tail, distinct rust coloured lores whereas Red-backed female has very pale cream lore. **Status:** Common to very common in the Kimberley. **Where to find:** Throughout the Kimberley grasslands then patchily distributed south to Cape Keraudren.

Purple-crowned Fairy-wren *Malurus coronatus* 14 cm

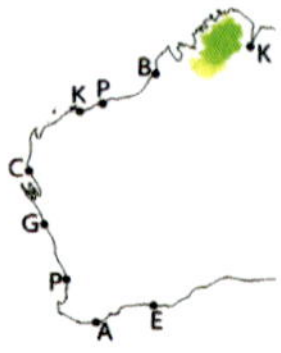

The largest of the fairy-wrens, its colouring quite different to the other fairy-wrens and even though there is less variety of colour range on the male, it still is a stunning bird, even Gould described it as his favourite of all the many fairy-wren species. The male Purple-crowned Fairy-wren has a bright lilac cap with a small black central black area on top of the crown , bordered by a broad black eye stripe broadest on the ear coverts, narrowing at the nape and base of bill, white under the throat and under-body merging into whitish-buff on the flanks, back beige-brown, wings darker, long green-blue tail, dark pinkish legs, bill black and longer than all fairy-wrens. Female lacks the distinct facial pattern and colours of the male, has grey-beige crown darker on forehead and becoming paler on mantle, wings brown but has distinct chestnut patch on the lores. Male in eclipse plumage lacks all the bright lilac and black colouring on the head, head colouring looks similar to female but ear covert develops into a plain black-brown patch. Requires a distinct type of habitat, thick pandanus lined creeks mostly where the river or creek opens into a large permanent pool. Populations have been affected by cattle over grazing on creek lines and damaging pandanus as they get down to water. Also changing fire regimes have had a major effect on many species, especially Purple-crowned Fairy-wrens. I have seen them disappear from some pools where they were once common such as on Ellenbrae Station. Luckily they still can be found on many major rivers throughout the central Kimberley. **Nest:** Bulky dome nest often woven paperbark located in thick pandanus. Clutch size: 2-3. **Similar species:** None. **Status:** Uncommon and classed as vulnerable. **Where to find:** Drysdale River. Mornington Station. Isdell River. Parts of the central Fitzroy River. Carson River there are other locations but territory relatively small and very vulnerable.

Photo: Simon Nevill

Rufous-crowned Emu-wren *Stipiturus ruficeps* 13 cm

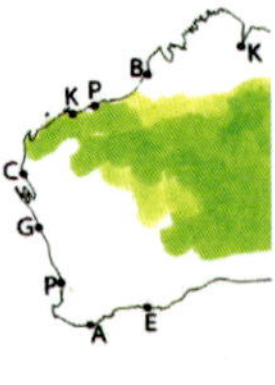

A very small bird, found over a vast range of the arid lands of WA. I never know when I'm going to encounter this diminutive little bird, usually as it flies past my bullbars in some remote desert region. A lover of big domed spinifex where it spends much of its life and I'm always amazed as it zooms in and out of the sharp spikes of the Spinifex, that it never impales itself. You often encounter it by hearing its very high-pitched delicate call (if you can hear such high pitches). The reason why these tiny birds of the genus *Stipiturus* are called emu-wrens was that early naturalists thought the tails resembled the structure of Emu feathers. The tail is certainly unique in the birding world having only six feathers and the feathers do not lock together having no barbicels. I am sure you have in the past tried to separate the individual elements on a feather and noticed how well they are interlocked, this is because each filament has barbs that bind the feather together but emu-wrens lack this in their tail feather. The male Rufous-crowned Emu-wren has a rich rufous crown very distinct with a bright pale blue throat, brown back streaked with darker brown lateral stripes, cream under-body darker on the flanks, a very long tail nearly twice the body length. Female has no blue throat but does have rufous crown more subdued colouring, body colour similar to male, greyish ear coverts streaked white. Inquisitive and will come to soft 'pishing' calls. **Nest:** Dome nest built within clump of Spinifex. Clutch size: 3-4. **Similar species:** None. Several fairy-wrens co habit different parts of their range and they include Splendid, Variegated and White-winged Fairy-wren but only the females and non breeding males could possibly be confused but emu-wren tails and body size so different. **Status:** Common in optimum habitat although feral cats now throughout the desert region are taking their toll. **Where to find:** Throughout the Pilbara. Cape Range. Isolated population in the Kennedy Range. Much of the desert regions of WA. I have encountered the species as far north as the Canning Stock near the Breaden Hills to as far south as the Anne Beadell Highway. In the Pilbara high up on the plateau country near the mountain the 'Governor' to lower down near Millstream.

Photo: Simon Nevill

Southern Emu-wren *Stipiturus malachurus* subspecies *westernensis* 14-18 cm

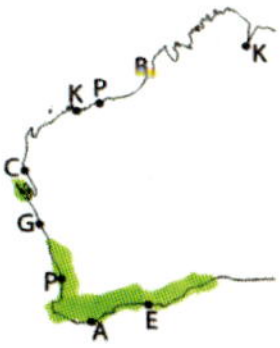

Eight subspecies throughout Australia and 2 within this region, *westernensis* in the lower South West and *hartogi* on Dirk Hartog Island. The Southern Emu-wren requires thick low heath, where it manoeuvres through the densest vegetation only occasionally popping up to check territory. A very small-bodied wren, the male having a distinct pale blue throat and chest as well as a pale blue supercilium eye stripe. The subspecies on Dirk Hartog Island has a much paler blue chest and the flanks are a lighter orange-cream, black streaking on the crown not as bold. Their call is so highly pitched that people who cannot hear high notes just can not hear them and that includes me but hay, after 40 sessions of radio therapy to the throat and jaw, one tends to lose the hearing for very high pitched calls but at least I'm still alive. Their range follows much of the coast from Jurien Bay all the way down to the Capes and along the southern coast to Cape Arid. They are not found in the main Darling Range except in the very south of the range. From Albany north they can be found in several of the lower Wheatbelt reserves. **Nest:** Is a small round dome with topside entry often woven into grasses or small plants growing alongside. Always well concealed and the inner nest lined with grasses or fur as a soft internal lining. **Similar species:** Rufous-crowned Emu-wren, not found in this region and territory does not over lap (furthest southern population on the top of the Kennedy Ranges). **Status:** Moderately common in closed heath throughout its range except the lower

Wheatbelt where it is scarce. **Where to find:** Heaths near Cervantes and Lancelin. Remnant population near Ellis Brook on the Darling scarp. Locally common between Cape Naturaliste and Cape Leeuwin and all along the southern coast to Esperance and Cape Arid. Locally common in the following parks: The Leeuwin-Naturaliste NP Walpole-Nornalup NP West Cape Howe NP Waychinicup NP and Fitzgerald River NP Cape Le Grand and Cape Arid NP.

Photo: Simon Nevill

Western Grasswren *Amytornis textilis* 18 cm

Most grasswrens are associated with spinifex (genus *Triodia*) but the Western Grasswren (previously named Thick-billed Grasswren) can be found in saltbush and bluebush country as well as acacia thickets on sand dunes. The Western Grasswren *Amytornis textilis* one of the smaller grasswrens, with a shorter tail than most and as the old name implies has a stout bill. The sexes are almost identical, the female showing a small but distinct rufous patch on the flanks. Western Grasswren are no longer found throughout much of their former territory, which included a large area on the Nullarbor and into the Great Victoria Desert. It is believed with the introduction of the rabbit, the impact on bluebush country greatly affected their survival and now the subspecies *textilis* can only be found on the Peron Peninsula and east of Hamelin Pool and parts of Woodleigh Station. They are omnivorous not only taking insects but also feeding on fruits and seed. They have two basic calls, a contact call which is just a single high pitched 'zsit zsit zsit' and strange mewing call. **Nest:** One of the few grasswrens that does not nest in Spinifex, nesting mainly in bluebush and samphire. In the Monkey Mia region there are several other low shrub species that it also nests in. The nest can be a deep cup of dense matted grasses and twigs without a hood, or with a partial hood or a complete hood with side entry. Clutch size mostly 2 sometimes 3. **Similar species:** Dusky Grasswren which is a long way from both subspecies. **Status:** Uncommon throughout most of its range. Locally common at Monkey Mia and adjacent dunes. **Where to find:** Easiest location

Photo: Simon Nevill

is in the parking area and adjacent dunes opposite the resort and camping area as they have become accustomed to people There is a remnant population of Western Grasswren to be found near Whyalla and in Pinkawillinie NP in South Australia these are the last remaining populations that once stretched across the Nullarbor into WA.

Striated Grasswren *Amytornis striatus* of subspecies *striatus* subspecies *whitei* 15-17 cm

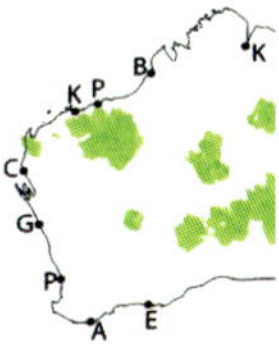

Serious bird watchers always want to find the endemic grassrwrens in Australia, I guess because so many have restricted and distinct locations and are sometimes hard to find which is always a challenge. The Striated is one whose distribution range is extensive and there are 4 disjunct populations (the Flinders Range species has been separated now known as the Short-tailed Grasswren). The range stretches from the high gullies in the Pilbara where the most western form, the subspecies *whitei* occurs to across WA's central deserts to the south east populations in Wyperfield National Park and then to far north east where the remnant population of subspecies *rowleyi* in the Bladensburg Ranges south of Winton occurs. Whether these fragmented populations will be split again remains to be seen. In WA as mentioned, we have the two subspecies, *whitei* in the Pilbara and *striatus* throughout much of our central deserts. The Pilbara grasswrens are the most attractive of all the subspecies and they may in fact one day be split as their plumage is quite distinctive and their habitat requirements are very different however this is only conjecture on my part. The male *striatus* has a white under-body with buff-cream flanks brown wings and upper-body with lateral white streaking on the mantle, crown and forehead, small chestnut-orange patch on the lores, distinct sub-moustachial black stripe, long brown tail. Female very similar but has slight rufous patch on flanks. The male subspecies *whitei* has a rich rufous-brown upper-body and crown, not brown, has strong buff lower under-body with larger rufous patch on flanks, merging into buff belly, outer tail feathers rufous-brown not plain brown. The subspecies *striatus* in WA lives and breeds mostly in plains and valleys where large domed Spinifex occurs. The subspecies *whitei* lives and breeds in much different habitat, mostly steep gullies and valleys in the Pilbara, the valleys often strewn with large boulders. They can run fast with short bursts of flight when alarmed. **Nest:** Large domed oval nest made of woven Spinifex on outside, lined with softer grasses inside, built in clump of Spinifex or base of dense bush. Clutch size: 2-3. **Similar species:** None, only young could be confused with fairy-wrens but still bulkier birds different jizz. **Status:** I believe *striatus* is getting less common in the desert regions as see them less on my desert trips compared with 30 years back. The Pilbara population appears to be more stable as they live where boulder country occurs, so can retreat from feral cats. **Where to find:** The subspecies *whitei* much of the Pilbara but also an isolated population in the Cape Range. The subspecies *striatus* covers large range. I have seen them on the Anne

Photo: Simon Nevill

Beadell, Connie Sue, Great Central Road and Gary Highway. I have never seen them on the Gunbarrel or the David Carnegie or Hunt tracks but that certainly does not mean they are not there, just indicating that on those long runs no birds have ever been sighted by me. Some authorities are calling both the sub sp *whitei* of the Pilbara and the sub sp *striatus* of our desert regions as the Rufous Grasswren and attributing the populations of sub sp *striatus* in SA and Victoria as Striated Grasswren. For this publication I will leave it as previous as there may be a change to the Pilbara sub sp *whitei* as it's colours and habitat requirements are very different to our desert sub sp *striatus*.

Dusky Grasswren *Amytornis purnelli* 16 cm

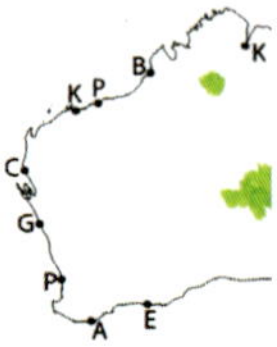

Formerly this species was grouped with Thick-billed Grasswren, then a disjunct population near Mt Isa was considered the same species but now that too, has been classed as a separate species as well, the Kalkadoon Grasswren. Dusky Grasswren is one of the most accessible grasswrens to see, as it is relatively easy to find in the Macdonnel Ranges in Central Australia particularly at several of the well known gorges like Emma and Serpentine Gorge. A fairly confiding grasswren, often seen in large family groups of 5-8 birds. Dusky Grasswren has a rufous-brown body, crown and mantle darker with typical grasswren white spiked streaks passing laterally from the head down the upper body stopping at the middle of the back and belly, darker rufous brown wings and flanks, brown long tail normally cocked. The female, like all the entire subspecies of this group have a distinct but diffused rust-orange patch on flank. **Nest:** Typical grasswren nest made in Spinifex with softer intertwined grasses and finc twigs. Clutch size: 1-2. **Similar species:** None. Female Splendid Fairy-wren should not confuse. **Status:** Locally common. **Where to find:** In WA not too many localities. Found in the Rawlinson. Wingellina Hills and Mt Davies (permit required for these two localities).

Photo: Rohan Clarke

Black Grasswren *Amytornis housei* 19 cm Endemic

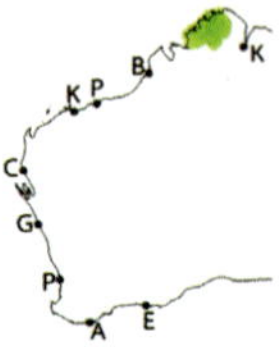

A large grasswren looking generally dark brown but if you are lucky to get close and in full sunlight the subtle changes in plumage are rather beautiful. Its home is the deep dissected sandstone gorges of the north-west Kimberley. Most of its territory is relatively inaccessible except for the tourist destination of the Mitchell Plateau where most people go to find this restricted grasswren. Much of the time, when looking for this species you think they are not around, but suddenly out they emerge from the deep fissures in the rocks. Their call is powerful, like the White-throated Grasswren and easy to pick up from a reasonable distance. They run fast if alarmed negotiating the rocks with ease, rarely flying. The male has an all black body and chestnut back which merges into the black, distinct lateral white streaks cover the head and pass down the throat to the breast and on the back to the mantle, tail is black, stout black bill. Female is similar but under-body is rich rufous not black and rufous plumage covers the shoulder and back. **Nest:** Large dome nest with side entrance having an entry platform made from Spinifex, leaves and twigs. Clutch size: 1-2. **Similar species:** None.

Photo: Simon Nevill

Variegated Fairy-wren subspecies *rogersi* co habits the same territory but should certainly not confuse the observer. **Status:** Locally common. Changing fire patterns have had an adverse effect on habitat, even in the sandstone country. **Where to find:** Sandstone country between the Charnley and west of the lower King Edward River. Common in the Harding, Edkins., Elizabeth and Catherine Range. Small populations found as far south-east as the Caroline Range. Common on the Gardener and Mitchell Plateau region.

Western Bristlebird *Dasyornis longirostris* 17 cm Endemic

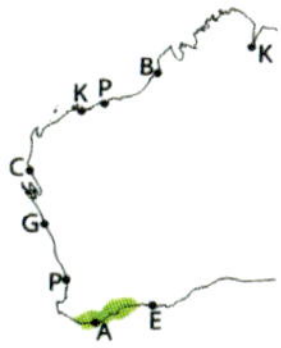

Now restricted to some of the dense low heaths from Two Peoples Bay to Hopetoun, although there are some translocated birds in the Walpole area. A little more widespread than the Noisy Scrub-bird in its distribution with pockets of populations throughout the Fitzgerald River NP. Cheynes Beach area. Mt Manypeaks. Betty's Beach and Two Peoples Bay area. This is a scarce bird even more so than its eastern states counterpart the Eastern Bristlebird. Increased protection of the Noisy Scrub-bird has certainly helped the Two Peoples Bay Bristlebird population. For years, the author has seen populations of bristlebirds come and go in the Fitzgerald River NP. Western Bristlebirds require low heath mostly very dense, occasionally more open in stands of Baxter's Banksia (*Banksia baxteri*) and Southern Plains Banksia (*Banksia media*) heath on the eastern side of the Fitzgerald. Wild fires particularly in the Fitzgerald and Mt Manypeaks area have affected many small populations. However, its status is nowhere near as tenuous as the Western Ground Parrot's. It spends most of its time on or near the ground, taking insects from low shrubs or in leaf litter. If approached

Photo: Simon Nevill

will sometimes give a warning 'click' call similar to a Noisy Scrub-bird. It has a plaintive but strong 'deeee..da...diditdit........deeee..da...diditdit'. In the Fitzgerald River NP, it often gives this after the main call then a rolling trill hard to put into words. The Western Bristlebird is a plain brown ground dwelling bird with short dark chestnut wings and with a long tail. Slight scalloping on the mantle and crown. Slight off white throat, not as extensive as Rufous Bristlebird or Noisy Scrub-bird but very similar to the Eastern Bristlebird. Has a small white eye stripe above the lores. Pale grey-brown underbody with scalloping far more pronounced than in the Eastern Bristlebird. Its species name *longirostris* comes from the Latin '*longus*' meaning long and '*rostrum*' meaning billed. However, its bill is only slightly longer than the Eastern Bristlebird's. **Nest:** The nest is a large oval ball constructed of dry grass or sedge stems placed close to the ground at the base of a tuft of sedge or other low shrub. There is a large entry hole at the top side of the oval. Inside it is lined with finer grasses. Clutch size 2. **Similar species:** The 'jizz' and colouring has some similarities to Noisy Scrub-bird but even though they share the same general area, their territory is different. The Scrub-bird inhabits dense riverine gullies or taller dense scrub with an understorey of thick sedges whereas the Bristlebird inhabits mostly low vegetation with no tall over storey as it prefers thick low heath. The Noisy Scrub-bird has a bulkier body being a heavier slightly larger bird. Its back is darker and tail is shorter in relation to body. Bill more down curved. Throat on male very white, contrasting with the dark crown. White on underbelly. Female less white on throat and no white on underbelly. Calls are totally different, the Bristlebird sounding plaintive, the Noisy Scrub-bird being just that, with a powerful strong resounding and far carrying call. **Status:** Scarce to rare. **Where to find:** Mount Manypeaks. Two Peoples Bay. Waychinicup. Cheynes Beach. Fitzgerald River NP.

Spotted Scrubwren *Sericornis maculatus* 11-15 cm

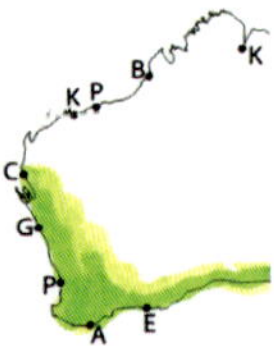

Authorities have now split what was formerly the White-browed Scrubwren sub sp *frontalis* in Western Australia naming it now Spotted Scrubwren and the eastern Australian populations retain the original name White-browed Scrubwren. The West Australian birds have far stronger streaking on the underparts, differing calls and occur in general in drier habitat. In this region, three subspecies occur. *maculatus* occurs in the lower South West and is illustrated here with slightly more yellow on the flanks and belly. *mellori* is found along the southern coast across the southern Nullarbor and partly into the Wheatbelt; it has a white underbody with only pale yellow on the lower flanks and vent, strong black streaks on the chest. *balstoni* occurs from Jurien all the way up to Shark Bay, has less bold streaking on a plain white chest and white flanks and only the vent and lower flank have slight pale yellow. The White-browed Scrub-wren is bold bird that will defend territory and scold intruders with its rasping chatter. Spends most of its feeding time within the foliage but will go into the lower strata of small trees rarely the canopy of tall trees. **Nest:** The nest is a loosely woven dome structure built in the centre of a low dense bush, with a side

Photo: Simon Nevill

entry hole mostly low to the ground. It is made from grasses or strips of bark and lined with soft materials. Clutch size 2-3. **Similar species:** None really. The only possibility is the Shy Heath-wren, as both often occur in the same area. The Heathwren does have a white supercilium eye stripe and also chevrons on the wing but does not have the dark facial patch. Has a bright rust coloured rump. Streaking on the underbody is far more extensive and bolder than on the Scrubwren. The Heathwren is a slightly larger bird and has a totally different call and as the name implies does not readily show itself nearly as much as a SpottedScrubwren. **Status:** Common throughout range, *balstoni* less so in the north. **Where to find:** Not in the inner part of Perth but in the surrounding country and foot hills and throughout the rest of the South West across to the mallee on the southern coast of the Nullarbor.

Shy Heathwren *Hylacola cauta* subspecies *whitlocki* 13 cm

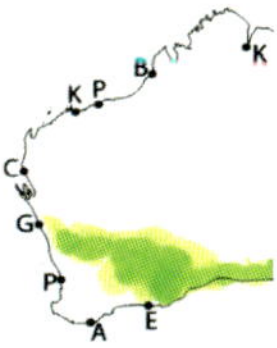

There are four subspecies in Australia and only one in this state, subspecies *whitlocki*. Their range is very habitat based, requiring semi-dense shrub-heath but not as dense as the low heath that Bristlebirds require. Mostly it occurs wherever there are shrubs under mallee such as Broom Bush (*Melaleuca uncinata*). Their range is confined to the South West WA, basically from Bullfinch in the north, across to Buntine Reserve near Wubin south to Dongolocking Reserve and the Stirling Range. East to the Fitzgerald River N.P. then Grass Patch and east to Eyre on the southern Nullarbor. It is a ground dwelling bird getting nearly all its food needs from the vegetation litter. Shy by name and certainly shy by nature, not easy bird to approach and typically, will run away rather than fly. The male has a distinct white supercilium eye stripe with black line through the eye from lores to rear of head. Very strong black streaking on the throat and underbody. Has black and white chevrons on the wings. Rust coloured rump seen clearly. Female is very similar but lacks the intensity of facial colours, the eye strip is brown not black. The streaking on the throat not so bold and underbody streaks narrower. Call is very musical, a rapid series of up and down notes. When on the ground always cocks its tail at a steep angle. **Similar species:** Rufous Fieldwren which can occur in close proximity to the Heathwren. It lacks the black eye stripe. Streaking marks not quite as broad. All subspecies have a rufous crown not dark brown as in the Shy Heathwren. No white chevrons on the wings. Rump colour is a more subdued chestnut colour. Has mottled dark brown streaking on the brown back. White-browed Scrubwren, but has no rust colouring on rump and rarely cocks tail. Chevrons are smaller, black eye stripe broader. **Nest:** Is a loosely structured dome nest, flattened at the base with a low side entry, invariably placed directly on the ground and normally at the base of a mallee or casuarina bush or under a dense shrub. Clutch 2-3. **Status:** Locally moderately common. Throughout much

Photo: Simon Nevill

of the range uncommon. Western Wheatbelt populations threatened with inadequate areas to expand and gene pool weakened. **Where to find:** Dongolocking Reserve. Wongan Hills. Buntine Reserve. Milton McNeil Reserve. Stirling Range NP. Corakerup NR. Fitzgerald NP. Hyden NR. Throughout the Transitional Woodland from the Vermin Fence east of Lake Cronin, all the way across to Dundas Reserve. Common in the mallee around Eyre Bird Observatory. Some small and rarely visited reserves like North Tarin Rock have amazingly high densities of Heathwren, showing their tenacity to survive in the most drastically changed environment.

Rufous Fieldwren *Calamanthus campestris* 12 -13.5 cm

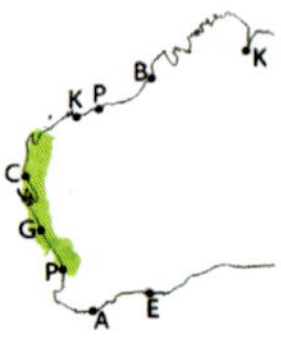

Since writing the first edition of this book the IOC and other authorities have decided to revert to the original split in the *Calamanthus* complex and re establish the species name 'Western Fieldwren' as a separate full species. I am not commenting in great depth in this publication but I will raise just a few points. There is a great range of clinal coloration change from the northern rufous headed, brown backed form of the Rufous Fieldwren to the lower south-west olive backed lighter crowned form of the Rufous Fieldwren. The point I raise is at what point do you make the split in species. It a very important thing that authorities do when establishing a full species status to a given bird and it should be clear and well defined in establishing that status. The Rufous Fieldwren is complex group with eight recognized subspecies. Six subspecies occur in WA. The nominate race *campestris* found from the SA/WA border to Israelite Bay *rubiginosus* occurs along the central west coast from Kalbarri to Exmouth; *wayensis* on salt lakes from Lake Austin in the west to Yeo Lakes in the east and Lake Disappointment in the north and Lake Barlee in the south; *montanellus* south of line from the lower Nullarbor across to Geraldton to the lower west. *hartogi* on Dirk Hartog Island; and *dorrie* on Dorre Island. Like the Shy Heathwren, this is a shy ground dwelling bird that occurs in low dense heaths feeding mainly on insects gleaned from the ground. When disturbed it will run along the ground with tail cocked high moving fast away from intruders. In the breeding season it will perch on a low bush singing its lovely melodious song. The South West subspecies *montanellus* is normally found in low Kwongan Heath, sometimes in adjacent samphire. The west coastal bird *rubiginosus* has a preference for samphire and low bushes in low dunes that hug the coastline. The inland birds *wayensis* are solely found around the perimeters of salt lakes with samphire and adjacent low shrubs. The sexes are almost impossible to split by viewing, if anything the female's facial markings and black streaks are less prominent than the males. The male has a rufous crown, most rufous in the

Photo: Simon Nevill

inland birds and darkest on the central west coast birds and least rufous in the South West birds. All have a pale rufous rump but not as pronounced or bright as Shy Heathwren. Mantle and lower back varies from pale rufous-brown with pale dark rufous-brown streaking in inland and west coast birds, to plain olive-brown with darker streaking. Call is a plaintive series of notes, not as rapid as Shy Heathrwen and softer. **Nest:** The nest is a round dome structure with finer strands of grasses than Shy Heathwren and is lined inside with soft fine grasses and leaves. Sometimes built in a depression made by the bird, occasionally built off the ground in the centre of a very dense bush. Clutch size 2-3. **Similar species:** Refer Shy Heathwren. **Status:** Inland subspecies *wayensis* locally moderately common, west coast subspecies *rubiginosus* locally common, south west subspecies *montanellus* is threatened according to some authorities, but the author believes is vulnerable if any more habitat is lost, though where they occur locally they can be moderately common. **Where to find:** West coast species at New Beach. Red Bluff at Kalbarri. Coral Bay. Many coastal heathlands from Jurien Bay all the way to Exmouth. Inland species: Lake Annean. Lake Disappointment. Lake Carnegie. Yeo Lakes and possibly Plumridge Lakes. South West species: Dongolocking NR. Coblinine NR. Tarrin Rock and North Tarrin Rock NR. Fitzgerald River NP. particularly on coastal heath. *Campetris* very common in the samphire and blue bush on the Nullarbor near the WA border.

Western Fieldwren *Calamanthus montanellus*

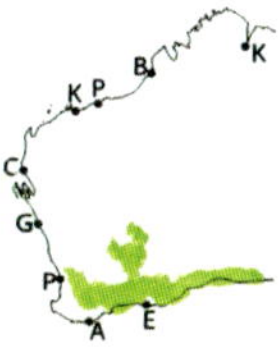

The southern range birds that occupy the southern heathlands have been grouped as a singular species by the IOC. I include a Photo: of a fieldwren that occurs in this range taken at Dongolocking Nature Reserve. If I had taken photographs of birds on the south coast say in the Fitzgerald River National Park or the Stirling Range, then they would show an even more olive back coloration. If one looks at the extreme range of the *Calamanthus* complex in WA and discusses the differences in plumage it could be summarized as follows: The northern birds (Rufous Fieldwren) those birds found north of Carnarvon have a strong rufous crown, nape and ear coverts. Their upper-body is pale brown and bill slightly thicker. Those birds on the southern coast (Western Fieldwren) have less rufous on crown and no rufous on the ear coverts. Bill finer. Those birds found just north of Perth would show a blending of the two extremes through natural clinal change.

Photo: Simon Nevill

Redthroat *Pyrrholaemus brunneus* 10.5 -12.5 cm

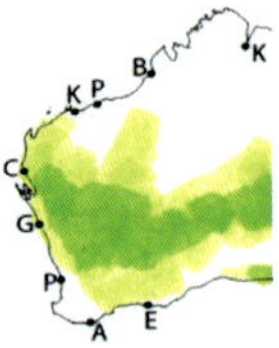

A bird of the semi-arid regions of Australia. Found from as far north as the Ashburton River in the Pilbara right across into the eastern deserts, then south east to the northern part of the Great Victoria Desert and south to the Transitional Woodland Line and most of the drier areas of the Wheatbelt. Its habitat requirements are varied and although common in much of the Mulga Woodland and Transitional Woodland belt it is patchily distributed and found in distinct locations like thick acacia scrub along creek lines in stony country or thick acacia in open mulga. In the Transitional Woodland zone particularly south of Coolgardie, they are found in the understorey bushes of the Salmon Gum Woodland, mostly medium height bushes of the melaleuca group. Its call is well known to many and is an absolute joy to listen to. It is also a great mimic bringing other bird calls into its repertoire, particularly scolding calls of thornbills, or parts of the Pied Butcherbird call. It is a plain brown bird bigger than thornbills or gerygones, the male having a small orange-brown circular throat patch not always easily seen. The young juvenile males do not have this coloured patch but look very similar to the female. Both sexes have a grey-brown back, grey breast and white belly with buff flanks, the female having no orange-brown throat patch. **Nest:** The nest is a dome structure made from bark, fine twigs or grasses; it is quite a compact ball not a loose structure, with a side entry, rarely on the ground like Shy Heathwren but placed in thick bush above ground usually not more than a metre off the ground and placed in the centre of the dense shrub. It is a sedentary bird. Clutch size normally 3. **Similar species:** None, possible confusion with Gilbert Whistler as they share the same habitat in the southern Transitional Woodland. The Gilbert Whistler is larger totally different jizz. Has a larger throat patch, distinct black lores and a thicker whistler type bill. Calls totally different. **Status:** Moderately common throughout range. Now scarce to rare in the inner Wheatbelt. **Where to find:** Nearest to Perth, Wongan Hills area. Some northern Wheatbelt reserves like Billyacatting. Chiddarcooping NR. Buntine NR. Reserves east of Dalwallinu. Low coastal dunes along Gladstone coastline. Throughout the Mulga Region down to the eastern Wheatbelt in reserves like Lake Gaunter near Hyden. In the southern Pilbara and Gascoyne Basin and southern deserts. Southern Nullarbor in mallee.

Female

Photos: Simon Nevill

Weebill *Smicronis brevirostris* subspecies *occidentalis* 8.5 cm

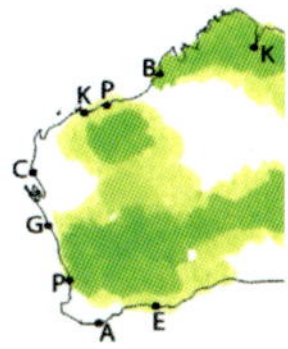

Australia's smallest bird, although if the emu-wren did not have such a long tail it would be the smallest. Weebills are found throughout WA although rarely in the central Great Sandy. There are four subspecies with three occurring in WA. The south west birds *occidentalis* have a darker upper body, slight faint streaking on the throat and upper breast and overall are darker than the subspecies *ochrogaster* of the Pilbara and the upper Murchison. The Pilbara birds have no faint streaking at all on the throat and upper chest. Further north into the Kimberley, the subspecies *flavescens* are very different having a bright yellow under body. All weebills have an orange-chestnut tinge to the lores. When feeding, Weebills spend nearly all their time gleaning lerps and insects off leaves. You can actually hear them peck at lerps on leaves as they make their way through the foliage. Weebills chatter as they communicate with other members of the group as they feed with their call often described as 'I'm a weebill' with the first notes starting high and dropping down. Their preferred habitat in the South West is areas with taller eucalypts and middle storey shrubs. In the Mulga Woodland they can be anywhere working creek lines or moving through open mulga. **Nest:** They build a suspended nest, normally from the upper branches of eucalypts or acacias. Is a pear shaped domed nest with a top side entry, neatly woven from fine grasses and held to the branches by spiders webbing. Is a more dense structure than the gerygone's or thornbill's nests. Located in the dropping outer branches of eucalypts, similar location to Yellow-rumped Thornbill but harder to find and much smaller. The inside is lined with fine grasses and feathers. Clutch size 1-3. **Similar species:** Possibly Western Thornbill, which has longer tail and body dumpier body than the Weebill. Slight flecking on forehead. No orange flush to lores and no yellow tinge on flanks but buffy colour. Bill slimmer and longer. Call different. **Status:** Common throughout range. **Where to find:** Varied habitat throughout the state. Common in eucalypt and acacia woodlands. Mallee in the southern Nullarbor.

Photo: Simon Nevill

Dusky Gerygone *Gerygone tenebrosa* subspecies *christophori*

11.5 cm Endemic

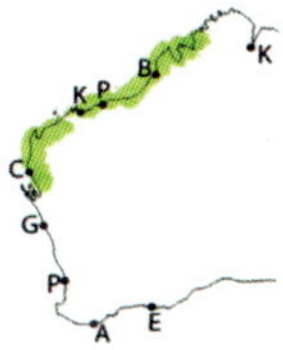

There are two subspecies in Western Australia and the differences are very slight, mainly based on wing and tarsus size. Totally restricted to mangroves and in Broome they are the dominant mangrove gerygone, surprisingly not the Mangrove Gerygone which often frequents paperbark woodland near or away from the coast. Dusky Gerygone subspecies *christophori* range from Gladstone to Cape Keraudren south of 80 Mile beach. Subspecies *tenebrosa* occurs from La Grange to the Walcott Inlet. It is a grey-brown backed gerygone, with a split white eye ring and importantly, a white eye. Slight buff colouring to flanks. Plaintive whistle 'de..dude..de'. **Nest:** Nests within mangroves; the typical gerygone nest is a neatly woven suspended structure with trailing beard at base, bound to a mangrove branch with spiderwebs and woven bark or plant strips. Clutch size 2-3. **Similar species:** Mangrove Gerygone has white at the base of the tail and red eye, Dusky has no white and white eye. Western Gerygone has two bands of white in the tail and also red eye not white. Totally different calls. **Status:** Common in all mangroves. **Where to find:** New Beach. Miaboolia Beach Carnarvon. Cape Range. Onslow. Karratha. South and north of 80 Mile. Common Roebuck Bay.

Photo: Simon Nevill

Western Gerygone *Gerygone fusca fusca* 10.5 cm

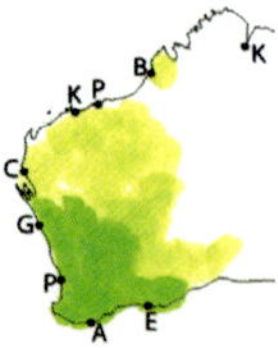

Three subspecies occur in Australia with two occurring in WA. Subspecies *fusca* found in the south west and migrates north to the Pilbara and Kimberley. Subspecies *mungi* occurs in the central deserts and also moves north out of the breeding season. A very common bird found in a variety of woodland and forest habitats. Is considered partly a passage migrant, leaving the deeper colder South West region in autumn and winter and returning in spring to breed. Found in the Jarrah and Karri Forests and Wandoo Woodland and peppermint groves in the south west. When attending nest to feed young, does varying twists and turns as if chasing insects, never flying straight to the nest. Adult bird has slightly white supercilium eye stripe. Dark red eye and a split orbital eye ring. White to the outer tail feathers, seen well in flight from underneath particularly if it is fanning its tail. Plain grey-brown upper body and

Photo: Simon Nevill

off white under body. Call melodious, starting high pitched continuing in a descending scale which sounds as if it is not finished, 'ceeeeee…ce..ce..ce. **Nest:** A suspended dome nest typical of gerygones is placed in outer foliage of a eucalypt or mulga tree. Made with fine grasses bound by spiders webbing and lined with plant down or feathers. Clutch size 2-3. **Similar species:** None really. Possibly Western Thornbill which is smaller and has no white eye ring or white in the tail. Buff-fawn under body while the Western Gerygone's is off white. Shorter tail. Mangrove Gerygone less white in tail and call different only overlaps in summer in the Kimberley. **Status:** Common throughout the South West in winter and spring. Uncommon to moderately common in the arid regions, particularly in winter. Where to find. Anywhere in the south west where forest and woodlands occur and in the mulga woodland.

Mangrove Gerygone *Gerygone levigastor levigastor* 11 cm

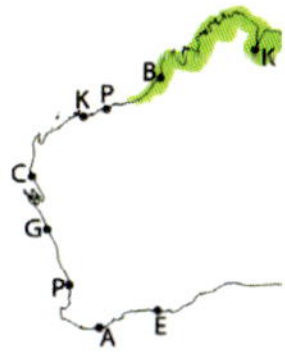

As the name implies, a bird of the mangroves, although strangely enough, around the region of Broome it is often found more often in paperbark woodland and yet when I am in the northern territory or Queensland the species is very common in mangroves and I wonder sometimes if the similar species Dusky Gerygone has become the dominant gerygone to feed in mangrove habitat on the north west coast. In the mangroves at Wyndham the Mangrove Gerygone is the common gerygone. Calls, beside habitat and other factors, are so important when assistance is needed when identifying very similar plain looking species. This applies to the entire gerygone genus. The Mangrove Gerygone has a plaintive musical song, similar to Western Gerygone. I can only describe the Dusky Gerygone call as a fragmented up and down jerky call, with longer gaps between calls. Best to hear a recorded call, you will easily split the calls. Feed actively in canopy often hovering momentarily to get insects. The Mangrove Gerygone is plain grey brown on upper body with pale buffish under-body, pale brown ear coverts, faint white eyebrow and faint orbital eye ring, Importantly, dark red eye and in flight shows white to base of outer tail feathers. **Nest:** Dome nest interwoven around stem of mangrove small branch in canopy, neatly woven grasses and leaves with topside entrance with hood. Clutch size: 3-4. **Similar species:** If seen closely the Dusky has paler under-body and most importantly a white eye, no white in outer tail feathers that can be seen well i.e. faint and diffused, call distinctly different. Plumage colour more like Western Gerygone but it is not found in this region. **Status:** Moderately common in the west Kimberley and common in Derby and Wyndham mangroves. **Where to find:** Adjacent to coast in paperbark woodland and sometimes mangrove near the coast at Roebuck Bay and along parts of Cape Leveque coast, away from coast around wetlands like Lake Eda and Taylors Lagoon, also Derby and Wyndham mangroves.

Photo: Simon Nevill

Large-billed Gerygone *Gerygone magnirostris magnirostris* 11 cm

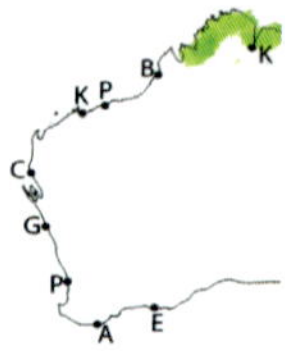

Largest of the gerygone genus, having a larger bill, seen readily when you are acquainted with the various gerygone species. Also an olive tinged brown back although not as olive tinged as the east coast subspecies *cairnsensis*. It frequents paperbark woodlands and lusher forested woodlands adjacent to mangroves. Also found inland monsoonal vine forests particularly if water pools are present. Can share habitat with Green-backed Gerygone. Large-billed has a pale olive-brown back, white under-body and throat, no white in tail just slight pale brown spots on tip of tail not really discernable in flight like Western or Mangrove Gerygone, red eye, importantly, broken eye ring not easily seen in subspecies *magnirostris*, thicker slightly bigger black bill than other gerygones. Call very strong and varied at times plaintive like Mangrove but more forceful and then at times goes into a rapid repeated note a little similar to Green-backed Gerygone. Feeds mainly in the canopy of mangroves but also in the lower strata of rainforests. Large-billed are very much a dual habitat feeding bird found both in mangroves and rainforest but not in riverine or rainforest habitat as much in WA as they do in the NT and the east coast. **Nest:** Long suspended dome nest almost reminiscent of a Sunbirds' nest and you know they are in the area if you find a nest. Although be careful both Bar-breasted Honeyeater and Green-backed Gerygone make a similar suspended nests but more untidy and rarely suspended on a long vine or thin sapling. Clutch size: 2-3. **Similar species:** Mangrove Gerygone is smaller, call different, Mangrove has clear white eyebrow that goes to base of bill whereas Large-billed has small white lateral split eye ring. Dark loral black line in Mangrove. Dusky the white eye not red is the best feature to look for. Has a pure white under-body no buffy tones to white again, different call. **Status:** Moderately common, north west Kimberley coast north of Walcott. **Where to find:** Mangrove habitats of the north west Kimberley coast particularly the larger mangroves of the *Rhizophora* genus, from north of the Walcott Inlet around the coast to Cambridge Gulf. Normally require boats to access these areas. Has been found in inland monsoonal vine forests and rainforests. More accessible areas are on the Mitchell Plateau in vine thicket and rainforests and lush paperbark woodland.

Photo: Graeme Chapman

Green-backed Gerygone *Gerygone chloronotus* subspecies *danwini* 10.2 cm

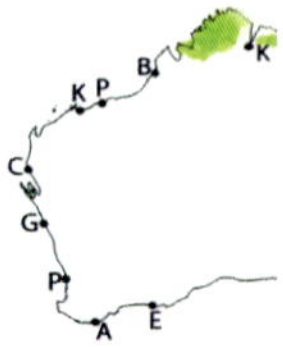

A small short tailed dainty gerygone. Luckily fairly easy to identify as no gerygone in the same habitat in the Kimberley has such an olive-green back and dark brownish-grey head. I love finding Green-backed Gerygone, particularly in the Kimberley as you can walk into the smallest patch of monsoonal vine thicket even less than a couple of acres in area and there they are. Then you can walk only a few metres into open eucalypt woodland and you would never get them unless your there in the middle of the wet. I have seen them forage at all heights within these small pockets of remnant forest. The Green-backed Gerygone has an olive-back, pure white under-body with yellowish tinge on flanks, grey-brown head, red eye, there is a clearly defined demarcation between the dark grey head and the pure white throat as in White-throated Gerygone, no white in tail. Call is unmistakable, 3 stages of rapid calls starting with a fast repeated note, ending in the most rapid single noted call I know of any passerine in Australia. The subspecies *darwini* call is still a very rapid call but is slightly less intense and softer in tone than subspecies *chloronotus*. **Nest:** A suspended dome nest similar to Large-billed Gerygone but has trailing tail from the base of the dome, the nest secured directly below the branch and not suspended on a woven cord like a Large-billed Gerygone. Clutch size: 2-3. **Similar species:** No other gerygone has the olive-green upper-body and White-throated does not share same habitat and has bright lemon-yellow under-body. **Status:** Uncommon but widely distributed in small pockets of lush forest or dense riverine vegetation. **Where to find:** Resident in Windjana Gorge particularly below the south facing wall. Found in any of the small remnant vine thickets throughout central and the northern Kimberley. El Questro Station in the spring fed palm forests.

Photo: Simon Nevill

White-throated Gerygone *Gerygone olivacea* subspecies *rogersi*

10.5 cm

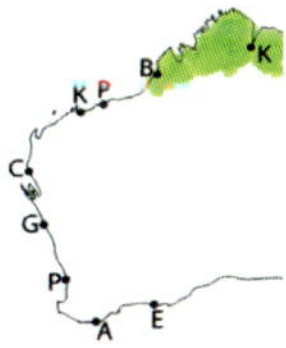

An attractive gerygone found through the drier open eucalypt woodlands but particularly along riparian habitat. Its plaintive call is reminiscent of a Western Gerygone with descending notes. Moves through eucalypt foliage fairly quickly gleaning insect as it goes. The White-throated Gerygone has a grey-brown upper-body and head, lemon-yellow belly, white vent and white throat, loral white spot at base of bill, red eye stands out from darker grey lore area, in flight shows clearly its white spots on the tip of the tail showing as a band in flight. **Nest:** Typical gerygone suspended dome nest but tightly woven bark to form outside structure. Clutch size: 2-3. **Similar species:** No other gerygone has the

Photo: Simon Nevill

combination of white throat and yellow underbody in WA. **Status:** Common throughout the Kimberley. I have seen them as far south as the large paperbark woodlands near Samphire Road House and as far north as the King Edward River and to the east on the lower Ord River. **Where to find:** Stop at any of the woodlands from Shamrock all the way to the Willare Bridge and good chances of seeing this common gerygone. Found around Broome Bird Observatory. Found through out the Kimberley and south to the Fitzroy River. Common along the Ord south of Parry's Lagoon.

Inland Thornbill *Acanthiza apicalis* 10.5 cm

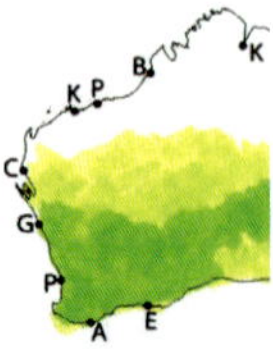

Named 'Inland' because the eastern states birds are found on the eastern side of the great divide while the Brown Thornbill occupies the more humid habitats but the name is a little misleading for the South West birds as they frequent the wettest areas as well as the semi-arid regions of Western Australia. The lower South West birds are the subspecies *apicalis* and their range goes as far north as the central Wheatbelt; the subspecies *whitlocki* covers the mulga and Pilbara region and like most subspecies hybridisation occurs were the ranges overlap. The South West subspecies *apicalis* are the darkest of all the Inland Thornbills and the scalloping on the head is heavier, also the streaking on the breast is much bolder. The subspecies *whitlocki* have paler backs, finer scalloping on the forehead and streaking on the breast. Throat streaking not as contrasting or bold, also the flanks have a paler buff colouring. In the South West, the birds feed in the thick understorey of the forests as well as working the canopies of taller trees. The drier country subspecies whitlocki frequent the mulga woodland and will feed in mixed flocks with Chestnut-rumped and Slaty-backed Thornbills. Inland Thornbills in the lower South West tend to cock their tails most of the time, a characteristic that does not occur with Brown Thornbills on the east coast. The calls are pleasant soft whistles with occasional grating notes. Surprisingly it is a good mimic and combines calls of other birds in its repertoire. The adult bird has heavy streaking on the throat and chest. A red eye. Faint horizontal flecking on the forehead. Distinct chestnut rump. Buff flanks. Diffused dark band on tail. **Nest:** Is a loosely structured dome nest with a slightly hooded entry woven to vertical branches or placed in the centre of a prostrate bush. Nests are mostly from 50-120 cm off the ground. Lined with plant down, fine grasses or feathers. Clutch size 2-4. **Similar species:** Chestnut-rumped Thornbill has no streaking on the throat and chest and has white eye not a red. Does overlap territory with in the South West, so be careful. Slaty-backed does have a red eye but no streaking on throat or chest and has lateral small streaks on crown. All species have distinctly different calls although Slaty-backed and Chestnut-rumped Thornbill

Photo: Simon Nevill

have similarities. **Status:** Common in South West. Moderately common in the arid region of the South West. **Where to find:** *apicalis* anywhere in the heavier forested region, less so in the Wheatbelt. *whitlocki* along creek lines and in the denser parts of the Mulga Woodland and mallee country north of the Nullarbor.

Chestnut-rumped Thornbill *Acanthiza uropygialis* 9.8 cm

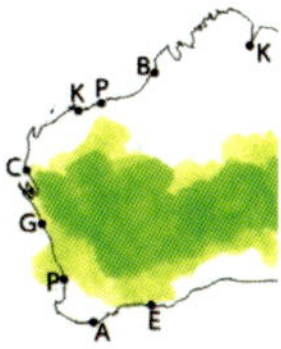

A thornbill of the inland regions of Australia not the deep South West forests. It ranges from the central Wheatbelt from Corrigin avoiding the west coast until just north of Geraldton up to Manilya and across to central Australia. In the south from Corrigin, sometimes lower, its range runs north east into the desert regions across to the centre. It is the most common specie of thornbill in the Mulga region, sometimes occurring in mixed flocks with Southern Whiteface and occasionally Slaty-backed Thornbill. Out of the breeding season, feeding parties of thornbills move through the bush accompanied by different species like Rufous Whistler and Red-capped Robins. Adult birds have a grey-brown upper body, red eye, chestnut rump, black-brown tail with a white tip, horizontal black and white flecking on the forehead, and plain off white under body. Feeds in trees but also on the ground. **Nest:** Mostly in hollows in trees, often in larger mulga trees or dead stumps. The nest is still a typical dome structure but the entrance is placed where the hole in the tree is located. Clutch size 2-4. **Similar species:** Slaty-backed Thornbill, which is very similar and requires clear observation but once you know the Slaty-backed you can pick up the pure grey back rather than grey-brown back. The Slaty-backed has a distinct plain dark grey back no hint of brown. Very fine lateral black streaks are set against the grey crown, unlike the Chestnut Thornbill that has fine horizontal black and white mottling on the forehead, which could be described as scalloping rather than streaks. The eye of the Slaty-backed Thornbill almost looks black, in fact it is a dark red whereas the Chestnut-rumped is white. Slaty-backed has two sets of calls, one is soft and high pitched not dissimilar to Chestnut-rumped so one must be careful if relying on calls. **Status:** Common to very common in the Mulga and Mallee. Moderately common on creek lines and acacia in the desert regions. **Where to find:** Throughout the Mulga station country and Pilbara. Transitional woodland out to the central desert ranges.

Photos: Simon Nevill

Slaty-backed Thornbill *Acanthiza robustirostris* 10.5 cm

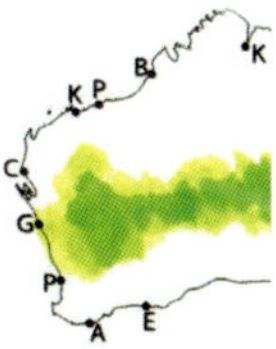

Slaty-backed Thornbill virtually covers the same range as the Chestnut-rumped Thornbill but is far less common. Tends to feed less on the ground than Chestnut-rumped but still accompanies them through mulga woodland. It's said that they are not easy to pick from Chestnut-rumped but once you are familiar with them, the plain slate grey back stands out from the lighter brown grey of the Chestnut-rumped, and the darker head the red eye confirm the difference. Has a slightly thicker bill than the Chestnut-rumped. It has two types of call, one a high pitched soft twittering similar to the Chestnut-rumped and the other stronger more agitated grating call, that is uttered rapidly 'chirp chirp chirp chirp'. The nest location is also different to Chestnut-rumped not built in a tree hollow but a dome nest built in a low bush such as an eremophila bush or sometimes in a dense mulga tree. **Status:** Can be moderately common locally but generally uncommon throughout its range. **Where to find:** Stations in the southern Mulga Region from Tallering Station in the west right across to Yeo Lakes in the east. Specific stations. In the Paynes Find-Yalgoo region the following stations: Maranalgo. Thundelarra. Muralgarra. Gabyon. Carlaminda. In the Cue region: Nallan and Austin Downs stations. Found also in mulga country in the Pilbara region. Acacia woodlands in the central ranges.

Slaty-backed Thornbill Photo above:: Graeme Chapman

Photo: Simon Nevill

Western Thornbill *Acanthiza inornata* 9 cm Endemic

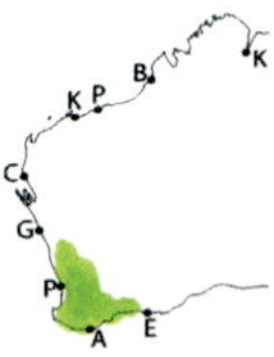

Found in the heavily forested regions of the South West. Feeds in the middle strata of tall eucalypts and occasionally in lower shrubs. Has an olive-brown back, slight darker brown on wings, white eye, fine flecking on the forehead not always clear, orange-brown tinge on the lores, under body cream-olive white, short tail. Call varies and is generally softer than most thornbills with also a rapid 'seep..seep..seep..seep'. Its range extends from Hill River in the north and follows the edge of the Jarrah Forests and inner Wandoo Woodland through

Photo: Simon Nevill

to Boyagin Rock, down to the Stirling Range and across to Pallinup River as its furthest east location. Often not as easy to get as many of the other smaller passerines, many visitors on brief trips miss out or overlook this small endemic. **Nest:** Is a dome structure with top side entry, lined with soft bark or fur or feathers, placed in various locations, within drooping foliage on eucalypts but more often wedged behind loose bark or in a crevice between vertical limbs, which is a common location in the Darling Range. Clutch size 2-3. **Similar species:** None. **Status:** Generally uncommon but can be moderately common in some areas. **Where to find:** Darling scarp within the Jarrah Forest, Karri Forest and Wandoo Woodland.

Slender-billed Thornbill *Acanthiza iredalei* subspecies *iredalei*

10 cm

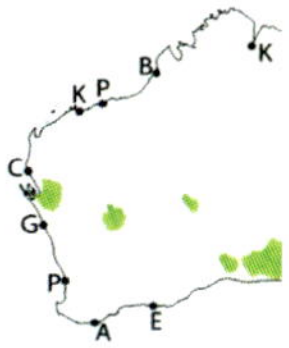

There are three subspecies in Australia and only one in WA, namely *iredalei*. This thornbill lives its life entirely in samphire and bluebush habitats. In the Nullarbor it is found throughout the region where bluebush has not been heavily grazed but alas much of its habitat on the West Australian side of the Nullarbor has been greatly affected by the grazing of sheep. The other populations in Western Australia are confined to some of the larger ephemeral salt lakes and a band of coastal samphire on the mid west coast. The range is fragmented, with populations on the west central coast from Lake MacLeod down to Hamelin Bay and a population on Peron Peninsular. Other populations occur on Lake Annean and Lake Austin. Also Lake Barlee, Lake Ballard, Lake Way and Lake Throssell. Also in the Yalgoo area. None of these however match the numbers that can be found on the central eastern Nullarbor, particularly on the South Australian side. The Slender-billed Thornbill is one of the smallest thornbills. It has a longish bill for its size, pale under body tinged buff-lemon yellow on the flanks, strongest on the rump but far paler yellow than Yellow-rumped Thornbill. Pale grey-brown back with faint mottled throat almost not visible. Black and white mottling on forehead and lores is the most distinct feature. Has a white eye. Out of the breeding season forms small feeding parties moving through the bluebush or samphire. The Slender-billed Thornbill found in mid western Australia have a yellower under body and a greener rump. The upper body is pale olive-grey rather than dark olive as in eastern subspecies *rosiinae*. **Nest:** Is round dome structure of very loose grasses sometimes seaweed when near the coast, placed near the top of a samphire or bluebush shrub. Clutch size normally 3. **Similar species:** Buff-rumped Thornbill subspecies *reguloides*, but that does not occur in this state. **Status:** Uncommon south of Carnarvon. Uncommon in all other Western Australian locations. Moderately common on the central and eastern Nullarbor.

Photo: Graeme Chapman

Yellow-rumped Thornbill *Acanthiza chrysorrhoa chrysorrhoa* 11 cm

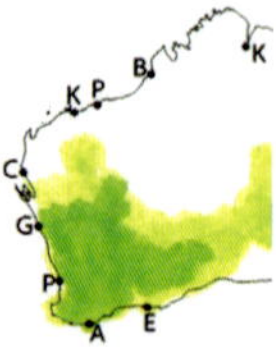

Mainly a ground dwelling thornbill although rarely found on wide open plains; requires open woodland where it will occasionally forage in trees but most of its time is spent feeding on the ground. One of the easiest thornbills to identify with its bright yellow rump, black crown with white spotting, diffused supercilium stripe and black eye stripe with white spotting on ear coverts, slight buff flanks, olive-brown back. When in flight yellow rump stands out clearly against dark black-brown tail and brown back. It is the largest of the thornbills, and is a sedentary species rarely moving far from its territory except in arid zones where it will move after bad seasons. Often its tinkling contact call leads one to the birds and, out of the breeding season, one can see the small feeding parties hopping across the ground, they don't walk. Found throughout the South West from the tall forests of the lower South West, through the Wheatbelt and into the Mulga Region. Their range stretches to the Pilbara and across central Australia but avoids the Nullarbor. There are four subspecies in Australia with two in Western Australia subspecies *chrysorrhoa* in the South West and mid west WA and *normantoni* in the eastern parts of central Western Australia. The western birds tend to be pale olive-brown rather than olive-brown and the rump is paler yellow but they have more buff colouring on the under body than the eastern birds *leighi* and *leachi*. **Nest:** The nest is extremely large for a thornbill, unlike any other and perhaps one of the easiest nests to find. Various trees are used in the South West especially marri trees, and in the Wheatbelt the sheoak trees are a common location. In the Mulga Region often in Kurrajong trees; the common denominator with all these trees is that they have dense foliage that often droops down in thick clusters of leaves. The Yellow-rumped Thornbill builds at least one but often two false entry holes and false nesting chambers. The main chamber is at the base of the structure. They may work on the nest in the pre-breeding season taking up to 2-3 weeks to complete. Clutch size 3-4. **Similar species:** None. **Status:** Common in the South West. Moderately common in parts of the Wheatbelt and Mulga Region. **Where to find:** Throughout the South West. The Mulga. The central desert ranges.

Photo: Simon Nevill

Southern Whiteface *Aphelocephala leucopsis* 11.5 cm

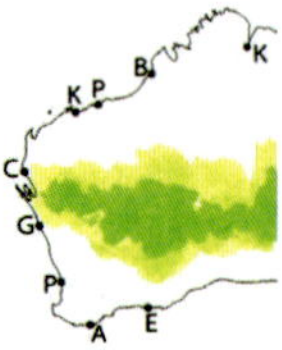

The most common of all the whitefaces, found in the semi-arid and arid areas of WA. There are two subspecies both occurring in Western Australia, *castaneiventris* and *leucopsis* with a hybrid zone between the two subspecies. *castaneiventris* is found from Carnarvon across to the edge of the Victoria Desert and the Nullarbor and *leucopsis* is found through the rest of central Australia. The subspecies *castaneiventris* that covers the bulk of the Mulga in the southern WA is the more attractive of the two subspecies, having a distinct chestnut flank and the head markings are cleaner with more demarcation around the head with its pure white throat, forehead and under body, whereas *leucopsis* has pale grey-buff underparts. Feeds almost entirely on the ground, hopping all the time and picking insects as it goes although seeds make up a reasonable part of their diet, hence the finch-like bill, which can crack small seeds. **Nest:** They nest in a variety of locations but the vast majority of nests in Western Australia are placed in a hollow or a crevice between trunks, mostly in mulga trees. Close to the entry hole they wedge grasses and twigs to form a base for the nest, and line it with feathers. Clutch size 3-4. **Similar species:** Banded Whiteface, refer to text on Banded Whiteface. **Status:** One of the most common birds of the Mulga Region. **Where to find:** All stations from north and east of the Transitional Woodland line to the Gascoyne catchment across to Warburton and south to the Nullarbor.

Photo: Simon Nevill

Banded Whiteface *Aphelocephala nigricincta* 10 cm

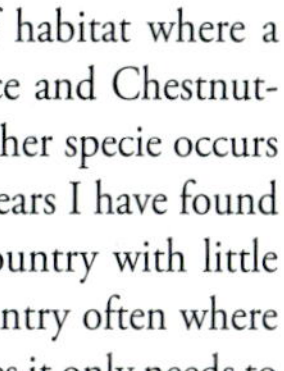

The Banded Whiteface has, as its name implies, a black band running through the centre of the breast. It also has a chestnut mantle, which is clearly seen in flight. On the flanks there is also a diffused area of chestnut. The head is similar to Southern Whiteface having the white forehead and grey-brown crown, but the colour is lighter and bleeds into chestnut on the nape. The tips of the tail feathers are white. The type of habitat where a specific species occurs, can be intriguing, no less so than Banded Whiteface and Chestnut-breasted Whiteface. We will just concentrate on Banded Whiteface as the other specie occurs in SA. Banded Whiteface can be found in a few types of habitat. Over the years I have found them in three types of habitat in Western Australia. One is open stony country with little shrub growth mainly north and south of Meekatharra. Another is open country often where chenopod plants (samphire etc.) are growing with scant vegetation; sometimes it only needs to be a small area such as on the southern boundary of Nallan Station, others are more extensive such as on the northern part of the David Carnegie track in the northern Victoria Desert. The third type of habitat is in recently burnt country particularly in desert regions, for example on the Canning Stock Route and also the Gary Highway. However in this state they are not as prevalent as in the Simpson and Strzelecki Deserts. **Nest:** The nest is a bulky dome structure with a wide entry hole on the side, placed in thick foliage typically the thorny Kurara bush. The nest is lined with soft feathers. Clutch size 2-3. **Similar species:** Southern Whiteface, which has a plain grey-brown back not a chestnut back. This is often all you see of the Banded Whiteface as it flies away from you but its back shows up clearly as being different from the more common Southern Whiteface. The Southern Whiteface has no black band on the breast. However their territories can overlap so one must be aware of that. **Status:** Uncommon throughout WA. **Where to find:** West of Cue. North and south of Meekatharra. East of Gascoyne Junction. Throughout parts of the Little Sandy and Great Sandy Desert north to the Breaden Hills.

Photo: Simon Nevill

Spotted Pardalote *Pardalotus punctatus* subspecies *punctatus and xanthopyge* 9 cm

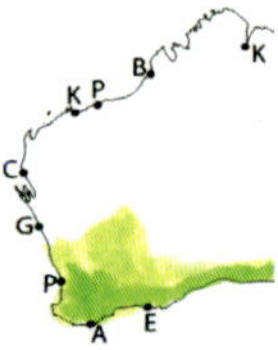

There are three subspecies in Australia and two occur in the west, although there is hybridisation in the south east region. The two subspecies were once separated as two distinct species, one being a bird of the wetter forests of the South West and one a bird of the mallee. Spotted Pardalote can be found in the heavier wooded regions of the South West as far north as Badgingarra and then east across to Coolgardie and down to the coast at Esperance in all areas within this boundary line. The subspecies *xanthopyge* has a preference for the mallee regions. The birds in the far eastern mallee of WA call differently than the South West forest birds *Punctatus* and have a slight variation in their call, whereas *xanthopyge* the mallee bird has a single high pitched but soft call repeated several times and easy to imitate to call up the bird. Both subspecies nest in hollows in the ground unlike Striated Pardalotes that nest in tree hollows, although I have seen Striated nest in a hole in a limestone wall in Perth. Both sexes play an active role in building the nest but the female does the final nest chamber and nest building. Their nest site often gives them away as one is walking or driving slowly down a track; they will fly out of the nest hollow when there is imminent danger. The female has a more subdued black-grey crown and spotting does not stand out as much. The throat

Photo: Simon Nevill

is not as bright yellow as the male. Overall colouring very similar but colours less contrasting nor as bright. **Nest:** Is built at the end of an excavated tunnel, the length normally 30–80 cm long. The end nest chamber consists of a ball of dry grasses or narrow strands of bark woven into a tight ball. Clutch size 3-4. **Similar species:** Striated Pardalote, which has no spots on the black crown just faint white streaking mostly at rear of crown. No white spotting on wing but shows a single red dot at the top of the white edging on the wing. No red or yellow on rump, just beige colouring. Larger supercilium white eye stripe with frontal part near the bill yellow. Mantle is plain grey-brown not scalloped brown. **Status:** Moderately common throughout range, more common in the heavier forested areas. **Where to find:** Subspecies *punctatus* within the Jarrah Forest but more so in the Wheatbelt and Wandoo Woodland. Found in Dryandra Woodland Reserve. Subspecies *xanthopyge*: Common in the mallee in Fitzgerald River NP. Lake Magenta. Cape Arid. Transitional Woodland. Mallee along the southern Nullarbor.

Striated Pardalote *Pardalotus substriatus* 9-12 cm

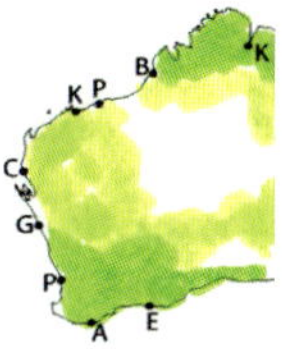

There are six subspecies in Australia. In Western Australia there are two subspecies *substriatus*, throughout the southern parts of WA and the Pilbara and *uropygialis* In the Kimberley. The differences are slight but *uropygialis* has more white in the wings. Distribution is broken by the Great Sandy Desert where Red-browed Pardalote is the commoner pardalote. The Striated Pardalote is slightly larger than the Spotted Pardalote and nests mostly in tree hollows. It breeds not only in spring but can breed in middle and late summer, one of the few passerines to do so. Often seen calling at the very top of a tree when holding territory. It feeds on lerps and other insects gleaning them off leaves and bark. Often feeding high in the eucalypt canopy which makes them hard to see but they do come down to lower shrubs. The call of each of the subspecies throughout Australia varies a lot and anyone visiting this state will hear marked differences to the eastern states birds. The female is hard to split from the male and the only feature is that the crown is black-grey rather pure black and slightly blotched. **Nest:** Is placed in a hollow and prefers hollows that are only just wide enough for pardalotes to enter. The actual nest is a mass of dry grasses placed at the base of the hollow. Clutch size 3-4. **Similar species:** Refer to Spotted Pardalote. **Status:** Common throughout most of WA except the Nullarbor Plain. **Where to find:** Perth region. The Darling Range. Jarrah Forest and Karri Forest. Throughout the Wheatbelt. Pilbara. Deserts and the Kimberley.

Photo: Simon Nevill

Red-browed Pardalote *Pardolotus rubricatus rubricatus* 9-11 cm

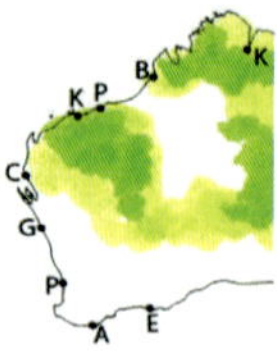

A pardalote of the arid regions, particularly along river and creek systems where it builds its nest and feeds in the eucalypts and other trees that line the river; it can be found away from rivers but that is the common location for this species. There are two subspecies in Australia but only the nominate race *rubricans* occurs in WA, from the Gascoyne River across through the Yeo Lakes to the SA boarder. North of this line to the Kimberley region. It has a plaintive call 'daaa de de dit….daa de de dit….daa de de dit', the calls are softer in the west and the centre than those of the east coast birds. It will call even when the temperatures are very high. Red-browed Pardalote have a black crown with small white spots. The yellow supercilium eye stripe has an orange-red dot near the bill. Its back is olive-brown with mottling in the feathers. The rump and upper tail is olive-yellow and there is a small faint patch of yellow on the breast. **Nest:** Is similar to Spotted Pardolote but often on higher banks alongside dry river beds. **Similar species:** Striated Pardalote, which has a black crown but no white spots just a few faint white streaks. Has a red dot on the side of the wings and clear white band on the primaries. Red-browed does not have but has yellow tinging in the centre of the wings. Lacks the bright yellow throat and has a small red patch on the front of the eyebrow. Spotted Pardalote is not found in the same region as the Red-browed. **Status:** Common in the Kimberley, the Pilbara and northern deserts regions less so the southern Mulga becoming more common towards the central ranges. **Where to find:** Easily found in the Pilbara and the Kimberley and northern Great Sandy Desert, along creek beds and river systems.

Photo: Simon Nevill

Red Wattlebird *Anthochaera carunculata* subspecies *woodwardi* 35 cm

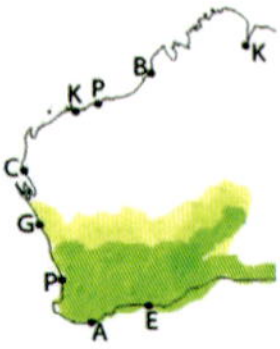

The Red Wattlebird is the largest honeyeater in the South West, having two distinct red wattles protruding from the base of the lores and a yellow belly and vent. Feeds mostly on flowering plants for nectar but also eats invertebrates. The range of flowering plants is diverse including banksias, bottlebrush, flowering eucalypts and grasstrees. Found throughout much of the South West, its northern limit basically following the outer limits of the Transitional Woodland and the limit of most banksias, hugs the south coast till the populations almost meet in SA. There are records as far as Goongarrie north of Kalgoorlie and beyond, in the east of their range and as far north as Billabong Roadhouse. House. Red Wattlebirds are pugnacious birds driving off smaller honeyeaters from their feeding territory. Some remain sedentary but with many there is partial migration to the north leaving the South West Region in April and May when the greatest number of banksias are flowering, and then move south in early spring in August and September when a greater range of flowering plants are in bloom in the south. Their calls are varied from noisy, harsh and grating calls to loud squeaks uttered regularly. There are 3 subspecies in Australia, *woodwardi* in South West Australia is smaller than the other subspecies, has a smaller wing length and the yellow underbelly is brighter. **Nest:** Is a shallow cup of larger twigs on the outside and smaller twigs on the inner lining, being finally lined with soft grasses or wool or feathers. Located in various bush or tree species, often in banksias and dryandras. Clutch size 2 to 3. **Similar species:** Western Wattlebird, which is smaller. No red wattles. Finer and longer curved bill. No yellow on the belly. Has chestnut in the wings, seen very well in flight. Longer down curved facial patch. Call similar but certainly different not as rasping or harsh, more like gurgling chatter that continues for a longer period. **Status:** Common throughout the lower South West, not so common in the outer regions to the Transitional Woodland. **Where to find:** Found throughout Perth in gardens and parks. In the Jarrah Forest. All along the southern coast to Cape Arid and beyond, although nowhere near as plentiful in the Esperance region as Western Wattlebird, range passes east of Kalgoorlie.

Photo: Simon Nevill

Western Wattlebird *Anthochaera lunulate* 30-33 cm Endemic

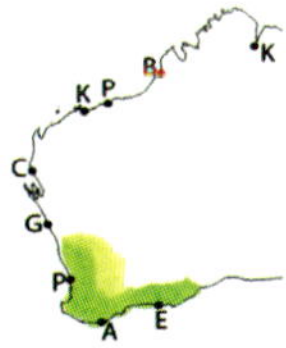

Now classed as a separate species from the Little Wattlebird of the east coast. The Western Wattlebird has a longer stouter bill, much darker and less streaked crown and back, the eye is distinctly red not brown. Has a slightly larger cheek patch that contrasts with the dark crown, more so than the eastern Little Wattlebird. Feeds on nectar and insects. Can share the same territory as Red Wattlebirds but generally avoids flowering areas where Red Wattlebird occur, as they are the dominant bird and extremely pugnacious. Does not travel to the drier areas in the north or the north east as does the Red Wattlebird: its limit is restricted to the wetter zone which follows the eastern edge of the Jarrah –Marri Forest and most of the dense dryandra heaths to Eneabba down to Tutanning and further down to the Stirling Range and across to Cape Arid and beyond. Throughout most of their range they are not a common bird. In the Darling Range they have a preference for *Dryandra sessilis* and can be locally common where it grows. However, Western Wattlebirds can be common from the Fitzgerald River National Park through to Esperance and Cape Arid where they are the dominant large honeyeater and extremely common in the township of Esperance. One of the major plant species that attracts them to this region is the Showy Banksia (*Banksia speciosa*), which has an extensive flowering period. There are of course many other flowering plants that they feed off, particularly the Dryandras. **Nest:** Mostly in dryandra and banksia bushes. Nest is similar in construction to Red Wattlebird but smaller. Clutch size 1-3. **Similar species:** Refer to Red Wattlebird. **Status:** Uncommon throughout most of its range except the southern coast from

Bremer Bay to Cape Arid. **Where to find:** Kings Park occasionally in the botanic gardens section. Yanchep NR, Bungendore NR, in late spring. Heathlands in Dryandra NR in late spring. The Fitzgerald River NP, The town of Esperance where it is very common. Cape Le Grand and Cape Arid NP.

Photo: Simon Nevill

Spiny-cheeked Honeyeater *Acanthagenys rufogularis* 23-26 cm

Photo: Simon Nevill

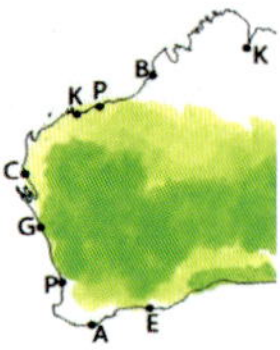

A large honeyeater of the semi-arid regions its distributionsextending from the inner Wheatbelt way out across the Transitional Woodlands into the central deserts. Found in a variety of habitats including Mulga Woodland, dry river/ creek lines, breakaway country, and heath. However it is most common throughout the mulga region feeding on a variety of flowering plants particularly the eremophilas, amyemas (mistletoes) and grevilleas. It is a semi-nomadic species moving to areas where plants are flowering. Little is known about their partial migrations but flocks can be seen moving in late November south through the northern Wheatbelt from the drier regions north. They do not occur in the wetter South West. The Spiny-cheeked Honeyeater's name relates to the small feathers that grow from the side of the neck, not easily seen. The important features are the distinct bi-coloured bill, black at the tip and pink at the base, orange-pink throat, pink eye ring, contrasting white spines on side of neck with dark brown crown, crown and mantle olive-brown with black scalloping. Black flecking on the white lower breast and belly. Long tail with white tip. The call is a pleasant series of gurgling and bubbling notes normally given perched on bare branches at the top of a tree. **Nest:** Is a deep thin walled cup suspended from branches. Constructed with long dry grasses and fine twigs bound with spiders webbing and lined with any soft material available. Clutch size 2 to 3. **Similar species:** None. **Status:** Uncommon in the Wandoo section of the Wheatbelt. Moderately common in the central, eastern and northern Wheatbelt. Moderately common in the Transitional Woodland. Locally common in the Mulga particularly along creek lines. Less common in the desert region. **Where to find:** Northern Wheatbelt reserves like: Buntine. Billyacatting. Mollerin. Chiddarcooping. Walyahmoning. Goongarrie. Most stations north of the Transitional Woodland line into the desert and central ranges.

Yellow-throated Miner *Manorina flavigula* 28 cm

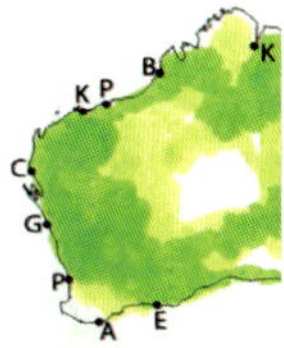

There are 5 subspecies of Yellow-throated Miner in Australia. In WA, there are two obscura in the SW and *wayensis* from the outer Wheatbelt into the deserts and covers an area excluding the wetter South West and running as a line from Moora across the central northern Wheatbelt through a rough line from Kellerberrin across to the Australian Bight, but it's a subjective line and obviously hybrids occur. The differences involve subtle changes in the colour of the back and upper breast but in the context of this book it really is not important; suffice to say that *obscura* occurs in the lower part of the Wheatbelt to Esperance and *wayensis* in the semi-arid and arid zone. We will discuss the range as combined. On the Swan Coastal Plain before European settlement they were more common but numbers have reduced greatly. They are found as far south as Bunbury. They avoid the heavy Jarrah Forest area but come around its eastern boundary where it abuts the Wheatbelt, down to the coast just west of Albany and throughout the rest of the South West. Its distribution stretches all the way north to the Kimberley where the subspecies *lutea* occurs. Yellow-throated Miners are gregarious birds rarely found alone. They have little colour save the bright yellow bill and distinct yellow eye ring, which extends to the rear of the eye. They have a dark cheek patch that contrasts with the yellow eye patch. They are highly aggressive to intruding birds and group together in defence of territories, giving their raucous calls. **Nest:** Is a thin walled cup suspended between a horizontal fork consisting of dry grasses and fine twigs bound with spiders' webbing and lined with soft materials. Clutch size 2. **Similar species:** None. **Status:** Uncommon in the Swan Coastal Plain. Uncommon in the Mulga (subspecies *wayensis*) occasionally locally common. Common in the northern, central and eastern Wheatbelt.

Photo: Simon Nevill

Extremely common in the Esperance to Cape Arid area and would be the most common bird of that region. Common in parts of the Pilbara and also the Kimberley. **Where to find:** In the Perth region locally common in parts of Pinjarra. Occasionally in the Armadale Golf Course and Whiteman Park. Found throughout the Wheatbelt particularly moving along roadside verges. Common in Esperance (subspecies *obscura*). Common throughout most of the arid zone and also the Kimberley.

Singing Honeyeater *Lichenostomus virescens virescens* 19 cm

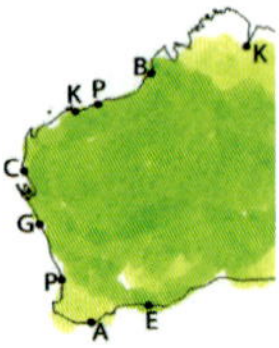

The Singing Honeyeater can be found throughout most of the semi-arid and arid parts of Australia. There are 4 subspecies in Australia all with small subtle differences; subspecies *virescens* covers much of the South West hybridising with *forresti* from a line roughly from Shark Bay across through the Wheatbelt and across and down to Eyre. *forresti* distribution spreads from the mulga, through the deserts all the way to the Kimberley. They are very common in Perth and have adapted well to the urban environment. They rarely enter the central parts of the Jarrah and Karri Forests, preferring more open woodland. They are at home on the Swan Coastal Plain as they are in our driest desert regions. In the South West there is little movement with this species being mainly sedentary. Like most honeyeaters, they feed on both nectar and insects but their percentage of insect intake is much higher than nectar, which is a reversal of many of the honeyeaters. **Nest:** Is a shallow cup of twined bark, grasses and spiders webbing bound from the rim to a fork. Clutch size 2 to 3. **Similar species:** No other species in this State. On the east coast both Mangrove and Varied Honeyeater are similar. **Status:** Common to very common throughout most of the state although less common in the central and eastern Kimberley. **Where to find:** Throughout Perth and the Swan Coastal Plain. Most of the Wheatbelt, the Mulga, the deserts and central ranges. In parts of the Kimberley.

Photo: Simon Nevill

White-eared Honeyeater *Lichenostomus leucotis* subspecies *novaenorciae* 17-20 cm

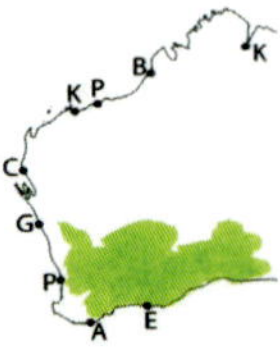

Three subspecies in Australia only one in WA, subspecies novaenorciae. The differences are minor with the western birds having shorter wings and the ear patch not as large. On the east coast White-eared Honeyeaters not only occur in dry areas but also in wet-sclerophyll forests, which they certainly do not in Western Australia. In the South West they occur in distinct habitats. In the Wandoo Woodland they occur in patches of heath but less so within pure stands of open Wandoo and Powder Bark Wandoo. Also in the Wandoo Woodland they occur in sheoak woodland and also common in areas of mallee and heath. Common in the Transitional Woodlands. Its strong scolding 'cheeew…cheeew…cheeew', is often the first sign they are in the area. The WA birds are a nervous bird and will skulk and hide when approached. They are a medium sized honeyeater bigger than say Yellow-plumed Honeyeater. Their range extends from inland of Geraldton south to Broomehill and across to Twilight Cove, and then north and east of these areas to the edge of the main eucalypt belt. **Nest:** Is thicker walled than many honeyeaters and quite deep, often not far from the ground in either shrubs growing in open heath or taller shrubs in mallee. Clutch size 2 to 3. **Similar species:** None. **Status:** Common in some southern central Wheatbelt reserves but not on the coast. Common in the Transitional Woodland east of the Vermin Fence east of Hyden. **Where to find:** Various reserves including: Wongan Hills. Buntine. Billyacatting. Tutanning. Boyagin. Dryandra. Tarin Rock, both north and south blocks. Dragon Rocks. Lake Magenta. Dunn Rock. Great Western Woodlands and across the southern the base of the Nullarbor where the mallee occurs.

Photo: Simon Nevill

Purple-gaped Honeyeater *Lichenostomus cratitius* subspecies *occidentalis* 16-19 cm

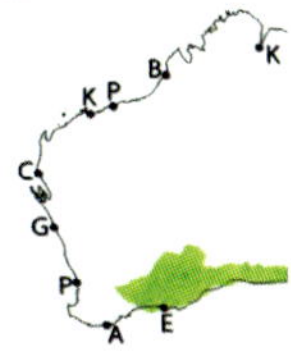

Two subspecies, one in South Australia cratitius on Kangaroo Island, the other in Western Australia, Victoria and south west NSW. In the South West their preferred habitat is mallee. They also have a close association with granite outcrops, feeding on the mallees that surround the bases of many of the larger granite rocks, particularly those within the Transitional Woodland belt between Hyden and Norseman and up to Coolgardie. The purple gape that their common name is derived from is not easy to see but it contrasts well with their plumed lores and yellow throat streak. They are a very active bird and their rapid chatter around flowering mallees, particularly in the breeding season when defending territories, fills the air. Their range is now quite fragmented as much of the mallee has been cleared but even in small patches of just say 600 acres they can still survive and remain there as sedentary birds. Their range has now been reduced, a few birds remain in the northern Wheatbelt mostly centred in the mid and southern zone of the Wheatbelt. **Nest:** The nest is a small delicate deep cup made of woven grasses and fine bark held together with spiders webbing with eggs sacks often surrounding the cup. Lined with any available soft material such as plant down or wool. Normally suspended between two vertical branches in a dense bush, mainly from 1 to 2 metres above ground. Clutch size 2 to 3. **Similar species:** None although for those starting birding maybe Yellow-plumed which has

Photo: Simon Nevill

no grey crown, has black streaking on a fawn breast not yellow throat and breast, no dark lores. **Status:** Common throughout its range. **Where to find:** The following Wheatbelt reserves: Chinocup NR. Corackerup NR. Northern Stirling Range NP. Fitzgerald River NP. Lake Magenta NR. Dragon Rocks NR. Dunn Rock NR. Also from Hyden all the way across to Dundas NR mainly at the base of granite outcrops like McDermid Rock and Disappointment Rock. Along Salt River road to the northern Stirling Range boundary. Mallee around Eyre Bird Observatory.

Grey-headed Honeyeater *Lichenostomus keartlandi* 13-16 cm

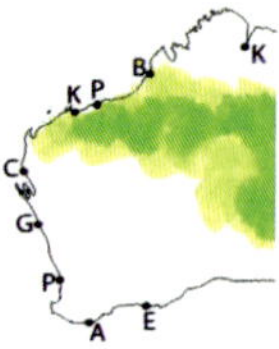

A bird of the sub-tropical arid deserts of Australia, coming to the coast only in Western Australia. A nomadic species of the deserts although more sedentary in the Pilbara region. In the deserts, moving to areas where there has been abundant rain. Very common in the Pilbara particularly along creek lines. Found in the desert regions when the Holly-leaved Grevillea (*Grevillea wickhamii*) is in full flower, then hundreds will congregate in those areas. An attractively plumaged honeyeater. Has a blue-grey head, black lores with bright yellow submoustachial stripe at the base of the lores. Yellow throat and upper breast and grey mantle. Pugnacious around flowering plants, chasing off rival feeders. It does occasionally come to the far north corner of the South West. **Nest:** Is a fine thin walled, deep, suspended nest normally placed between two horizontal branches. Clutch size 2. **Similar**

Photo: Simon Nevill

species: Possibly Grey-fronted that overlaps the territory but has no grey on the crown, but olive green. No dark facial patch the most obvious difference. No yellow on throat but beige with pale yellow streaks. **Status:** Locally very common. **Where to find:** The Pilbara. The Great Sandy Desert. Southern Kimberley. Central Ranges. occasionally south to the Gascoyne River catchment.

Grey-fronted Honeyeater *Ptilotula plumula* 14-16 cm

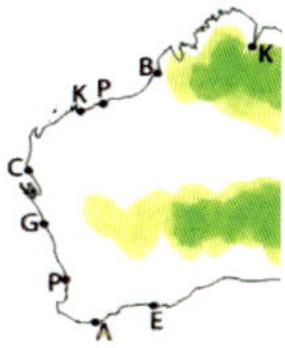

An interesting distribution for this species. The subspecies plumulus occurs from the western Transitional Woodland (roughly White Wells) across to the SA border through the Victoria Desert. Its numbers increase markedly east of Laverton. Its southern boundary starts roughly north of the Trans Line at Haig and Nurina north to Warburton and through the central ranges. The population then blends into a wider band of the subspecies planasi that spreads through Lake Gregory region into the Kimberley but not so much the western Kimberley. Their habitat is diverse in the west from flowering tall eucalypts to mallee as well as hakeas, eremophilas and grevilleas. **Nest:** Is similar to Grey-headed Honeyeater. Clutch 2. **Similar species:** Yellow-plumed Honeyeater. Now these 2 species can be tricky, as their territory in the north and north-east part of the Wheatbelt region almost connects, although few Yellow-plumed get to the northern Wheatbelt. The Yellow-plumed Honeyeater's plume turns to a point and is more raised from the neck and longer. The black border shows clean and contrasting against the yellow plume. Yellow-plumed has stronger, darker black-grey streaking on a white underbody. The Grey-fronted has weak grey-brown streaking to a grey white underbody, not always yellow tinged as discussed in many books. The calls are different but both have a range of calls, the dawn call is clearly different. The Yellow-plumed is more melodic but still has harsh calls at times like the Grey-fronted. **Status:** Uncommon in the eastern Transitional Woodland become more common until very common in parts of the Great Victoria Desert. **Where to find:** Between the 40 km and 80 km range north of the Murchison Bridge but scarce. From Yalgoo to White Wells scarce. Further west, better chances along the Leonora to Laverton road and then very common on the Anne Beadell Hwy and parts of the Warburton Road. Common in the Fitzroy Basin and the Bungle Bugle region.

Photo: Simon Nevill

Yellow-plumed Honeyeater *Litchenostomus ornatus* 15-18 cm

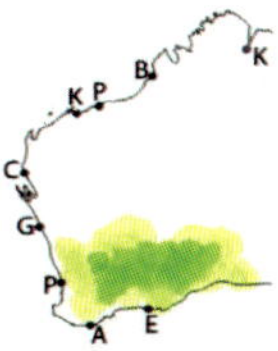

In the South West Region the Yellow-plumed Honeyeater has lost much of its former territory. It was known to be common in the Wongan Hills region and the northern central Wheatbelt particularly around Kellerberrin but now it frequents the central Wheatbelt reserves. One of its favoured habitats is the Wandoo Woodland, also reserves like

Dryandra Woodland Nature Reserve. It is by far the most common honeyeater in this area with high population densities. It also ranges across the eastern Wheatbelt through the Transitional Woodland to Dundas NR where it is common throughout much of that range. In the Wandoo Woodland it feeds both in the canopy of the Wandoo and on the low understorey like the Sandplain Poison, although interestingly it feeds less in the heath where other honeyeaters are dominant like New Holland, White-eared, Tawny-crowned and White–cheeked Honeyeaters. Yellow-plumed Honeyeater's calls are varied and territorial disputes are common, with family groups defending breeding and feeding territories strongly. Like New Holland Honeyeaters and other honeyeaters, they will warn all birds of their kind when hawks or falcons are flying into their territory, the fast staccato sound spreading across the woodland from one bird to another. Warning calls vary and a snake alarm call is different again, not so high pitched, a slower harsh pitching. **Nest:** Is a small shallow cup of matted dry grasses and fine twigs lined with soft plant down and bound with spiders webbing. The nest is suspended from rim from thin branches. Height can vary markedly even in the same area, one nest can be in a low gastrolobium bush and then another nest a few metres away can be on the outer suspended branches of a tall Wandoo tree some 8 metres high. Clutch size 2 to 3. **Similar species:** None in the central and eastern Wheatbelt. On the north and eastern periphery of their range the very similar Grey-fronted Honeyeater occurs, refer to that species for details. **Status:** Overall moderately common. Locally very common. **Where to find:** Throughout the eastern boundary of the Darling Range where the Wandoo and Powder Bark Wandoo abut the Jarrah Forest. Found in Boyagin Rock NR. Dryandra Woodland NR. Dongolocking NR. Mission Road NR, Kojonup. Northern Stirling Range. Transitional Woodland to Dundas NR.

Photo: Simon Nevill

White-plumed Honeyeater *Ptilotula penicillata* subspecies *carteri*

14-18 cm

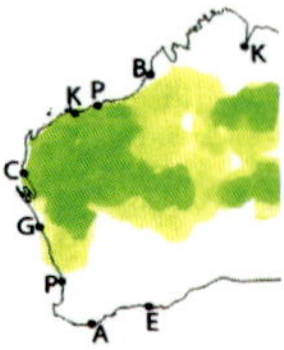

A honeyeater of the mid-west and the Pilbarra. When one looks at distribution maps particularly for honeyeaters, one could easily think that they all occur in the same habitat but that is not always the case. White-plumed are an example, you rarely find them in open dry mulga where other honeyeaters occur. White-plumed are found wherever there are tall eucalypts normally along watercourses and where ever there is a body of permanent water. Their favoured habitat is riparian vegetation where the eucalypt Red River Gum (*E. camaldulensis*)

Photo: Simon Nevill

is the principal tree, in fact so much so that its distribution throughout much of western and central Australia is governed by the distribution of this tree and also the Coolabah tree (*E. victrix*). The subspecies *carteri* is found as far south as Mingenew. It is the common town bird of Geraldton and is very common along the Murchison and Gascoyne River systems. It is not noted as being a migratory bird but remains fairly sedentary along the watercourses and Coolabah (*E. victrix*) flats that it inhabits. Their pleasant calls of 'wheat-a-wit….wheat-a-wit' greet you whenever walking along the mid west rivers. It must be remembered that eastern states birds have different tones and timing in their calls, so an eastern states White-plumed Honeyeater does not have the same melody. Only the alarm call seems to remain a constant. **Nest:** Can be located in a variety of plants at varying heights, also the materials to make the suspended nest vary from plant stems to dry grasses bound with spiders webbing and lined often with hair, feathers or wool. Clutch size 2 to 3. **Similar species:** None, it is the only honeyeater with a white plume contrasting with the black moustachial stripe. **Status:** Very common along river systems, uncommon away from rivers. **Where to find:** Murchison River. Gascoyne River. All the Mulga Region. All the Pilbara and along the Fitzroy Basin in the Kimberley also the Central Ranges.

Brown-headed Honeyeater *Melithreptus brevirostris* subspecies *magnirostris* 12-14 cm

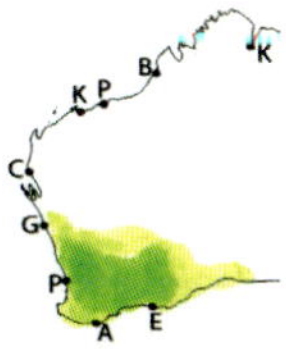

There are 5 subspecies throughout Australia, only one in the south west, subspecies *magnirostris*. It is less common than the similar plumaged White-naped Honeyeater. It often reminds one of a bird which has not come out of the immature plumage stage, having no strong contrasting colours and being over all-dull in coloration. Often seen in small flocks passing over the canopy to new feeding areas, keeping in contact with a continuous clicking rasping call not unlike their territorial call but more rapid. More arboreal than most honeyeaters, feeding mostly in the higher strata of eucalypts, although will come down to heath plants to feed. Found in a variety of habitats especially heath with extensive stands of mallee, Salmon Gum Woodland and Wandoo Woodland far less so in the thicker Jarrah Forest. Particularly common in stands of Mallet (*E. astrigens*) which occurs on stony breakaways in the inner and central Wheatbelt. Common in parts of the Transitional Woodland. **Nest:** The nest is nearly always located on the outer edge of a eucalypt where the outer branches droop down. It is a neat heavily woven structure made from flexible grasses and lined with soft plant down and bound with spiders webbing interwoven around the stems of suspended branches. Clutch size can be 1 to 2. **Similar species:** White-naped Honeyeater, which is larger and has a black head not brown. Important feature is the eye ring, which is a distinct cream colour whereas on the White-naped is pale white-blue. The white band on the White-plumed passes from the nape to centre of head but stops before the eye leaving the black of head showing, whereas on the Brown-headed it bleeds into a cream eye ring. Bill is finer and slightly longer on White-naped. Mantle is bright olive–green on White-naped but dull olive-green on the Brown-headed. The black collar goes down to shoulder on White-naped but on Brown-headed the brown head colouring passes around neck leaving thin white gap. Calls of White-naped less grating than brown headed, more whistle-like, markedly different from the eastern states subspecies. The difference between juveniles is a little tricky as the head is brown and does not develop into the black head for the White-naped for a couple of months after leaving the nest. However all the other features discussed above are still relevant. **Status:** Generally uncommon but can be locally common particularly east of Hyden. **Where to find:** Scarce in the Jarrah Forest. Best on the eastern edge of the Darling Range in the Wandoo Forest. Uncommon in its northern range as far as Hamelin Pool. Best to see in the central and southern Wheatbelt reserves Dryandra Woodland NR. Dongolocking NR. Southern coastal heaths like the Fitzgerald River NP. Ravensthorpe region. Throughout the southern Transitional Woodland all the way east to Dundas NR.

Photo: Simon Nevill

Gilbert's Honeyeater (formerly White-naped Honeyeater)

Melithreptus chloropsis 13-15.5 cm **Endemic**

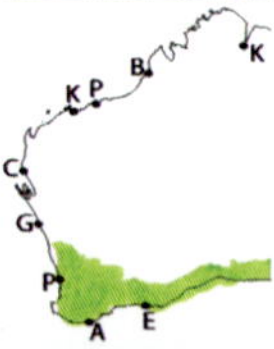

The Gilbert's Honeyeater differs from the Eastern Australian White-naped Honeyeater being slightly larger and has a different coloured eye ring, pale cream instead of bright red as in subspecies

Photo: Simon Nevill

lunatus on the east coast. The IOC has now designated the WA subspecies to the status of a full species. The Western Australian birds prefer the wetter regions of the South West and are the dominant honeyeater in the Karri Forest and southern Jarrah Forest. Their range almost follows the distribution of Jarrah stopping west of New Norcia and south west to Yanchep and covering all the Jarrah and inner Wandoo down to the Stirling Range and across to Esperance and all areas below this border line. Very common in the deep South West and southern coastal belt, less so past Two Peoples Bay. Far more common than the similar Brown-headed Honeyeater and is more sedentary in its feeding habits whereas Brown-headed tends to move through areas. Now uncommon on the Swan Coastal Plain and north of Moora. **Nest:** Is very similar to Brown-headed Honeyeater but not as thick walled. Clutch size 2 to 3. **Similar species:** Brown-headed Honeyeater, refer to that species for similarities. **Status:** Very common in Karri, Jarrah and Wandoo Woodland. Moderately common in heathlands to Esperance. **Where to find:** Throughout the Darling Range. Lower Wheatbelt reserves particularly the Stirling Range. All the parks of the lower South West. Nowhere in the semi-arid and arid regions.

White-throated Honeyeater *Melithreptus albogularis albogularis* 13.5 cm

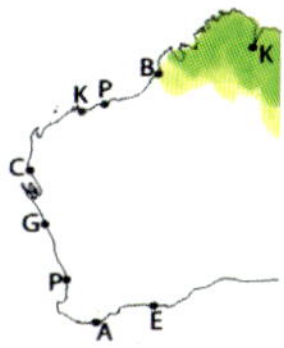

Found throughout the sub-humid and semi-arid regions of the Kimberley. It takes the place of its close relative, the White-naped Honeyeater (Swan River Honeyeater) but that species population is nearly 2000 km south, so no problems with overlap or hybridisation. Like many Kimberley honeyeaters it is found in eucalypt woodland, riverine habitat and rainforest, but unlike some can be found further away from water in more open eucalypt woodland. White-throated has an olive-green upper-body and tail, black crown that stops below eye, White under-body, white crescent on black crown widest on nape narrowest near eye, eye wattle white to very pale blue. Nest. Nest average honeyeater depth, that is not deep and not shallow, a well woven nest of leaves and fine twigs bound with spiders' webs and often littered with spiders egg sacs, nest normally suspended from horizontal fork. Clutch 2-3. **Similar species:** In its range in north west is the Black-chinned Honeyeater which is larger more robust, has distinct bright lime-green back not drabber olive-green, dark patch on throat, crescent touches wattle, under-body greyish-white not pure white, call totally different. One bird that is almost identical is the very closely related south west population of the White-naped Honeyeater. Agreed their population is many hundreds of kilometres apart but if you are looking at White-throated in the Kimberley sub-species

Photo: Simon Nevill

Melithreptus albogularis and the south-west White-naped Honeyeater sub-species *Melithreptus chloropsis* and you are having trouble splitting these species take heart as they are one of the closest species in terms of colouration. The subtle differences are that on the White-naped the black hood encroaches onto the white throat below the bill, with White-throated it runs from the bill itself, the bill on the White-naped is slightly thicker, the wattle on the White-throat is slightly larger importantly colouring on both species is white which does not help in identification. There is comment about the crescent being closer in White-throated but this is variable. If we were looking at the south-east White-naped subspecies *lunatus* it would be easy, as they have a red eye wattle not white. Talking to Ron Johnstone, he confirmed from skins, that the White-throated of the Kimberley has a brighter olive back than the south-west White-naped. Calls of all sub-species vary, compounded by having 3 main call types including agitation calls but in general terms, calls of the north-west White-throated Honeyeater are softer than the other two southern sub-species of White-naped Honeyeater. **Status:** Common throughout the Kimberley's sub-humid zone less common in the semi-arid southern region. **Where to find:** Broome across to Fitzroy Crossing and across to Halls Creek and Kununura and throughout central and northern Kimberley to Kalumburru.

Brown Honeyeater *Lichmera indistincta* 12-15 cm

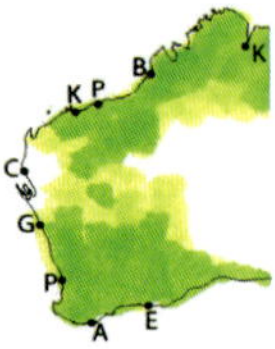

Three subspecies throughout Australasia but only one in Western Australia the nominate race indistincta. The word 'indistinctus' in Latin means 'obscure' and relates to the plain colouring of this small honeyeater and it is certainly that; its only features besides its overall grey-brown colouring are the yellow gape and the yellow rear orbital eye ring. However, it certainly makes up for its drab plumage with a great array of chirping musical calls as well as its common scolding call. Found throughout the WA, except for part of the mid west coast and much of the lower central deserts. A small honeyeater, highly active and very common in the Perth region. Most diverse in its choice of feeding habitats including wet sclerophyll forests, mangroves, dry arid woodland and heath. Found in town gardens throughout the Wheatbelt and South West. Although its distribution shows it covers much of the state, it is more common in the South West and the far north in the Kimberley and far less common in the semi- arid and arid regions. This is understandable as it requires a high intake of nectar and is very common in heath country particularly where flowering grevilleas occur. **Nest:** The nest is a relatively deep cup suspended from thin branches mostly in low bushes, particularly in calothamnus or dryandra bushes. Clutch size 2. **Similar species:** None that occur in this region. **Status:** Very common to uncommon throughout the region. **Where to find:** Anywhere there are flowering plants.

Photo: Simon Nevill

New Holland Honeyeater *Phylidonyris novaehollandiae* subspecies *longirostris* 16-20 cm

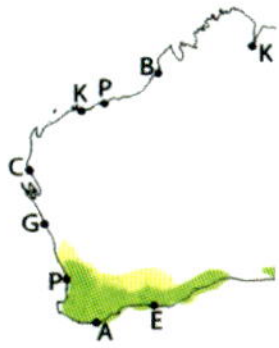

Five subspecies throughout Australasia with only one subspecies *longirostris* in Western Australia. Very similar to the eastern states birds but bill is finer and longer with minor plumage variation. A bird of the southern part of Western Australia, occurring in areas with blossoming plants. Certainly there are partial migratory movements in the state as it is not uncommon to see moving flocks of birds in autumn and early spring. Found throughout the Swan Coastal Plain, Jarrah Forest, Karri Forest and southern heaths. Moderately common in the southern Wheatbelt, in reserves that contain Kwongan heath. Can be found in all areas south of a line from Dongara through Northam, Corrigin and across to Pingrup, Ravensthorpe to Twilight Cove. **Nest:** Deep cup but untidy outer structure of intertwined small twigs but internally neater, often lined with the down from banksia cones. Nests often in dryandras or banksias. Clutch size 2-3. **Similar species:** White-cheeked Honeyeater which has much bigger white facial patch on the lores. Complete long eyebrow, which is not broken. Dark brown iris. Call distinctly different. **Status:** Common in Karri and Tingle Forest. Common on coastal heaths and wheatbelt heaths. Less common in the Jarrah Forest except on the scarp. **Where to find:** Throughout gardens in Perth and in suburbs on the Darling scarp. All inner Wheatbelt reserves with heath including Tutanning NR. Dryandra Woodland NR. Dongolocking NR. Stirling Range NR. Throughout the coastal heath from Dunsborough all the way to Albany and Esperance and Cape Arid.

Photo: Simon Nevill

White-cheeked Honeyeater *Phylidonyris nigra* subspecies *gouldi* 18 cm

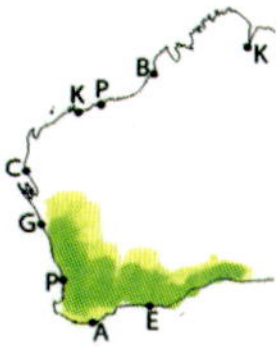

There are two subspecies in Australia, *nigra* on the east coast and *gouldi* in the south west of Western Australian. The Western Australian birds are slightly larger but the white facial disc is not as large and narrower, quite evident for those that know both subspecies. The other features regarding the breast and forehead are so subtle they will confuse and are difficult to see. The western bird's call is louder and more intense, not as soft as that of eastern subspecies. White-cheeked are not as common overall but can be very numerous locally, particularly in the north of their range in the Eneabba area. Their distribution goes further north than New Holland. The distribution is unusual and slightly fragmented, mainly centred on areas with rich Kwongan heath or shrubland heath. They occur as far north as the Kalbarri coast then there is a break south until the flora-rich heaths of the Eneabba sandplain commence. Their range then continues down the coast through Perth to north of Mandurah and there are populations on the Cape Naturaliste to Cape Leeuwin coastal heaths. The Jarrah Forest breaks their range as far as the eastern side starting as far north as Calingiri and running south along the eastern side of the Darling Range where there is heath in the Wandoo Woodlands, all the way down to the coast at Denmark then along the coast to Cape Arid and inland to Hyden. **Nest:** Nest and clutch size as for the New Holland Honeyeater. **Similar species:** New Holland Honeyeater, refer to that species for details. **Status:** Uncommon to locally common throughout their range particularly in the Kalbarri, Eneabba and Badgingarra sandplain. **Where to find:** In Perth particularly the botanic gardens but seasonally more common in the northern Perth suburbs. Yanchep NP. Bungendore Reserve. Kalbarri NP. Eneabba NP. Badgingarra and Mt Lesueur NP. Tutanning NR. Dryandra Woodland NR. Heathland 65 km due east of Hyden on the Norseman road.

Photo: Simon Nevill

White-fronted Honeyeater *Phylidonyris albifrons* 14-18 cm

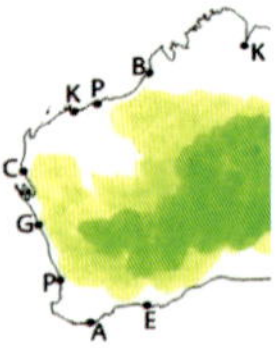

An unusual plumaged honeyeater, almost like a New Holland or White-cheeked Honeyeater without clear head markings. It has a huge range extending from the Pilbara in the north, south to the northern Wheatbelt and from Shark Bay west across through the central deserts into the NT and SA. Not found in the lower South West. However they are scarce in most of the southern Wheatbelt and do not reach reasonable numbers until the northern and eastern Wheatbelt. From the Transitional Woodland zone out to the mulga and deserts, they become more plentiful. They are highly nomadic and in summer may come deep into the heathlands to feed on banksias and other flowering plants in the northern Wheatbelt. In the mulga they feed on eremophilas, grevilleas, hakeas and mistletoe. They are common in the eastern Transitional Woodland east of Hyden and Lake King. Their calls are plaintive and varied with chuckles, whistles and the typical 'tuk..tuk' call. The important feature to note is an all white forehead running around the eye with a long white malar stripe. Black throat, lores and crown, the crown being faintly streaked with white. Pale lime-green on the wings. Strong black streaks to the white underbody giving way to an all white belly and vent. They have an unusual bright red patch of bare skin just behind the eye but it is so small you can almost miss it. **Nest:** Can be either suspended from a horizontal branch or wedged in a vertical fork and bound to the stems with spiders webbing. Neat soft lining woven with plant down or wool. Clutch size 2. **Similar species:** None in this state. **Status:** Uncommon in the Wheatbelt, Moderately common in the Transitional Woodland and Mulga Woodland. **Where**

to find: Heaths due east of Hyden. Within the Transitional Woodland from Eurardy Station down to White Wells across to Mt Jackson and particularly the extensive woodland between Coolgardie and Peak Charles. Also the lower Victoria Desert and the mid central deserts and mid Great Sandy Desert.

Photo: Simon Nevill

Tawny-crowned Honeyeater *Phylidonyris melanops melanops*

15-18 cm

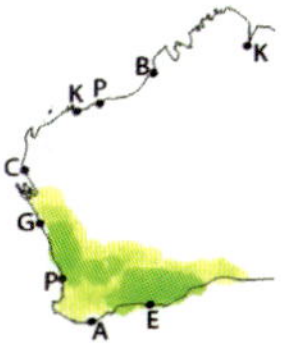

Two subspecies in Australia, *melanops* on the mainland both in the south west and south east and subspecies *chelidonia* in Tasmania. In Western Australia it virtually follows a line from the Transitional Woodland all the way to the lower South West. It is a bird primarily of the species-rich flowering heaths, and wherever flowering heaths occur Tawny-crowned will normally occur. Its range is the same as White-cheeked Honeyeater extending to the sandplain heaths of Kalbarri NP, then with a break to the Eneabba and Badgingarra sand plains the range becomes fragmented through the Wheatbelt down to the coast. It occurs between Cape Naturaliste and Cape Leeuwin and follows the southern coastal heaths all the way east to Cape Arid and north to Hyden. Will feed on eucalypts but primarily feeds in low heath where it also nests. The nest is mostly located very low sometimes only 20 cm off the ground, often placed in low heath shrub. The outside is an untidy meshed set of twigs but internally neatly woven with down or soft foliage inside. Clutch size 2 to 3. The bird will rarely fly directly to the nest but will perch on a nearby shrub then drop to ground and run to the

Photo: Simon Nevill

nest in a similar manner as the chats. **Similar species:** None. **Status:** Uncommon overall but locally common in species-rich Kwongan heath. Now uncommon on the Swan Coastal Plain. Is not found in the dense Jarrah except heaths and Wandoo woodland that occur in the Darling Range. **Where to find:** Kalbarri NP. Eneabba sandplain. Mt Lesueur and Badgingarra sandplain. Various reserves in the Wheatbelt including Gardner NR. Boolanelling NR. Tutanning NR. Boyagin NR. Dryandra Woodland NR. Dongolocking NR. Tarin Rock South and North NR. Stirling Range NP. Fitzgerald River NP. Sandplains east of Hyden and Lake King.

Western Spinebill *Acanthorhynchus superciliosus* 12.5-15 cm

Endemic

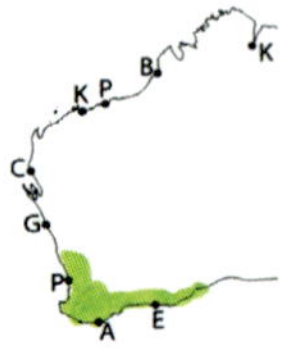

Requires areas with sufficient flowering plants to support its very high nectar intake, so its range becomes restricted to wetter South Western Region and parts of the state that have extensive Banksia Woodland and Kwongan Heath. Found from Badgingarra south east through to the inner part of the Wheatbelt and across from Ravensthorpe to Esperance. On the south coast they may be sedentary only moving locally but the Jarrah Forest birds and Wheatbelt birds have seasonal movements. They occur along the Swan Coastal Plain including Perth and into the Jarrah Forest, feeding mainly on Bull Banksia (*Banksia grandis*) and lower storey plants like Hairy Jug Flower (*Adenanthos barbigerus*), Fuchsia Grevillea (*Grevillea bipinnatifida*), Wilson's

Male

Photo: Simon Nevill

Grevillea (*Grevillea wilsonii*) and One-sided Bottlebrush (*Calothamnus sanguineus*). In the Wheatbelt Region, they are found where there are remnant patches of heath, which occur in the central and lower Wheatbelt. The sexes differ markedly when adults but when they are juveniles they are very similar. The female lacks the varied markings of the male particularly on the throat, having a plain brown back and white underbody like the male but the only colour is a chestnut rear collar band. **Nest:** Can vary in its location from higher in the outer foliage of tall banksias to low shrubs only 40-60 cm off the ground. The nest is a neatly made thick sided cup and always lined with soft materials from banksia down to fine bark or feathers. Clutch size 1 to 2 . **Similar species:** None. **Status:** Moderately common on the Swan Coastal Plain and Jarrah Forest. Common in heath in the south west Wheatbelt Region and southern coast less so east of Hopetoun. **Where to find:** In Perth at Whiteman Park. Seasonally Kings Park. Bungendore Reserve. Many suburban gardens on the Darling scarp. Jarrah Forest when Bull Banksias in full bloom. Wheatbelt reserves with heath including Tutanning NR. Boyagin NR. Dryandra Woodland NR. Dongolocking NR. Tarin Rock NR. and Stirling Range NP. and along the coast between Cape Naturaliste and Albany.

Black Honeyeater *Sugomel nigrum* 11 cm

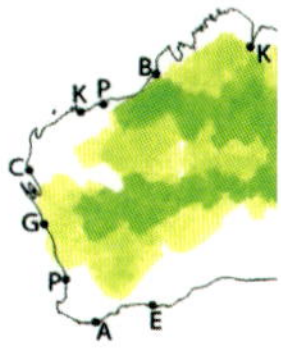

A nomadic species of the drier regions of Western Australia can be found from just south of the Transitional Eucalypt all the way to the Mid Kimberley. Not found in the deep south west only rarely. Its normal range when the inland region is not experiencing drought, is the Mulga Woodland and northern desert regions and less so the Pilbara and the northern Kimberley. Occasionally when the deserts and mulga are experiencing drought some birds will enter the Wheatbelt. I have seen them in Boyagin Rock Reserve and Dryandra Woodland Reserve several times; in fact on a tour one gentleman found the nest in a calothamnus bush in Dryandra Nature Reserve, it's very much an irruptive species. One interesting thing we do know is that Black Honeyeaters will move into regions after burns; I witnessed this just east of the Vermin Fence east of Hyden, where the heath was burnt and only the grasstrees were sprouting and yet there were many pairs in the area. Another interesting aspect with Black Honeyeaters is that they have a propensity to eat carbon like a few other species. I have seen this on a couple of occasions when they have come to my campfire early in the morning to take carbon. The two sexes are quite different, the male has the contrasting black and white markings with the black throat band passing down the underbody as a thin black stripe. The female has a plain brown back and white underbody with slight brown streaking on the throat and down the top of the breast. Both have a long fine down curved bill. The call is a single extended high pitched note, higher pitched than the Red-eared Firetail 'seeeee….. seeeee…..seeeee' and if you get the tone right, you can whistle them in. **Nest:** In a variety of locations but often in very exposed low shrubs, often in burnt bushes in the desert and yet not easily seen, as the nest is very small and normally woven to vertical branches. The outside is a rough tangle of fine twigs, lined

Photo: Simon Nevill

with fine soft twigs and grasses. Clutch size 2. **Similar species:** None. Some similarity with the Pied Honeyeater, which is larger and bulkier and has white on the shoulder and down the wing at rest. The black throat does not form a small band down the centre of the breast like the Black Honeyeater. The Pied Honeyeater has a distinct blue crescent mark below the eye. Bill is bulkier and shorter in proportion. Call totally different. **Status:** Can be locally common but overall uncommon. **Where to find:** Best areas to see are within the Mulga Region and even more so, the flowering northern deserts.

Pied Honeyeater *Certhionyx variegates* 16-19 cm

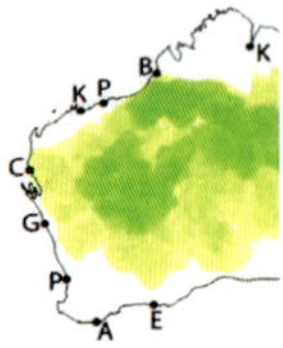

A bird of the semi-arid to arid regions of Australia. Its range starts from the mulga Eucalypt line to the northern deserts. In the mulga it is very much reliant on rains and will move out of areas in severe drought but most years can be found in the Carnarvon region and inland along the Gascoyne catchment. They rarely pass the Transitional Woodland zone into the northern Wheatbelt, occasionally far south as Mullewa. In the breeding season when the Mulga Region is ablaze with flowers the Pied Honeyeaters can be everywhere. The males fly up into the sky with their aerial displays while making their repeated high pitched whistle, confirming and holding their small territory with other males as close as one hundred metres away doing the same thing, flying up and down. Feed mostly on the eremophilas and grevilleas in the mulga. After good rains in late autumn through to winter and early spring they will gather in large groups and pair bonds may be established before nest sites are located. This occurred one year on Muggon Station when a concentration of maybe over 60 birds were connecting and then flying through. The female has a plain brown upperbody and hood with a off white throat and underbody with brown streaking on the throat and breast, with some diffused white in the brown wings. Her blue eye patch is more subdued in colour. **Nest:** Are often placed in the centre of eremophila bushes, particularly the Turpentine Bush (*Eremophila fraseri*) or the thick Kurara bushes (*Acacia tetragonophylla*). The nest is a loose suspended nest with rough outside of loose twigs bound with spiders webs but the cup is lined with finer dry grasses or very thin twigs. Clutch size 2 to 3. **Similar species:** None, refer to Black Honeyeater for possible similarities. **Status:** Locally very common in parts of the Mulga Region. **Where to find:** Any of the mid western and lower western sheep station properties. They become even more common in the northern and central deserts in winter.

Photo: Simon Nevill

Grey Honeyeater *Conopophila whitei* 11-13 cm

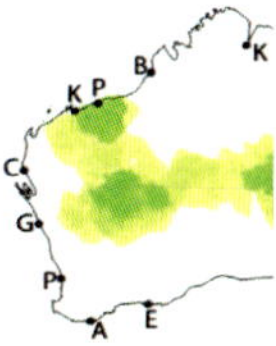

A fascinating scarce to uncommon bird with still scant information on its behaviour. Found in a band of mulga country in Western Australia from west of Yalgoo through to Wiluna in the east and north into the Pilbara and the to the central ranges around Alice where I maintain it is far commoner. In WA, sightings are few and far between and most are from the Tom Price area. I have seen them a few times in the Pilbara but more so in the mulga on Nallan Station, Yalgoo, Pindabunna and more times in the Northern Territory in the MacDonnell Ranges out of Alice at least once in every three to four tours there over the last 20 odd years. The bird is not only seen in mulga woodland but also in dry creek beds with scant acacia cover along the sides of rocky hillsides. They often frequent areas in the mulga that have a lot of mistletoe. Partly the reason for many people overlooking this bird, is the fact that it rarely calls, remains fairly unobtrusive and its bland colouring, which does not attract the observer but if it does birdwatchers often think it is a Western Gerygone or female Redthroat. It is most un-honeyeater-like in its behaviour choosing often to follow thornbills and gerygones through the mulga, gleaning insects more than feeding off nectar. When it does take nectar, its bill not being long and down curved like most honeyeaters, makes it have to pierce the side flower heads, particularly eremophilas, to get at the pollen and nectar. There are some important features to note. Firstly the important word jizz applies to this bird. Its shape is closer to the birds of its genus *Conopophila*, being the Rufous-banded and Rufous-throated Honeyeaters, which do not occur in the Grey Honeyeater's territory. It does not have the jizz of gerygones, which are slimmer bodied and smaller and proportionally look to have a longer tail. One reason for confusion with gerygones is that the tail length is shorter than any other conopophila and most other honeyeaters, but it is still a longer and bigger bird than any gerygone. The upperbody is a plain grey with slight fawn tingeing in the wings. The bill is stout and short for a honeyeater but still larger than any thornbill or gerygone. It has white tips to the tail, which Western Gerygones also have. Around the eye is a small pale cream coloured split eye ring not easily observed. **Nest:** Is one of the frailest of all nests and it is a wonder that the eggs do not fall from the nest. The shallow cup is suspended from branches using spiders webbing. Clutch size 1 to 2. **Similar species:** Western Gerygone arid subspecies *mungi* which is a slightly smaller and slimmer bird weighs 6 g, the Grey Honeyeater weighs 10 g. Bill straight on the Western Gerygone, thicker and slightly down curved on Grey Honeyeater. White short supercilium stripe to eye plus dark loral band from bill to eye on Western Gerygone, white eye ring split front and rear of eye on the Grey Honeyeater. Eye distinct dark red in Western Gerygone, black in the Grey Honeyeater. More white in tail of Western Gerygone. **Status:** Scarce but not rare. **Where to find:** Yalgoo region including Gabyon, Carlaminda, and Thundelarra Station. In the Paynes Find area Maranalgo Station and Pindabunna. In the Cue region Nallan and Austin Downs Stations. Possible in any of the mulga in the Pilbara, particularly when mistletoe is in bloom. Can be found near Tom Price close to the townsite.

Photo: Simon Nevill

Kimberley Honeyeater *Meliphaga fordiana* 19.5 cm Endemic

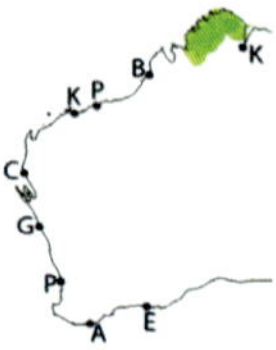

Only in recent years has this species been split from the White-lined Honeyeater *Meliphaga albilineata* which was never surprising as to have two species that are basically sedentary and their core populations over 600 km apart would always produce some genetic variation. It was not changed for the simple reason that they look so alike and the differences are very subtle *fordiana* lacks citrine edges on wing feathers and the under-body is whiter. This species is found in the pockets of denser riverine vegetation and the monsoonal vine thickets of the sandstone country of the north west and central Kimberley. The Kimberley Honeyeater has a darkish brown upper-body, pale grey-brown under-body with distinct dark brown mottling on the throat and chest reducing in intensity towards the lower body, vent plain pale grey-brown, distinct white eye stripe below the eye that starts from the ear coverts and terminates at the gape, gape pale yellow, eye iris pale blue. Call fairly strident that echoes off the sandstone cliffs where it often occurs just like the White-lined Honeyeaters territory. A piercing regular, 'teww…tew…tew.teww…tew-tew'. Has another two variations on that call. **Nest:** Typical honeyeater, open cup with interwoven twigs lined with fine twigs and bound together with cobwebs and tied to branches with spiders' webbing, normally on a horizontal fork. 2-3. **Similar species:** Only bird in region similar is White-gaped Honeyeater, which is a bigger stockier bird, different call, larger round cream-yellow loral patch, plain dark brown all over body, no mottling on breast. **Status:** Locally common in larger pockets of vine forests and thick riverine vegetation in the sandstone gorges. **Where to find:** Mitchell Plateau area is a reliable area and fairly common in the area. Found throughout the coastal gorges of north west coastal belt where monsoonal vine forests occur. Has also been seen off the Gibb River road at various locations such as Manning Gorge. Found on the Carson and Drysdale River. Mostly north of the Charnley River.

Photo: Simon Nevill

White-gaped Honeyeater *Lichenostomus unicolor* 21 cm

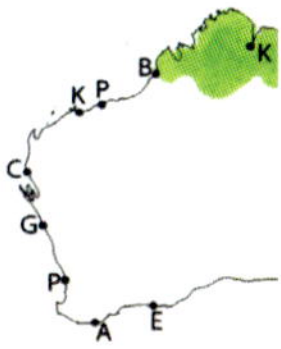

A common honeyeater of the northern tropics, found through much of the Kimberley from Broome to Kununurra. Has a preference for riverine habitat, closed gallery forests, and vine thickets but also found in mangroves. Can be pugnacious with other smaller honeyeaters when feeding off blossom. Has a strident, rollicking call a rapid 'whit...whit... a-whit...whit...whit... a-wit' then often a piercing 'cheew' often one of the first birds to call pre dawn. Does not flock like many honeyeaters at most normally groups of 4-6. A plain brown bird with the only prominent feature being a crescent shaped cream patch on the gape. **Nest:** Typical honeyeater, open cup with interwoven twigs lined with grasses walls can be fairly open but bound well together with cobwebs and tied to branches with spiders' webbing. 2-3. **Similar species:** Only all brown bird in region is the Brown Honeyeater and it is much smaller with longer curved finer bill, song totally different. **Status:** Common only in specific habitat, not common in open woodland. Found through out much of the Kimberley. **Where to find:** Very common in Kununurra in the lush riverside vegetation, found in Broome and along the Fitzroy River system and Gibb River Road where lush creeks occur.

Photo: Simon Nevill

Yellow-tinted Honeyeater *Lichenostomus flavescens flavescens* 14.5 cm

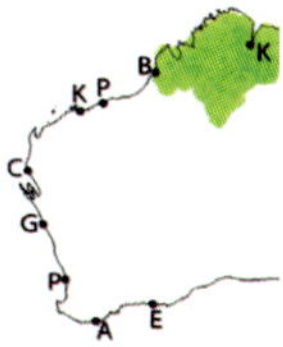

One of the most common honeyeaters in the Kimberley, found in open woodland and riverine habitat across tropical Australia. A particularly active honeyeater often feeding in small groups. Very common in cadjeput trees (*Melaleuca leucadendra*) along rivers and flower bearing eucalypts. Yellow-tinted has olive-grey nape and back, yellow on primaries, pale yellow head merging into nape, yellow stronger on cheeks, important diagnostic feature is the black crescent from the base of the bill and curves below the eye and stops at the rear of the ear coverts, the black contrasts with bright yellow plume below black, breast is pale buffy-white with faint fawn lateral streaking, belly and vent grey-white, black bill down curved, dark olive-brown tail with yellow edges. Many varied calls but the most common is a rapid tew...tew..tew..tew. **Nest:** Typical honeyeater, shallow open cup with interwoven twigs lined with grasses walls bound well together with cobwebs and tied to horizontal branches with spiders' webbing. 2-3.

Similar species: White-plumed Honeyeater very similar but look for a black coloured crescent above a pure white plume rather than Yellow-tinted Honeyeaters black above yellow. Both species range overlaps in many areas such as Halls Creek, Fitzroy Crossing and at the northern end on 80 Mile beach. It is always interesting to see where a species overlaps with another similar species and which of the two are the more common and at what point is there a reversal in numbers, for example in Halls Creek there are greater numbers of Yellow-tinted than White-plumed Honeyeater but go a short distance south to the Lake Gregory and White-plumed by far out number Yellow-tinted Honeyeater. **Status:** Moderately common in the northern Kimberley but far more common in the dryer region particularly along the Fitzroy River system. **Where to find:** Most areas of the Kimberley particularly the Kununurra area and Fitzroy River.

Photo: Simon Nevill

Bar-breasted Honeyeater *Ramsayotnis fasciatus* 14 cm

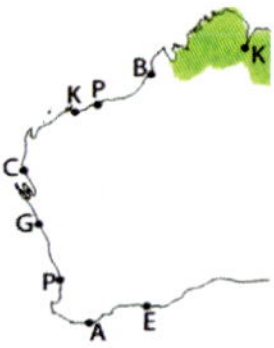

A small, unmistakable honeyeater with distinct horizontal barring on its breast, similar jizz to the closely related Brown-backed Honeyeater. Seems to have a very close association with woodlands adjacent to wetlands and riverine habitat. Occurs more in the tropical part of the Kimberley with its normal distribution range going no further than Derby, Geikie Gorge

Photo: Simon Nevill

and Bow River north of Warmun. Commonly occurs along permanent watered creeks along the Gibb River and Kalumburu Roads and wetlands near Kununurra. Rarely found in large group, often found in cadgeput woodlands around water holes and creeks where it also nests. Feeds on flowering grevilleas. Bar-breasted Honeyeater has black barring on breast finishing at lower belly but passes lower down to flanks, barring across crown, distinct fine dark brown moustachial stripe, dark pinkish bill, short brown tail, brown back with dark brown streaks, pink legs. **Nest:** Unusual for honeyeaters to build a suspended dome nest, covers nest with bark strips and leaves that help camouflage it, has top side entrance with twigged hood forming an entry tunnel. Clutch size 1-3. **Similar species:** None in region. **Status:** Moderately common in the central and northern tropical part of the Kimberley, ranges from Derby in the west to Fitzroy Crossing in the south and across the WA boader at Kununurra. **Where to find:** Common around Kununurra. Creeks near Parry's Lagoon. Drysdale River. Many creeks along Gibb River Road. Manning Gorge.

Rufous-throated Honeyeater *Conopophila rufogularis* 13 cm

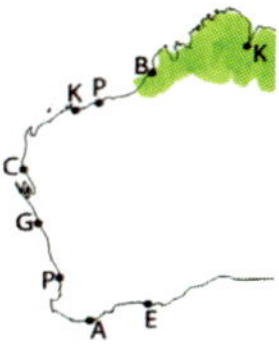

A small plain brown honeyeater with a slight down curved bill, its only main distinguishing feature is a rufous throat but if you see an immature bird the throat patch may not have developed and you may think briefly you are looking at a Brown Honeyeater. A slightly larger distribution range than some of the more tropical honeyeaters like Bar-breasted Honeyeater. This species enters the semi-arid zone as far south as the Edgar Ranges. The Rufous-throated adult male has a bright rufous throat, female rufous less intense and more diffused, immature may not have rufous at all, brown upper-body, brown head and crown, short dark brown tail, greyish-white under-body from throat below eye to vent, distinct yellow margins on flight feathers seen at rest, yellow on outer tail feathers, grey-black legs. When feeding in blossom they continually chatter to one another while feeding maintaining contact. A bill is often diagnostic feature of what a honeyeater may predominantly feed on, so the stouter more robust and less curved the bill, often the greater the intake of insects compared with the consumption of nectar by down curved billed birds, so Rufous-throated is more of an insectivorous feeder than a nectar gatherer. Call scratchy and raucous with harsh contact call 'chzip'. **Nest:** A beautifully deep woven suspended bag attached by weaving around fine branches, made of leaves, paperbark, insect cocoons, and lined with fine twigs or grasses. Clutch size 2-3. **Similar species:** Immature bird could be confused with Brown Honeyeater, which has different jizz, finer longer down curved bill, less yellow on wing margins, under-body much darker. **Status:** Common in the central and southern sub-tropical part of the Kimberley less so far north Kimberley, ranges from mid 80 Mile Beach, south to the Edgar Ranges and across to Derby and WA border at Kununurra. **Where to find:** Common throughout the Kimberley but more so in the semi-arid regions. Common in Broome. Kununurra. Fitzroy Crossing.

Photo: Simon Nevill

Dusky Myzomela (Dusky Honeyeater) *Myzomela obscura obscura* 14 cm

Vagrant

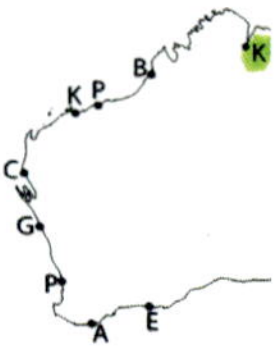

An all, chocolate-brown coloured bird with a long down curved bill with a classic *myzomela* jizz. Its distribution is interesting, as it is very common in the NT and very common in sub-tropical areas like Katherine with a similar climatic and vegetation type that occurs around Fitzroy Crossing but they are certainly not there. They have been sighted in the eastern part of the Kimberley near Kununurra, which is not surprising as they visit Keep River National Park just over the border in NT. They do require a high intake of nectar and that is why they are common on the margins and canopy of rainforests where flowering plants are often prolific, as in the rainforests on the east coast. Dusky Honeyeater has an all chocolate-brown body, darker on crown, back and tail, throat and breast has a reddish tinge in the chocolate-brown, pale brown vent, black eye and legs, long down-curved bill. Needs to feed on blossoming plants. **Nest:** A delicate thinly woven nest woven between thin sapling horizontal fork, has very open sides, but base of nest more leaf and grass interwoven with spiders' webs. Clutch size 2-3. **Similar species:** The only bird in the Kununurra region to compare, is the common Brown Honeyeater with shorter, thicker, less down curved, paler brown, yellow in wings, different jizz. **Status:** Scarce but not rare on the WA border at Kununurra. **Where to find:** Kununurra region. Common central and northern NT and east coast rainforests.

Photo: Simon Nevill

Red-headed Myzomela (Red-headed Honeyeater) *Myzomela erythrocephala erythrocephala* 12 cm

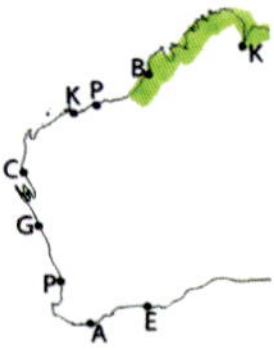

A very small honeyeater, almost entirely restricted to mangroves, particularly those of the larger *Rhizophora* and *Avicennia* genus. There have been sightings away from the coast but this is very uncommon and will occur when the mangroves are not flowering well. I have seen them feeding in flowering melaleuca over a kilometer away from the mangrove edge but this is because they are always adjacent to their prime feeding and nesting area the mangroves, although in some towns like Darwin they have understandably expanded into the city gardens to take advantage of flowering shrubs. Red-headed Honeyeater have a dark brown body and breast that merges into a paler brown under-body and vent. Tail is dark brown, head a vibrant bright scarlet-red that breaks sharply at the neck line with the dark body, rump bright scarlet-red to middle back seen well in flight, fine down curved bill, black loral stripe from base of bill to eye and surrounds eye as an orbital ring. Female lacks bright red head, has reddish tinge only around the forehead and chin, back paler brown than male and pale grey-brown under-body merging into very pale vent. A very territorial species, chasing off other Red-headed Honeyeaters from its territory. Call is a high pitched fast four-noted repetitive 'tseep tseep..tseep..tseep' then repeated, also a harsher lower toned 'tsoop..tsoop'. **Nest:** A shallow woven nest, woven between thin horizontal branches, lined with leaf and grasses at base interwoven with spider's webs. Clutch size 2-3. **Similar species:** The only bird in the Kimberley that could be confused with the female Red-headed Honeyeater would be the Brown Honeyeater. Brown Honeyeater bigger, thicker bill, no reddish facial tinge, has small cream-yellow rear eye plume, different jizz. **Status:** Common throughout the tall mangroves of *Rhizophora*. **Where to find:** All around the Kimberley coast. Seen in the mangroves adjacent to the town site of Broome. Mangroves along the Diggers Rest road about 5 km along where the road runs close to creek off the Wyndham-Kununurra Road. A good location to see them is Port Smith mangroves, camping/caravan facilities excellent there at the northern end of 80 mile Beach.

Photo: Simon Nevill

Banded Honeyeater *Certhionyx pectoralis* 12 cm

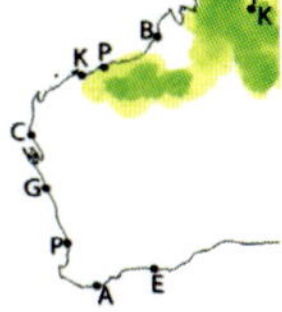

A distinctive honeyeater commonly found feeding in flowering eucalypts such as Darwin Woolybutt (*E. miniata*). Scarlet Gum (*E. phonicea*). Silver Box (*E. pruinosa*) and also the grevilleas particularly Silky Grevillea (*G. pteridifolia*). It's common in the central Kimberley but being semi-nomadic it will move its range to where the eucalypts and melaleucas are flowering well. Unlike some of the Kimberley Honeyeaters, it will spread out to the open woodlands away from water in the sub-tropical zone where mass flowering occurs, as it is a very nectar dependent bird. When it is near coastal waters, it will feed in mangroves. Banded Honeyeater has a black crown that passes through the eye, has a black upper-body and tail, pure white under-body from chin to vent including, white rump, diagnostic black band passing across the breast, fine down curved bill. Juvenile birds have a brown back and crown and breast band not black. Call: 'cheet..cheet.. churpi..cheet..cheet' also a mousy 'cheerr'. **Nest:** A compact tightly woven nest placed between a horizontal fork, fairly deep but certainly not a shallow cup like many honeyeaters but then not as deep as Rufous-throated

Honeyeater's nest. Clutch size 2-3. **Similar species:** Black Honeyeater does come into the Kimberley region. I have seen them on the road into Ellenbrae Station. Black Honeyeaters are even smaller with a longer, finer bill, have a complete black hood passing down as a lateral band, so should not confuse and call very different. **Status:** Moderately common throughout the Kimberley, locally very common in high concentrations of flowering eucalypts. **Where to find:** Anywhere in the Kimberley, found on the Gibb River Road. Drysdale Road and Kununurra area.

Immature

Photo: Simon Nevill

Black-chinned Honeyeater *Melithreptus gularis* subspecies *laetior*

16 cm

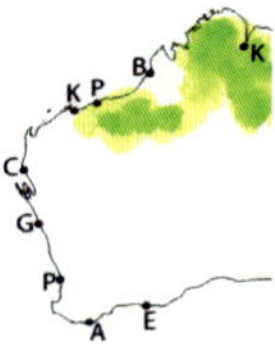

The largest species of honeyeater in the *Melithreptus* group. It has a widespread distribution from the northwest of WA across the northern parts of Australia and down the eastern parts of Australia to south east, with the Strong-billed Honeyeater taking its place in Tasmania. Mostly a bird of the open eucalypt woodlands of Australia. In WA its distribution covers a wide range but mostly above the Tropic of Capricorn. I have found them in woodland on the edge of Parry's Lagoon, along the Canning Stock Route in Sand dune Bloodwood (*Corymbia chippendalei*), in the high Pilbara plateaux area in Hamersley Bloodwood (*Corymbia hamersleyana*), at Carrowine Gorge in tall Cadjeput (*Melaleuca leucadendra*), and creeks north of Halls Creek, so a wide range of country but never common throughout just locally common when found. They are semi-nomadic and will move where trees are flowering but they must have a high intake of invertebrates as well as nectar. Their call is their giveaway for me, their strong call is relatively far carrying, a rasping call but still pleasant. The subspecies *gularis* is slightly different to the WA subsp *laetior*, which is an up and down 'cheew..chit..cheew..chit... chew', then a more rapid 'chew.chew.chew.chew.chew'. Feeds mainly in the canopy. Generally feeds higher in the canopy unlike other honeyeaters, except in the desert regions. The subsp *laetior* has much brighter yellow-olive back and rump hence its old name of 'Golden-backed Honeyeater'. The eye wattle above the eye, varies from lime-green to bright yellow but never blue as in *gularis*. subsp *laetior* has black head with a white crescent that narrows at the eye and broadest on the nape, white throat and grey-white under-body, black narrow lateral throat patch hence name, stout black bill, black iris. **Nest:** Often hanging in drooping foliage of a eucalypt neatly woven deep cup nest made of thin woven twigs and hair or similar, leaves of tree often woven close to nest to hide it. Clutch size 2-3. **Similar species:** White-throated Honeyeater has some similarities but is a smaller bird, dull olive-green back not bright lime-green, eye wattle small, whitish blue not as bright lime-green, no black throat patch, totally different call, bill finer. **Status:** Moderately common throughout the Kimberley, parts of the northern deserts, less common in parts of the central and northern Pilbara. **Where to find:** Gibb River Road. Drysdale River. Bungle Bungle. Around Kununurra and Wyndham region. Fitzroy River System. Halls Creek area. Hamersley Range east to Carrowine Gorge. Much of the Great Sandy Desert but nomadic in this region.

Photo: Simon Nevill

Blue-faced Honeyeater *Entomyzon cyanotis* subspecies *albipennis*

28 cm

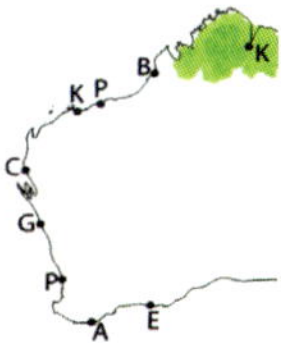

A large, colourful honeyeater with a vibrant, iridescent blue facial mask. Generally a gregarious bird, often moving through woodlands to feed in parties 6 -12 birds. Can be aggressive to other smaller honeyeaters but tolerates Yellow-throated Miners. Feeds on nectar but also larger insects where it will peel back bark and glean for insects on foliage, also will take small reptiles and has been known to take eggs, fruit is also a major food sauce. Blue-faced has an unusual distribution, it's rare in the pindan country of the west Kimberley and does not frequent the Fitzroy River system which one would think was an ideal habitat, however it gets south to Halls Creek and the Bungle Bungle. Its stronghold is the central Kimberley but is also common in the Kununurra, Lake Argyle region. Blue-faced Honeyeater has olive-green back and tail, black crown to name and below eye, large black patch on the throat and upper breast, white narrow malar stripe that begins as a narrow lateral band at the base of the bill and broadens on the side of the breast to form an all white under-body and vent, small black shoulder patch,

Photo: Simon Nevill

Bright iridescent blue facial mask covering the eye merging from bright blue to turquoise at the top of the patch, immature birds show a lime-green patch, small white crescent runs from rear of blue facial mask around nape, bill robust and slightly down curved. **Nest:** Nest location varies, sometimes it builds its own nest a shallow untidy bowl shape of intertwined twigs but messy, said to also use the nest of other birds particularly babblers but will open up entry hole. Call is noisy and varied but all calls are a repeated high pitched call, one contact sounds like 'wheeek.... wheeek...... wheeek' screech. Clutch 2. **Similar species:** None, should not be confused with any other species in the Kimberley. **Status:** Locally moderately common in the semi-humid central Kimberley and Kununurra, less so the semi-arid southern region. **Where to find:** Gibb River Road from the King Leopold Range right through to Kununurra but mostly between Mt Bell and Ellenbrae and north to the Drysdale. Moderately common on stations like Mt Elizabeth.

Silver-crowned Friarbird *Philemon argenticeps argenticeps* 29 cm

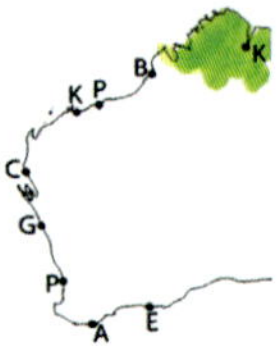

A medium sized friarbird, gregarious and noisy although less noisy than Little Friarbird, will mix with other honeyeaters and also Little Friarbird when flowering eucalypts are prolific. Found in varied habitats such as riverine, mangrove, open woodland and particularly the sandstone plateaux habitat, almost taking the place of the NT escarpment friarbird, the Helmeted Friarbird that for some reason does not occur in the Kimberley. Silver-crowned Friarbird feeds on a variety of foods, particularly figs in the sandstone country, also feeds on insects, seeds and takes nectar. Its distribution and numbers are not as great in the Kimberley as the Little Friarbird. Little Friarbird for example are found in Broome but not Silver-crowned also Little Friarbird goes further south into the semi-arid zone. Silver-crowned Friarbird has a grey-brown back, darker tail with fine white tip, throat white with dark streaked feathers stopping above belly, bare black facial mask that peaks behind eye, slight decurved bill, short white tufts protrude from bare skin on the nape, slight raised black casque at base of bill, shorter length than Helmeted, bright red eye in adult bird, immature black. Call difficult to describe and very varied but almost a gurgling, cackling sound. One common sequence sounds like 'wareecha...wareecha... warecha' has a single noted high pitched contact call with long pause 'shweeek.........shweeek ....shweeek'. **Nest:** Deep cup made of woven twigs, grasses or bark bound with spider webs, lined with fine grasses, bound to a hanging fork of thin branches. Clutch 2-3. **Similar species:** Little Friarbird which lacks the casque on the crown, facial mask is blue-grey not black, crown darker grey and lacks the silvery tufts on nape, call different but same scratchy noisy tones. **Status:** Moderately common in the semi-humid central Kimberley and Kununurra, not common in semi-arid southern region. **Where to find:** Gibb River Road, Drysdale to Kalumburu, particularly common in the sandstone plateaux country in the central and north west region of the Kimberley.

Photo: Simon Nevill

Little Friarbird *Philemon citreogularis* 27 cm

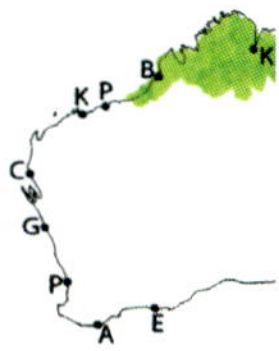

Easily distinguished from other friarbirds by its smaller size, lack of a casque on base of bill and the only friarbird to have a blue-grey facial patch. Even though it is the smallest of all the friarbirds it still has that pugnacious quality and will drive off other honeyeaters. Found often in pairs or small groups to 6-10 birds being semi-nomadic, moving to flowering areas out of breeding. Feeds mostly on insects, beetles and grasshoppers and is not so dependent on nectar. Found in open woodland, riverine habitat and mangroves but not enclosed forests. Little Friarbird has grey-brown and tail with white tip, under-body pale grey with darker lateral speckling, crown grey with lighter streaks, diagnostic grey-blue facial mask surrounding eye to bill, lacks casque on base of bill. Calls similar in tone to Silver-crowned Friarbird but more rolling raucous calls, with short contact repeated calls. **Nest:** Nest similar to Silver-crowned Friarbird but shallower more bowl shaped and tighter weaving of grasses, suspended either from a horizontal fork or placed in vertical fork lined with grasses. Clutch 2. **Similar species:** Silver-crowned refer to Silver-crowned Friarbird. **Status:** Very common throughout the Kimberley and the semi-arid region as far as Lake Gregory and Edgar Range and south to mid 80 mile beach. **Where to find:** Common in the Kimberley towns like Broome. Fitzroy Crossing and very common in Kununurra region to Halls Creek.

Photo: Simon Nevill

Crimson Chat *Epthanura tricolor* 11-13 cm

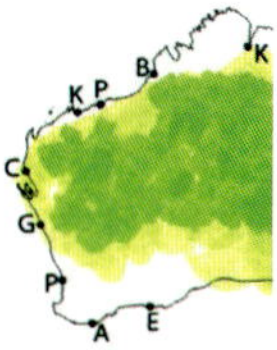

A brightly coloured ground bird of the semi-arid and arid regions of WA. Its favoured habitat is the Mulga woodland and it can be very abundant in good seasons. When the low shrubs and everlastings are in full bloom in the Mulga woodland, Crimson Chats can be one of the most common species and the nesting sites can be extremely close. Their distribution covers all the desert areas to the southern Kimberley. The female is rather plain but in breeding develops small patches of crimson on lower breast and flanks and, like the male, has a bright red rump, seen well in flight. The male in non-breeding plumage lacks the intensity of the red on the crown and the breast is patchy and less intense. Being highly nomadic, they sometimes venture into the northern and north eastern Wheatbelt even as far south as Ongerup. Most nests are only 20 -50 cm off the ground sometimes in ephemeral plants that are short lived like Tall Mulla Mulla (*Ptilotus exaltatus*) or Cotton Bush (*Ptilotus obovatus*) but retain their dense foliage right through the brief breeding season. When approaching the nest, it will land on a nearby bush or dead twigs on low bushes and check surroundings before dropping to the ground and running to the nest site. **Nest:** The nest is a shallow neat cup made from grasses and twigs, lined with fine grasses and wool bound with spiders' webbing.

Clutch size 2-3. **Similar species:** None. **Status:** Locally common. **Where to find:** Remains fairly sedentary in the mulga unless extreme droughts occur then it may come to the northern Wheatbelt. Common in station country, but it's always best to check where there has been winter rains before one travels. Can be common in the central ranges and Great Sandy Desert.

Below: Male

Photos: Simon Nevill

Orange Chat *Epthianura aurifrons* 11-12 cm

Female

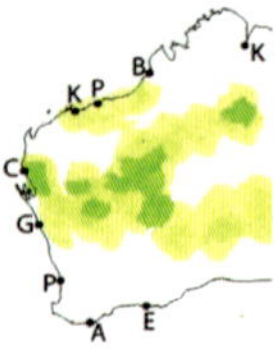

The Western Australian birds are almost exclusively found in samphire (*Halosarcia*), saltbush (*Atriplex*) and bluebush (*Maireana*) country, normally on low-lying areas often adjacent to ephemeral salt lakes. Easily overlooked as they feed, working one bush then running to another; but every now then, they will sit on the top and survey the area and if the male does this, the bright orange contrasts really well against the dark greens of the samphire. The female is less colourful but when the female bird flies the bright yellow-orange rump can be clearly seen, similar to the brightness of the male. Like all chats they nest low to the ground, building an open nest. In the region of the South West, they are found from Carnarvon to Hamelin Bay along the coast at various points in samphire flats. Inland their range is fragmented and the locations include Oyster

Creek Road, Carnarvon. Also 9 km south east of Murchison Settlement on the Meeberrie -Wooleen road. On the Wooleen Lake system. 4 km south east of Yalgoo. Lake Austin. Lake Annean. Lake Barlee. 2 km south east of Mullewa. This is the furthest south west that I have seen them although they have been recorded at Lake Hinds near Wongan Hills. The furthest north I have seen them is on Lake Tobin on the northern Canning Stock Route. They can also be found on the northern samphire areas of Lake Gregory and also on the Nullarbor Plain. **Nest:** The nest is a small neat cup lined with fine grasses and placed only 20-40 cm in samphire or bluebush on wide open flood plains of samphire or bluebush. Clutch size 2-4. **Similar species:** Yellow Chat, but not found in these localities, as it requires different habitat with freshwater nearby. **Status:** Locally common. **Where to find:** All the above ephemeral lake systems after good rains. Can often be found around the town of Carnarvon particularly on the dirt track to the racecourse and in good years around McNeill Claypan.

Photos: Simon Nevill

Yellow Chat *Epthianura crocea* 11 cm

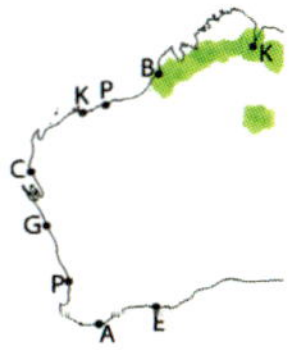

An interesting species with fragmented populations throughout central and northern Australia. The Kimberley is blessed with reasonable populations of this generally rare but sometimes locally common bird. I have seen many Yellow Chats in the drying lignum in shallow waters on the Parry's Lagoon floodplain nesting in close proximity to one another. I have also seen high concentrations of nesting birds along the irrigation channels at Camballin Station and on the Roebuck Plains when in flood. Like all chats this is an attractive bird. The male in full breeding plumage shows well the black crescent bar on the breast contrasting well with the bright yellow body. Yellow Chats occur in large numbers when Lake Gregory is in flood and could potentially hold the largest populations in Australia when conditions are right. **Similar species:** Orange Chats both male and female are similar in coloration. The male Orange Chat has a black throat patch with no black breast band. The adult female Orange Chat is very similar to the female Yellow Chat and difficult to split but if seen closely the feature that distinguishes the two species is that Orange Chats have a dark red eye and the Yellow chat has a white eye. In the

Photo: Simon Nevill

Kimberley unlike the central desert regions of QLD and SA, Orange Chats never occur in close proximity to Yellow Chats so confusion should not occur. However at Lake Gregory both species can be seen in close proximity to each other. **Status:** Locally common but overall rare in the Kimberley. **Where to find:** Parry's Lagoon when conditions are optimum. Parts of the Fitzroy floodplain. Sections of Lake Argyle. Lake Gregory when in flood. Roebuck Plains and Lake Eda.

White-fronted Chat *Epthianura albifrons* 11-23 cm

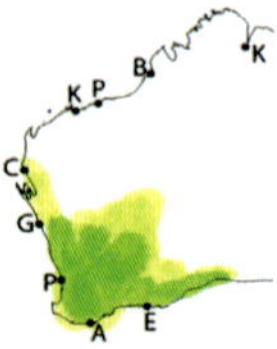

The distribution of the White-fronted Chat in Western Australia is almost a reversal of that of the Orange Chat but far more common in WA. Both can be found in samphire and saltbush but the White-fronted Chat will also occur in heathland particularly in the Wheatbelt. It does range as far north as Coral Bay and then hugs the coast where samphire occurs particularly south of Carnarvon to Gladstone. It is found throughout parts of the Shark Bay and Dirk Hartog region, particularly around Hamelin Bay. In the inland it then becomes fragmented, reaching as far north as Lake Moore area and east as far as Yeo Lakes on the edge of the Great Victoria Desert and east to Cape Arid. However the greatest concentration are on the Swan Coastal Plain and southern coastal belt to Esperance. They would have been far commoner on the Swan Coastal Plain before much of their territory of samphire was cleared and land reclaimed. Like all chats and many other birds they do the typical broken wing distraction display if you approach a nest. **Nest:** Is a little more substantial than Orange Chats but still neatly bound with grasses and soft lining of tightly woven grasses, placed low in samphire or in Kwongan heathland. Sometimes nests near water but other times at higher elevations such as the heaths in Tutanning NR. Clutch size 2 – 3. **Similar species:** None.

Above: Male

Status: Moderately common on larger expanses of samphire and saltbush as well as Kwongan Heath. **Where to find:** Common on Rottnest. Vasse Estuary. Peel Inlet. The coastal belt from Lancelin to Dongara. The coastal belt from Hopetoun to Esperance. Throughout the Wheatbelt around salt lakes and in heathland, often seen moving along roadside verges to new feeding areas.

Photo: Simon Nevill

Grey-crowned Babbler *Pomatostomus temporalis* subspecies *rubeculus* 25 cm

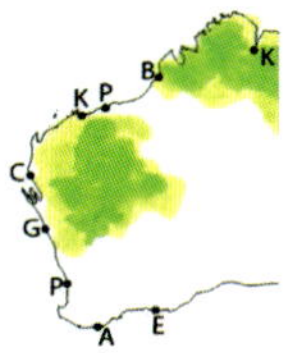

There are two subspecies in Australia and only one *rubeculus* in WA. Its range covers the mid west, Pilbara and Kimberley. A gregarious bird having 6 to 12 members in a family group. Like all babblers, they have a complex family structure with a dominant male and dominant female who have a life long pair bond. Then there are non-breeding support members, the young of previous years. In bigger groups there can be two dominant breeding pairs which share nesting and the raising of young. All support each other by mutual preening, or defending group members from danger. Will congregate together when agitated. They are highly active birds; when feeding they call regularly, keeping contact while turning over litter or pecking into bark removing insects. The sexes look alike. Found mostly along river systems but occur in semi-open acacia woodland. The similar White-browed Babblers occur in a large overlap range mainly in the Murchison and Gascoyne regions, for example Talisker and Muggon Stations inland from Overlander have both species. The Grey-crowned Babbler has a fine grey-brown lateral line running across the crown. A darker grey-brown eye stripe runs from the bill and broadens at the lores and into the upper-body. The southern subspecies *rubeculus* has a paler chestnut coloured chest and belly, hence the Kimberley birds are called the red-bellied form of the Grey-crowned Babbler. **Nest:** Their nests are bulky and there can be several in the one territory and the family group will roost in one or two of the non-breeding nests. One nest is solely used for breeding and is well lined with plant down and feathers. Clutch size usually 2-3. **Similar species:** White-browed Babbler which is noticeably smaller (length 19 cm whereas Grey-crowned is 25 cm). The White-browed has no chestnut on the chest or belly but white chest and belly with only the flanks having a grey brown tinge. The supercilium eye stripe is much narrower. Calls are different. **Status:** Moderately common throughout its range. **Where to find:** All stations from the central Murchison north to the Kimberley but not east of Carnegie Station. Both Grey and White-browed Babblers overlap at Woodleigh Station east of Hamelin Bay.

Photo: Simon Nevill

White-browed Babbler *Pomatostomus superciliosus ashbyi* 19.5 cm

There are currently four subspecies in Australia and two occur in WA. superciliosus extends from the Victoria Desert down to Esperance and in areas north and north east of a line cutting across the south west from Ravensthorpe in the south to Geraldton on the coast and into the Pilbara. *ashbyi* can be found in all areas within that line i.e. mainly the whole of the South West Region. The differences between the two subspecies are minor and unless you have the bird in the hand there is little variation that can be seen, which includes wing size etc.

The range of the White-browed Babbler is extensive; the only area in the South West that it no longer occurs, is most of the Swan Coastal Plain. In the Wheatbelt even though the country has been largely cleared the White-browed survives in most areas, even in the most cleared farming areas. It is one of the few birds that can be seen working the fragmented roadside verges, mainly out of the breeding season, as it will use the remnant reserves and patches of uncleared bush as the main breeding territory. In the South West Region it occupies a more varied habitat than in the north being found in thick mallee, woodlands adjacent to heath and throughout the Transitional Woodland. In the Mulga Region it occurs mostly in dense stands of mulga in the southern Mulga Region. Like all babblers it is very gregarious and is Australia's smallest babbler foraging mostly on the ground but will go up into low trees picking off bark trying to get at insects. **Nest:** The nest is similar to the Grey-crowned Babbler's but is smaller and the entrance hole can be more exposed and wider. The location tends to be more concealed in denser shrubbery. Clutch size 2-3. **Similar species:** Grey-crowned Babbler. For differences refer to Grey-crowned Babbler text. It will not be a problem misidentifying the babblers south of the Transitional Woodland or throughout most of the Goldfields, as the Grey-crowned does not occur there. **Status:** Moderately common throughout the Wheatbelt, Transitional Woodland and Mulga Region. **Where to find:** Most large Wheatbelt reserves mostly in northern, central and southern belt. Common in the larger reserves like Magenta NR. Dragon Rocks NR.

Photo: Simon Nevill

Black-throated Whipbird (formerly Western Whipbird)

Psophodes nigrogularis 23.5 cm — Endemic

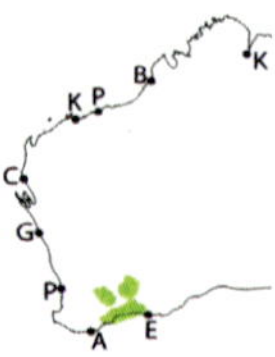

The Black-throated Whipbird has now been identified as being a separate species from the South Australian White-bellied Whipbird based on DNA research. In Western Australia there are two subspecies of the Black-throated Whipbird subspecies. *Psophodes nigrogularis nigrogularis* and 'western heath' subspecies *Psophodes nigrogularis oberon*, 'western mallee'. In South Australia there are two subspecies of the White-bellied whipbird subspecies: *Psophodes leucogaster leucogaster*, 'southern mainland' subspecies: *Psophodes leucogaster lashmari*, 'Kangaroo Island'. The Black-throated Whipbird is a ground dwelling bird that occurs in a variety of vegetation habitats including: peppermint (*Agonis flexuosa*) in the Two Peoples Bay region, dense stands of Tallerack (*Eucalyptus tetrogona*) and peppermint at Waychinicup NP. Amongst stands of Cayley's Banksia (*Banksia cayleyi*) and Southern Plains Banksia (*Banksia media*) and a variety of mallees in the north and eastern parts of the Fitzgerald River NP including Tallerack. In remnant reserves like Chinicup NR in the dense stands of mallee. In the Stirling Range it's mostly found in Tallerack and a variety of mallees and stunted Jarrah. The Black-throated

Whipbird has an olive-brown back and head with a small crest, black throat with white sub-moustachial stripe. The under-body is light grey-brown with a tinge of olive on the flanks. A long tail with black and white bars to the outer tail feathers seen best in flight. It is not a strong flyer and rarely flies more than 10 - 15 metres and rarely over the canopy of trees. Very defensive of territory from other Whipbirds. Its call is totally different to the similar east coast species the Eastern Whipbird; the female of that species gives her loud whip cracking call at the end of the pair's duet. The Black-throated Whipbird has a far more musical up and down scaled call, starting on a high note then dropping down to an extended note deee...do...dididee' repeated 'deee... do... dididee'. Its call has a ventriloquial quality so it sometimes takes a while to get onto the location of the bird. If patient, you will see this bird in the breeding season come up to the top or near the top of low mallee to give its territory call and that is when you will get the best views. **Nest:** The nest is a broad bulky cup with fine weaving of dry grasses in the centre, placed in dense bushes often in banksias. Clutch 2. **Similar species:** Eastern Whipbird not in this region. **Status:** Scarce throughout its range except parts of the Fitzgerald River NP where it is locally common. Vulnerable in some of the small Wheatbelt reserves, some surviving on bush blocks no more than 250 hectares. **Where to find:** Best area without too much disturbance is the Fitzgerald River NP near the north east entry to the park. Also the southern end of Hamersley Drive in the eastern section of the Fitzgerald River NP I have seen the subsp *oberon* in possibly the most northerly location recorded in WA at Dragon Rocks NR north of Newdegate in mixed banksia mallee woodland.

Photo: Simon Nevill

Chiming Wedgebill *Psophodes occidentalis* 20.5 cm

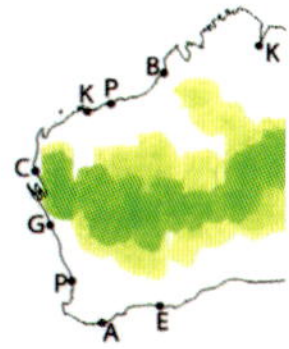

A bird of the same genus as the whipbirds and as the name implies has a very robust bill. Found in the semi-arid and arid regions of WA, particularly the mulga country north and south of Carnarvon and just north of Meekatharra at Karalundi where high densities of this species occur. However, their range is well spread out through most of the central belt of mulga and way out into the remote central deserts along the Gary Highway; less common in the southern Mulga Region. It is a plain coloured bird having a plain fawn head and mantle, darker wings with white on part of the primaries. A long brown tail with white on the tip of the outer tail feathers. Under-body is a cream-brown. The main feature is a very long crest curving slightly forwards. The song is one of the first things you may encounter with this bird as it can skulk in thick thorny acacias. The call is a monotonous rapid undulating call which has been given the vernacular wording of 'did you get drunk-did you get drunk'. This song is repeated over and over again, starting faintly almost inaudible, and then reaching full strength minutes latter. It has another call which it also sings late into the night and the notes go up then down 'did you - did you –did you –did you' repeatedly at some length. **Nest:** The nest is a neat open cup nest with finely woven grasses and plant stems, lined with finer soft twigs in the centre. Clutch size 2. **Similar species:** None, the very similar Chirruping Wedgebill is not found in this state. **Status:** Common in the central Mulga. **Where to find:** Carnarvon region. Lower stations adjacent to Gascoyne River. Other stations Boolathana. Edagee. Callagiddy. Near Meekatharra, Yoothapina. Near Cue, Nallan and Austin Downs. One of the most southerly populations occurs along the Warne River near Paynes Find. On the Gary Hwy way out in the Gibson Desert.

Photo: Simon Nevill

Copper-backed Quail-thrush

Cinclosoma clarum 24 cm

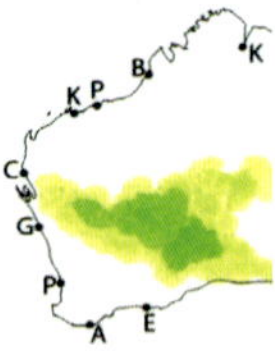

Three main subspecies occur in Australia. Two of which are found in the WA *clarum* occurs from Shark Bay right across to central Australia to South Australia at its northern limits and south via the Transitional Woodland and through the Mt Jackson to Goldfields region. It is found in varied habitats mostly under acacia woodland in sandy soil in the western areas south east of Shark Bay. In the Mt Jackson area under thick acacia and other dense shrubs, often in rocky terrain but not open stony country like the Western Quail-thrush in the north of this region. The subspecies *fordianum* occurs in the southern areas of the South West, in mallee and most now survive in the Transitional Woodland east of Lake Cronin across to Dundas NR and then in an outlying band along the southern Nullarbor coastline between Eyre and Eucla. From Coolgardie through to Balladonia is the main crossover where the subspecies intergrade. Copper-backed Quail-thrush males have chestnut mantle and rump that varies from the Shark Bay area right through to Dundas NR. All males have a black throat, chest and flanks with a white sub-moustachial stripe that borders the black throat. White scalloping on the shoulder. White supercilium stripe to almost the rear of the head. Grey-brown tail with white tips. Female quite different, having a blue-grey throat and chest not black, so the throat lacks the strong contrast between black and white as in the male. The male Copper-backed Quail-thrush will sit 4-8 metres up in low branches of trees and give their territorial contact call particularly an hour or so after dawn, a very high pitched single piping note with an even scale repeated for about 10 seconds and then repeated again and again 'te..te..te.. te..te..te te te'. Feeds entirely on the ground and only flies up into low branches to give its contact calls. **Nest:** The nest is an open cup nest sometimes in a depression scratched out by the bird, often placed alongside the base of a dense bush. The southern birds *fordianum* nest mainly near Blue Bush (*Maireana sedifolia*) or Grey Bush (*Cratystylis conocephala*). Clutch size 2–3. **Similar species:** Western Quail-thrush. There are a few overlapping ranges, for example on the Butchers Track before the start of Toolonga Nature Reserve their habitat are close but the Copper-backed Quail-thrush occurs where there are low breakaways with stony terrain. Also there is an overlapping area in the Mt Jackson region. The main differences are that the male Western Quail-thrush has a complete orange-rust upper-body which goes down to the tail. It has a similar black throat and white sub-moustachial stripe but below the chest is a distinctive chestnut band running across the body splitting the black chest and black upper belly; also has orange-chestnut flanks. The main thing to register if you see it briefly, is that the bird appears all over orange-brown compared with the Copper-backed Quail-thrush which is a greyer-brown bird overall. The female Western Quail-thrush still has an orange-rust back but it is far paler with colours more subdued. The belly is all white with a pale beige band on the chest, and the head is lighter and greyer with cream throat not grey. **Status:** Uncommon in the northern parts. Locally common in the Mt Jackson area. Common in the Transitional Woodland from 20 km east of Lake Cronin all the way to the eastern boundary of Dundas NR. **Where to find:** A close and easy location to visit is on the Mc Dermid Rock to Coolgardie Track and also even better the old track to Norseman that runs off the Coolgardie track and enters the Coolgardie Norseman road about 20 km north of Norseman (be careful, this track is rarely used and requires 4WD). Of course, for best results, support Eyre Bird Observatory on the Nullarbor and stay there, it's a magic place and you can walk to see the birds.

Above: Male

Photos: Simon Nevill

Western Quail-thrush (Chestnut-breasted Quail-thrush)

Cinclosoma marginatum 23 cm

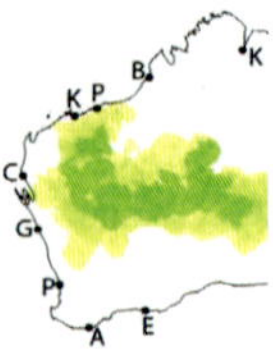

There were two subspecies of Chestnut Quail-thrush, *marginatum* (shorter wings, males only) in Western Australia and the nominate *castaneothorax* in central eastern Australia. BARC and IOC has given the subspecies *marginatum* a full species status and now have called it the Western Quail-thrush. Understandable as the two populations of the previous combined species were 1000 km apart and certainly no chance of interbreeding and this distinct plumage differences. Western Quail-thrush are slightly smaller than Chestnut-backed Quail-thrush with a slimmer body. Western Quail-thrush is found mostly in stony country on stony hillsides and open stony plains with often scant vegetation, sometimes along creek lines that run through stony country. This is a bird of the arid regions as far south as 40 km north of Mt Jackson and across to Thundelarra Station. They are common north of the Murchison Settlement region, Gascoyne Junction region and into the Ashburton catchment. Also way out in the northern Victoria Desert and Little Desert on the Hunt and David Carnegie tracks. The male can be recognised by his chestnut band between the black throat and the black on the lower breast. Has a complete chestnut crown, back and upper tail. In flight shows distinct white and black on the tips of the outer tail feathers. When surprised say by a vehicle, will fly low off the ground with an undulating flight and seek retreat under a shaded bush, so register where it flies. It has a few calls, one a high pitch and piping not too dissimilar to the Chestnut Quail-thrush but in the breeding season it has a soft and plaintive call almost mournful. This bird blends in very well with the stony country where invariably the soils and stones are so similar in colour, so seeing it is not always easy. **Nest:** Its nest is placed below a bush on the ground. A small depression is made by both the male and female and a shallow cup of loose mulga leaves is laid, often with twigs and leaves radiating out from the edge of the nest, which makes it well concealed. **Similar species:** Refer to Chestnut-backed Quail-thrush text. **Status:** Uncommon in the northern part of the South West. Common in the central Gascoyne River area to the central Ashburton River. **Where to find:** Common in the Murchison Settlement area, including stations: Muggon, Mt Narryer, Wooleen, Murgoo, Twin Peaks, Specific locations: At the junction of the Gascoyne Junction-Mullewa road and the Butchers Track. On the Beringarra-Pindar road 3 km north of the turn-off to Twin Peaks. 16 km south east of Wooleen on the Mereberrie–Mt Wittenoom road. Also east in the Tukanarra area north of Cue and 10 km east of Cue. Near Alexander Springs Little Sandy Desert. 10-15 km north of Everard Junction.

Photos: Simon Nevill

Nullarbor Quail-thrush *Cinclosoma alisteri* 18.5 cm

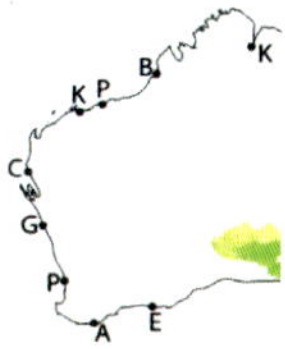

The photograph shows a Cinnamon Quail-thrush for reference only. The Nullarbor Quail-thrush male's mantle and crown are a rich rufous not grey-brown, supercilium stripe brighter cream not dull beige, importantly, the breast has clearly defined black chest running from belly to the throat, not broken by white as in Cinnamon. The female are almost identical in colour except the mantle is more rufous than grey-brown. Both sexes are smaller than all other Quail-thrush. The Nullarbor Quail-thrush has once again been re-established as a full species and so it should be, not only is it a smaller bird but also has marked different colour characteristics from its previously grouped species the Cinnamon Quail-thrush. This is a bird of the wide open limestone plains of the Nullarbor. It lives amongst the drought resistant chenopod bushes, which include Bluebush *Maireana sedifolia*, Greybush *Cratystylis conocephala* and Saltbush A*triplex vesicaria*. The first sign you know the bird is near you if you hear its delicate high pitched contact call 'stzee...stzee.stzee stzee…stzee.stzee'. It will normally see you first, as you stand well above the 60 cm tall bushes and slowly walk off out of sight but with patience you will get them with binoculars. The male has a rufous crown, rump and tail, distinct black breast band that forms a black vertical throat band with white panels contrasting on side of neck, has a distinct cream supercilium stripe stopping at nape, white under-body, black and white on shoulders, in flight white and black on tail tips shows very well. Female plumage over all duller, upper-body paler grey-rufous, no black on breast just grey-brown, white belly and white throat, also supercilium stripe but duller cream than male. **Nest:** Cup nest of twigs and grasses often placed alongside base of bush. Clutch 2-4. **Similar species:** Only bird that possibly could confuse in the region if running from you, is Inland Dotterel but a totally different bird and different jizz. **Status:** Locally common in pristine habitat, alas very uncommon in much of previous habitat where sheep grazing occurs. **Where to find:** In WA getting harder to find but still on stations like Rawlinna Station but grazing, wildfires and feral predation have taken their toll. Best seen in just over the border near Nullarbor Road House, SA.

Photo: Simon Nevill

Varied Sittella *Daphoenositta chrysoptera* subspecies *pileata* 11.5 cm

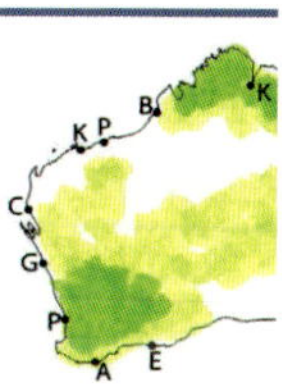

Certainly varied in subspecies coloration. There are at least five subspecies recognised, with two in WA. *pileata* in the south west and *leucoptera* in the Kimberley. There is little hybridisation in WA as the Great Sandy Desert breaks their distribution. *pileata* is slightly larger than *leucoptera* and has brown band in the wings not white, seen well in flight and the mantle is less streaked. Both males and females have black caps but the males only have a white eye patch. All have white under bodies and black wings. Vent is barred black and white, tail short and black with white tip. Bill is fine and upturned with yellow at the base of the bill. Pale yellow legs. Sittellas are gregarious birds often in a family party of 4-6 birds with two primary breeding adults and 2-4 helpers who can be adult birds or the young of last season.

They help with building the nest and feeding young. When they fly it is normally as a group constantly twittering to keep contact. Sittellas can be found in a variety of woodlands. *pileata* in the South West occurs in all the well forested areas and open woodlands, including all the primary eucalypts. Their range extends from the lower South West up into the Pilbara and east through the central deserts avoiding the Nullarbor and Great Sandy Deserts. They do occur in the Mulga Woodland but more so in the Transitional Woodland *leucoptera* is less common and frequents a varied selection of eucalypts not always the taller species. **Nest:** The nest is one of the exquisitely crafted nests, made with loose bark placed on the outside in small patches to imitate the trunk and branches of the host tree. Nearly all nests are at the junction of a steep fork, sometimes a three-way junction, mostly well over 5 metres up in trees. The inner nest is lined with plant down and spider's webs and egg sacs. Clutch size 2-3. **Similar species:** None. **Status:** In the South West moderately uncommon. Uncommon in the outer Wheatbelt. Transitional Woodland and Mulga Woodland. Moderately common in the central Kimberley. **Where to find:** *pileata* in the Darling Range. Larger areas of Wandoo. Often in Yate woodland. *Leucoptera.* Also a range of other eucalypts, particularly those with rough bark, not always tall eucalypts but in mixed woodland with acacia such as on the Gibb River road near Ellenbrae Station.

Photos: Simon Nevill

Crested Shrike-tit *Falcunculus frontatus* subspecies *leucogaster* 17 cm

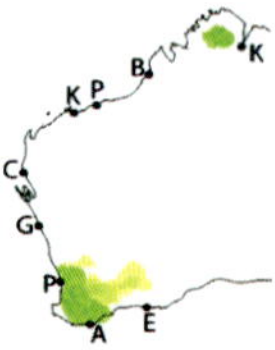

There are three subspecies of Crested Shrike-tit in Australia, two in WA, *whitei* in the Kimberley and leucogaster in the South West. Unlike the eastern states subspecies *frontatus*, which is relatively common, the Kimberley and South West birds are certainly not common. The Kimberley subspecies *whitei* is in fact fairly scarce and I have only seen them myself off the Gibb River Road near Mt Elizabeth and the Drysdale River Station. However the Kimberley birding expert George Swann certainly knows where to find them in specific habitats. The Kimberley birds are slightly smaller than the south west birds but do not have a white belly but a complete yellow under-body. They often feed on the eucalypt species with dissected bark. Like the south-west birds their mournful call gives them away as otherwise they can be difficult to see in the canopy. The south-west birds have lost much of their prime feeding habitat the Wandoo Woodland, and they require adequate stands of mature Wandoo to support a pair's territory. In the main block of Dryandra Woodland Reserve I know of only five territories within an area of about 8,000 acres. They also occur in the Karri Forest. Crested Shrike-tits have a mournful undulating territorial call 'suewit... sueweee... suewit... sueweee'. They also have a grating agitation call. Their strong bill allows them to lever off bark to get at insects, one of the few birds in the South West able to do this. They are totally arboreal in their feeding habits, often spending time in the canopy of trees, and can be easily overlooked as out of the breeding season they call infrequently and only the cracking of bark alerts one to their presence. **Nest:** The nest is placed in a fork normally above 6 metres. It is an exquisite construction made of thin woven strands of bark with copious amounts of spiders' webs. The inner cup is lined with finer bark strands or twigs and grasses. Clutch size 2-3. **Similar species:** Possibly male Golden Whistler which has a similar combination of colours. The Golden Whistler has no crest and has a white throat not black. Head markings totally different. **Status:** Uncommon in the Karri Forest. Vulnerable in the Wandoo Woodland. **Where to find:** *leucogaster*: Pemberton region particularly Warren National Park. Boranup forest in the Leeuwin-Naturaliste NP. Northern and central areas of wandoo in the Stirling Range. Dryandra Reserve. Wandoo Conservation Park. Whitei: Found in various eucalypt woodlands but Darwin Stringybark. Kimberley Yellow Jacket and Rusty Bloodwood give a good chance (per. G. Swann). 50 km radius around junction Kalumburu/ Gibb River Road strong possibilities.

Photo: Simon Nevill

Crested Bellbird *Oreoica gutturalis gutturalis* 21 cm

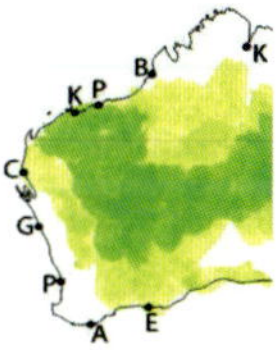

There are two subspecies in Australia. Both occur in WA, with *gutturalis* in southern WA, and *pallescens* in northern WA, through to the Northern Territory; the subspecies overlap in the Gascoyne River region. Both have little variation save that the southern birds *gutturalis* are darker backed birds which is typical of a birds being smaller and lighter the further north one goes. The Crested Bellbird has a distinct crest, white face with surrounding black border, bright orange eye, grey head with black crest and darker brown upper body and pale underneath. Has a stout bill. The female lacks the white throat and black surrounding mask, does have a black crest but smaller. Has the well known bell-like call with a ventriloqual quality that makes it hard to place as it can sound close and then far away. If one is very close to the bird, you can actually hear the bill clatter when the call reaches its highest pitch. Birds in Western Australia have a marked difference in their call from the eastern birds. It is one of the few mulga species that remains sedentary even through the worst droughts when other nomadic species will move. Found in semi-arid and arid regions mostly past the Transitional Woodland belt, having a preference for open mulga country. It also is found in the Wheatbelt region but has declined in the south west Wheatbelt. Dryandra Woodland Reserve lost its last male Crested Bellbird in about 1992, the same bird had been there for at least 5 years, calling for a mate that never came. In the late 1970s I knew of at least 2 pairs. Even though scarce in the south west Wheatbelt, they do come down to the Stirling Range and Fitzgerald National Park and some of the larger eastern Wheatbelt Reserves, Most of its feeding is done on the ground. **Nest:** It builds a bulky untidy nest, the outside twigs being fairly robust and loosely gathered; the centre cup is a little more finely woven with thin grasses and plant stems usually placed in the fork of a dense mulga bush only 1 -2 metres off the ground. Clutch size 2 – 3. **Status:** Scarce in the inner Wheatbelt. Uncommon in the southern Wheatbelt. Moderately common in the eastern and northern Wheatbelt. Common in the Mulga Region. **Where to find:** Stirling Range. Central Fitzgerald River NP. Magenta NR. Dunn Rock NR. Chiddarcooping NR. Throughout the Mulga Region north and north east of the Transitional Woodland belt. In the Pilbara and lower deserts to the border.

Photo: Simon Nevill

Gilbert Whistler *Pachycephala inornata* 19 cm

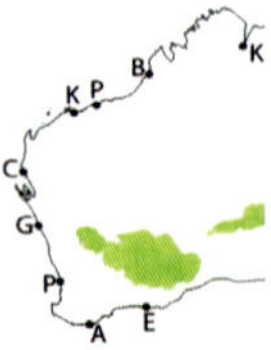

A whistler of the semi-arid regions of the Transitional Woodland belt. Found mostly east of Lake Cronin 90 km east of Hyden to Balladonia in the east. It frequents the Salmon Gum woodland and wherever the Goldfields Teatree (*Melaleuca pauperiflora*) occurs, that is where you will find them even if the dominant tree is not Salmon Gum. The Gilbert Whistler's call is one of the most powerful calls of all whistlers in fact of any Australian passerine. The call starts off slowly and then finishes with the loudest penetrating 'chop chop' calls getting higher near the end 'soooee.... soooee... soooee., chop... chop... chop... chop.... chop...chop' The male is recognised by its orange throat and dark lores. The rest of the body is a dull grey-brown with a lighter under-body. It has a robust bill just a bit shorter than White-breasted Whistler. The female lacks the orange throat patch. The crown is lighter and the lores much lighter. Gilbert Whistlers were once relatively common in the western Wheatbelt but by the 1930s were almost extinct in that region and only now survive on a few of the larger north-eastern Wheatbelt reserves. There is a small outlying population on the Warne River east of Paynes Find, in a habitat not typical of most of their range in Western Australia. **Nest:** The nest is a deep cup shape with twigs getting finer to the centre mixed with slivers of smooth bark all bound tightly together, placed only 1-3 metres high. Clutch size 2-3. **Similar species:** None for the male. Red-lored Whistler, not in Western Australia. The female can be confused with the female Golden Whistler. The Golden Whistler is a smaller bird with different 'jizz', tail not as long. Has an olive-grey look to the back. Bill is not as robust. The territory of the north western population is highly unlikely to have the Golden Whistler and they are not common in the Transitional Woodland east of Hyden although they tend to frequent denser vegetation, sometimes at the base of granite outcrops in that region where Gilbert Whistler do not occur. **Status:** Common from Jaurdi north east of Coolgardie south to Peak Charles and Lake Cronin in the east to Balladonia in the west. **Where to find:** Easiest anywhere just east of Mc Dermid Rock. The Warne River population is just north of where the Warne River passes over the Paynes Find and Sandstone road.

Photo: Simon Nevill

Western Whistler (previously Golden Whistler)

Pachycephala fuliginosa 17 cm

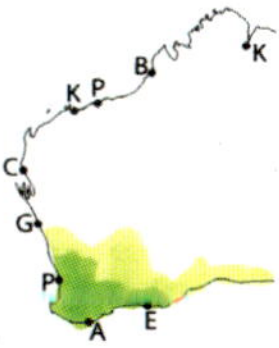

Western Whistler is now considered a full species, although I am interested where the colour variation changes as at the Eyre Bird Observatory there are Western Whistler sightings. A medium sized whistler with a slim bill, stocky body and large head in relation to body. Sexes totally different in adult birds. The male has strong head markings with a white throat, black hood and yellow collar and yellow under-body. Olive back, black tail tinged dark green. The female has no yellow tingeing on the vent unlike eastern states birds. The back is olive-brown. Slight supercilium stripe. Faint scalloping on pale brown throat. Buff under body darkest on chest. The Western Whistler is a bird of the wetter South West being commonest in the tall eucalypt forests and thick understorey of the lower South West and common in the Karri Forest, and southern Jarrah and Tingle Forests. Found feeding in the middle and lower stratum and on the ground of forested areas. It occurs in the drier woodlands particularly Wandoo Woodland but mostly in sheoak thickets and mallee and dryandra over dense heath, less so in open Wandoo. Its range belies its population density as it occurs as far north as Shark Bay and south east to Kalgoorlie and down to Balladonia but generally past the northern Darling Range and lower wheat belt it is not common, however it is more so in the lower Transitional Woodlands south of the Merredin to Kalgoorlie highway. Its pleasant whistling call is familiar to many 'choo… choo… choo… whip'. **Nest:** The nest is shallower than most whistlers with the outside constructed of thin twigs, which get finer and more compact in the centre often with the addition of leaves bound to the twigs with spiders webbing. The inner lining is of fine rootlets or grasses. Clutch size mostly 2. **Similar species:** Mangrove Golden Whistler that does not occur in the same region. **Status:** Very common in the lower South West. Common in the Darling Range and southern coast and Transitional Woodland. Uncommon past the Transitional Woodland belt. **Where to find:** Throughout the lower South West and Darling Range. Common in wetter gullies particularly in peppermint stands on the southern coast. Common on some islands like Woody Island off the Esperance coast. Most inner Wheatbelt reserves. Moderately common in the southern Transitional Woodland. Uncommon in northern Wheatbelt and northern Transitional Woodland.

Above: Female

Photos: Simon Nevill

Mangrove Golden Whistler *Pachycephala melanura melanura* and subspecies *robusta* 16 cm

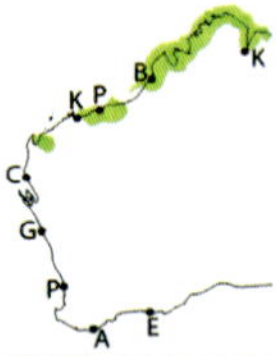

Mangrove Golden Whistler nominate melanura is found from Onslow to Cape Keraudren, then further north from Port Smith and then around the northern Kimberley Coast to Joseph Bonaparte Gulf where the nominate species intergrades with the subspecies robusta. Although Johnstone et al 2004 recognizes only two subspecies in Australia, *melanura* and *spinicauda*. Mangrove Golden Whistlers are generally confined to

Above: Female

Note yellow vent

Photos: Simon Nevill

mangroves although in the northern parts of the Kimberley they will enter adjacent vine thickets and rainforest. The male is so similar to Golden Whistler that in the field you will almost not pick the differences. The bill is slightly longer and more robust in the Mangrove Golden Whistler, the black hood has more of a gloss sheen, the yellow neck collar is slightly broader and brighter. The good news is the two species do not overlap in WA as they do in central Qld. For male colouring refer Golden Whistler. Female Mangrove Golden Whistler certainly differs from female Golden Whistler. In the nominate race *melanura* found through most of WA, the female has a bright yellow vent as the birds intergrade from Port Warrender around to the Cambridge Gulf the yellow on the vent increases further up the belly and in the full subspecies *robusta* the yellow merges with to the lower breast and easily identifiable as female Mangrove Golden Whistler. Female Golden Whistler in the SW subspecies *fulginosa* has no yellow on the under body whereas female Mangrove Golden Whistler does. Calls vary for the various subspecies although not as variable as Golden Whistlers. The call of Mangrove Golden Whistler call is a strong more robust call than the Golden Whistler, although similar in tone and notes. It feeds not only on insects but also small molluscs and small crabs. **Nest:** Well woven, deep cup of fine twigs and mangrove rootlets, lined with softer grasses placed between vertical Clutch 2–3. **Similar species:** Golden Whistler as described above but not in same habitat except on east coast. **Status:** Common wherever there are areas of substantial canopy mangroves mainly of the genus *Bruguiera* and *Rhizophora*. **Where to find:** Barred Creek. Port Sampson. Cape Keraudren. Port Smith. Roebuck Bay. In Derby and Wyndham be careful here-crocodiles a problem. All around the Kimberley coast and some larger islands to the WA border in NT and east coast.

Rufous Whistler *Pachycephala rufiventris rufiventris* 16 cm

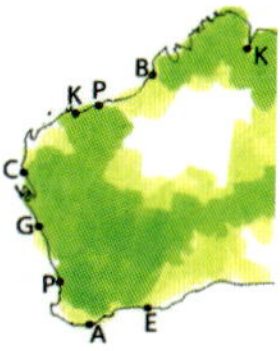

Found throughout the WA. Least of all the wetter south-west, the Nullarbor and most of the central Great Sandy Desert. A common bird with a preference for semi-arid habitat, not the wetter areas of the south west, including the Karri Forest and heavier forested Darling Range. A master songster with one of the finest calls of any Australian bird given with so much power, particularly in the breeding season. Sexes very different, the male has a white throat, black surrounding mask, grey crown and back. Buff under body. The female has no black or white on the head. Back grey-brown, under body light buff with fine streaking on the throat and chest. There are four

Above: Female

Photos: Simon Nevill

subspecies in Australia with two in WA, one *rufiventris* in the South West and *falcate* in the far north intergrading where they overlap. The South West birds typically are darker, with darker backs and richer buff underparts getting lighter in the northern regions where there are in greater numbers. **Nest:** The nest is slightly bulkier than the Golden Whistler with larger twigs spiralling into finer twigs in the centre, sometimes lined with a few feathers sometimes not. Clutch 2–3. **Similar species:** Female Golden Whistler has a pale throat with slight mottling, grey brown chest bleeding into a light beige belly. Grey-brown back. The Rufous Whistler female has pale cream under body with long fine black streaks and grey back. **Status:** Common throughout the state particularly the Mulga Region. Common in the Pilbara and central and lower Kimberley. **Where to find:** Near Perth, Bungendore Reserve and Wandoo Conservation Park. Any of the larger Wheatbelt Reserves. Throughout the Mid West and Pilbara and much of the Kimberley south to the northern Great Sandy Desert.

White-breasted Whistler *Pachycephala lanioides* 19 cm

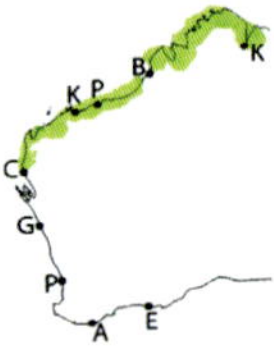

A whistler that only inhabits mangroves. Found from Bushy Bay (not common) south of Carnarvon all the way to the northern Kimberley and then populations in the Cambridge Gulf. Has the largest bill of any whistler, which is used to open small molluscs. Feeds in the mangrove trees but when the tide goes out, will take food from the mud floor before returning to the mangroves with the onset of the high tide. The sexes are different. The male has a white throat, black crown and facial mask with a chestnut collar below. The back is pure grey with a black tail. Under-body white. Very robust bill with down curved tip. Female is a plain coloured bird with beige upper body and pale cream under body but with streaking from the throat to the lower belly. Grey tail. Same robust bill as male. Male has several whistling calls, the main territorial call is hard to put into words but is very varied undulating, not as robust a call as the Gilbert or Rufous Whistler's call, more a plaintive song but still carrying far through the mangroves. Has long pauses between bursts of song. There are three subspecies in Australia. All three occur in Western Australia but only *carnarvoni* in this region reaching just south of Carnarvon, then north in fragmented populations until Cape Keraudren where Eighty Mile Beach becomes the barrier to the next subspecies *lanioides* in the Kimberley. The subspecies *fretorum* is found from Wyndham around the coast to the NT. **Nest:** The nest is not dissimilar to the Rufous Whistler's nest but twigs are taken from mangroves or samphire bushes. It is placed near the tops of a large mangrove well above the high water mark. Both sexes build the nest. Clutch size 2.

Above: Female

Photos: Simon Nevill

Similar species: None. The female is similar to the female Rufous Whistler but not found in the same habitat. The White-breasted female is larger than the Rufous Whistler with a more robust body. **Status:** Not common in the Carnarvon region. Becomes more common in the Exmouth Gulf region particularly in Gales Bay on Bullara Station. Common throughout the northern Kimberley mangroves. **Where to find:** Bushy Bay. Giralia Bay. Near Onslow. Point Sampson. Balla Balla near Whim Creek. South of Roebuck Bay. Derby Mangroves. Wyndham Mangroves now hard to access.

Grey Shrike-thrush *Colluricincla harmonica* 25 cm

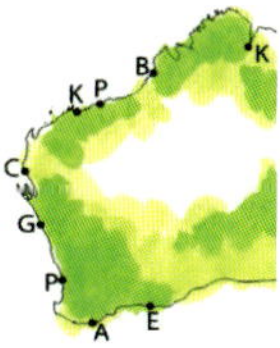

A large thrush-like bird, larger than whistlers. An overall grey bird with a grey head, dark grey back with only a slight chestnut tinge on the mantle and not as mush as many books illustrate, being much greyer than eastern birds. The head is totally grey with a small white patch on the lores. Throat grey-white and under body slightly darker. Tail long and dark grey. A long heavy bill with distinct hooked end. The female is similar in appearance but has fine streaking on the throat and belly. Feeds mostly on larger insects but will take small reptiles and also eggs and young birds, although not to the same degree as currawongs or ravens. Feeds both in trees and on the ground. The song is strong but very melodious with great variance throughout Australia, one of the master songsters. Found throughout the state except the central Great Sandy and much of the Nullarbor, found in varied habitat also in the wetter forests of the South West, throughout the Wheatbelt, Transitional Woodland and Mulga Region. Deserts and the Kimberley. There are five subspecies in Australia and it is a complex taxonomy. Two in WA subspecies *rufiventris*, in the south west. Pilbara and Central deserts. The Latin name *rufiventris*, relates to the slight rufous colouring on the vent, *brunnea* occurs in the Kimberley. Nests: The location for Grey Shrike-thrush nests can vary greatly, in the Mulga Region located in thick mulga in the centre only a metre or two high. In South West forested areas, can be found in tree hollows, not inside a trunk but where the trunk is broken or heavy bark peeled back, normally above 4 metres. The cup nest is a combination of woven twigs and strands of bark with an inner lining of fine rootlets and grasses. Clutch size mostly 3. **Similar species:** Sandstone Shrike-thrush which is a longer, slimmer bird with darker brown back and rufous lower under body and no white lores. Do not share the same habitat but comes close. **Status:** Common throughout their distribution. **Where to find:** Anywhere there is woodland or thick mulga. Riverine vegetation even rocky hillsides.

Photo: Simon Nevill

Arafura Shrike-thrush *Colluricincla megarhyncha* 19.5 cm

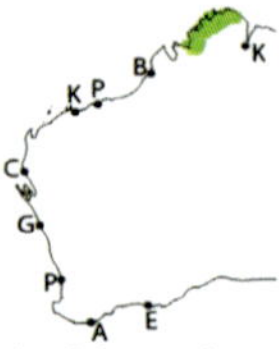

This species of Shrike-thrush has justifiably been spilt from the Rufous Strike-thrush having distinct differences in colouration, having a pale face and a pale off white supercilium eye stripe. It also lacks the rufous colouration of the underbody which the Rufous Strike-thrush has being a pale off white cream colour. Also there is a distinct call varation. Arafura Shrike-thrush in the Kimberley is found mostly along the coastline from north of the Walcott to Port Warrender but I am sure there are many populations around the remoter north-east coast. They live in the remnant rainforest patches and thick monsoonal vine thickets, although I have found them in heavily forested gullies but not rainforest that run down to the coast, so presumably they survive moving from one patch of thicker rainforest to another. Arafura Shrike Thrush is distinctly smaller than Grey Shrike-thrush. They can be found inland but their stronghold is the north-west rainforests close to the coast and on the larger islands. Mainly an arboreal feeder, gleaning insects. If you are quiet and remain still, they will sometimes approach as they are rather inquisitive angling their head as they look, one of the joys in walking in remote country where few people go. Their calls are powerful and variable, ringing through forest and their main contact a long 'tseeee…' then a rapid and powerful 'chuk..chuk'. Their main song is beautiful series of whistles that are varied ending in a 'cheee' not unlike that of Grey Shrike-thrush. **Nest:** Well woven, deep cup of fine twigs lined often with soft vine stems, the out side having leaves woven into the nest as camouflage, nest placed in vertical branches. Clutch 2-3. **Similar species:** Grey Shrike-thrush that generally does not share same habitat. It is a larger bird with no rufous-beige under-body and a grey back not brown, song similar tones but easily identified as being different. **Status:** Moderately common in larger canopy forests of the north-west. **Where to find:** Mitchell Plateau. Prince Regent River. Sale River. Roe River. Glenelg River. Upper Calder River. Wren Gorge.

Photo: Simon Nevill

Sandstone Shrike-thrush *Colluricincla woodwardi* 25 cm

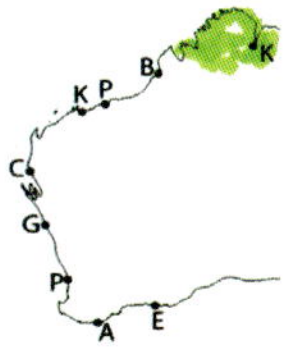

A long tailed Shrike-thrush with a beautiful song that echoes across the gorges and rock country where it lives. A very specialized shrike-thrush, finding its niche in the rocky hillsides, normally where there are deep crevices in the tropical region of Australia. Sandstone Shrike-thrush has an all brown upper-body, longish black bill, longest tail of all shrike-thrush, whitish-grey lores and throat with faint dark streaking, buffy under-body becoming more rufous at vent, Kimberley birds have more rufous on vent than NT birds. Female is very similar but bill dark grey rather than black, streaking on breast less distinct. Sandstone Shrike-thrush spends its time feeding amongst spinifex clumps and into rock crevices looking for insects however it will feed in adjacent woodland below rocks. When it sings its beautiful territorial song, it does so high up on an exposed rock ledge, making its presence well known. Patchily distributed throughout the Kimberley from Windjana Gorge in the west to Kununurra in the east and from King George River in the North down to the Emanuel Range in the south. It not only inhabits sandstone country but also limestone rocky areas. **Nest:** Untidy shallow cup placed on rock ledge under canopy or in rock crevice, made of fine twigs lined soft spinifex or grass stems. Clutch 2-3. **Similar species:** Grey Shrike-thrush generally does not share same habitat. Tail not as long, more upright jizz and back greyer, body of Sandstone Shrike-thrush slimmer. **Status:** Locally common but often populations patchily distributed. **Where to find:** Kununurra, easy to get at Mirima NP (Hidden Valley). The Grotto north of Kununurra. Common in the gorges of Purnululu NP (Bungle Bungle). Manning Gorge. Windjana Gorge and across to Geikie Gorge.

Photo: Simon Nevill

Restless Flycatcher *Myiagra inquieta* 19-21 cm

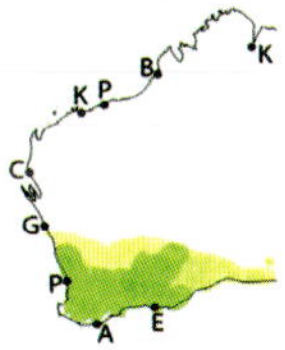

Well named, as this is certainly an active feeder frequently twisting and turning in flight to gather insects. It also has the unusual habit of hovering just 1-2 metres off the ground almost in Kestrel fashion, looking for movement of insects. It often does this while giving a continuous grating call, hence its popular name 'scissor grinder bird', derived from its calls resemblance to the sound of steel being ground on a stone grinder. Besides the grinding call, its contact call is a rapid 'chewe..chewe..cheweeee'. The male has dark blue-black head, nape, wings, upper body and tail with a slight metallic sheen only seen in direct sunlight. The under body and throat are all white. The male has a small crest and a long fine bill. The female is almost identical but has slight buff on the side of the white shoulder. The crest is not as pronounced and the upper body lacks the metallic sheen of the male. South west birds are particularly at home in open woodland, namely Wandoo, Salmon Gum and Mallee. Much of the Wandoo and Mallee has gone from the Wheatbelt, so they tend to occur in fragmented areas in the Wheatbelt Region. Their preference is for open woodland where they can both forage in the canopy or hover close to the ground. They rarely feed in the thicker Jarrah Forest but as soon as it mingles with the Wandoo, they occur. They do occur in Karri Forest but mainly after burns for a few years, while the understorey is clear. The northern limit is basically a line from Lancelin in the west across to Zanthus on the eastern side of the Nullarbor down to the southern coast. The Kimberley birds mainly frequent open areas near swamps where paperbark occurs. **Nest:** All flycatchers build exquisite nests and the Restless Flycatcher is no exception. A tight moulded mass of shredded bark and grasses all bound with spiders' webbing and camouflaged on the outside with bits of bark or sometimes spiders egg sacs. The inside of the cup is made of fine woven dry grasses. Normally placed on a horizontal limb where it forks. Clutch size 2-3. **Similar species:** Willie Wagtail, which has a different jizz and a black throat not white. Has small white supercilium stripe. Tail slightly longer and broader when perching. Call different. **Status:** Uncommon deep south west. Moderately common in the Transitional Woodlands. Common in the wetland areas of the Kimberley. **Where to find:** Wandoo Conservation Park. Dryandra Woodland Reserve. Dongolocking Reserve. Stirling Range National Park. Magenta Reserve. Main stronghold is the eastern Transitional Woodland from Lake Cronin through to Balladonia. Kimberley. In the Kimberley, the Fitzroy Basin and the wetlands around Kununurra.

Photo: Simon Nevill

Paperbark Flycatcher *Myiagra nana* 17-18 cm

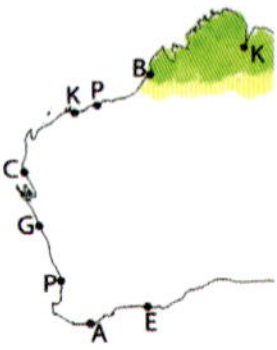

It was only a matter of time that *nana* was given full species status. The original two subspecies did not overlap in distribution. Paperbark Flycatcher has several differences from the southern Restless Flycatcher. It is substantially smaller and also a slimmer bird with a shorter, broader bill and smaller crest (see photograph). The species weight is at least 30% lighter than the Restless Flycatcher. Found along creek lines and wetland areas, although can be found in open dry woodland. Very common in paperbark woodland hence the name. **Similar species:** Willie Wagtail, which has a different jizz and a black throat not white. Has small white supercilium stripe. Tail slightly longer and broader when perching. Call different.

The Paperbark Flycatcher does not use the 'scissor grinder' call that the Restless Flycatcher has. It is also less prone to hovering over the ground that is typical of the Restless. **Status:** Common along river margins, paperbark swamps throughout the Kimberley. **Where to find:** In the Kimberley along the Fitzroy Basin and the wetlands around Kununurra. Also around the Broome region.

Photo: Simon Nevill

Magpie-lark *Grallina cyanoleuca* 26-28 cm

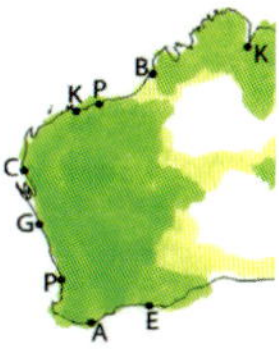

A well known bird found throughout most towns in the South West. It is hard to imagine this is a member of the 'monarch flycatcher' group, it certainly is Australia's largest monarch flycatcher. The strong black and white markings make it a striking bird. The sexes have different head patterning. The male has a black face and throat, the female has white face and throat. The black band passing through the male's eye runs horizontally and in the female vertically. It is a ground dwelling bird and one of only three birds in Australia that builds a mud nest, hence one of its common names is 'Mud Lark'. The other two mud nest building birds are the Chough and the Apostle Bird. Magpie-larks occur throughout the state except the deserts and Nullarbor. Mostly found in open ground near lakes or dams as they require water for their nest building. Adult birds defend their territory strongly and remain in their territory permanently. When you see small moving flocks particularly through the Wheatbelt at the end of summer and autumn, these are mostly younger birds congregating to travel to other feeding areas for winter before returning to form territories themselves in spring. There are two subspecies in Australia, *cyanoleuca* basically south of the 20 degree latitude and *neglecta* north of 20 degree latitude. The tropical northern birds are smaller, typical of 'Bergman's Law' having shorter wings and tail and less weight. **Nest:** The mud nest is held together with dry grasses and the inner cup line with finer grasses and sometimes feathers added. Mostly placed high in trees directly on top of a larger horizontal branch. **Similar species:** None. **Status:** Common throughout the state. Adapted well to urbanization. **Where to find:** Common in towns and throughout the south west, mid west and Kimberley where water is available.

Photo: Simon Nevill

Grey Fantail *Rhipidura albiscapa* 15-16.5 cm

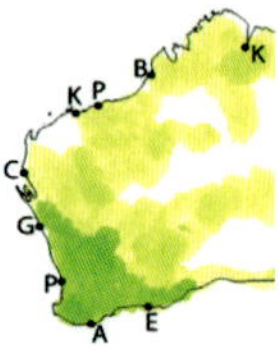

A small fantail, familiar to many as it is quite a confiding bird, in fact will take advantage of people while they walk by taking insects when disturbed. There is a great variance in colouring with this species. There are five subspecies in Australia with two occurring in WA: *preissi* in the lower South West and *albicauda* in the semi-arid and arid regions of central Western Australia. Besides subtle differences in tail length and tarsus length, which you are unlikely to see in the bush, what you may notice is that the inland birds *albicauda* have a distinct white tail and only the central tail feathers are dark grey, also the body is distinctly paler. This is one bird that may become a separate species in the future, with the South West bird's subspecies *preissi* moving through their territory travelling north in autumn and winter but *albicauda* remains in the same territory. The calls are slightly different the soft tinkering seems more grating with the inland birds. The calls of Grey Fantail are very soft with a high frequency, so people with hearing problems may not always pick up their calls. The South West subspecies *preissi* is found in the wetter regions particularly along riverine habitat, also Karri Forest, Jarrah Forest and open Wandoo Woodland. When one enters the outer Wheatbelt regions particularly the Sheoak areas, the white-tailed subspecies *albicauda* becomes common and is the dominant Grey Fantail in the central Mulga Region; it is nomadic moving to where weather conditions are favourable but may be sedentary in the Pilbara. **Nest:** They build the most exquisite nest which is moulded onto a thin branch of a horizontal limb with countless fine grass stems tapering to a point at the base of the nest giving it a wine glass appearance, the whole structure being bound by countless spiders webs. Clutch size 2 – 3. **Similar species:** None. **Status:** *preissi*, moderately common throughout the South West. *albicauda* moderately common locally, uncommon in the Mulga Region. Moderately common in the southern Pilbara uncommon in the Kimberley. **Where to find:** Throughout the South West, less so outside the south west.

Photo: Simon Nevill

Mangrove Fantail *Rhipidura phasiana* 14-15 cm

Very similar to the Grey Fantail but, as the name implies, a bird purely of the mangrove habitat. Found from the northern tip of Peron Peninsula north to the Kimberley mangroves. Note that, particularly in winter the Grey Fantail subspecies *preissi* can move through the coastal mulga belt but rarely feeds in mangroves. The Mangrove Fantail is much paler than the Grey Fantail especially a*lbicauda* which possibly does not reach the far west coast south of Carnarvon. The Mangrove Fantail has the palest upper body and also a paler very narrow chest band (see photograph). When seeing the bird for the first time, you will notice that it is distinctly smaller than Grey Fantail. Feeds in the mangroves and on the mud or sand. Its technique of dislodging insects is often to twist and turn through foliage disturbing insects and making them move. It occasionally will fly out over adjacent samphire hawking for insects away from the mangroves. **Nest:** The nest is similar to the Grey Fantail, although the tail of the nest not as long, and it is built straight onto a horizontal mangrove branch above high water mark. Clutch size 2-3. **Similar species:** Grey Fantail, see notes above. **Status:** Common in mangroves from Carnarvon to the northern mangroves of the Kimberley. **Where to find:** At Miaboolia Beach and Oyster Creek near Carnarvon and beaches north and south of Eundoo Creek and New Beach. Also northern Peron Point in Herald Bight. Giralia Bay. Near Onslow. Point Sampson. Balla Balla near Whim Creek. South of Roebuck Bay. Derby Mangroves. Wyndham Mangroves now hard to access.

Photo: Simon Nevill

Northern Fantail *Rhipidura rufiventris* subspecies *isura* 17 cm

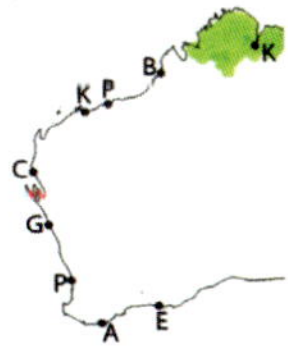

Found in variety of habitats, rainforest, mangroves, paperbark woodland, riverine or any dense canopy forest. What it does not like is open woodland and drier riparian woodland, preferring shade of the canopy forests. You will see them perched in their typical upright posture, different from a Grey Fantail that perch more horizontally with their tail cocked or fanned upwards. Flies less erratically than Grey Fantail and rarely fans tail and does not have jerky fanning movements of the tail. Northern Fantail has a wide grey breast band with white speckling, white throat, dark grey back, fine supercilium stripe, broad white tail bar tipped dark grey, white belly and vent. They cohabit the margins of mangroves with Mangrove Grey Fantail, particularly where vine thickets

Photo: Simon Nevill

or rainforest merge with mangroves but Mangrove Grey Fantail tends to seek food at a lower stratum even settling on mangrove roots to look for food. Call is similar to Grey Fantail but slightly more powerful and not as high pitched, its song is a sweet but jerky call up and down. It also gives a few very short sharp harsh notes with long pauses between 'kik……… kik. **Nest:** Typical neat 'wine glass' nest, tightly woven bark and grasses bound with spiders' webs, moulded around a branch, tail or stem looser not as fine as other fantails. Clutch 1-3. **Similar species:** Both Mangrove Grey Fantail and Grey Fantail are smaller, far more active in movements both swaying a fanned tail, breast band broader, Northern Fantail has a distinct jizz with upright posture and slightly larger head. **Status:** Moderately common and more wide spread than other species that are restricted to wetter rainforests and vine thickets. **Where to find:** Found in drier zones like Broome but still need denser woodland and vine forests. Found in lusher riverine habitats along the Gibb River Road and Kalumburu Road. King Leopold Range, particularly in Silent Grove. El Questro. Geikie Gorge. Kununurra and Ord River. Bungle Bungle. Throughout the northern Kimberley rainforest margins.

Arafura Fantail *Rhipidura dryus* 16 cm

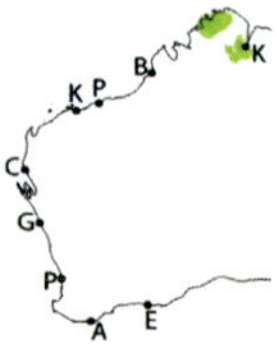

Only in recent years has this species been split from Rufous Fantail *Rhipidura rufifrons*. Johnstone et al 1990 had long seen this species as a separate species naming it Wood Fantail, but it was named by Christidis an Boles et al in 2008 using the geographic location 'Arufura', as the species is also located on the Lesser Sundas islands. Arafura Fantail is scarce in the Kimberley and mostly located in the mangroves and rainforest in the far north west Kimberley coast but more common in the NT. There have been sightings east of Kununurra but most records held by the WA Museum are located in mangroves from the north-west coast and the Cambridge Gulf but there may be some dispersal into lush riverine habitat. I have never managed to see this species in WA but have seen it in the NT whilst conducting tours. Arafura Fantail is very similar to Rufous Fantail, the variances are subtle based on wing, tail and bill length, breast markings be seen as different in the field, what will be obvious is when the tail is fanned as Arufura Fantail shows only rufous on the lower part of the tail in the rump region and white tail tips contrast well with the dark brown tail, as seen well in this photograph, where as Rufous Fantail the rufous colouring on the tail covers half the tail length well past the rump and the tail tips are diffused beige and hardly contrasts at all with the dark tail, this is quite noticeable in the field. Call is a very high pitched but soft delicate ' chip..chip…chip' also delicate ' seeup… seeup…seeup'. **Nest:** A neat cup moulded to a horizontal branch with a long wine glass tail, made of fine twigs or rootlets bound with spiders webbing. Clutch 2. **Similar species:** None. Mangrove Fantail totally different colouring a much faster more active feeder. **Status:** Uncommon in mangroves has been sighted in lush riverine vegetation away from coast, could be immature birds. **Where to find:** The north-west coastal belt of the Kimberley and rainforests close to the coast. Mangroves in the Cambridge Gulf.

Photo: Simon Nevill

Willie Wagtail *Rhipidura leucophrys leucophrys* 19–21 cm

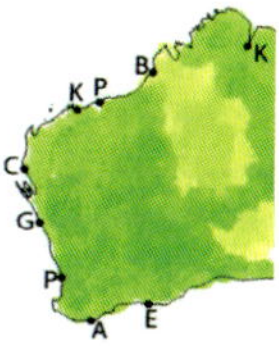

Well known to most people, this bird can be seen throughout Australia except in rainforests. Found in cities, towns, heath, woodland, desert dunes and rocky hillsides, it certainly has adapted well to all terrains and weather conditions and most human invasion and fearless of no bird except the *Accipiters.* This pied flycatcher is an active feeder constantly wagging its tail from side to side dislodging insects. As mentioned above can be found in all habitats but generally prefers areas where it can feed in the open and does not like enclosed thick woodland, as much of its feeding takes place on the ground. Its call is a sweet scratchy sound, often given well into the night particularly on full moon nights. The Noongars of the South West aptly called them the 'Chitty Chitty Bird' which sums up their call exactly. There are three subspecies, *leucophrys* in southern Australia, *picata* in northern Australia and *melaleuca* on islands in the Torres Straight and Pacific. **Nest:** Builds a compact neatly woven nest, often on a dead exposed horizontal branch of a fallen tree and out in the open. When the chicks are fully developed, it's amazing they never fall out of the small cup nest, which seems built for a much smaller bird. The nest is moulded onto the horizontal branch, being constructed from dry grasses woven with copious amounts of spiders webbing and lined with fine grasses. Clutch size 3-4. **Similar species:** Refer to Restless Flycatcher. **Status:** Very common throughout most of the state, less so the Nullarbor and arid deserts. **Where to find:** Throughout the state including towns.

Photo: Simon Nevill

Broad-billed Flycatcher *Myiagra ruficollis* subspecies *mimikae* 15.5 cm

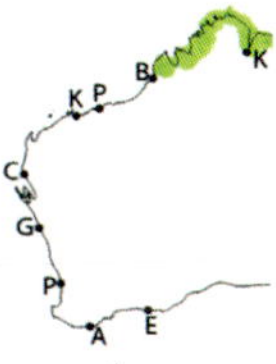

Mostly a species of the mangroves but it can be seen inland particularly around wetlands with extensive areas of paperbark but not as common further inland as the NT. When observed clearly, the broad bill is readily identifiable. It is an active feeder moving through the mangroves, perching and then darting out to take insects in flight or hover over mud to take them. Broad-billed Flycatcher has an all slate-grey back having a metallic sheen, with orange-rufous throat from below the eye to the mid breast, white under-body and vent, has typical flycatcher raised crest on rear of crown, fine pale orbital ring, barblets well defined at base of bill, broad bill, grey blue tail lighter below. Sexes are very alike, female upper-body slightly paler with less sheen, this is not easily seen unless very close, the orange-rufous throat is paler and you can pick this. Female lores have a pale margin near bill. Call similar in tone to other flycatchers with the most common call a rapid, repetitive 'chewee.chewee.chewee.chewee' also a longer drawn out grating call 'toweee….. toweee ….. toweee'. When it gives these calls it often quivers its tail from sided to side. **Nest:** Neat cup moulded to a horizontal branch, made of fine twigs or rootlets bound with spiders webbing but the outside often covered in bark or leaves to camouflage as they are often very exposed in the mangroves. Clutch 1-3. **Similar species:** Leaden Flycatcher can be found close to mangroves, the female is very similar to both sexes of Broad-billed Flycatcher, the only reliable or easy way to split these two species is the broad bill of the Broad-billed Flycatcher, which when seen below shows clearly. It's almost frogmouth in bill shape. It's not an ideal way to split but if you are in open eucalypt woodland which is a common habitat for Leaden, it would be extremely rare to see Broad-billed in that environment in WA. Leaden females colouring tends to be a bit more subdued. **Status:** Moderately common in large areas of mangrove. **Where to find:** Port Smith. Barred Creek. Cape Boileau.

Above: Female

Throughout the north west coast of the Kimberley where ever there are canopy mangroves, also in the Cambridge Gulf. Not found in the Kununurra area.

Photos: Simon Nevill

Leaden Flycatcher *Myiagra rubecula rubecula* 15 cm

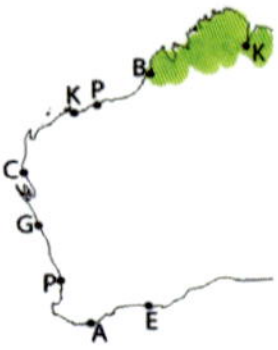

Can be found in variety of habitat, from rainforest, riverine to open woodland. In the Kimberley. I have found this species more often along river sides which is unusual, as on the east coast and NT is fairly common on ridge tops and open eucalypt woodland. The male is similar to a male Satin Flycatcher but that species does not occur here. There is no other flycatcher in WA with the same colour combination, however the female is almost identical to the female and male Broad-billed Flycatcher, see notes under Broad-billed Flycatcher. The male upper-body colouring is as the name suggests, a leaden colour or as some would say, a gun metal-blue. It has a leaden hood that stops abruptly at the middle breast, wings and tail leaden colour with a metallic sheen, has typical raised crest at rear of crown but when agitated crest is raised even higher, typical for most of the *Myiagra* genus. Female has a paler leaden crown and upper-body, orange-rufous breast and throat, grey under tail. Both sexes quiver tail frequently particularly when landing. Sit on perches and sally out to take insects, feed in canopy higher than most flycatchers. Many calls, one a high pitched repeated 'situ..situ..situ..situ' almost like a Restless Flycatcher, another sweeter, metallic high pitched contact call 'seeeat.seeeat. seeeat'. Unlike subspecies *yorki* in the south east of Australia that migrates north to PNG the WA and NT subspecies *rubecula* is sedentary. **Nest:** Similar to Broad-billed Flycatcher, a neat cup moulded to a horizontal branch, made of fine twigs or rootlets bound with spiders webbing but the outside almost always covered in bark that match perfectly the host tree often above 5m. Clutch 1-3. **Similar species:** Leaden Flycatcher can be found close to mangroves, the female is very similar, the only reliable or easier way to split is the broad bill of the Broad-billed Flycatcher which when seen below shows clearly. **Status:** Moderately common in riverine woodland or open eucalypt woodland. **Where to find:** Not common in Broome. Found on Cape Leveque. Willare Bridge. Geikie Gorge and along Fitzroy River. Silent Grove. Not so much in the outskirts of Kununurra but found in riverine woodland. Main stronghold central sandstone plateaux of the Kimberley particularly from Mt Elizabeth to the far north-west coat.

Above: Female

Above: Male

Photos: Simon Nevill

Shining Flycatcher *Myiagra alecto subsp melvillensis* 18 cm

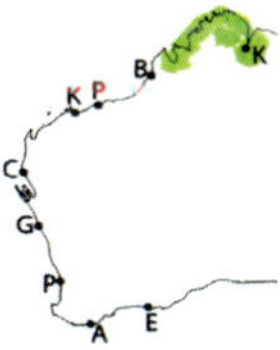

Found mostly in mangroves and rainforest in the Kimberley but also on a few more luxuriant riverine habitats, like the creeks at El Questro Station (habitat sadly over populated now with tourists). Their favoured habitat in the Kimberley is where mangroves link with rainforests. The male Shining Flycatcher is unmistakable with a metallic blue-black body with a distinct sheen, the only bright colour to be seen is when it opens its mouth while calling having a bright vibrant red throat which contrasts well with its black head. The female plumage colouring is totally different, having chestnut back and tail, the crown is a blue-black colour passing below the eye, the throat and under-body is white. There is no other flycatcher with this colour combination. It is an arboreal feeder but also feeds in lower strata more than many other flycatchers, often along creek lines where it hovers dislodging insects. In mangroves it seems happy to quickly hover over mangrove roots to the mud seeking small crustaceans

Above: Female

and insects. Its varied calls are unmistakable, from rapid whistle 'chew.chew.chew.chew' and a distinct buzzing, nasal type call 'zzzzzcheew....zzzzzcheew'. **Nest:** Similar to Broad-billed Flycatcher, a neat cup moulded to a horizontal branch, made of fine twigs or rootlets bound with spiders' webbing but the outside almost always covered in bark that matches perfectly the host tree often above 5m. Clutch 1-3. **Similar species:** No other species has the same coloration of either the male or female. **Status:** Uncommon in riverine woodland or open eucalypt woodland close to gorges. **Where to find:** The north-west coastal belt of the Kimberley and rainforests close to the coast. Derby. Along the Durham River. El Questro Station. In the mangroves between Wyndham and Home Valley. Along the river margins of the Ord River in Kununurra. Cambridge Gulf.

Male

Photos: Simon Nevill

Black-faced Monarch *Monarcha melanopsis* subspecies *pallida*

18 cm

Vagrant

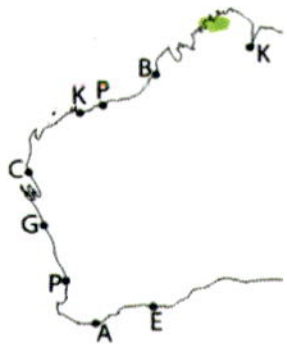

A migrant flycatcher that mostly migrates from the south-east and eastern coast to New Guinea but has been recorded coming to the Kimberley. Found mostly in rainforest in the Kimberley but also on a few more luxuriant riverine habitats. Their favoured habitat in the Kimberley would be evergreen vine forests and rainforest. Typical of the monarch flycatchers the male is very attractive with an orange under-body, grey mantle and head and black face. Female has the same orange under-body and grey head but lacks the black face and a dark bill not pale grey as in the male. **Similar species:** No other species has the same coloration except the Black-winged Monarch and that is a very remote chance it would overpass from its wintering ground of New Guinea. It has the same coloration except has a black on the wings not dark grey, also the female. **Status:** Rare possibly in Kimberley rainforest. **Where to find:** The north-west coastal belt of the Kimberley and rainforests close to the coast.

Photo: Simon Nevill

Lemon-bellied Flycatcher *Microeca flavigaster* subspecies *tormenti* 13 cm

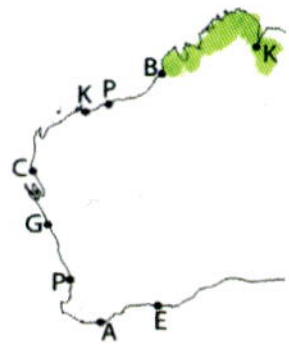

There are two subspecies of Lemon-bellied Flycatcher in the Kimberley, *tormenti* is a mainly a species of the mangroves, although is found inland mainly in paperbark wetlands. Lemon-bellied Flycatcher subspecies *tormenti* has a grey-brown back, pale under-body and no white in outer tail feathers, just slight pale white tips to outer 3 tail feathers. Lemon-bellied Flycatcher subspecies *flavigastor* is easier to recognize, as it has a pale yellow under-body with white throat. Jacky Winter subspecies *pallida* which is also found in the Kimberley is very similar to the brown-tailed form of the Lemon-bellied Flycatcher subspecies *tormenti*. In the field it is not easy to pick up the differences. Lemon-bellied Flycatcher subspecies *tormenti* has an all brown tail and lacks the white outer tail feathers of the Jacky Winter, easily seen when Jacky Winter does its typical alighting tail wag, which Lemon-bellied do not do. There are other anatomical difference but not really reliable for field observation. Some say calls are very different but both have the same pitch and tone, with slight differences in variation. Jacky Winter is certainly a more powerful call. Lemon-bellied Flycatcher subspecies *tormenti* is softer than *flavigaster* 'terwit..terwitittoo'. Jacky Winter subspecies *pallida* is a much stronger and rapid 'chewee. chewee. chewee' also shorter 'chip.chip.chip.chip'. **Nest:** A neat cup made of fine twigs with spiders webbing, placed on horizontal fork, bigger than Jacky Winter nest. Clutch 2. **Similar species:** As discussed above, Jacky Winter. **Status:** Uncommon in mangroves, has been sighted in lush riverine vegetation away from coast, some could be immature birds. **Where to find:** The north-west coastal belt of the Kimberley in mangroves and adjacent woodland. Mangroves in the Cambridge Gulf.

Photo: Simon Nevill

Jacky Winter (Brown Flycatcher) *Microeca fascinans* subspecies *assimilis* 13 cm

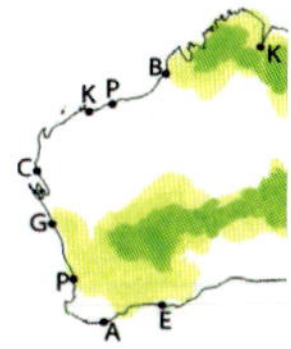

There are three subspecies in Australia and one in New Guinea. Two occur in WA. The subspecies assimilis range takes them from the Stirling Range in the south west north to Billabong Road House adjacent to Meadow Station, then south east down to Kalgoorlie and Dundas NR and east across through the southern boundary of the Nullarbor. They also occur in the southern Great Victoria Desert. However their main populations occur in the Transitional Woodland and some of the larger Wheatbelt reserves where there is open woodland particularly Wandoo. They do not occur in the thicker Jarrah and Karri Forests or the Swan Coastal Plain or southern coastal districts. Jacky Winter require open areas to feed and hawk for insects but must have trees adjacent to the open areas and are particularly common in also in Swamp Yate woodland (*E. occidentalis*). The subspecies *pallida* occurs in the Kimberley, having more white in the outer tail feathers. There is no overlap in the two subspecies range. They call strongly in the breeding season from high up, normally on dead exposed branches, with their well known high-pitched rapid 'peter, peter, peter' call. Jacky Winter will always shake their tail when alighting after flight, revealing the white in their tail. Often hovers for insects although flight is not as prolonged as say a Restless Flycatcher. **Nest:** The Jacky Winter's nest is one of the smallest, shallowest cups of any passerine, placed mostly high up on a horizontal branch where a branch forks. The nest is constructed of fine grasses and decorated on the outside with slivers of bark all tightly bound with copious amounts of spiders' webbing. Clutch size 2-3. **Similar species:** Possibly female Hooded Robin, a bulkier bird with no supercilium stripe and white on the tail only in the middle of the outer tail feathers, it has an all grey hood coming down the breast while the Jacky Winter has a white chest and throat. Hooded Robin never wags its tail like the Jacky Winter. Calls are very different. **Status:** Uncommon to moderately common in open woodland in all locations it occurs. **Where to find:** Wandoo Woodland in the eastern Darling Range. Dryandra NR. Stirling Range NP. Dongolocking NR. North Karlgarin NR. Throughout the Transitional Woodland. Woodlands in the Great Victoria Desert. Open woodlands often adjacent to water in the Kimberley.

Photo: Simon Nevill

Scarlet Robin *Petroica multicolour* subspecies *campbelli* 12.6 cm

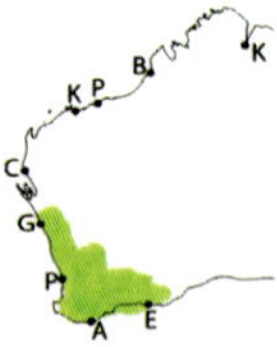

There are four subspecies in Australia and at least 13 on Pacific Islands. In Western Australia there is only one *cambelli* which is slightly smaller than the eastern mainland subspecies *boodang* and the Tasmanian subspecies *leggii*, which has a longer bill. It is a robin of the wetter South West although can be found in the lower south west Wheatbelt. Most prolific in the larger forests of the Karri, Tingle, Jarrah Also along the southern coast. In Dryandra Woodland N R where the dominant tree is the Wandoo, it still selects those areas that mostly have Marri or Jarrah. When larger stands of Jam trees (*Acacia acuminata*) occur adjacent to Jarrah or Wandoo, Red-capped Robins hold territory, but in the Jarrah, Scarlet hold a territory but I have found that the Scarlet will dominate on the margins and confront the Red-capped Robin which retreats back into the Jam woodland. The Western Yellow Robins do not have the same conflict, as they prefer the open Wandoo Woodland with often little or no understorey, which is not the preferred site for Scarlet Robins and anyway totally different robins. Like most robins, the male is a stunning bird with its distinct white forehead spot, set against the black hood. The breast is scarlet as the name implies, with white flanks and vent, white to the outer tail feathers and white on the shoulders on the black wings. The female breast lacks the intense scarlet, being a more subdued pale scarlet over a smaller area the scarlet only going to the lower chest. The hood and back is a dark grey-brown. Has a small white patch on the forehead above the bill and white eye ring broadest at the rear of the eye. The call is a delicate tinkling, often uttered high in a eucalypt. **Nest:** They can nest in low shrubs like sheoak but mostly higher in the mid strata of the taller eucalypts like Jarrah and Marri. The nest is a neat thick walled open cup lined with loose bark bound together with spiders' webbing. The inner lining is woven with fine grasses and soft feathering woven into sides. Almost invariably placed in the fork of a horizontal branch. Clutch 2-3. **Similar species:** Red-capped Robin is smaller and a red patch on the forehead rather than a white. Its call is totally different. The female Red-capped Robin has a pale scarlet patch on the forehead. The Scarlet has a smaller white patch above the bill. **Status:** Common in the Darling Range and Karri and Tingle Forest, less common in the Wandoo. **Where to find:** Throughout the Darling Range. Dryandra Woodland NR. Stirling Range. Throughout the Margaret River and Augusta region all the way along the coast to Esperance hugging the coastline.

Above: Female

Photos: Simon Nevill

Red-capped Robin *Petroica goodenovi* 11 cm

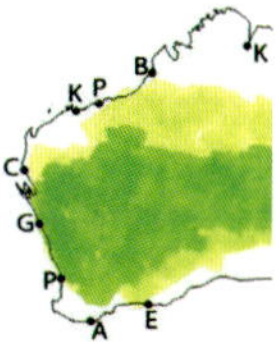

A small robin and it certainly is an attractive one, with its bright red forehead spot, set against the jet-black hood and the bright red breast to match. It prefers more semi-arid to arid areas in WA, than the Scarlet Robin, avoiding much of the lower South West forests. Its range almost mirrors the limit of the Scarlet Robin extending north of the Porongurup's all the way to the Pilbara and southern parts of the Kimberley and east through all the desert regions except the central Nullarbor. There is a marked movement of some South West birds to over winter in the north. They can be found throughout much of the Wheatbelt, preferring stands of sheoak woodland where they also nest. Further north they become the dominant robin as one enters the Mulga. They are not so common in the Transitional Woodland, there the Yellow Robin is the dominant robin. On Rottnest there is a remnant population that is quite common on the island. They often give themselves away with their so called telephone call 'drrrrrit drrrrit.......drrrrit drrrrit........drrrrit drrrrit' reminiscent of a phone ringing. If you come too close to its nesting area it will give rasping clicks and flick its wings repeatedly. **Nest:** The nest is a small compact cup of tightly woven grasses and bark heavily moulded with spiders' webbing and the outside bark strands matching the branches it is attached to, normally a fork in the centre of an acacia bush. Clutch size 2-3. **Similar species:** Refer to Scarlet Robin. **Status:** Moderately common in the Wheatbelt.

Common in the Mulga and less so the deserts. **Where to find:** Rottnest Island. Eastern side of the Darling Range. Most large Wheatbelt reserves particularly with good stands of sheoak. Found throughout the Mulga and desert regions and gullies in the stony country of the north.

Photo: Simon Nevill

Hooded Robin *Melanodryas cucullata* subspecies *westralensis* 16 cm

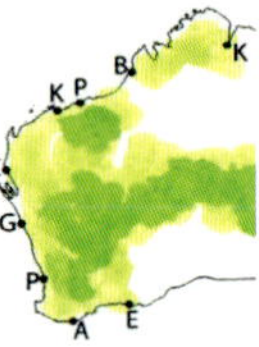

There are four subspecies in Australia with two occurring in WA, subspecies *westralensis* covers most of WA from the Wheatbelt in the south through the Pilbara and across through the central desert regions into SA and the NT subspecies *picata* is found in the Kimberley and typical of Bergmann's Rule, the subspecies is smaller than those in the south. Hooded Robins have an unusual distribution in the South West. They can be found quite low down in the South West in areas like Boyup Brook but then patchily through the Darling Range wherever small patches of Wandoo occur as in areas like North Bannister. They can be found on the eastern side of the Darling Range to the east of Yarra Road between Brookton Highway and Great Eastern Highway. I feel that the population in the Wheatbelt may be declining. In the 1970s it was not difficult to see 5 to 6 pairs of Hooded Robin in the main block in Dryandra Woodland Reserve but by the early 1990s it was difficult to find two pairs. By 2000 only one pair, then for a while none and only recently has a pair returned. The male has a complete black hood with a white under body, black wings with a white bar, and a black

Above: Female

Photo: Simon Nevill

tail. The female has a dull grey hood and back with darker grey wings and the under-body is grey-white. Its call is a short series of semi-harsh notes 'churpede churpede churrrp' often uttered from a high dead branch. It feeds like many robins by sitting on lower branches of trees and waiting before pouncing on prey. **Nest:** The nest is a larger version of the Red-capped Robin's nest but more loosely structured on the outside and in more diverse sites, sometimes placed behind peeling bark or wedged in the fork of a mulga tree. Clutch size 1-2. **Similar species:** None. **Status:** Uncommon in the South West. Moderately common in the mulga and uncommon in the Kimberley. **Where to find:** Reserves like Tutanning and Dryandra. Near Perth in the Wandoo Conservation Park. Easiest to find in the mulga sheep station country. Moderately common in the central ranges. In the Kimberley, patchily distributed in dry open areas.

Western Yellow Robin *Eopsaltria griseogularis griseogularis* 14.5 cm

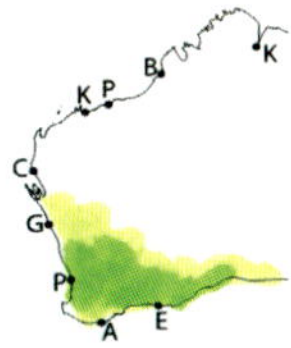

There are two subspecies in Australia and both occur within WA. The differences are not so much in terms of plumage variations but mainly in size of wing, bill and tarsus. The subspecies *griseogularis* can be found within a general line from Lancelin in the north down to Northam and to the coast near Albany. The subspecies *rosinae* can be found outside the lower South West beyond a line from Lancelin north to Tamala Station south of Shark Bay, then a line in a south easterly direction via Coolgardie and down to the Eucla coast. It has a longer wing and tail than the lower South West subspecies *griseogularis*. The Western Yellow Robin requires open woodland where it can perch and drop down to take insects. It can be found throughout the Wandoo Woodland often where a grove of young saplings are growing and will fly from trunk to trunk looking for insects. The sexes are alike. Has a lemon-yellow rump, vent and lower belly, white throat, grey crown and mantle with dark grey wings, faint white supercilium stripe and a stout bill. Has a great variety of calls from its agitation call, a rasping 'chip chip chipcheeeee', to the delicate high pitched 'pip...pip...pip...pip' of its contact call. **Nest:** It builds an exquisite well camouflaged cup nest with trailing bark that matches the trunk and branches where the nest is located. The internal nest is made of tightly woven fine grasses and lined at the base with feathers. Clutch size 3. **Similar species:** None. **Status:** Locally common in Wandoo. Moderately common through the Wheatbelt. Common in the Transitional Woodland east of the Vermin Fence. **Where to find:** Darling scarp at Bungendore NR. Julimar Forest. Wandoo Conservation Park. Boyagin Rock NR. Dryandra Woodland NR. Stirling Range NP. Peak Charles NP. Dundas NR. Mt Jackson region. Along the Lake King and Hyden roads to Norseman. Mt Jackson region. Mallees along the southern Nullarbor.

Photo: Simon Nevill

White-breasted Robin *Eopsaltria georgiana* 15.5 cm Endemic

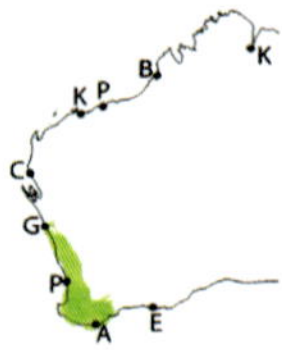

Found from Lake Indoon near Eneabba south to Lancelin. Then a break till the northern end of the Darling Range in Helena Valley and south to the heavily forested region of the lower South West and east to the Gairdner River on the eastern side of the Fitzgerald River National Park. A bird that requires densely vegetated or a lush understorey, as in the Karri Forest. Surprisingly, it still occurs in dense vegetation gullies in the semi-arid regions of Hill River area up to Lake Indoon; may be relic populations from when the coastal belt was wetter. Sexes alike, with a grey crown, back and lores, white supercilium stripe and white throat and under-body. In flight show distinct white wing bars. Like most robins they have a variety of calls including a musical soft whistling call but the commonest call is a contact territory call of a series of single harsh notes 'cheuw.....cheuw.....cheuw'. **Nest:** The nest is similar to a Scarlet Robins but has often ferns or other plant foliage woven into the sides disguising the nest. Placed in a fork often only a metre off the ground. Lined with fine grasses and soft bark. Clutch normally 2. **Similar species:** Jacky Winter, but territories do not overlap. **Status:** Common South West. Uncommon in riparian habitat in the Darling Range. Uncommon in dense thickets between Jurien and Eneabba. Common in the Karri and south coast. **Where to find:** Near Perth. Wungong Valley. Upper Bickley Brook. Piesse Brook. Jarrahdale. Common in the Karri Forest and throughout the coastal scrub from Margaret River region to Waychinicup NP. Porongurup NP.

Photo: Simon Nevill

Mangrove Robin *Eopsaltria pulverulenta* subspecies *cinereiceps* 15 cm

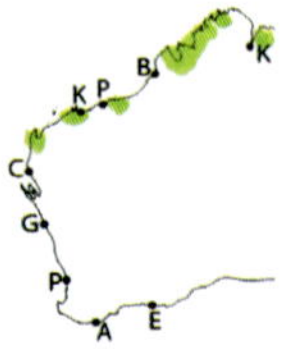

Only found in mangroves. Very territorial. Each family group defending its feeding area from other Mangrove Robins. It has a fragmented range in WA, sadly having gone from several locations. There was a small population on the north west of the Cape Range near Exmouth I knew many years ago but alas I think they have gone. I knew of several pairs at Cape Keraudren but I have had no luck over the last few years so we see changes to populations in our own life time which is always sad knowing the time span of this earth of ours. There are populations at Giralia Bay. Point Sampson. Balla Balla. Port Hedland. Then a long gap at 80 Mile Beach until Cape Leveque and Derby. Found in several locations throughout the mangroves of the north-west and north of the Kimberley where the populations are healthy, also in the mangroves of the Cambridge Gulf. In coloration, it is similar to the south-west White-breasted Robin, having a dark grey crown and mantle, black-brown wings, brown tail with two central bars to the outer tail feathers, white throat and under-body, most robust bill of all robins with a hook end, barblets at base of bill, black eye and legs. Clings to the side of mangrove stems waiting patiently like all robins for movement of insects or small crustaceans. Feeds mostly on the ground or up to 2 m rarely higher. It is said that they prefer *Rhizophora* forests but I have also seen them in different lower mangroves species. Call is a faster 'choowupi. Choowcheee' and also a slow drawn out mournful low-noted 'chooooow .....cheee'. **Nest:** Nest is placed in a vertical fork fairly deep cup camouflaged with leaves intertwined with grasses or rootlets. Clutch 2-3. **Similar species:** None in the mangroves. **Status:** Sometimes common but often uncommon depending on extent of mangroves. Locally common in a few locations in the Pilbara but generally scarce. Greater populations off the north west Kimberley coast. **Where to find:** All the locations above. A boat trip on the north west coast would almost guarantee you seeing this species on a bird watching tour. Derby one of the better locations to find or south at Point Sampson.

Photo: Simon Nevill

Buff-sided Robin *Poecilodryas cervinivetris* 15 cm

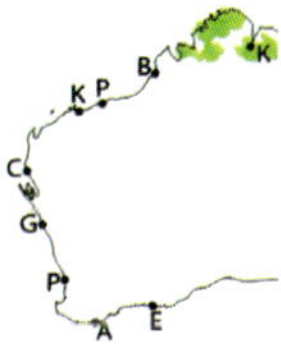

Buff-sided Robins require a distinct type of habitat normally thick riverine forest, not necessarily rainforest. In the north west sandstone country, even the smallest patches of monsoonal forest sometimes only one hectare in size will still hold a pair of Buff-sided Robins. They can be found in many disjunct locations throughout much of the Kimberley, although not found in the drier pindan country of Broome and Cape Leveque. Mostly found in the vine thickets and edges of thicker rainforests. Common on the banks of the Ord River in Kununurra. Annie Creek and upper Fitzroy on Mornington Station. Found in riverine thickets, as on the Behn River south of Kununurra and several other creeks leading into Lake Argyle. Buff-sided Robin has dark brown back, crown and tail with white tip, White supercilium

Photo: Simon Nevill

stripe, black-brown eye mask, white throat and under-body, white chevrons on wings, the diagnostic feature is buff on flanks and vent which the east coast White-browed Robin lacks and population do not connect. Call is very distinct, powerful but not musical, a harsh repetitive 'tittytoo… chook!… tittytoo…chook!' also a plaintive 'sissy.see..chew….sissy.see.. chew' **Nest:** Nest is an untidy loosely woven nest, with loose twigs spreading out from nest. Clutch 2-3. **Similar species:** None in their habitat. **Status:** Locally common but can be just one pair occuring in small singularly lush area. **Where to find:** Geike Gorge, Upper Fitzroy on Mornington Station. Charnley Station. Mt Elizabeth Station. Lower Pentecost River. Common in Kununurra and around Lake Argyle where heavier riverine vegetation occurs.

Southern Scrub-robin *Drymodes brunneopygia* 20.5 cm

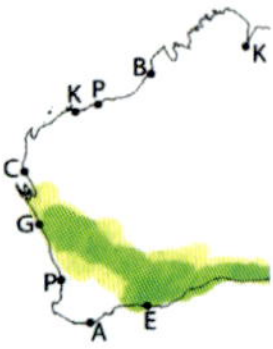

A ground-dwelling thrush-like bird, with a long tail that it will cock at times, particularly when agitated. Has a grey body with a chestnut rump and slight chestnut tinge in the tail feathers. Has a distinct white eye ring broken by a small black vertical stripe. Has white chevrons on the wing. Long legs. Has a soft contact call rising at the end of the two note call 'cheeeew... we... cheeeew....wee' also a warning scolding pitching call for intruders 'pipipitcheeee......pipipitcheeee'. It is an inquisitive bird and it is not hard to imitate its contact call; if you sit quietly and just give a few extended 'picheeeee' calls, it will come and check you out, obviously do this once only. It ranges from Peron Peninsula down through the transitional Woodland to Coolgardie and south east to Dundas NR Along parts of the southern coast through the Fitzgerald River NP and then north through to the Stirling Range. Wherever it occurs, it requires thicker vegetation to 2 metres high. In the southern region it is mainly in dense thickets of melaleuca under mallee. In the Wheatbelt say at North Tarin Rock, it is not only in melaleuca thickets but also in low Kwongan heath as there is limited habitat for them to survive in. Below Mc Dermid Rock east of Hyden it is found in broom bush (*Melaleuca uncinata*). Above the cliffs at Quoin Head in the Fitzgerald River NP. it occurs in dense wind swept thickets of Lemann's Banksia (*Banksia lemanniana*). In the northern regions just north of the Murchison Bridge it occurs in heath with banksia woodland. In the Kalbarri NP. in heath with an over storey of mallee. **Nest:** The nest is always placed on the ground built into a depression excavated by the bird. It is a neat cup made of concentric fine twigs but with large twigs woven into the outside. Often placed at the roots of a mallee or mulga tree and invariably laid between a fallen branch or up against a tree root. Clutch size 1.

Photo: Simon Nevill

Similar species: None. **Status:** Moderately common through most of its range. Scarce to extinct in many of the remaining small western Wheatbelt reserves. **Where to find:** All the locations mentioned above. Very common in the southern part of the Fitzgerald River NP.

Black-faced Cuckoo-shrike *Coracina novaehollandiae* 30 -38 cm

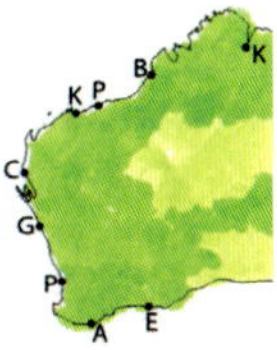

A medium sized Cuckoo-shrike with a long tail and long wings found throughout the South West. Generally prefers drier open woodlands but can be found in the lower South West although tends to avoid the dense Karri Forest and Jarrah Forest. Feeds mostly in the canopy of woodlands and rarely feeds on the ground, unlike the Ground Cuckoo-shrike. Feeds on wide range of insects, including caterpillars, beetles and grasshoppers but also will take fruit particularly in the semi-arid regions where native figs grow. The sexes are similar only the immature birds can be recognised as different. They lack the full black facial mask but instead have a black broad eye stripe and the throat has not developed the pure black but has grey mottling that runs all the way down to the belly. Black-faced Cuckoo-shrikes have the interesting habit when landing after a flight, of always shuffling their wings. There is a distinct movement after the breeding season from the South West in mid January to March to the northern regions, mainly the immature birds. Call is rather cat-like rasping. **Nest:** They nest mostly above 5 metres on a horizontal branch where it forks, and build a shallow nest, constructed of fine twigs and grasses and on the outside bound together with spiders webbing. Clutch size 2-3. **Similar species:** Possibly Ground Cuckoo-shrike which is a bigger bird and deeper bodied. Has black wings that contrast with pale grey upper body, barring on the under body and rump. No facial mask. The Ground Cuckoo-shrike's flight is different as it is a direct flight that does not undulate. White-bellied Cuckoo-Shrike in the Kimberley could be confused: smaller bird. No large black facial patch just small black patch on lores. There are two subspecies in WA: *subpalllida* in the Pilbara region has a paler crown and upper body, paler under body than *melanops*, which occurs throughout the rest of the state. Status. Common throughout range. Moderately common the Mulga Region. **Where to find:** South West. Pilbara. Kimberley. Less so the desert regions.

Photo: Simon Nevill

Ground Cuckoo-shrike *Coracina maxima* 32-36 cm

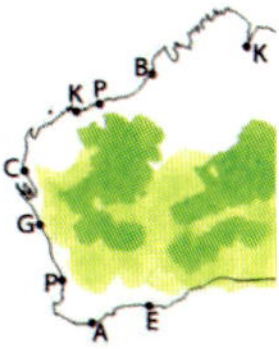

This is the largest of Australia's Cuckoo-shrikes, with a long tail, long pointed black wings, a grey head and upper back, distinct barring on the breast and rump. Yellow eye stands out from the grey head when close. A ground dwelling bird that walks on open ground with a jerky head movement searching for large insects, including spiders and caterpillars. Has a preference for feeding on stony ground often between spinifex or hummock grass.

Normally a very nervous bird, when approached will fly off at a steep angle in the same manner and at the same speed as magpies, with a straight fast flight unlike other cuckoo-shrikes that are slower and undulate their flight. When it does fly, it shows the distinct contrast between its long black pointed wings and a very pale grey back, with paler white and black barred rump and long black tail making it easy to identify. It also utters its warning two-noted contact call, a high pitched 'peep peep…peep peep' as it ascends. Often feeds in family groups of 5-6 birds. Found throughout the Mulga Region, not a common bird but becomes more prevalent in the central Mulga Region centred on Meekatharra region and out to the desert regions and up to the Pilbara. I have seen it a few times in the north eastern Wheatbelt near Chiddarcooping NR and in ploughed fields north of Bonnie Rock and also north of Beacon. Many times in the mulga and deserts. Nests: Normally between 5-10 metres up in a fork not necessarily on a horizontal limb like Black-faced Cuckoo-shrikes. The nest is deeper and more substantial, made from fine grasses and strands of bark bound with lots of spiders' webbing, often still with spiders egg sacs surrounding the outside. Clutch size 2-3. **Similar species:** Black-faced Cuckoo-shrike (refer notes Black-faced Cuckoo-Shrike). Calls quite different. **Status:** Uncommon throughout the Mulga Region. Eastern Pilbara. Deserts and Nullarbor. Moderately common in the southern Kimberley. **Where to find:** Anywhere in the Mulga Region, can be seen as soon as one clears the Transitional Woodland. Seen more so in the southern deserts and central ranges.

Photo: Simon Nevill

White-bellied Cuckoo-shrike *Coracina papuensis* subspecies *hypoleuca* 18.5 cm

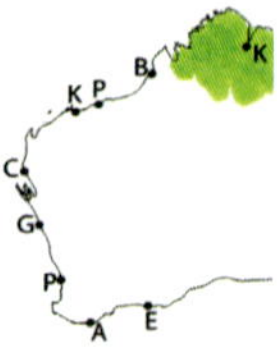

Previously called Little Cuckoo-shrike. There are five subspecies in Australia with only one in WA, namely subspecies *hypoleuca*. There is a clinal variation in colouring from the Kimberley birds to those of the lower south-east subspecies *robusta*. Typical of Bergmann's Rule the birds of the tropical north are smaller, colouring differences are compounded by the fact that there are differing morphs. The pale morph and by far the most common, is smaller and has an all white under-body not grey, facial lores are narrower than *robusta*, mantle is lighter. Dark morphs have a complete black hood that merges with the white under-body and mantle is darker grey. Found mostly in the sub-tropical areas of the Kimberley but go to the edge of the semi-arid regions such as Broome and Halls Creek. Found in a variety of habitats but common in riverine areas, gallery rainforest, also found on sandstone plateaux country and open eucalypt woodland, common in mangroves and adjacent paperbark woodland, tends to occupy wetter habitats than Black-faced Cuckoo-shrike. Mostly feeds in canopy but also seen perching looking for prey and then dives from perch, like most cuckoo-shrikes will shuffle wings when just alighting on a branch. Staple food is insects, including spiders, grasshoppers and beetles. Will also eat fruits. Main call different from Black-faced Cuckoo is a harsh, rasping call made between long pauses ' keesick….. keesick……keesick'.

Photo: Simon Nevill

Nest: Shallow broad cup can be in a vertical triple stemmed cluster of branches or on a fork of a horizontal branch, made of woven bark and lined with grasses and fine twigs. up nest of twigs and grasses often placed alongside base of bush. Clutch 1-3. **Similar species:** Black-faced Cuckoo-shrike particularly immature birds. Firstly call is very important as very different. Majority of adult birds very different mainly facial markings with Black-faced having an all black face merging into breast whereas White-bellied has black eye stripe on the lores. Black-faced is not only bigger but is more robust so the jizz of both birds different. **Status:** Common in the sub-tropical parts of the Kimberley less common in the semi-arid zone. **Where to find:** Central Kimberley easy to seealong Gibb River Road and Kalumburu Road and very common around the Kununurra region. Range extends from Broome in the west to Halls Creek in the south and east to the WA border and north to Kalumburu.

Varied Triller *Lalage leucomela* subspecies *macrara* 19.5 cm

Photo: Simon Nevill

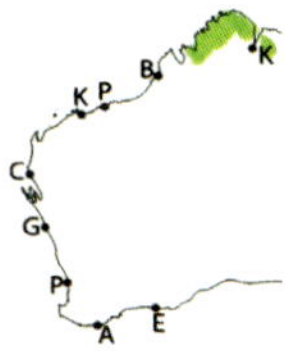

A sedentary species of the sub-humid zone found mostly in canopy rainforest forest and monsoonal vine thickets in the winter months but in the wet will move into adjacent woodlands and mangroves in search of fruiting trees. Slightly bigger and thicker bodied than the White-winged Triller. Most remnant rainforest pockets larger than a couple of acres will often have a pair of Varied Trillers. Mainly a canopy feeder feeding on fruits, insects and seeds. The male's upper-body, crown and upper wings are black with white barring on wing coverts, white under-body with heavy grey-brown scalloping, vent orange-rufous, white superscilum stripe contrasted with black stripe. Female is similar but instead of black back has a dark grey-brown most important difference, and barring much heavier on under-body and white is tinged with rufous-brown, supercilium eye stripe dull cream. So basically, even though patterning almost alike the female colouring over all looks browner, particularly the upper-body and crown, lacks the strong black and white contrast. Call is often the first thing you will know of the bird's presence, their contact call is what I call a reeling call, the same notes repeated regularly 'drreeel....drreeel....drreeel'. **Nest:** Very shallow cup, moulded with spiders webbing to a fork on a horizontal branch, like a larger Jacky Winter's nest lined with fine twigs and often the external sides camouflaged with bark and greenery. Clutch 1. **Similar species:** White-winged Triller can be found in the same region but not same habitat, needs open eucalypt woodland. Major differences, White-winged Triller has no white supercilium stripe, no black eye stripe to contrast with white supercilium, plain white on under-body not barred, larger white shoulder patch and lastly smaller, slimmer bird. **Status:** Locally, moderately common but habitat of rainforests and vine thickets is limited. **Where to find:** Mitchell Plateau, many rainforest pockets on far north west coast and the hinterland as in the Sale, Calder and Prince Regent Rivers.

White-winged Triller *Lalage tricolor* 17-19 cm

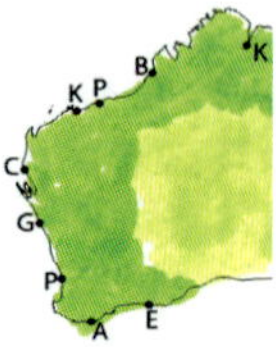

A small, slim black and white cuckoo-shrike. The male has a black cap, neck and mantle, a white throat and under body and white shoulders on black wings. Female has a brown crown, neck and mantle, black and white in the wings and a slightly mottled fawn on white throat and under body. In non breeding plumage, the male is almost identical to the female but rump is greyer and mantle darker brown with more black and white in wings but difficult to split. White-winged Trillers are a passage migrant moving into the South West in the spring to breed and more nomadic in the Mulga Region, travelling to areas that have had favourable rains where populations can be quite dense. When moving into areas to breed the male becomes really active calling loudly with low display flights between trees. The call is very powerful, a strong rapid 'chur chur chur chur chur' that can last a good 15 seconds or more. In the Mulga Region when an area has high densities of breeding birds after good rains, the triller may nest close to other birds and the calls of trillers, Pied Honeyeaters and Chiming Wedgebills asserting territories can be very noisy. Their nest is flimsy, normally only 1 – 2 metres off the ground in the mulga and higher up in trees in the South West. They prefer an open understorey below trees in the breeding season and do not enter the deeper south west in any reasonable numbers having a preference for the Wheatbelt and eastern southern coastal areas. Feed mostly on the ground. Two subspecies occur in Australasia only one on the mainland tricolor. **Nest:** The nest is similar to a Black-faced Cuckoo-Shrike's nest, smaller in diameter but proportionally deeper. The outside is loosely bound with spiders webbing. Clutch size 2-3. **Similar species:** Pied Honeyeater, which is slightly smaller and has

Photo: Simon Nevill

a complete black hood and mantle, black coming down to the middle of the chest. Distinct blue eye ring at base of eye. Much longer, finer, down-curved bill. White rump not grey-white. Calls totally different. **Status:** Can be very common in the Mulga in the breeding season, disperses after breeding. Moderately common in the Wheatbelt. Uncommon on the Swan Coastal Plain south of Perth. Scarce in the heavy forested lower South West. **Where to find:** Wandoo Conservation Park (uncommon). All Wheatbelt reserves. Northern Stirling Range NP. Throughout the Mulga, less so in the Transitional Woodland. Found in the Pilbara and the Kimberley.

Common Cicadabird *Coracina tenuirostris* subspecies *melvillensis*

25 cm

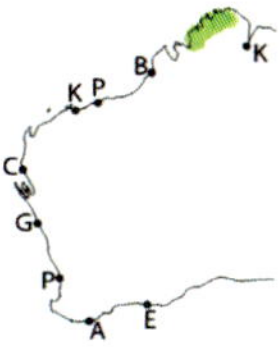

Smaller and slimmer than a White-bellied Cuckoo Shrike, a bird of the rainforest and heavy canopy woodlands and paperbark woodland of the sub-humid zone. We know that the east coast subspecies *tenuirostris* migrates north to the tropics from the south-east in our winter but its not entirely clear if Kimberley birds are mostly migratory or sedentary. What we do know is they are scarce in the Kimberley, becoming more common as you move east via Darwin and relatively common in the wet tropics of north Queensland. One of the problems regarding sightings is that few people get to the far north east of the Kimberley in the wet season, although there are few records in late winter mainly from rainforest patches on the Mitchell Plateau area. The call is unmistakable and is normally how one locates these birds as they mostly feed in the canopy of thick rainforest foliage. The complete male body is slaty-blue-grey, with a darker facial patch and dark edges on wing, in flight shows a pair of white bulls-eyes contrasting with dark grey wings. Female totally different having a dark-brown upper-body, beige under-body heavily barred with horizontal dark brown stripes, has cream orbital ring. The call as the name implies is like a cicada, a buzzing call. The subspecies *melvillensis* has slower more drawn out buzzing call than the east coast subspecies *tenuirostris* a grating buzz 'zeeee…zeeee…zeeee…zeeee' it also has a sharp clicking warning call almost like one of the calls of a Buff-sided Robin. Food is the same as Cuckoo-shrikes taking insects, fruits and seeds. **Nest:** Shallow cup can be in a vertical stemmed cluster of branches or on a fork of a horizontal branch, made of woven bark and lined with grasses and fine twigs, often camouflaged with rainforest materials like lichen and bark. Clutch 1. **Similar species:** None in region, although female could be confused with female Varied Triller, which is a smaller bird having a black-grey back and strong black and white wings with stronger black eye stripe contrasting with white supercilium stripe. **Status:** Scarce in the far north west Kimberley. **Where to find:** Rainforest patches on north west islands, possible in the few rainforest patches on the Mitchell Plateaux and along the Sale. Prince Regent and Roe River.

Photo: Simon Nevill

Australasian Figbird *Sphecotheres viridis* subspecies *ashbyi* 28 cm

Not a common bird in the Kimberley, patchily distributed through parts of the central Kimberley and remoter parts of the north-west areas. Its distribution is governed by the location of fruiting plants, which form a major part of its diet. Over the many years of travelling to the Kimberley I have often found this species in Windjana Gorge and few other areas and it is simply due to the fact that many ficus trees growing in the gorge. The subspecies *ashbyi* found in the Kimberley and the NT has marked contrast in coloration to the lower east coast subspecies where the under-body is bright yellow whereas the Kimberley birds have an all green under-body and upper-body. The subspecies merge in northern Qld.

Photo: Simon Nevill

Similar species: Adult males could not be confused with any other species but both females and immature males could be confused with Olive-backed Oriole. Plumage colouring is amazingly similar but this is where the jizz of both species plays a vital role in splitting species, The Figbird is a stockier bird with a shorter tail. Orioles look longer and slimmer. It's not easy to see but Figbirds have a jet black eye whereas Orioles have a dark red eye. Calls differ markedly. **Status:** Uncommon in the Kimberley. **Where to find:** Parts of the central and northern Kimberley particularly north of the Gibb River Road.

Olive-backed Oriole *Oriolus sagittatus* subspecies *affifinis* 28 cm

Photo: Simon Nevill

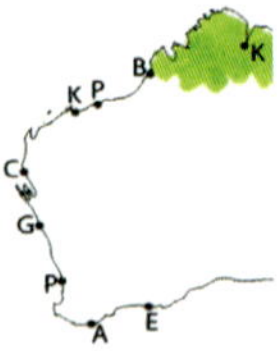

A much wider distribution than the Yellow Oriole, found in open eucalypt woodland but mostly along lusher, riverine habitat as it is a fruit eating bird, it will however frequent the edges of rainforests. Moderately common from Port Smith north of 80 Mile Beach across to Kununurra and south to Fitzroy Crossing and Halls Creek and north to Kalumburu. Basically sedentary in the Kimberley although its said that there is some movement away from the coastal areas with the onset of the wet. Olive-backed Oriole has dull olive-green to dull olive-brown back, under-body white with dark green-brown flecking, dull pinkish bill, eye is fairly bright red, tail grey-brown tipped white. Feeding habits like Yellow Oriole mainly on fruits and then secondly insects. Call softer, less powerful than Yellow Oriole a 'chew.chew.chew. chuweee'. **Nest:** Similar to Yellow Oriole but tighter weaving of grass stems, placed on a fork, made with loose twigs and pieces of bark. Clutch 2. **Similar species:** Female figbird does not mainly overlap but the figbird has a subtle different jizz particularly as figbirds head slightly larger, more thicker set at chest, streaking on the under-body bolder streaks wider, has dark eye not red, shorter tail. **Status:** Common throughout the central Kimberley from Broome to Kununurra. **Where to find:** Broome. Derby. Kununurra. Fitzroy Crossing. All along Gibb River Road where riverine vegetation occurs.

Green Oriole (Yellow Oriole) *Oriolus flavocinctus* 28 cm

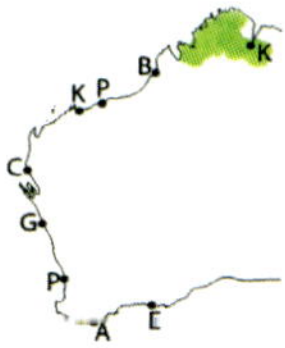

An unmistakable bird of the northern tropics. It is always interesting in evolution, how one species splits into two separate species. The Yellow Oriole and the Olive-backed Oriole are classic examples of this, so similar in morphology and colouring but both requiring different habitats. The Yellow Oriole is bird of the wetter rainforest and thicker canopy forests of the tropics, found mostly in the north west of the Kimberley, then a gap through the drier central region to Kununurra where it is moderately common, most probably due to the increased flooding of the Ord River and the subsequent lush growth that has now developed around its margins. An attractive, all lime-green bird, with darker olive wings and tail tipped light yellow, bright orange bill and conspicuous orange-red eye. Its primarily a fruit eater but also takes insects, foraging in the mid to high canopy. It has a powerful repertoire of calls, some almost ventriloqual, always far carrying. Its common territorial call is a strong, guttural series of calls 'chop. chop. chop. choowp'. **Nest:** A loosely woven deep cup, often suspended from hanging branches or placed on a fork, made with loose twigs and pieces of bark, often paperbark. Clutch 2-3. **Similar species:** Olive-backed Oriole has a dull olive-brown back, under-body white with dark green-black or dark brown streaking, bill paler salmon pink. Inhabits dry open eucalypt woodland. **Status:** Locally common but often populations patchily distributed. **Where to find:** Rainforests and vine thickets of far north-west and along the Ord River. Moderately common in the town of Kununurra particularly alongside the river front in lush forests.

Photo: Simon Nevill

Spangled Drongo *Dicrurus bracteatus* subspecies *baileyi* 30 cm

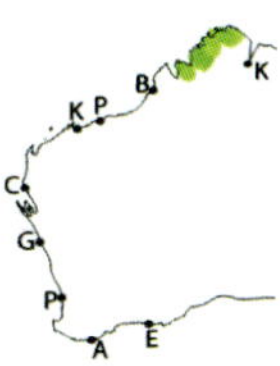

Very much a bird of the north-west sub-humid part of the Kimberley from Collier Bay to Cape Bougainville, found in the remnant rainforest pockets and vine thickets. It can often be found in open areas under the canopy sitting waiting to hawk insects or pounce on prey. It needs open areas to feed but still prefers the rainforest habitat in WA, if not within the forest often sits near the edge flying out to open woodland if it sees prey then flies back into the cover of the rainforest. Spangled Drongo has an all satin black body with a metallic green-black sheen, long forked tail used to manoeuvre when chasing flying insects, has a distinct red eye, robust bill, grey scalloping on the vent, and fine pale grey spots on the black breast. Perches patiently with a vertical posture, never horizontal, showing its

Photo: Simon Nevill

splayed fork tail well when at rest. There are 3 subspecies, all basically same markings but vary marginally in size, the WA birds are sedentary. Omnivorous eating insects, fruit and will even kill chicks of small birds. Call is unique hard to describe but basically a nasal, scratchy, rasping series of calls. **Nest:** Very open nest, normally suspended between two horizontal stems, loosely woven with twigs lined with finer twigs. Clutch 2-4. **Similar species:** No other all black bird similar in shape or size. **Status:** Uncommon in rainforest and vine forests of the far north west, comes inland to a few denser rainforest pockets but most on the coast. **Where to find:** Rainforests and vine thickets on Saint George Basin and Prince Regent River. Prince Frederick Harbour rainforests. Rainforest pockets on Bigge Island. Osborn Island. Mitchell Plateau. Camp Creek.

White-breasted Woodswallow *Artamus leucorynchus* 16-18 cm

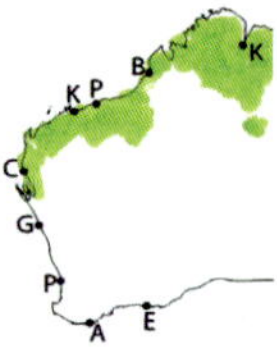

The largest of all the wood swallows, although only marginally so. Has a stocky body and short tail. Dark grey-black head and upper body and wings. Distinct break from grey-black hood to white under body. White rump seen well in flight, with grey-black tail. Grey-blue bill, small black tip. Found in coastal region around Carnarvon north to the Kimberley. In the Kimberley region the species not only occupies the coastal belt but also can be seen throughout the Kimberley mainland. Often found along coastal beaches with a preference for mangrove areas and also near inland wetlands with paperbark woodlands. In the north west does not move as far inland as it does in the Kimberley. Often perches in groups or singly, on power lines when at rest flying out to take prey on the wing. Will take prey from the ground. **Nest:** Nest in a variety of locations, mostly in mangroves or paperbarks near the coast. Can be in hollow tree trunks, broken off trunks, or a fork in branches. The nest is a broad cup shape of rough twined twigs getting finer to the centre with an inner lining of fine twigs and grasses. Clutch size 3-4. **Similar species:** Possibly Masked Woodswallow which occurs in the same area. They have a black mask and dark grey crown. Upper body is grey not black-grey. Under body pale grey not pure white. No black in tail. **Status:** Moderately common in Carnarvon region. Far more common further north of Carnarvon from Exmouth and north to Wyndham. **Where to find:** Carnarvon. Onslow. Karratha. Port Hedland. Roebuck Bay. Broome. Cape Leveque. Fitzroy Basin. Kununurra region.

Photo: Simon Nevill

Masked Woodswallow *Artamus personatus* 19 cm

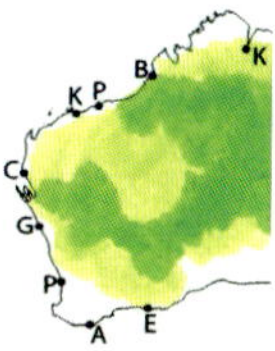

Similar size and shape as the White-browed Woodswallow, which it occasionally flocks with but in very low numbers in WA. As the name implies, has a mask on the face. The male has a black mask with a white border at rear of mask. Dark-grey crown, grey back and pale grey underbody. Fine down-curved bill, longer than White-breasted Woodswallow. Female has similar markings but mask is grey not black and under body more buff-grey than pale grey. Lacks the white border to the mask. Found in the semi-arid and arid regions of the WA, mostly in open mulga. Occasionally comes into the northern and eastern Wheatbelt. Will fly over in flocks in the Transitional Woodland both east of Hyden and north east of Wubin. Forms large groups after the breeding season, in some areas very large flocks that can be heard calling in the sky several hundred metres high hawking for insects and constantly chattering keeping contact. Watch out for White-browed Woodswallow's with their dark bodies, often in the north they will fly together with Masked Woodswallows but they are in far less numbers. Masked Woodswallows will also feed on nectar, descending in hundreds on the Honey Grevilleas (*Grevillea eriostachya*) and in the deserts on the Holly Grevillea (*Grevillea wickhamii*). When near colonial nesting sites, pairs will rise up give a rasping harsh chatter. **Nest:** The nest is a shallow cup of twigs and bark quite frail at times, normally placed in the fork of a tree or end of a hollow, can be quite exposed. Clutch size 2-3. **Similar species:** None. **Status:** Can be locally very common. **Where to find:** Patchy distribution throughout the Mulga Region, can be on any of the stations where feeding conditions are good. Found in large numbers maybe 100 plus in the Great Sandy when grevilleas are in full bloom.

Right: Male Masked Woodswallow

White-browed Woodswallow *Artamus superciliosus* 18-20 cm

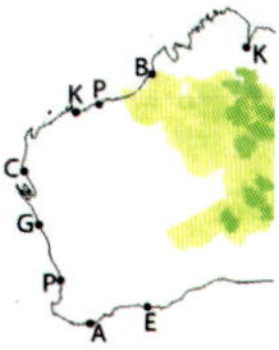

Can occasionally come into the north-east desert region from the NT, mostly coming with large flocks of Masked Woodswallows from the eastern deserts and crossing into the Great Sandy Deserts and occasionally into the Kimberley. The most colourful of all the woodswallows. The male has a rich chestnut-red under body with blue-grey upper body. Bold white supercilium stripe hence its species name. Grey tail with white tips. The female has a less rich coloured under body being chestnut-brown. Back and head grey not blue-grey. Supercilium stripe diffused off grey-white. Easy to pick in flight with the Masked

Photo: Simon Nevill

Woodswallow, having dark under body and white wings whereas the Masked Woodswallow has a pale grey almost white under body with pale grey wings, no dark colours at all. **Status:** Very uncommon north and east, but can be locally common in Lake Gregory region. Common in central eastern and south eastern Australia.

Black-faced Woodswallow *Artamus cinereus* 17-19 cm

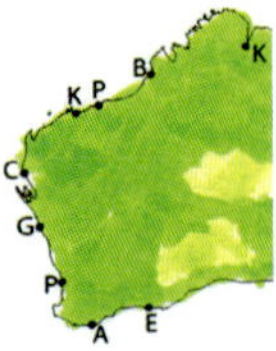

A medium sized Woodswallow that covers most areas of Australia and all of the WA, except in the wetter South West. Sexes alike with a small black facial mask on the lores and upper throat. Pale grey, slightly down curved bill with black tip. Grey back. Black tail with white tip. Like the Dusky Woodswallow, when birds alight on a branch they twist their tail in a figure eight movement before resting or after each time they call from a perch. A bird of varied habitat but not dense forests. Can be found on stony plains, in dry river courses, and mulga woodland, roadside verges alongside open fields, heathland. Does not like enclosed woodland even wandoo woodland. Before European settlement, the Black-faced Woodswallow did not occur in the lower South West but with land clearing the species spread into the agricultural region. It is interesting to note the preferences for certain habitats of species and specifically woodswallows: If you drive say to Dryandra Woodland Reserve in the inner wheatbelt you will often pass the occasional Black-faced Woodswallow along the roadside verges but as soon as you enter the uncleared Wandoo Woodland of the reserve the populations of Black-faced Woodswallows cease and immediately Dusky Woodswallows

Photo: Simon Nevill

become the dominant woodswallow. There are four subspecies in Australia and two occur in Western Australia. *melanops* that covers most of the arid part of Australia including the Kimberley down to the Transitional Woodland line, and the subspecies *cinereus* that occurs in the South West. Differences are only slight and not easily visible in the field and integration occurs along the Transitional Woodland Line. **Nest:** Nests are a woven mesh bowl of twigs in varied locations from low shrubs, grasstrees in heath to the end of tree hollows. The nest is similar to a Masked Woodswallow's but not so exposed being placed in the centre of bushes or in vines or other thick foliage. Clutch size 3-4. **Similar species:** Ducky Woodswallow, same size and jizz but all dark chocolate-brown with no contrast between upper body and under body; also has white in the wings which can be seen at rest. **Status:** Very common in the Wheatbelt and common in the Mulga. Common in the Pilbara and lower Kimberley. **Where to find:** Swan Coastal Plain. The Darling scarp. Throughout the Wheatbelt and Mulga Region. Pilbara. Parts of the desert region and the Kimberley.

Dusky Woodswallow *Artamus cyanopterus* subspecies *perthi*

16 -19 cm

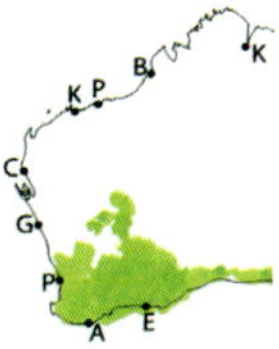

A medium sized woodswallow found in the forested South West. There is evidence of some movement in autumn and winter away from the lower South West. Dusky Woodswallows have an all dark chocolate coloured plumage and a gunmetal blue bill with a small black tip. The tail has a white tip and white on the primaries shown as a thin white line on the wings when at rest. In flight shows its light coloured grey underwing and from above an all dark brown body and upper wing, except the outer primaries which have white shafts. Dusky Woodswallows hawk from perches to gain insects, sometimes going to ground to feed. They are particularly prevalent in the Karri Forest and remnant Wandoo Woodland. There are two subspecies: *cyanopterus* in the east coast region and *perthi* in the South West. **Nest:** They nest mostly against the trunk of trees where bark is dislodged, a small branch juts out from the trunk or a small trunk is broken off but rarely build at the end of a horizontal branch as do many birds. A frail bundle of twigs is constructed with little or no lining, placed 3-8 metres high. Clutch size 3-4. **Similar species:** Refer to Black-faced Woodswallow. **Status:** Moderately common in Karri and Wandoo and some Jarrah woodland. Scarce in the open Wheatbelt. Common in the eastern part of the Transitional Woodland but not the northern part. Rare in the mulga although some movement north at the end of autumn and through winter. **Where to find:** Julimar Forest. Wandoo Conservation Park. Boyagin Rock Reserve. Dryandra Reserve. Dongolocking. Cherry Pool Reserve. Stirling Range NP. All of the Karri Forest including the eastern population of Karri in the Porongurups. Transitional woodlands east of Hyden.

Photo: Simon Nevill

Little Woodswallow *Artamus minor* 13-15 cm

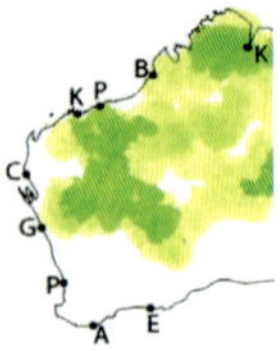

The smallest of the Woodswallows, found in the Mulga Region mainly near rocky breakaways. Darker black-chocolate-brown than the even grey-brown Dusky Woodswallow. The whole body is black-brown with black-grey wings. Small white inside tip to the tail. Fine blue-grey bill. Mostly found in woodland near rocky terrain, including steep rock faces or granite boulders. A gregarious Woodswallow that will roost and rest in a group mainly on cliff faces. During the day will perch on dead saplings or mulga and fly out hawking for insects. There are two subspecies: *minor* which is found in the mulga and central ranges and *derbyi* found in the Kimberley and eastern northern Australia. **Nest:** The nest is normally placed on a rock ledge or in a crevice but occasionally in a broken off part of a tree or hollow. The nest is a small cup shape of twigs with fine grasses or twigs on the inner lining. Clutch size 2-3. **Similar species:** Dusky Woodswallow which is noticeably larger, is a lighter chocolate-grey-brown and has white in the outer wings. Rarely found in the same location except in the Mt Jackson region. **Status:** Uncommon and patchily distributed in the Mulga Region and central ranges. Common in the Kimberley. **Where to find:** Walga Rock east of Cue. The 'Granites' north of Mt Magnet. Tallering Peak north of Mullewa. Wooleen Station. The base of the Kennedy Range. Parts of the Hamersley range. Throughout much of the Kimberley.

Photo: Simon Nevill

Grey Butcherbird *Cracticus torquatus* 27-30 cm

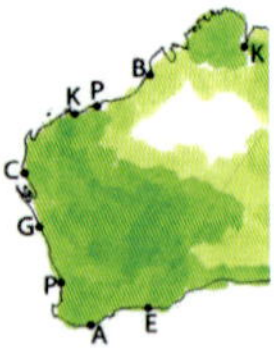

The subspecies *leucopterus* is found throughout most of WA to the northern Pilbara. The Grey Butcherbird has a black head with white throat, grey back, black and white in the wings, all white under-body, black tail with white tip, white spot on lores. A massive bill for the size of the bird with a down-curved black tip. Is a smaller bird than the Pied Butcherbird. Feeds on insects, lizards and small birds mainly young. Like all butcherbirds, has a strong and varied call that sounds like chuckling but is still melodious; is an accomplished mimic, although I find the western subspecies *leucopterous* mimics less than the east coast subspecies *torquatus*. Varied in its choice of habitat, it is the only butcherbird found in the wetter forests of the deep South West and much of the lower South West, although it avoids much of the central Darling Range. Found in the Wheatbelt. Transitional Woodland. The Mulga. The Pilbara and the southern deserts. In the Mulga Region there is sharing or overlapping of territories but it is not totally understood how these habitat can support both Pied and Grey Butcherbirds with similar feeding requirements. The Grey Butcherbird is certainly more prevalent in the southern Mulga Regions and the Pied Butcherbird more so north. There are five subspecies in Australia, in the Kimberley a bird known as the 'Silver-backed Butcherbird' is a distinctive subspecies *argenteus*. The birds of South West *leucopteris*, vary only subtly from the east coast birds but the Kimberley birds are much lighter on the back than the South West birds and have a pure white under body and do not have a white loral spot, call is also different and very powerful when it echoes in the paperbark or eucalypt riverine woodland where it occurs.

Nest: Placed on a multiple fork from 2-8 metres above ground or in lower shrubs or trees but requires dense outer foliage; the nest is constructed of loosely placed twigs but the inner lining is neatly woven grasses and fine twigs. Clutch size 3-4. **Similar species:** Pied Butcherbird, which is distinctly larger and has a black bib going down to the lower breast. Grey Butcherbird has all white throat and under body, no bib. Pied has a black back not grey and its white collar extends completely around neck, the Grey Butcherbird's stops at nape. Pied has an even more massive bill. Calls very different. **Status:** Moderately common in lower South West. Common in parts of Perth. Uncommon in Wheatbelt. Moderately common in the southern Mulga. Uncommon further north. **Where to find:** subspecies *leucopterus* Perth. Darling Range scarp. More common in southern coastal areas particularly in the Esperance region. Less common in the northern and eastern Wheatbelt than the Pied Butcherbird. Found around Paynes Find, Yalgoo and Cue and Murchison Settlement. Patchily distributed through the Pilbara. The 'Silver-backed' subspecies *argenteus*. can be found in the taller cadjeput and tall river eucalypts in riverine locations, particularly in the central Kimberley as in the Hann, Drysdale, and Charnley Rivers.

Kimberley 'Silver-backed' form of the Gray Butcherbird.
Note: no white loral spot.

Photo: Simon Nevill

Pied Butcherbird *Cracticus nigrogularis* 28-32 cm

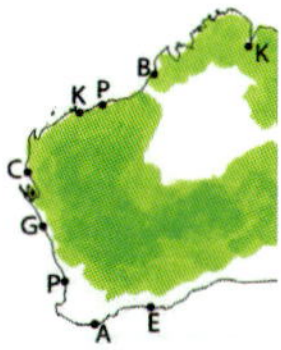

Found throughout much of the state except the wetter south west. A bird principally of semi-arid and arid regions of Western Australia. The sexes are almost alike. Both have a black bib and head; white collar, black back and wing, with white in the wings. Black tail with white under tail feathers near tip; white underbelly; massive bill. Female has slightly narrower white collar that is diffused with grey mottling but not always clear. The subspecies *pictatus* found in most of WA has slight differences to the east coast nominate race *nigrogularis*. Birds in the South West have a wider white collar, longer bill, bib does not come as low and the clean marked break with white belly not diffused. There are small differences with the female, which has a shorter tail and shorter tarsus. Feed on insects including beetles and grasshoppers. Some fruits, small reptiles, young birds. **Nest:** Large round twig nest with a deep centre placed in the fork of a tree, normally from 5–12 metres high up in eucalypts but in mulga obviously lower down. Call is astounding and as one American sound recordist stated, it was one of the finest songsters in the avian world he had ever heard. A very rich melodious call, quite varied. Unlike the Grey Butcherbird's rapid staccato type call, the Pied Butcherbird's notes are long and slow. They will vary the song and bring in different elements. Pied Butcherbirds do not come down to the deeper South West preferring drier more open country. They are particularly common in the northern and central Wheatbelt and since land clearing have extended their range from the Mulga Region. **Nest:** The nest is similar to the Grey Butcherbird but tends to be higher and sometimes more exposed. Clutch size 3-4. **Similar species:** Refer to Grey Butcher notes. **Status:** Scarce in the lower South West including the southern Swan Coastal Plain and most of the southern coast. Moderately common in the inner Wheatbelt, getting more common the further east and north one goes. Uncommon in the eastern Transitional Woodland. Moderately common in the northern Transitional Woodland. Common throughout the Mulga Region. The Kimberley 'Silver-backed' Butcherbird is locally common in prime habitat. **Where to find:** North of Moore River. Wongan Hills. Kellerberrin. Merredin. Reserves near Hyden. Kalgoorlie. Wubin and all stations north of the Transitional Woodland line. The deserts and Pilbara. In the Kimberley tall lush riverine woodlands.

Photo: Simon Nevill

Black Butcherbird *Cracticus quoyi* subspecies *spaldingi* 38 cm

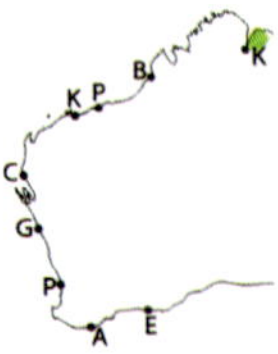

The largest of all the Butcherbirds, found in rainforests and canopy mangroves. The WA and NT subspecies *spaldingi* is slightly bigger weighing more than the east coast subspecies and the bill slightly longer. Most people will not get to see this species in WA, as it is confined to dense mangroves north of Wyndham, where there are large areas of mangroves in the Cambridge Gulf and around the coast to the NT which are very inaccessible except by boat. It inhabits the tall *Bruguiera parviflora* and *Rhizophora stylosa* mangroves. It feeds on small crustaceans and large insects, will also rob other birds' nests taking their chicks and wedge them between forks, I have seen this behaviour with Pied Butcherbirds a few times, once right in the middle of the hotel complex at Ayers Rock. Feed both on the ground and in the canopy of the mangroves. You often hear them before you see them as they will often slip away if approached if not used to human intrusions. Their call is reminiscent of orioles

being powerful and echoes through the rainforests or mangroves. Those birds in the rainforests of NT call differently to those in the rainforests of the east coast. Black Butcherbirds are an all blue tinged black with a slight sheen, robust bill with distinct hook at end grey at base and black half way to tip, black eye, long tail for a butcherbird. The immature birds of the WA subspecies *spaldingi* does not get the distinct rufous morph of immature birds that occur with some other subspecies, so no confusion possible. Very territorial said to often use the same nest the following year. **Nest:** Large bowl of loosely woven twigs mostly flexible mangrove stems, on a horizontal limb. In rainforests, built more often in vertical multi-forked stems. Clutch 3-4. **Similar species:** No other all black bird similar, both Torresian Crow and Spangled Drongo different sized birds and totally different jizz. **Status:** Moderately common in the Cambridge Gulf becoming more common in northern NT. **Where to find:** Rainforests and mangroves requiring a boat trip from Wyndham.

Grey Currawong *Strepera versicolor* subspecies *plumbea* 44–57 cm

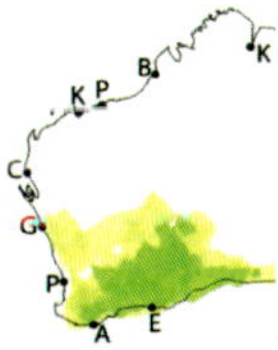

There are distinct variations in Grey Currawongs throughout Australia with six subspecies identified to date. In the South West Region there is one subspecies *plumbea* that overlaps with *intermedia* on the southern belt of the Nullarbor. A very large crow-like bird but unrelated to the corvids. Grey Currawongs have a massive bill almost conical in shape. An all dark grey body and wings when at rest. Distinct yellow eye. White vent. In flight shows clear white bull's eyes in wings and white vent and tip on under tail. Even though all currawongs in the South West are grouped under one subspecies *plumbea*, I find distinct variance in their grey colour, finding a clinal change from the lower south west birds being darker than those in the Transitional Woodland and getting lighter as one enters the mallee on the Nullarbor. The calls of birds in Jarrah Forest are distinctly different to birds east of Lake Cronin and also birds in the mallee in the Ongerup region, particularly the alarm calls. Grey Currawongs have varied feeding habits. They feed walking along in a Ground Cuckoo-shrike style, bobbing the head as they look for insects and small reptiles. They will cling to grasstree stems and pick out the many insects attracted to flowering grasstrees. They are a very successful hunter of small birds and I have seen them pull a Striated Pardalote from a small nest hollow or take out the complete clutch of very young Common Bronzewing Pigeons. They will work through bush well after sunset, trying to flush out small roosting passerines. In a nesting territory, they can have a drastic effect on the bird population immediately around a nesting site. **Nest:** Their nest is a bulky affair though not as big as a Ravens, about 5–10 metres in eucalypt or other large tree, normally more concealed or less exposed than corvids' nests. Clutch size mostly 3. **Similar species:** Could be confused with Australian Raven, which has no white in the wings or tail, best seen in flight. Different bill shape and no yellow eye. **Status:** Moderately common in the Jarrah and Karri Forest. Moderately common in the southern Mallee Regions. Uncommon in the Wheatbelt mainly found on larger reserves. Common in the Transitional Woodland east of Hyden. Sporadic in the southern Mulga Region. Does not occur much further north than the Murchison River. **Where to find:** The Darling Range and scarp particularly on roadside verges of the Brookton and Albany Highway Wandoo Conservation Park. Boyagin Rock Reserve. Dryandra Reserve. Stirling Range National Park. All the southern national parks from Leeuwin –Naturaliste in the west to Cape Arid in the east and north to Coolgardie and up to Mt Jackson. And the southern Nullarbor.

Grey Currawong juvinile hybrid – *plumbea-intermedia*, found on Nullarbor. Photo: Simon Nevill

Australian Magpie *Gymnorhina tibicen dorsalis* 37-43 cm

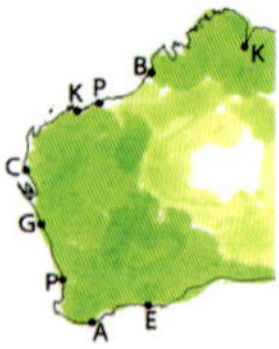

The Australian Magpie can be found throughout most of WA, being most common in the South West. In the arid regions their distribution is fragmented and one may see a family group and then not another for a long distance, sometimes as much 20 to 40 km in remoter country. They prefer open areas with woodland to roost and nest in but feed almost entirely on the ground. Have adapted well to European settlement and are found throughout cities and towns. Magpies have a long stout bill with down-curved hook at the end and long and strong legs with distinct pied plumage. There is great variation in patterning of plumage particularly the distribution of white on the back, and also in weight with a variance from 220 g to 350 g, which is a large disparity. Authorities identify 8 subspecies throughout Australia. Three can found in WA; *dorsalis* from the lower South West to the Murchison and then south east down to the edge of the Nullarbor, and above this line overlapping with the subspecies *longirostris* of the Pilbara. The subspecies *eylandtensis* is found through parts of the Kimberley. The South West birds are generally the largest of the subspecies and are likely to have a white back more than the 'black-backed' form. As the species goes north and overlaps with *longirostris* there are more black-backed magpies. The complexities of subspecies in Australia is compounded by the fact that there is partial migration of immature south west birds to other northern areas. However the South West population is the most cut off from external influences and retains a higher degree of consistency, with males mostly white-backed. The average reader should not get bogged down with the complex taxonomy but it is interesting to see the changes from South West birds to Pilbara birds and Kimberley birds. Magpies are extremely territorial and will defend feeding areas aggressively. Family groups will sit together with head and bill pointing skyward, giving their combined carolling calls asserting territory with the same technique as Kookaburras. The family relationships and their complex behaviours are interesting, like submissive displays where one member will hold its head down fully extended while fluttering wings and a dominant bird will stand over and lightly peck at the neck or rear of head. Many such behaviours with Magpies have been well studied. The Australian Magpie has a varied song which is linked to the social territorial group. They call well into the night during the breeding season, less so it seems in an urban environment. **Nest:** The nest is slightly neater than a Grey Currawong's but is a similar broad twig nest neatly lined with fine grasses and plant material. Often built quite high from 8-15 metres. Clutch size 3-4. **Similar species:** None. **Status:** Very common now in urban environments. Common throughout the South West both in the Wheatbelt and the open forested regions. Less so in the denser forested areas of the Darling Range and Karri Forest. Moderately common to uncommon in the Mulga Region. Locally common in the Pilbara and deserts. Less common in the Kimberley. **Where to find:** Throughout the state except most of the Great Sandy Desert.

Photo: Simon Nevill

Western Bowerbird *Chlamydera guttata* 27 cm

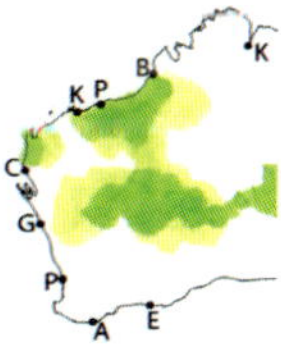

A stocky bowerbird of both central Western Australia and central Australia found in a variety of habitats, particularly stony country where the Rock Fig (*Ficus platypoda*) grows. It does however feed on a variety of fruits and other figs like Sandpaper Fig (*Ficus opposita*) in the northern range, the fruits of Sandalwood (Sorghum spicatum) and Northern Sandalwood (*Santalum lanceolatum*), Quandong (*Santalum acuminatum*) in the southern range. Has adapted to feeding around homesteads where there are fruiting trees and permanent water. The male's bower is normally built under an acacia tree with an umbrella covering, the bower being placed near the trunk of the acacia. This occurs in typical open mulga country. They also can be found away from rocky areas along creek lines that have thick riparian vegetation and will have their bowers in the dense vegetation where vines or dense shrubs are compacted together. One would never walk in these areas normally but their hissing, scalding and mimicry of other birds gives their bower's location away. Other typical locations particularly for the subspecies *carteri* can be in narrow gullies on the side of hills. As is well known, the male plays no part in the nest building and raising of young. His energies are entirely concentrated on maintaining a successful display bower. He is polygamous and will mate with any female that is attracted to the bower and will mate in or adjacent to the bower. The female locates the bower by hearing the male's contact calls that are made at the top of a nearby tree. It goes through its repertoire of chuckles, hissing and mimicry calls. Once the male sees that a female is attracted, he will descend from the tree and start a sideways hopping dance, near the bower and will try and lead the female into the bower. He will perform calling in the centre of the bower with his pink nape display plumes fully extended and the female may inspect the bower. If the female is impressed with the bower, she will allow the male to mate and in the breeding season the male maintains the bower, attending it most periods of the day and bringing new additions of white objects such as land snail shells, bits of white china, white quartz and small bleached bones. In addition, there will be some green objects, never as many as the white ones but nevertheless most bowers will have some. In the central Mulga Region they are normally non-ripened Sandalwood or Quandong fruit that have not turned purple and will last several weeks in a green state. The centre of the bower is an immaculate structure so finely woven with twigs that the passage of a bird moving through the bower is not hindered. Bowerbirds are unmistakable in flight being the only stocky orange-brown coloured bird to fly in the central semi-arid regions. Their flight is a long undulating flight only a few metres above the canopy of trees or ground. There are two subspecies of Western Bowerbird: *guttata* found in central Australia and central Western Australia and an outlying population in the Cape Range near Exmouth subspecies *carteri*. There are complex variations in tarsus and wing size but in the field Cape Range birds are darker, their plumage is richer orange-brown and particularly the under-body is darker. **Nest:** The female builds the nest and raises the young alone. The nest is an untidy affair of twigs in a shallow cup shape built normally in a three way fork in acacias or eucalypts almost always away from the male's bower area. Clutch size mostly 2 can be 3. **Similar species:** None. **Status:** Locally moderately common in the Mulga Region. Overall, uncommon in the Mulga Region. **Where to find:** Not far into the Mulga Belt just south east of Paynes Find on Pindabunna and Pullargaroo Stations and north of Kalgoorlie. Moderately common in the Mt Magnet, Meekatharra and Wiluna region. In the Pilbara in the Barlee Range. Shay Gap almost to the edge of the Great Sandy.There is a break east in the distribution until the central ranges.

Photo: Simon Nevill

Great Bowerbird *Chlamydera nuchalis* 36 cm

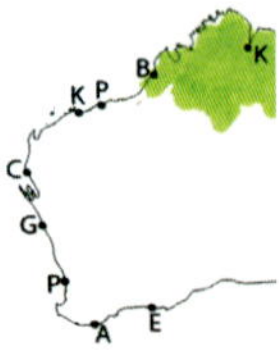

Largest of all the Australian bowerbirds, found throughout much of the Kimberley. Normally its bower is not far from water. It is often one of the easiest bowers to find as it can be fairly well exposed but always under the canopy of tree or bush. Of all the bowerbirds it is fairly accepting of human habitation. It is omnivorous in its feeding habits but like all bowerbirds it is particularly fond of fruiting berries. It occurs in very dry areas such as the Bungle Bungle region. Very common along the fringes of rivers where its hissing and scolding calls attract one's attention. If you are lucky to see it in full performance if a female is close, you will see its vivid pink head plume extend from the nape. It will circle the bower and hop almost side to side as it manoeuvres around. Then it will enter the bower with bill open uttering hissing calls while facing the female. The male as is well known plays no part in the building of nests or raising young. The energies are placed on maintaining the display bower and the performances that it routinely performs. **Similar species:** None. **Status:** Common in the Kimberley to the southern Fitzroy Basin. **Where to find:** Anywhere between Broome and Kununurra. Often found in mangrove fringes as well as eucalypt riverine habitat.

Photo: Simon Nevill

Australian Raven *Corvus coronoides* 46-53 cm

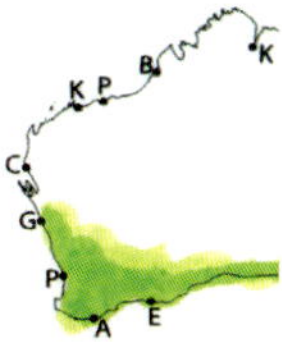

Most non birdwatchers think there is just one corvid and call them all crows. There are 5 corvids in Australia, 3 occurring in Western Australia, two crows, the Torresian Crow (*Corvus orau*) and Little Crow (*Corvus bennetti*), and one raven, the Australian Raven (*Corvus coronoides*). In the South West, the range of the Australian Raven extends as far as the Wooramel River and south-east through Kalgoorlie down to the base of the Nullarbor at Eyre and Eucla. One of the largest corvids, it is found throughout the South West. It least favours the dense forested areas although it does occur in them but is more common in the Wheatbelt, the Swan Coastal Plain and the open woodlands of the Transitional Woodland and parts of the southern Mulga Region. Ravens basically a carrion eater scavenging for food or killing prey and is primarily an opportunistic feeder. Often works in family parties or small groups. Identifying corvids is not always easy but the Australian Raven is one of the easiest to identify in the field, as it has long throat hackles that are visible most of the time

Photo: Simon Nevill

particularly when calling. Calls are the primary way birdwatchers differentiate one species of corvid from another. The Australian Raven has a few short loud calls finishing with a long extended call that reduces in volume 'aaahk, aaahk, aaahk .....aaaaaaah'. Calls vary and are complex amongst a family group. Sometimes meaning food is available, sometimes warning the group of danger, sometimes it is just a contact call, a single 'aaark'. There are two subspecies of Australian Raven with only one in Western Australia *perplexus*, which has slightly shorter hackles although still long compared with other corvids and slightly smaller than the east coast subspecies *coronoides*. **Nest:** The nest is bulky and loosely constructed, lined with finer twigs or grasses and also sometimes wool or plant down is wedged into the fine twig lining. The nest is placed normally well above 8 metres. Clutch size 4-5. **Similar species:** Torresian Crow and to a lesser degree Little Crow. Even though some authorities say Torresion Crow looks the same size as the Australian Rave, the raven certainly looks a bulkier corvid particularly around the head. If not extended the hackles hang from the throat giving the neck line a thicker appearance whereas the Torresian has only slight hackles that definitely do not extend from the throat to any great length. Also the bill is fractionally bigger and bulkier on the raven. It is not a diagnostic feature but the raven occurs mostly from the Transitional Woodland and the south west but the Torresian Crow rarely enters the South West Region, less so than even the Little Crow. The Torresion's call is different from the raven's it gives a short faster call 'ark ark ark ark' sometimes slightly extended at the end but never as long or low pitched as the Raven. It also makes a peculiar low-toned gurgling call difficult to put into words. The Little Crow is easier to differentiate as they are distinctly smaller and the bill is shorter, slimmer and more pointed. Has no hackles at all and the jizz of a Little Crow is of a sleek smooth corvid with proportionally longer-looking legs and less feathering on the thighs. In full sunlight has a distinct sheen. Its call has similarities to Torresian but is lower and mellower sounding 'naark, naark......naark, naark'. **Status:** Common in the Wheatbelt and Swan Coastal Plain. Uncommon in the lower south Jarrah Forest and Karri. **Where to find:** Perth and other large towns. Throughout the Wheatbelt. Roadside verges particularly on the Albany Highway through the Jarrah Forest.

Torresian Crow *Corvus orru* subspecies *cecilae* 46-51 cm

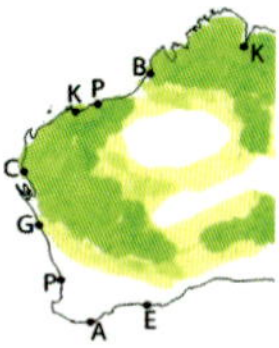

A corvid of the drier regions of central Western Australia and the Kimberley and other states. Does not come down to the South West and rarely comes past the line from Shark Bay south east to Paynes Find, Southern Cross and east through to Norseman then north to Zanthus and through the Great Victoria Desert to central Australia. Its

Photo: Simon Nevill

range virtually extends where Australian Raven does not occur. Does not favour the very open arid areas of the desert. Found more around watered areas or along river and creek systems. Will come around station properties scavenging for food. More omnivorous than the Australian Raven taking fruits as well as carrion or small, live prey. Less flocking than Little Crow. The Torresian's call is different from the raven's, it gives a short faster call 'ark ark ark ark' sometimes slightly extended at the end but never as long or low pitched like the ravens. **Nest:** Nests singularly not in groups, as does Little Crow. The nest is similar to that of the Australian Raven and placed normally higher than 6 metres. Clutch size 4-5. **Similar species:** Australian Raven and Little Crow, refer to notes on Australian Raven. **Status:** Common in the mid to northern region of WA not so common in the southern Mulga Region. Uncommon northern Transitional Woodland. Fairly common in the Kimberley and central ranges. **Where to find:** Carnarvon region, Gascoyne River. Mt Magnet. Throughout the Pilbara. Throughout most of the Kimberley.

Little Crow *Corvus bennetti* 44-48 cm

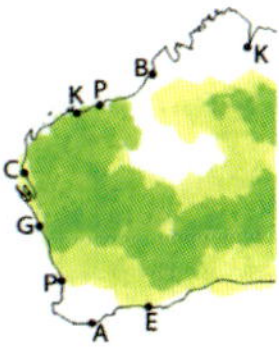

The smallest of the corvids. A sleek smooth plumaged crow often found in large flocks that fly high in spiralling flights, calling constantly and often performing aerodynamic displays in all directions. They do come into the Wheatbelt in summer and are often overlooked by birdwatchers, presuming them to be Ravens. They come further south than the Torresian Crow, which makes it easier to split from Australian Raven, as Torresian looks similar to Australian Raven. The Little Crow is more prevalent in the open mulga away from water than the Torresian Crow. Also, more likely to inhabit southern desert areas where sanddune country has some mallee and eucalypt woodland than Torresian Crow. Will often nest in small colonies. Found alongside outback roads feeding off road kills and when Wedge-tailed Eagles are feeding on a carcass, they will still come and try and take the meat, being a very agile corvid. Will shuffle wings when alighting. Its call has similarities to Torresian but is a slower more mellow 'naark, naark……naark, naark'. **Nest:** A smaller nest than the other two corvids and also can be lined with earth matted with the fine grass lining. The Little Crow's nest can be much lower and sometimes in denser mulga is only 4-6 metres off the ground. Clutch size 3-4. **Similar species:** Torresian Crow and Australian Raven, refer to notes on Australian Raven. Status. Common throughout the Mulga Region and southern desert regions. Uncommon in the Wheatbelt in summer. Very uncommon in the southern Kimberley and certainly not found in the northern Kimberley. **Where to find:** Any station property north of Paynes Find. Kalgoorlie and Carnarvon.

Photo: Simon Nevill

Horsfields Bushlark (Singing Bushlark) *Mirafra javanica*

12-15 cm

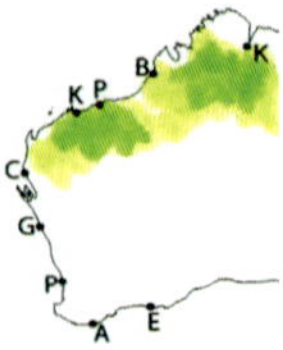

A short tailed stocky lark with a finch-like bill. Has a short crest when raised. Easily identified in flight by short triangular wings and short tail distinguishing it from the Australian Pipit and the song larks, all having longer tails and wings. A bird of the semi-tropical and tropical north found from as far south as Cobra Station and north across the grass plains of the Pilbara, east to Lake Gregory and into the Kimberley. Has a preference for short-grassed plains, some spinifex plains but mainly grassed country, particularly on black soil plains. Being a seedeater, it must always be not too far from water. The subspecies *woodwardi* is found in the Pilbara and reaches the base of the Dampier Peninsula. *forresti* is found in the Kimberley proper being one of the lightest fawn-coloured birds of the species. There is great variance within the *javanica* subspecies throughout Australia and much of this is related to the soils where they feed; obviously survival from raptors is increased by blending in with the soil colour as there are often no shrubs to retreat to. **Nest:** The nest is a well concealed, deep cup placed alongside grasses, often having a partial hood of grasses. The inner lining is fine woven grasses. Clutch size 3-4. **Similar species:** None really, possibly Australian Pipit which is a slightly longer bird with longer tail and longer legs and stands more upright. Outer white tail feathers are more prominent on the pipit, breast more streaked. **Status:** Common on grass plains north of the Gascoyne throughout the Pilbara. Northern deserts and lower Kimberley. **Where to find:** Any grass plains from the Gascoyne all the way to the central Kimberley.

Photo: Simon Nevill

Zebra Finch *Taeniopygia guttata* 9 - 11 cm

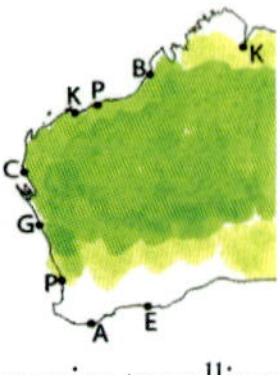

A small finch found throughout most of the semi-arid and arid regions of Australia. They do come into the northern Wheatbelt, particularly north of Beacon and Bonnie Rock mainly in the summer. In the mulga they become more common the further north and north-east one travels. They are a common sight at cattle and sheep troughs gathering often in hundreds particularly after the breeding season. They are a highly nomadic species travelling to regions where the winter or summer rains have been heavy. The sexes are different. The male has a large orange patch on the lores, the female just grey lores. The male flanks are orange with white spots, the female has no colour on the flanks just the same under-body fawn. The male has a grey throat and upper breast with fine grey-black barring and at the base of the chest a diffused black band. Zebra Finches must come to water and rarely are more than 10 km away from a water point. I once found a rock hole in the desert. Looking into the hollow could see no water at the bottom, but suddenly out of a small crack no more than 10 cm wide a few birds flew out, such is the need and ability to find water. They feed almost entirely on the ground sometimes in the company of Southern Whiteface. Unlike many species they can be sighted throughout all the desert habitats. **Nest:** The breeding habits of Zebra Finches are well known mainly from captive birds, but in the wild they breed in loose colonies making a ball nest. There will often be 2 or 3 nests in one thick acacia bush such as Kurara (*Acacia tetragonophylla*). Nest locations can vary, obviously most nests will be in dense bushes but it's not unusual to see them in the base of a larger raptor's nest such as a Black-breasted Buzzard's or a Wedge-tailed Eagle's nest. One station owner told me that he had one day's break from using his plane and come back a day later and they were building a nest in the engine intake grill, such is the opportunistic nature of these birds. They can breed at any time of year and will start mating within 24 hrs after heavy rains, although rain is not the only key factor; grass conditions and state and abundance of available seed are crucial. They also build roosting ball nests, which are untidier than true breeding nests. Clutch size 4-5. **Similar species:** None. **Status:** Common in the Mulga. The deserts and the southern Kimberley. **Where to find:** Throughout the Mulga Region, more prevalent north of Mt Magnet. Common in the Pilbara and deserts.

Photo: Simon Nevill

Star Finch *Neochmia ruficauda* subspecies *subclarescens* 11-12 cm

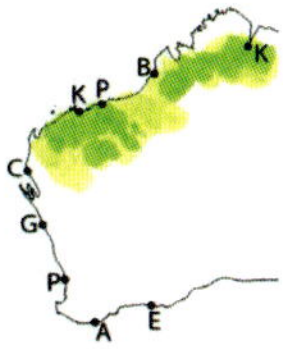

A finch that requires lush rank grass, which restricts its distribution, found in the Pilbara and southern Kimberley. A flocking finch rarely seen alone. The male has a large facial red disc and bill with white spotting over the lores. Rear of head, upper-body and wings olive-green. Breast and flanks green with extensive white spotting. Belly and vent pale yellow. Tail dark cinnamon brown with rump having white spots. Female very similar but facial red disc smaller and upper-body a duller olive-green. Under-body less contrasting green with white spots. When native grasses are lush and long, will cling to a stem dropping it to ground and then run bill along the stem taking the seed. In drier conditions will feed from ground. There are three subspecies in Australia, *subclarescens* found in the Pilbara, Kimberley and Gulf of Carpentaria, *clarescens* in the Cape York region and *ruficauda* in fragmented populations in northern Queensland. **Nest:** Nests in low thick shrubs rather than tall grasses, mostly near water. The nest is an oval shape ball with a side entrance, woven into the branches of a dense bush or located in thick tall grasses. The inner nest is made with fine grasses and lined with feathers. Clutch size 4-5. **Similar species:** None. **Status:** Uncommon in Gascoyne River region. More common north of the region in the Pilbara,. More common in the Kimberley like the Kununurra area and along the Gibb River Road. Nests: Often in a suspended nest from a hanging shrub, often exposed. **Where to find:** Millstream. Flood plains near Karratha.

Millstream and Opthalmia Dam. Lush grasslands and Typha on the Sherlock. Oakover and Maitland Rivers. In the Kimberley: Camballin Station. Tall grasses along the Gibb River Road in the King Leopold Range. Locally common in parts of the Kununurra region and around Lake Argyle.

Photo: Simon Nevill

Red-browed Finch *Neochimia temporalis* 10 -13 cm Introduced

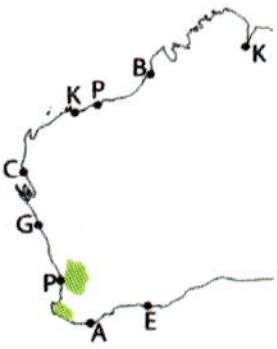

An introduced finch from the east coast, found in a few creeks on the Darling Range scarp surviving in riparian vegetation. A finch with a distinct red supercilium stripe and dark olive back. Dark grey head and under-body. Bright red rump and upper tail, very distinct in flight. Black lower tail, red bill. The bird on the east coast is mainly found in lush vegetation and in the west it survives in the few wetter creek lines close to Perth on the Darling scarp. **Nest:** The nest is a bottle shape, turned on its side with a drooping entrance hole. In the Darling Range mostly placed in thicker bushes like banksias and dryandras.

Photo: Simon Nevill

Clutch size 4-6. **Similar species:** None but could be confused with the endemic Red-eared Firetail which has barring on under body, not plain grey. Has no red supercilium stripe. Head and back is brown not olive-green. **Status:** Scarce in the Darling Range. **Where to find:** Piesse Brook. Helena Valley. Bickley Brook.

Red-eared Firetail *Stagonopleura oculata* 11-13 cm Endemic

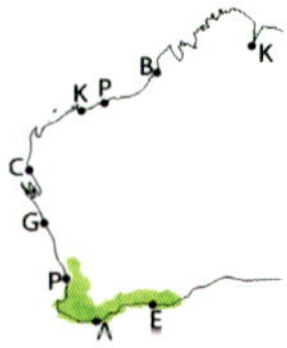

Found only in the wetter lower South West wherever there is lush thicker vegetation. Mainly found along creeks and rivers in riparian vegetation. Common in peppermint groves (*Agonis flexuosa*), also in coastal heath. The sexes are almost identical in the field, having small red patch on the ear coverts, small black area on the lores and a bright red bill. Chest and throat have faint light and dark brown barring. Belly bold black and white barring. Red rump and upper tail and black middle and end tail. Call is easily imitated, having a downward and then upward fine high pitched whistle 'wheeee.. wheeee … wheeee'. Feeds mostly on the ground. Its range extends from the Helena Valley in the northern part of the Darling Range east to Wandoo Conservation Park on Qualen Road Darkan, south to Cape Naturaliste and all along the southern coast to Cape Le Grand. Found on off shore islands in the Recherche like Woody Island. Inland to the Stirling Range and middle Fitzgerald NP. Easy to whistle up with its ascending 'wheeee….. wheeeee' call. **Nest:** Typical finch ball nest placed in various locations, mostly above 3-4 metres up in low tree often peppermint trees. I have seen a nest on the side of East Mt Barren wedged into the side of a Royal Hakea, not a typical location. The nest is dense ball of loose grasses with a side entrance tunnel. The inside is neatly lined with grasses and often with feathers or plant down; normally placed in the outer foliage of a drooping eucalypt or peppermint tree. Clutch size 4-6. **Similar species:** None. Refer to notes on Red-browed Finch. **Status:** Locally common in South West. **Where to find:** Wungong Gorge. Upper Bickley Brook. Ellis Brook. Along the southern coast from Cape Naturaliste to Cape Le Grand. Common between Walpole and Albany.

Photo: Simon Nevill

Painted Finch *Emblema pictum* 10-12 cm

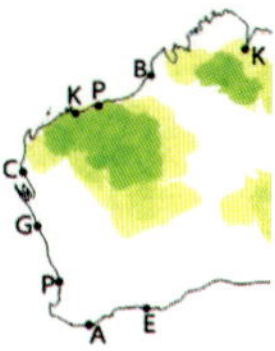

A bird of the northern parts of WA Two core population centres in WA. The Pilbara and the southern Kimberley. Frequents rocky habitats with spinifex (*Triodia*) and other grasses, normally never great distances from water. The male has a red facial patch from the forehead down to the upper throat. Brown upper-body and most of the head. The under-body is black with bold white spotting. Bright red rump and upper tail. Bi-coloured bill with lower mandible red. The female is very similar but the red facial patch is not as large. The white spots on the under-body are longer and cover more of the black under-body, appearing more like barring than just white spots. It has a plaintive, chirping call often given in flight. **Nest:** The nest is a ball shape of woven dry grasses almost invariably placed in spinifex either within the spinifex clump or wedged at the base on the ground below overhanging spikes of the spinifex. Clutch size 3-5. **Similar species:** None. **Status:** Common in the Pilbara and Kimberley. **Where to find:** In the Pilbara: Common in rocky hillside south of Tom Price. Common around Python Pool. Marble Bar and east out at Carrawine Gorge. In the southern Kimberley. Margaret River area. Halls Creek region. Wolfe Creek and the Breaden Hills.

Photo: Simon Nevill

Double-barred Finch *Taeniopygia gutata* subspecies *annulosa*

11 cm

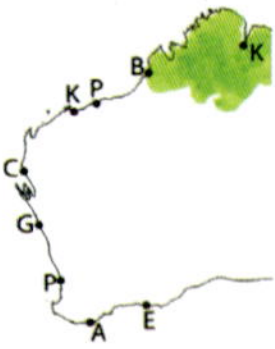

A very common, attractively marked finch, found from north of 80 mile Beach to Broome and Cape Leveque. South along the Fitzroy flood Basin, to the edge of the Great Sandy Desert. South to Halls Creek and throughout the Kimberley to the northern coastline. Double-barred Finch has black-brown forehead merging into grey-brown nape and mantle, Kimberley subspecies *annulosa* has a black rump not white as in nominate birds, white facial disc with fine black margin band, female's band is even finer, wings at rest show black base with lots of

Photo: Simon Nevill

small white dots, lower band on breast, white under-body, black vent, typical short triangular bill, blue-grey, black eye and bars contrast with white. Found in long grasses, feeding on seeds both from grass heads to fallen seed. Common along riverine margins. Normally seen in small flocks of 5-10 birds some times larger flocks. Call is a mousy single noted call similar to Zebra Finch, a soft 'zeow.zeow.zeow'. **Nest:** Oval ball of loosely woven grasses with short side entrance, normally placed at the extremity of a bush similar to Zebra Finch location. **Similar species:** No other finch as the combination of their markings. **Status:** Common throughout the Kimberley less common north of the Gibb River Road but still found in appropriate habitat all the way north to Kalumburu. **Where to find:** Very common in Kununurra and surrounding grasslands, reed-beds and plantations. Found throughout the Kimberley as far south as the edge of the Great Sandy Desert.

Long-tailed Finch *Poephila acuticauda acuticauda* 15 cm

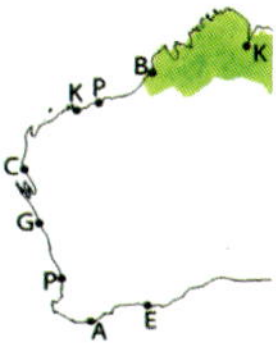

Moderately common in the Kimberley found from Broome north to Cape Leveque across to Looma and south east to Halls Creek and north-east to Kununurra. More common in the central Kimberley north to Kalumburu. Long-tailed Finch has a grey crown, fawn mantle merging into a brown back and wings, long black tail, white rump and vent, black bar on flank, pinkish-beige belly, black throat, grey cheek, black throat, pale yellow finch bill, black eye and small black lore spot, the Kimberley subspecies *acuticauda* most obvious difference is that it has a yellow bill rather than a red bill. There are other morphological differences but minor and not really useful for field identification. Male and female very similar. Long-tailed Finch is found in grassy country often sharing habitat with Masked Finch. Often found in recently burnt country where it is easier to gather fallen seeds. Often found in small groups 6–10, sometimes as many as 30. Call of subspecies *acuticauda* is a long drawn out ' ptseeueew….. ptseeueew' also other rapid shattering calls. **Nest:** A large oval ball, woven at the end of branches, located normally over 4m high. **Similar species:** Masked Finch which has a different facial pattern, black forehead mask covering eye with black chin, bill brighter yellow, black flank band much wider, tail shorter and lacks Long-tailed Finches two long central wire tails. **Status:** Common throughout the Kimberley. **Where to find:** Found through out the Kimberley, more so in the central Kimberley. Cape Leveque area. Kununurra region. Particularly coming to small pools.

Photo: Simon Nevill

Masked Finch *Poephila personata personata* 13 cm

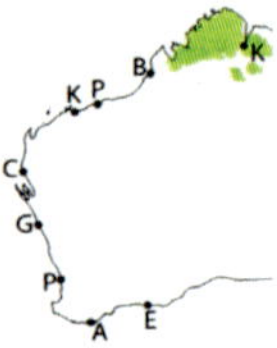

Less common than Long-tailed Finch and forms smaller feeding parties normally no more than 5-6 birds flocks of up to 10 uncommon but they do associate with Long-tailed Finch and one wonders why they evolved as different species and yet their feeding habits are very similar, feeding mostly on the ground and sharing the same habitat and often come into drink together. Their range is also less extensive as they are not found in Broome or Cape Leveque, mostly found in the central grasslands of the Kimberley where rocky hill are not far away. They become more common further east. Masked Finch has bright orange-yellow bill, black facial mask to eye, dark brown crown and fawn nape and back, pinkish-buff underbody with white rump and the largest black patch on the flanks of the two subspecies, long black tail with distinct white rump shows well in flight. They are primarily a grain feeder (granivorous) but during the breeding season will eat invertebrates to boost protein intake like all finches. It was interesting while I was watching finches and Mannikin's drink at a water hole, the Masked and Gouldian Finches sucked up the water whereas the Mannakin's had to scoop up the water and hold their heads back, trivial maybe but I found it an interesting behaviour. Masked Finch call is more harsh and grating than Long-tailed a more rapid 'chee… chittty…chittty'. **Nest:** A long oval ball with an extended entrance, nest not as bulky as Long-tailed Finch, woven at the end of branches, located normally over 4m high. **Similar species:** Long-tailed Finch which has different facial pattern having a black throat and grey head, not a black mask and brown head, tail is longer, black bar on flank narrower. **Status:** uncommon throughout the central Kimberley moderately common in the eastern Kimberley. **Where to find:** Gibb River Road and all the adjacent stations like Charnely River, Mt House, Mt Elizabeth, Ellenbrae, Mt House. Drydale River. Also Kununurra region north to Wyndham and south to Halls Creek.

Photo: Simon Nevill

Crimson Finch *Neochmia phaeton phaeton* 13 cm

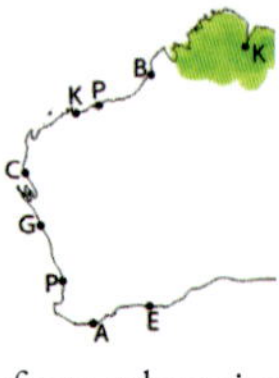

Can be very common locally but scattered populations through the Kimberley, although not found in Broome or the Dampier Peninsula, found as far south as the Fitzroy River, the Margaret River and Halls Creek and east to Kununurra and north to Kalumburu. Male Crimson Finch as the name implies, has a crimson facial patch, breast, flanks, wings and tail, buffy-grey back and nape, black belly and vent separates it from subspecies *evangelinae* which has a white belly and vent, has white spots on the crimson flank, red bill.

Female is similar with a crimson face but lacks the crimson under-body having buffy-grey under-body, olive-grey back with tinges of crimson, still has white spots on flanks but overlays the olive-grey. Immature birds of both sexes lack the full crimson colouring even the facial patch does not occur until the end of the first year, also lacks the white spotting on the flanks.

Photo: Simon Nevill

Very much a flocking bird, sometimes seen in large numbers as many as 30 birds. Feeds on grasses but far more prone to feeding on the seed heads of grasses rather than just on the ground like the majority of finches do. Often seen in riverine habitat where pandanus grows, as it is one of the commonest nesting location. Call is a repetitive high pitched and sharp 'zit.... zit.... zit.zit, zit.....zit'. **Nest:** A large ball oval with no spout, woven normally with wider grasses, with strips of fine bark, if nesting in pandanus always has a safe entry to nest. Some times builds in paperbark trees or eucalypts but when available prefers pandanus or palms in which to nest. **Similar species:** No other finch has the same coloration. **Status:** Locally common particularly around Kununurra, population more fragmented in the central Kimberley but normally near water where long grasses occur. **Where to find:** Gibb River Road and all the adjacent stations like Charnly River, Mt House, Mt Elizabeth, Ellenbrae, Mt House. Also Kununurra region north to Wyndham and south to Halls Creek. Very common in Kununurra.

Gouldian Finch *Erythura gouldiae* 13 cm

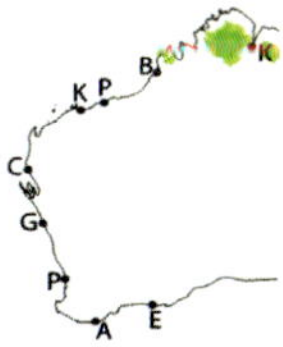

Such a beautiful finch, made even more interesting by the fact that there are 3 coloured morphs, although the yellow faced morph is extremely rare, in fact so rare that the probability of seeing it is estimated to be a ratio of between 500–1 and a 1000–1. The body chemistry that is involved in the genetic make up that leads to a red or black morph being selected is incredibly complex, involving a single sex-linked gene but the limitations of this book will not permit going into this though it is a fascinating study. The Gouldian Finch is not a common bird but is certainly not rare. Many people miss this bird thinking that they will find it near large pools, it actually has a preference for drinking from very small pools. After leaving the limestone Napier Range and starting to climb into the King Leopold Ranges, I take a few tracks I know that lead to a series of small pools, its these that attract the finches to drink. They prefer some adjacent trees near the pools from which to alight as a precaution against being taken by Collared Sparrowhawks. They mainly come to drink pre dawn and by the first hour of light the numbers dwindle to hardly any at all but they do return at different times during the day as being seed eaters they must drink. They mostly come down from the ridge tops and plateaux where the Northern White Gum *E.brevifolia* grows, it's one of their preferred nesting trees but also the ridge tops and plateau carry the grass seeds they prefer. Gouldians are mostly seen in small numbers, however the largest aggregation of birds I saw with one of my groups was over 60 birds. They dropped from the adjacent tree like confetti, on a small pool at Ellenbrae Station. There are reports of flocks of 200 but that is rare. With the changing fire regimes, numbers have changed over the years. I was talking to Peter Lacey, owner of Mt Elizabeth many years ago and he mentioned when he was a young lad they were in far greater numbers. They have several calls but the contact call is delicate 'zsit.zsit…. zist….zsit' also a grating call 'cheeip..cheeip'. **Nest:** Most nests are located in a hollow of a horizontal branch in a eucalypt, often on hill tops in snappy gums. An open cup nest made of grass at the base of a hollow. **Similar species:** No other finch has the same coloration. **Status:** Locally common but overall generally uncommon. **Where to find:** Small water holes off the Gibb River road in King Leopold Range. Mornington Station where staff will often assist with information regarding locations if they feel you are a genuine birdwatcher. Locally common on the Wyndham to Diggers Rest road. Sometimes come to dripping taps in the town of Wyndham, particularly in the camping ground. Seen on the northern margins of Lake Argyle and freshwater springs that occur in the Kununurra region.

Photo: Simon Nevill

Chestnut-breasted Mannikin *Lonchura castaneothorax* 12 cm

Locally common around wetlands in the Kimberley. Extremely common in the wetland areas around Kununurra, particularly along irrigation channels. Feeds on tall grasses and reeds. Always in groups when feeding, some times in large numbers in excess of 50 – 100 birds. **Nest:** Round ball of woven grasses woven to reed or grass stems having a side, hooded entry. Clutch: 4-5. **Similar species:** Yellow-rumped which has a pale grey head and yellow rump with no dark facial mask as in the Chestnut-breasted Mannikin. **Status:** Locally common within limited range. **Where to find:** Same locations as Yellow-rumped Mannikin. Kununurra irrigation plain. Kingston Rest Station. Along edges of diversion dam in Kununurra, in reed beds and on grassed areas. Found also in reed beds on Victoria River across the border in the NT.

Photo: Simon Nevill

Yellow-rumped Mannikin *Lonchura flaviprymna* 12 cm

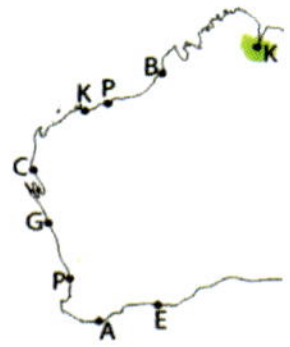

Have a restricted range in the Kimberley and some areas in the NT. Highest concentrations are to be found in the Ord irrigation area. A large headed, thick set finch with a grey head, grey bill, fawn breast merging into white belly, black vent, yellow-beige tail and rump, chocolate brown mantle and wings with a grey nape merging into grey head, grey-blue legs, sexes very similar. A flocking bird that often congregates with Chestnut-breasted Mannikin with which it also hybridises with, normally flocks of 5-20 birds but always outnumbered by Chestnut-breasted Mannikin. Feeds on seeds either feeding on ground or from seed heads. Seen regularly in Kununurra in the reed beds feeding on seed heads with Chestnut-breasted Mannikins. Said to extend range in the wet. Call is bell like similar to Chestnut-breasted Mannikin a 'seeeit…seeeit…seeeit'. **Nest:** Round ball of woven grasses woven to reed or grass stems having a side, hooded entry. Clutch: 4-5. **Similar species:** Chestnut-breasted Mannikin easily split having dark grey crown with black-brown face, breast orange-brown with distinct black bar and white under-body. **Status:** Locally common within limited range. **Where to find:** Kununurra irrigation plain. Kingston Rest Station. Along edges of diversion dam in Kununurra, in reed beds and on grassed areas. Found also in reed beds on Victoria River across the border in the NT.

Photo: Simon Nevill

Pictorella Mannikin *Heteromunia pectoralis* 12 cm

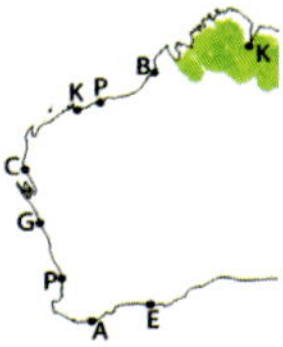

Found in the dryer regions of the Kimberley. Moderately common around Kununurra. A few sightings on the Dampier Peninsula and Roebuck Plains but scarce, most sighting in the grasslands of the central Kimberley as in Silent Grove, Charnley Station or driving into Mornington, not common further north of the Drysdale River but found further south than many Kimberley finches particularly along the highway between Fitzroy Crossing even south of Christmas Creek and across to Halls Creek. Certainly a flocking seed eater normally found in small groups of up to 20-30 birds, larger numbers occasionally seen. Over the years the species has been on the decline possibly by the changing fire regimes that now are lit in the winter months, affecting grass seeding, something that has also affected Gouldian Finches badly. Pictorella Mannikin has a black face, grey crown and upper-body, grey wings with white spots, white breast with black scalloping pattern, buffy under-body, grey-black tail. Male has beige margin that surrounds black mask female has hardly any noticeable margin and breast marking is less contrasty, over all the female appears to have a duller plumage. Contact call is soft 'tick…tick…..tick' but a flock makes a combined, constant chipping call. **Nest:** Oval ball of woven grasses, woven to reed or grass stems, sometimes stems of

Photo: Simon Nevill

Spinifex in dryer areas, always with a side extended entry, lined with soft grasses and a few feathers. Clutch: 4-5. **Similar species:** Chestnut-breasted Mannikin has plain orange-chestnut breast not white with black markings. **Status:** A nomadic species generally uncommon but locally, can be common. **Where to find:** Grasslands along the Gibb River Road. On the Mornington track. Track into Silent Grove. Possibility on the main highway from Fitzroy Crossing to Halls Creek. Grasslands in the Kununurra.

Australian Pipit *Anthus novaeseelandiae* 13-18 cm

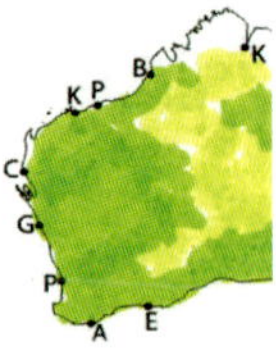

A very common pipit, found throughout much of WA, although does not frequent the heavily forested areas of the Darling Range and the southern Karri Forests. Its preferred habitat is open spaces often quite bare, sometimes stony, wherever there are short grassed or bare soils particularly in the Mulga Region, roadside verges, claypans, eroded areas near creeks, open areas in samphire, open bare areas in fields. Its bobbing tail is a feature of this small bird whenever it alights. Has a relatively long tail with a similar jizz to northern hemisphere wagtails but the tail is not as long and it does not wag its tail constantly like wagtails do. Feeds entirely on the ground with a typical run and pick technique. In flight white outer tail feathers are the main feature to quickly identify it over songlark's, and more extensive white than Singing Bushlark. There are thought to be eight subspecies in Australia but this is open to debate. Two subspecies: *australis* basically north and east of the Transitional Woodland line to the Kimberley and the central ranges. *bilbali* in the South West, both extremely similar *bilbali* having broader, darker streaking. Bergman's Rule applies with southern birds being slightly larger getting smaller in a clinal change the further north one goes. **Nest:** A small flat cup shape nest is built

Photo: Simon Nevill

at the base of grasses or against fallen timber with loose litter. The inner lining is woven fine grasses. Clutch size normally 3. **Similar species:** Singing Bushlark which has shorter tail, less white in the outer tail feathers and a more finch-like bill, not long and slender. Not as upright or as long legged. Has more chestnut in wings. Stockier jizz. **Status:** Common throughout the South West except the heavily forested regions. **Where to find:** Anywhere there is open ground and grassed areas, including playing fields in towns.

Yellow Wagtail *Motacilla flava* 17 cm

Scarce to WA but a regular vagrant to the Broome region and occasionally the Port Hedland area. A species with a large number of subspecies, various authorities identifying between 13-22 subspecies. Their range extends through Europe, Russia the Far East. Siberia and Alaska. Migrates to the southern hemisphere between October and March. The subspecies *tschtschensis* migrates from eastern Siberia and western Alaska down to China and Indonesia but often over passes to Australia. Other subspecies have migrated to Australia. Often found in northern towns wherever there are well watered grassed areas. Yellow Wagtails have a grey upper body and bright yellow under body in breeding plumage. The non-breeding plumage is mostly seen in Australia, the yellow under body is more diffused and paler. Has a white supercilium stripe. Long grey-black legs. Very long dark grey tail with white in the outer tail feathers. **Similar species:** Grey Wagtail which has a distinct black throat in the breeding season but moults to white in non-breeding. Yellow Wagtail always has yellow throat. Supercilium stripe not as long. Citrine Wagtail has all yellow head even in non-breeding plumage. **Status:** Very uncommon in the Kimberley except the Broome and Derby region where it is a regular visitor in very small numbers. Rare on the Pilbara coastline. **Where to find:** Have to visit Broome during the mid summer months from late October through to March–April.

Photo: Simon Nevill

Mistletoebird *Diccaeum hirundinaceum* 9 -11 cm

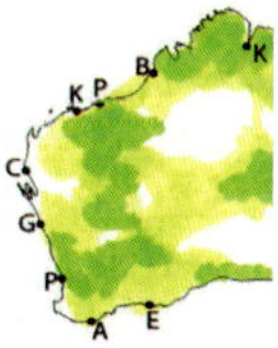

A small bird similar in shape to pardalotes. The male very distinctive with a bright red throat. Blue-black upper body with a slight metallic sheen. Under body off white with unusual feature of a black central band running from breast to vent. Red vent and under tail. Stout small bill. Female has plain grey-brown head and upper body, lacks the blue-black back. Has pinkish vent and under tail. Vertical band on belly is dark grey-black and under body diffused grey-white. Found in a variety of forested areas including open woodlands and mulga but always where there are trees that are parasitised by mistletoe. Where there are extensive stands of mistletoe the higher the probability of Mistletoebird's being present.

They range throughout the WA but are scarce in the thicker Jarrah Forest and Karri Forest. Their primary food is mistletoe fruit but they eat other fruits and also insects. In adaptation to feeding on such a high percentage of small fruits the Mistletoebirds digestive system has evolved into a unusual structure. They possess a rudimentary gizzard and a simple digestive tract with no enlargements and a long intestine, so berries pass from the oesophagus straight to the intestines; the berries passing through the body in as short a time as 4 minutes. They have the well known habit of defecating on branches, the seed being stuck to the branch by a sticky coating surrounding the seed. The bird is the primary transporter of mistletoe seed to other host trees. There are four subspecies of Mistletoebird throughout Australasia but only one in Australia. **Nest:** Their nest is unusual in as much as it is pouch-shaped with long oval entrance. It is a soft nest made from spiders webbing and fine grasses, the outside always is woven with either insect larvae sacs or plant down. Placed in the outer foliage of a eucalypt or acacia, often well hidden. Clutch size 3-4. **Similar species:** None. **Status:** Uncommon to moderately common throughout range. **Where to find:** Particularly prevalent in the inner Wheatbelt in Wandoo and York Gum woodland. Moderately common in the Mulga Region.

Above: Female

Photo: Simon Nevill

White-backed Swallow *Cheramoeca leucosternus* 13 -15 cm

Photo: Simon Nevill

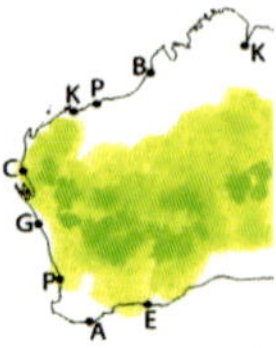

A solitary swallow that does not flock like other swallows, only flying with the immediate family. Has distinct patterning on the body, with a white crown, black eye band running from lores through the eye and around the nape. Triangular white patch on back. White throat and armpits. Long tail. A fast flyer. Found primarily in semi-arid and arid regions, near sand banks where it nests. Is a sedentary swallow so will remain in areas that afford a good nesting site and abundant food. One location due east of Hyden has been used for the last 20 years that I know. Found throughout the South

West, accept areas within the lower South West roughly from Perth south east around the Jarrah Forest down to Albany. If approaching a nest locality, the swallows will normally fly very high and sometimes almost out of sight watching their territory. Adults and dependent young leave nest together one after the other, flying high while making a high pitched chirp contact call; will circle for a while and then leave for feeding areas sometimes over heath or tree tops. They come back to rest during the day and then return to feed often not until just after the sun has set. **Nest:** A long tunnel excavated in a sand or clay bank with twigs at the base of a small nest chamber. Clutch size 4-5. **Similar species:** Can not be confused with any other swallow. **Status:** Uncommon throughout WA, a little more common in the desert regions. **Where to find:** Just north of Perth. Lancelin. Jurien. Hill River. Odd gravel pits in the Wheatbelt. On the Hyden Norseman road. Even sand banks within the Transitional Woodland. In the Mulga Region. Shark Bay region and all the central deserts.

Barn Swallow *Hirundo rustica* 15-18 cm — Vagrant

Photo: Simon Nevill

An uncommon but regular visitor to the north west of WA. Found in the Kimberley in Broome and Derby each year. Breeds throughout the northern hemisphere. Although normally in very small numbers in Australia flocks of 200 plus have been sighted in Broome. Often perches on telegraph wires particularly in Derby. **Similar species:** Very similar to the resident Australian Welcome Swallow but has longer tail streamers and a distinct black band at the base of the orange-chestnut throat. Underparts a pure white whereas Welcome Swallow has a slight greyish off white appearance. At rest say on wires, Barn Swallow shows longer tail streamers. **Status:** Locally common in parts of Kimberley between November and March. Uncommon but often seen in the Pilbara even records further south.

Welcome Swallow *Hirundo neoxena* subspecies *carteri* 14-16 cm

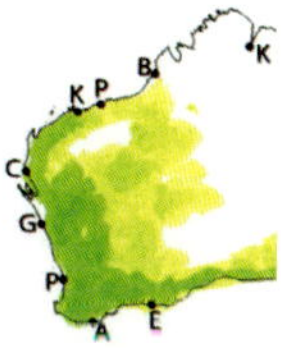

A very common Swallow found throughout the much of WA, although numbers diminish in the Pilbara and the desert regions. Sexes alike in the field. Has an orange-chestnut forehead and throat with black lores. Black-blue back with glossy seen in full sunlight. Tail passes wing tips when at rest. Found in a variety of habitats, mostly where there are open areas with woodland adjacent and often over water and tree canopies. Requires reasonable insect populations in flight which the semi-arid and arid zones do not have. Often around islands and coastal waters. In the southern Mulga Region around dams and troughs. Apparently was restricted to the coastal regions when Europeans first settled these shores but with land clearing spread to Kalgoorlie and further north. There are two subspecies in Australia *neoxena* on the east coast and *carteri* in the South West. One cannot split them in the field but the South West birds have a fractionally shorter tail and slightly longer bill. There are movements north to sub-tropical parts of WA in the winter months. **Nest:** Usually placed on a rock ledge often with a cover above nest. Since European settlement, has taken advantage of additional nesting sites particularly in towns under garage roofs or under eaves, particularly in the Mulga Region where nests sites are not so plentiful. If you are near a body of water and park a vehicle for few hours, the birds will begin to investigate the interior of the car as a possible nesting site almost oblivious of one's presence such is the strong instinct for the ideal nest locality. Their nest is a compact cup shaped nest, Consists of mud pellets mixed with grasses to make a dense hard nest. Clutch size 3-5. **Similar species:** Refer to Barn Swallow. **Status:** Common throughout the South West, Uncommon in the Jarrah Forest and Karri Forest. Becomes less common the further north and north-east one goes from the higher populations in the Mulga region. Normal range terminates about Port Hedland but has been recorded in the Kimberley.

Photo: Simon Nevill

Tree Martin *Peltrocheldon nigricans* subspecies *neglecta* 12 cm

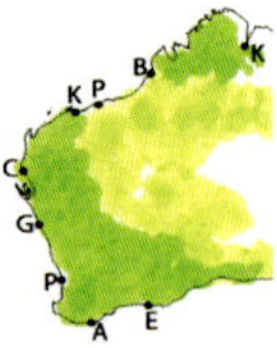

Found throughout WA but not in the treeless central deserts. Found wherever there are eucalypts. In the breeding season will form small groups within a woodland. nesting not far from each other and feeding often in groups. The South West birds are partially migratory. After the breeding season, they form larger groups and will disperse north mid summer to end of summer. Some birds do however remain in the South West. They are very common in Karri Forest and Wandoo Forest. There are two subspecies in Australia, *nigricans*

Photo: Simon Nevill

which breeds in Tasmania and migrates in late summer along the east coast with some birds flying to New Guinea and *neglecta* which breeds in all parts of southern Australia and moves north in late summer. **Nest:** They nest in tree hollows in live and both dead branches. Eggs are laid either on loose existing timber debris or leaves, twigs or grasses within the hollow. Prefer horizontal entrances on branches if available. Clutch size 4. **Similar species:** Fairy Martin, which has an orange-chestnut crown not black-blue. Does not nest in trees but makes a mud nest in caves or in culverts. **Status:** Common throughout the South West and Mulga, less so the northern Pilbara and the Kimberley. Where to find. Darling Range Karri Forest and much of the Wheatbelt and Transitional Woodland.

Fairy Martin *Peltrocheldon ariel* 12 cm

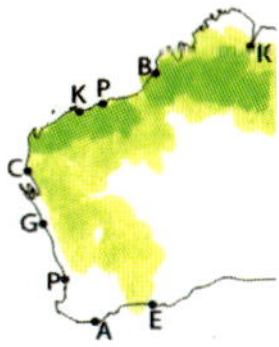

Found throughout the WA, except the heavily forested South West Corner. Not as common in the South West as the Mulga Region and even more common the further north one travels with the Pilbara and Kimberley having the greatest numbers; since European settlement with the introduction of culverts and bridges, the population has increased throughout its range. Fairy Martins are mostly found near water, and in the Mulga Region when good winter rains have fallen they become increasingly active in spring, particularly if there is plenty of surface water, as they require water to mix with earth to make their mud nests. Many of the earth nests found in caves can be many years old and the remoter the country the more caves are used, as culverts are few and far between. Fairy Martins have an orange-chestnut crown, black-blue back, black wings, white rump and white under body and throat. The throat is faintly streaked with fine pale brown markings. In flight shows a short square tail with a shallow wedge shape. Feeds in flight on insects but will come to ground to take insects but not as much as Tree Martins. **Nest:** A bottle-shaped structure with the neck of the bottle lying horizontal which is the entrance tube.. The nest cavity in the bulbous part of the bottle shape is lined with grasses. Fairy Martins nest communally, in fact throughout their whole life, they remain in a communal group. Clutch size 3-5. **Similar species:** Tree Martins which have a black-blue crown not orange-chestnut colour. Have a small orange-chestnut patch on the forehead just above the bill. Have the same jizz and white rump as the Fairy Martin so it is important to check the colouring of the head. In flight Fairy Martins make more of a soft-twittering call than Tree Martins. **Status:** Not common the South West. Moderately common in the Mulga Regions, Gascoyne River catchment and the northern Goldfields. **Where to find:** Can be common on the Carnarvon floodplains. Throughout the Mulga Region in areas that have had good winter rains. Common in the Fitzroy Basin, particularly near Camballin Station.

Photo: Simon Nevill

Australian Reed Warbler *Acrocephalus australis* subspecies *gouldi*

16 cm

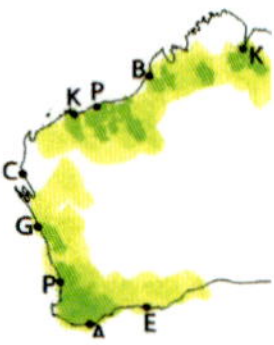

A medium sized reed warbler, found in fragmented populations throughout WA. The majority of birds occur in the South West region found in wetlands south west of a line from just north of Perth to Bremer Bay in the south, however there are good populations scattered throughout the Pilbara and particularly in the Kununurra Region. Mainly sedentary in WA but there is certainly movement to drier regions, establishing new territories in reed beds formed from either bores or new sewerage works. Their call is very strong and lets one know they are resident. They can exist in the smallest patch of reed beds as is often the case in the Mulga Region where excess water from bores forms only small wetlands. There are two subspecies in Australia, *australis* in northern and eastern Australia and *gouldi* in the South West and Pilbara region. The difference is related to wing size and bill length. **Nest:** They build a deep-cup shaped nest, woven between two or three reed stems mostly over water. Clutch size mostly 3. **Similar species:** Oriental Reed Warbler which comes occasionally to the Kimberley, particularly in the Kununurra area. It is a slightly bigger bird with a larger bill and faint fine black streaking on the side of the throat. Call very similar but can be distinguished as it is a far more grating, rougher call, Australian Reed Warbler's call not quite as fast. Oriental Reed Warbleris often overlooked. Little Grassbird is much smaller, with rufous crown and streaking on the breast. Call is totally different. Tawny Grassbird Almost the same size but slightly different jizz when perched. Mantle and crown has streaking that Reed Warbler does not have. Has chestnut buff on wings. **Status:** Locally common wherever reed beds occur. **Where to find:** In Perth Herdsman Lake. Lake Monger. Lake Gwelup. Bibra Lake. Other areas Benger Swamp near Harvey. Lake Seppings, Albany. Two Peoples Bay. Millstream. Sherlock River. Karratha Sewerage Ponds. In the Kimberley very common on the Ord River.

Photo: Simon Nevill

Oriental Reed-Warbler *Acrocephalus orientalis* 18 cm

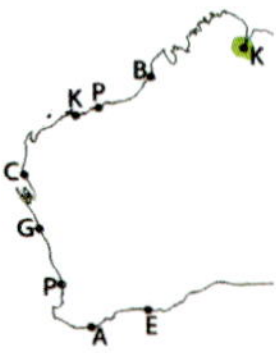

Originally considered a subspecies of the paleartic Great Reed-Warbler *Acrocephalus arundinaceus* but is now a full species. It certainly is a large reed-warbler, although still slightly smaller than the Great Reed-Warbler, which has not been recorded in Australia. It is larger than the Australian Reed-Warbler by at least 2 cm and is certainly a bulkier bird with a very robust bill. It's a scarce but not a rare vagrant to these shores there being at least 6-8 accepted sightings for Australia although many people do not submit sightings. It's however, a bird that can easily be overlooked and most probably is a regular visitor to the north, particularly in the Kununurra area where there is profuse amount of reed beds. It has been recorded in bulrushes, cumbungi and even mangroves. The species breeds in the northern summer in far eastern Russia, Mongolia, northern China and Korea. They move south in September, mainly over-wintering in southern China and Wallacea and some in PNG returning in May. However, a few over pass to northern and eastern Australia. **Similar species:** Australian Reed-Warbler. Oriental Reed-Warbler plain grey-brown upper-body, crown and tail, pale buff under-body and throat, pale, off white loral stripe, very robust bill, distinct brown streaking on throat although in moult which may occur in Australia, this may be less evident, light tip on tail shows better in flight. Australian Reed-Warbler has a plain brown back, not grey-brown, bill same length but finer, certainly not as robust, does not show any brown streaking on throat and breast. Calls different: Oriental Reed-Warbler does have same basic rhythm of notes and the jerky melody but is more guttural, being a deeper grating call. **Where to find:** Reed beds above diversion dam in Kununurra a possibility. Most likely to find between the months of September to May.

Little Grassbird *Megalurus gramineus* subspecies *thomasi* 12-15 cm

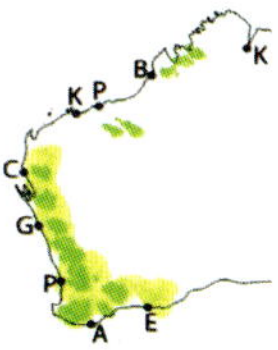

The majority of Little Grassbirds in WA are found in the south west and are the subspecies *thomasi*, There are outlying small populations in the west and more so, the east Kimberley namely subspecies *goulburni* both requiring reed bed habitats. Little Grassbird is a small reed dwelling warbler with a fine down-curved bill. Has a fine white supercilium stripe, dark rufous crown, faint black streaking on the breast, and bold dark brown mottling on the back. Relatively long tail for its size. Call is mournful soft whistle, the first note extended, then three rapid notes and final extended note 'weeee...wee, wee, wee...weeeeee'. Calls constantly in the breeding season but very little out of breeding. Will occupy even smaller bodies of reeds than Australian Reed Warbler, often being found in outback dams with reeds. It will also inhabit thicker samphire around estuaries such as at Peel Inlet or the wet samphire at Oyster Harbour, Albany. **Nest:** Similar to Australian Reed Warbler but smaller entrance that is more concealed, the nest chamber is lined with feathers. Sometimes uses old reedwarblers' nests. Not only nests in reeds but also melaleuca and samphire shrubs. Clutch size 3-4. **Similar species:** Similar to Tawny Grassbird which is only found in the Kimberley and Little Grassbirds are very uncommon in the Kimberley. They are a much bigger bird, lack the heavy flecking on the under body. No strong supercilium stripe. Call very different. **Status:** Common in all wetlands with reed beds. **Where to find:** In Perth. Eric Singleton Sanctuary, Bayswater. Herdsman Lake. Alfred Cove. Bibra Lake. Peel Inlet. Most wetlands in South West. Many bores throughout Mulga Region like at Cobra Station. In town wetlands at Carnarvon. Reed beds near the townsite of Kununurra.

Photo: Simon Nevill

Tawny Grassbird *Megalurus timoriensis* 17 cm

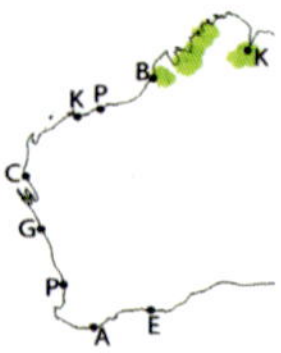

A species of the tall, rank grasslands found in the Kimberley especially in Rhodes Grass *Chloris gayana* and Pangola Grass *Digitaria decumbens*. Like so many reed or long-grassed frequenting species, its call is the first thing that will tell you it's in the area. It is a strong complex call with a descending, grating quality. Luckily the male often sings high on one of the tallest grass stems. The male will fly above the grasses with tail spread uttering its call, almost reminiscent of Rufous Songlarks as they ascend into the sky. It has the same jizz as a Spinifexbird, with heavy dark streaking on the back and a distinct rufous crown. **Similar species:** Little Grassbird which has been recorded in the Kimberley but very uncommon. A smaller bird with a totally different call being a soft mournful whistle. Lacks the rufous crown that Tawny Grassbird has. **Status:** Generally uncommon but can be locally common. **Where to find:** The best areas are in the Kununurra area on the lower Ord River and grasslands near Kingston Rest Station. Found in the few thicker grasslands such as those found in the Walcott Inlet.

Photo: Simon Nevill

Spinifexbird *Eremiornis carteri* 15 cm

Photo: Simon Nevill

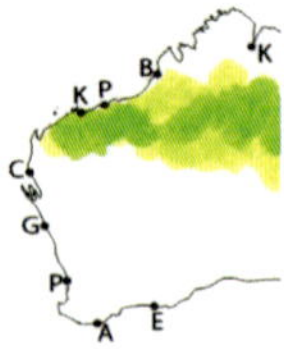

Certainly a bird that frequents spinifex habitat. Found in Spinifex either gorge country or the flat vast desert plains of the Western Deserts. They only occur in spinifex or acacia woodlands associated with spinifex but interestingly they do not occur in the dome spinifex habitat below latitude 25. Their range is from Cape Range across to McPherson's Pillar off the Gary Highway and north as far as Mornington Station and the Bungle Bungle Range. You normally never see Spinifexbird unless they sit up on branch or spinifex stem to call, you are more likely to hear them first, with their common sharp, rapid contact call given between long pauses 'chewychit.........chewychit ....chewychit'. They also have a harsh warning grating call, similar to many small passerines like Noisy Scrub-bird 'chuck...chuck...chuck. chuck' also a sweeter whistle reminiscent of the simple notes of a Grey Honeyeater 'sweeit.... sweeit....sweeit'. Spinifexbird is larger than Little Grassbird, has plain brown upper-body and nape, dark brown tail, distinct rufous crown and rufous rump seen well in flight, white supercilium contrasts with rufous crown, white throat and belly with buffy flanks and vent, slightly down-curved bill, large wedge-shaped tail as seen in the photograph. **Nest:** Large bowl type nest normally woven within a Spinifex dome, well protected but close to ground, lined with a few soft grasses. **Similar species:** None in their habitat. Tawny Grassbirds do not occur in same habitat. **Status:** Common throughout its range but easily missed. **Where to find:** Hamersley range particularly at the base of the ranges as below Wittenoom Gorge. Cape Range in gullies where dome spinifex occurs, Rudall River NP. Breaden Hills on the Canning Stock Route. Common south and east of Halls Creek. Common on the track into the Bungle Bungle.

Rufous Songlark *Cincloramphus mathewsi* 19-20 cm

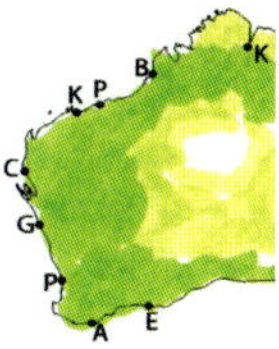

Found throughout the state but movements within the South West are dependent on climatic conditions. If the Mulga Region is experiencing bad droughts more birds will enter the South West, for example Dryandra Woodland NR will in some years have breeding birds and other years no birds at all. In the South West birds arrive in early spring and leave in early summer, however numbers are greatest in the Mulga Region particularly in areas when good rains have fallen. They prefer open woodland with long grasses, also creeks and river systems in the Mulga where grasses and ephemeral plants are at their peak in growth. Their call is in some ways similar to Brown Songlark but much softer and more musical. In the breeding season, the male calls from a perch and then will fly into the sky but not as high as a Brown Songlark, uttering its call as it ascends at a low angle and then descending into the grasses. The Rufous Songlark is a plain fawn-coloured songlark with a white supercilium stripe, slight flecking on the breast and a distinct chestnut rump. The sexes are very similar but the female is slightly smaller. **Nest:** The nest is an open cup structure placed on the ground either within a grassed area or at the base of a larger clump of grass with dry grasses leading from the top making it well camouflaged. Clutch size 3-4. **Similar species:** Brown Songlark female. The female looks very similar to both sexes of the Rufous Songlark but lacks the chestnut rump. The streaking on the back is stronger, bolder but is not an easy diagnostic

Photo: Simon Nevill

feature to pick. Australian Pipit is smaller and has a more upright stance and has white in the outer tail feathers and no chestnut rump. **Status:** Uncommon to moderately common in open woodland throughout much of the state. Not found in heavy forested areas of the South West. Can be very common in areas of the Mulga Region. Pilbara and throughout the central Kimberley. Central ranges. **Where to find:** Most open grassed woodlands in the Wheatbelt. In spring, early summer if conditions are good found in station country in the Mulga. Throughout the grasslands of the Pilbara, deserts and Kimberley but dependent on good conditions.

Brown Songlark *Cincloramphus cruralis* 19-25 cm

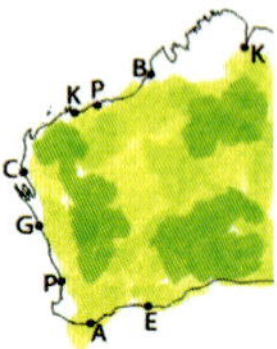

Found throughout most of WA, in areas of open treeless grasslands, farming areas, sparsely wooded areas and samphire flats. Does not enter the heavily forested parts of the South West although will occur in cropping areas as far south as the coast. I have seen them in the fields at Torbay. Rarely occurs in open woodland such as Wandoo, Salmon Gum or York Gum unless adjacent to large open areas. The male is far larger than the female, one of the largest variances between the sexes of any Australian bird. The male weighs on average 70 g, the female just 33 g, less than half the weight. In breeding, the male develops a very dark brown plumage mainly on the under body. In non-breeding is still darker than the female but the throat lightens to a patchy off white with dark brown flecking and the breast lightens. Immature males have similar colouring to the female but are still twice the size. The female is a lighter fawn-coloured bird with darker streaking on the upper body and proportionally shorter tail. In the breeding season the males fly almost vertically, while calling as they ascend in Skylark fashion, then they drop with wings held out extended almost in a 'parachute' drop, gliding to the ground. They will also call from the top of the tallest shrubs, with tail cocked. The call is a metallic grating discordant call 'pitchuweedeewe.... pitchuweedeewe..... pitchuweedeewe'. **Nest:** On the ground, under clumps of grass. In samphire or low bushes. Sometimes nesting semi-colonially in samphire in good years. Clutch size 2-4. **Similar species:** The female is very similar to the Rufous Songlark but lacks the rufous rump and the bill is thicker and the upper body a slightly lighter fawn colouring. **Status:** Very common in good seasons in the Mulga. Moderately common in open grasslands of the Pilbara. Parts of the Great Sandy Desert and grasslands in the Kimberley. **Where to find:** Samphire as at Wooleen, grasslands in the Mulga, Pilbara and the Kimberley.

Photo: Simon Nevill

Zitting Cisticola *Cisticiloa juncidus* subspecies *leanyeri* 11 cm

A cosmopolitan bird found in Africa, southern Europe, the Indian subcontinent and south-east Asia. Same size as Golden-headed Cisticola, having a much more restricted habitat range preferring grasses on the edge of swamp lands both freshwater and saline tidal flats. Confined in WA to the eastern side of the Kimberley mainly around the flood plains of the Ord River at Parry's Lagoon, wetlands south of Kununurra on grass plains adjacent to the upper Ord, found also at Kingston Rest. Zitting Cisticola has fawn back with heavy dark brown streaking, crown dark brown, whitish under-body and throat with buffy flanks, down-curved bill, under tail has a series of 'mirrors' at base of tail, shows well in flight contrasting with dark bars, has long legs, short tail, show faint off-white supercilium with darker brown patch on lores. In a nut shell, Zitting looks over all, a paler brown-beige bird whereas Golden-headed looks an orange-brown bird particularly in breeding. Call very distinct sharp clicking that it can keep up for a fair amount of time in the breeding season. **Nest:** Nest woven into grass stems on floodplains or swamps, a deep cupped nest with spider's webs that close it off except for small entry. Clutch 3-5. **Similar species:** Golden-headed Cisticola is very similar. In the breeding season identification is made much easier, the male and female Zitting has a dark crown heavily marked with black streaks over a brown base, Golden-headed male has orange-buffy crown with no streaking at all, however the female has streaking as well, but the face is a buffy colour where as female Zitting has a pale brown face. What you can notice both in and out of the breeding season, is that in flight, Zitting has a series of bars on the tail that are darker and contrast far more than those of Golden-headed. However the easiest way to split the two species is by their call but unfortunately it has to be in the breeding season as Zitting remain fairly quiet in winter. The call of the Golden-headed is far more varied, with a strong repetitive 'shewit… shewit…. shewit' then a totally different soft but nasal call 'zeeee…. zeeee…..zeeee'. Zitting as the name implies has a clicking, repetitive and constant zitting call 'zit…zit…zit…zit' mainly given in flight as it rises up in display flight in but will also call from a higher perch, only in the breeding season. **Status:** Locally common. **Where to find:** Parry's lagoon region and in grasslands above the diversion dam in Kununurra.

Photo: Graeme Chaplan

Golden-headed Cisticola *Cisticola exilis* subspecies *alexandrae* 9.5 cm

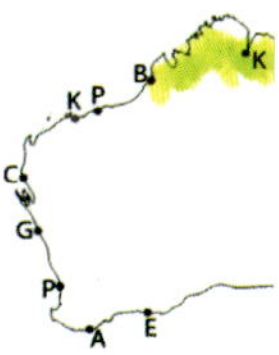

Far more common than the Tawny Grassbird in the Kimberley region. Found in the tall grasslands that cover much of the Kimberley. Often found in rank grasslands near water. The male has the same habit of perching in the tallest grass strands as the Tawny Grassbird does, singing a much softer squeaky song, repeated with short breaks. A small bird with a yellow-brown coloured head and dark streaking on upper-body with pale cream under-body. **Similar species:** In full breeding plumage the golden brown head makes it unmistakable

Photo: Simon Nevill

but out of breeding it could be confused with Zitting Cisticola. Zitting Cisticola has bold dark streaking on crown and darker loral patch that contrasts with off white supercilium stripe. Barring on tail more pronounced. Lacks the rufous crown that Tawny Grassbird has. **Status:** Throughout much of the grasslands in the Kimberley becoming scarcer the closer to the Great Sandy Desert. **Where to find:** Lush grasses off Gibb River Road. Along the Fitzroy. Many locations around Kununurra.

Canary White-eye (Yellow White-eye) *Zosterops luteus*

9.5 -13 cm

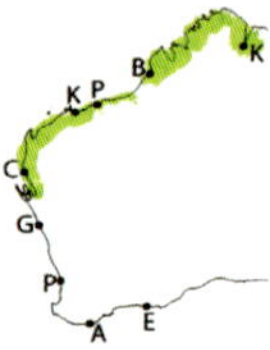

Subspecies *balstoni* found in mangroves from Long Point in the south to the Walcott Inlet the nominate *luteus* found from northern Kimberley across to QLD. A small White–eye with an all yellow under body. Lime-green head and upper body, fine bill, distinct white eye ring. Small yellow streaks on the lores. An active feeder moving along the coastline within its territory. Feeds on insects as well as nectar and fruit. Does not travel far from the coast except in a few localities in the Pilbara where they have been sighted as far inland as Millstream. Normally in ones and twos but occasionally groups of up to ten birds work through the mangroves together mainly out of the breeding season. **Nest:** Small cup nest placed in fork of mangrove 2-3 metres high. The nest is made of grasses and bound together with spiders webbing. Clutch size 2-3. **Similar species:** Silvereye, which has a grey-white under body with faint buff flanks but no yellow. No yellow patch on lores. Has grey-green mantle not lime-green. Territories can overlap. **Status:** Common in all the mangroves from throughout the WA coast. **Where to find:** New Beach. In Carnarvon, Mangrove Point. Oyster Bay, Babbage Island and Miaboolia Beach. Onslow. Port Sampson. Port Smith. Roebuck Bay. Willie Creek. Wyndham.

Photo: Simon Nevill

Silvereye *Zosterops lateralis* subspecies *chloronotus* 11-13 cm

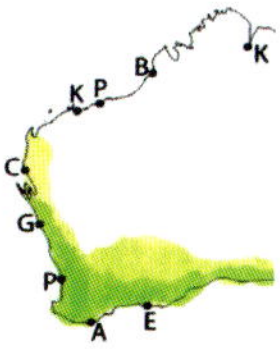

A common bird throughout much of the South West with distribution as far north as Shark Bay. There is a marked movement of Silvereye in the late summer and through winter along the coast moving north, then returning to breed in late spring. They range as far north as Shark Bay and then not far from the coast until Moora and then across to Kalgoorlie and down to the Eucla coast. Silvereye are extremely common in the lower South West Region and along the southern coast with large numbers in the Esperance region. Sexes are alike. The South West subspecies *chloronotus* has an olive-green mantle not a grey mantle, shorter wings and tarsus than the east coast nominate subspecies *lateralis*. **Nest:** Is neat woven cup nest made from fine grasses and bound with spiders' webbing. The base of the nest has a layer of feathers or dry grasses, placed in the centre of a low shrub. Clutch size 2-4. **Similar species:** Besides Yellow White-eye discussed under that specie, the only other small greenish coloured bird it could possibly be confused with is the Weebill which has an olive-brown rather than olive-green back. Yellow under body not grey. Bill smaller and thicker. **Status:** Very common in the lower South West. Uncommon inland away from the coast. **Where to find:** Throughout the South West.

Photo: Simon Nevill

Christmas Island White-eye *Zosterops natalis* 12.5 cm **Endemic**

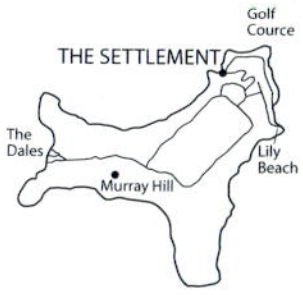

Found throughout Christmas Island, relatively common especially on the north east coast of the island. The species was also introduced on the Cocos Keeling Islands. A typical White-eye moving quickly through the lower stratum gleaning off leaves. There are no other species to confuse with on the island. Its difference from mainland Australian white-eyes

Photo: Simon Nevill

is marginal. Christmas Island White-eye has a fine black loral stripe from the eye to the base of the bill, that contrasts slightly with a white lower forehead. Bill is slightly longer and slightly down-curved more so than mainland Silvereye. **Similar species:** None on the Island. **Status:** Moderately common. **Where to find:** Around the settlement and along the many coastal tracks throughout the island.

Common Blackbird *Turdus merula* 25 cm — Introduced

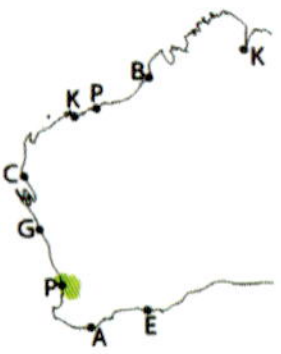

An introduced species to Australia. Has been recorded seen in WA but luckily a vagrant. WA certainly does not need this species. It has taken over many niche areas in the urban environment of the eastern cities and towns. It also has spread throughout much of New Zealand. A species found naturally in Europe, northern Africa and parts of the Indian sub-continent and the far-east. Do report this bird to the authorities if sighted.

Photo: Simon Nevill

Island Thrush *Turdus poliocephalus* 21 cm — Endemic

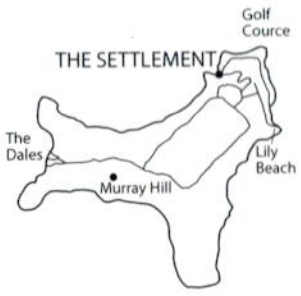

Wide spread throughout the islands in south-east Asia but in the Australian territories it can only now be found on Christmas Island. It is well named as it occurs on many islands, although found on mainland areas like Sumatra. Java and New Guinea. It was originally found on Norfolk and Lord Howe Island but the introduced Black Rat soon eradicated those populations. It was also introduced onto the Cocos Keeling Islands but has also become extinct there. Forages for insects turning over the leaf litter. **Similar species:** None on Christmas Island. **Status:** Locally common near the settlement. **Where to find:** Around the settlement. Tracks down to Lily Beach and also tracks near Steep Point.

Photo: Simon Nevill

House Sparrow *Passer domesticus* 15 cm

Introduced

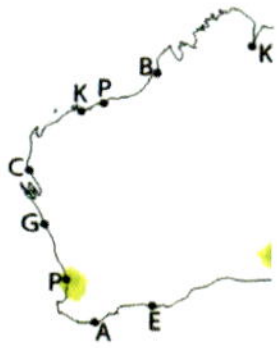

This species has always lived in close proximity to human habitation. It has spread far and wide in the east coast since its introduction. They have occurred in this state but luckily not firmly established. I recall being horrified to see them nesting at Nullarbor Roadhouse only a few kilometres from the WA border many years back. If you do see this bird, DO report it to the authorities for eradication, as we certainly do not need them in this state. **Nest:** Ball-nest wedged often in to the eaves of a house or in guttering. Clutch size large 4-5. **Similar species:** Tree Sparrow but highly unlikely in this state. **Status:** Rare. **Where to find:** Around harbours and ships or on the Nullarbor around buildings.

Photo: Simon Nevill

Common Starling *Sturnus vulgaris* 21 cm

Introduced (NSW) Vagrant (WA)

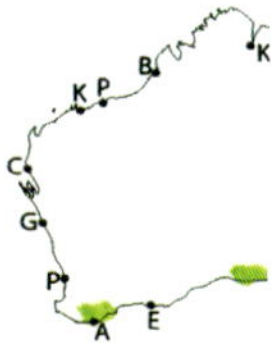

Introduced to eastern Australia from Europe, the Common Starling has now spread to much of Australia including NSW, Victoria, South Australia, Tasmania and Queensland. Over the last twenty years or so, small flocks have been entering across the eastern border of Western Australia. A multicoloured bird with a metallic tinge to its plumage. Highly gregarious, flocking in huge numbers when flying to roosting sites. Has preference for urban environments but also has spread to farming areas. A pugnacious bird forcing many birds from their nesting sites particularly parrots. In fact once starlings have nested in a tree hollow other native birds will generally not return to the nest site, such is the odour left by these birds. This species is a major problem in the eastern states and is commonly seen feeding in fields, has a preference for urban environments with open grassed areas or cultivated fields nearby. They are a major pest to orchardists damaging fruit. Birds have been expanding into Western Australia for many years now and birds are eradicated when sighted by the agricultural authorities. The author has seen first hand small flocks flying past Nullarbor Roadhouse just before the Western Australian border flying west towards WA. They are a problem in the Esperance region at this very time.

Photo: Simon Nevill

NOTE TO THE READER: Anyone sighting these birds in Western Australia should report these sightings to the Department of Agriculture and Food at their nearest regional office. They have an internet site: www.agric.wa.gov.au and a free call number 1800 084 881. Many people in Western Australia still have no idea what a major problem these birds can be so do phone in any sightings, you will be doing both farmers and birds a big favour.

GLOSSARY

Axillaries (axillars): the feathers at the base of the wing where the wing connects with the body. Example Grey Plover in flight is identified from Golden Plover by these feathers.

Arboreal: dwelling in trees.

Bergmann's Rule or Bergmann's Law: in simplistic terms, Bergmann's Rule refers to those species that reside both in warm regions as well as cold regions and the variation in their size and weight. Those individuals found in the warmer tropical regions of the world are nearly always smaller than the same species that occur in the colder regions of the world.

Breast: the upper region of the chest area.

Carpal: the region of the wing where it bends (anatomically at the wrist).

Cline: a gradual change in the plumage or size of a species over its range.

Colonial: relates to species that nest or roost in groups colonially. Example: Banded Stilt.

Cheek: the area on the side of the head, technically the 'jaw'.

Cosmopolitan: in this context a species found throughout much of the World.

Crepuscular: species that are active at twilight. Example: Letter-winged Kite.

Decurved: a bill that is downward curving. Example: Eurasian Curlew.

Diurnal: active by day.

Dorsal: feathers on the upperbody.

Ear coverts: the feathers to the rear and below the eye that cover the hearing system of the bird.

Eclipse plumage: the dull plumaged stage in a male's plumage following the breeding period. Example: All male waders particularly those from the Arctic that over winter in Australia go through this stage, or male fairy-wrens.

Endemic: species that are restricted in distribution within a country or region. Example: Western Spinebill is restricted to south Western Australia.

Family: the taxonomic order above genus, containing one or more genera.

Flight feathers: the longest wing feathers consisting of primaries and secondaries.

Gape flange: soft skin extending from the mouth often yellow as seen in young birds but occasionally is part of an adult bird's plumage as in Purple-gaped Honeyeaters.

Hackles: extended throat feathers.

Hosts: the species that incubates the eggs of a parasitic bird (typically the eggs of cuckoos) and raises the young of the parasitic species, mostly at the detriment of their own offspring.

Invertebrate: an animal lacking a spinal column or backbone such as molluscs, insects and worms.

Iridescence: the metallic or shiny effect when looking at certain species feathers like the Shining Bronze Cuckoo. Feather colouring is created by two main pigments 'melanin' which produces black or dark brown colouring or carotenoids that produce red, orange and yellow feathers. Also contained in feathers is a protein known as 'keratin' and the translucent keratin reflects certain colours particularly the blues, greens and violets typically seen on the wings of bronzewing pigeons. Lore: (pl. lores adj. loral) the area on the head between the bill and the eye. Sometimes it can be bare skin as in Great Cormorant but mostly feathered typically seen in the female fairy-wrens.

Mantle: upper part of the back below the neck line.

Mandible: the lower part of the bill sometimes called lower mandible although we tend to use the plural to describe both upper and lower portions of the bill.

Monotypic: having only one type of species in a genus.

Morph: a colour variation in certain species such as Brown Falcon but is not consistent and has no taxonomic standing.

Moult: the occurrence when species shed their old feathers and replace them with new. This occurs with many birds and some, e.g. swans cannot fly during this period until they have developed new feathers.

Nape: the rear of the head above the neck line.

Nocturnal: active during the night.

Palaearctic region: the biogeographical region in the northern hemisphere that includes North America, Eurasia south to the Himalayas.

Passerine: pertains to the largest Order of birds the Passeriformes, the 'perching birds'.

Pelagic: the term referring to the 'open seas' so pelagic birds are those that spend most of the lives on the high seas.

Plumage: refers to all the feathers on a bird and is also used to describe the colour changes that a species may go through when in moult or ageing.

Primaries: the largest outer wing feathers.

Raptor: refers to birds that have powerful claws and sharp talons used in tearing flesh from prey. Mostly refers to the Falconiformes the hawks and their relatives but also includes the owls.

Secondaries: the wing feathers between the primaries and the body, which are attached to the ulna.

Secondary coverts: the feathers that cover both the primaries and secondaries consisting of the greater, median and lesser coverts.

Species: a single bird or other animal that collectively breeds within its own group and isolated from other groups.

Subspecies: a geographic population of a species that has developed marked differences either in plumage or size than the nominate race and retains their differing characteristics when breeding. However, when one population of subspecies overlaps with another they will interbreed producing intermediate characteristics known as hybrids.

Supercilium: a lateral area above the eye that often looks like an eyebrow known as a superciliary stripe and is common in many birds including several honeyeaters and babblers.

Tarsus: a term that loosely describes the leg of bird.

Taxonomy: (adj. taxonomic) a scientific hierarchical classification of all plants and animals in terms of their relationship to each other.

Underbody: a loose term to describe all parts of a bird from the chin to the undertail coverts.

Upperbody: a loose term to describe all those parts of the upperbody but does not normally include the head or tail but includes the mantle, scapulas and rump.

Vagrant: a species that is irregular or rare to a geographical area.

Vertebrate: all those animals that have a vertebral column known as a 'backbone animal'.

Wattle: a fleshy appendage that hangs from the head or neck and is usually unfeathered bare skin often brightly coloured as in the Red Wattlebird.

INDEX – Latin

A NOTE TO READERS:

Due to the many changes to species classification and nomenclature by the IOC, it has been extremely difficult to list Latin names accurately in this index. Some species have been drastically changed being given totally new genus names as well as species names. This is going to lead to some contentious issues until stability in nomenclature stabilizes. It is important to double check the latest current Latin names with other more authoritative bodies rather than relying on this index as a definitive record.

INDEX – Common English Names (IOC)

Selected references

Barrett, G. et al. 2003, *The New Atlas of Australian Birds*, Birds Australia.

Beruldsen, G. 2003, *Australian Birds their Nests and Eggs,* G. Beruldsen, Kenmore Hills, Qld.

Brooker, M. I. H. & Kleinig, D. A. 2001, *Field Guide to the Eucalypts, Volume 2,* Bloomings Books.

Christidis. L. and Boles.W . E. 2008, *Systematics and Taxonomy of Australian Birds,* CSIRO.

Geering, A., Agnew, L. and Harding, S. 2007, *Shorebirds of Australia.* CSIRO.

Hayman, R., Marchant, J. and Prater, T. 1986, *Shorebirds. An identification guide to the waders of the world* Croom Helm.

Higgins, P. J. et al. 1990–2006 *Handbook of Australian New Zealand and Antarctic Birds Volumes 1 to 7,* Oxford University Press, Australia.

Jaensch, R., Vervest, R. and Hewish, M. J. 1988, *Waterbirds in Nature Reserves of South Western Australia 1981-1985,* Report No. 30, Royal Australasian Ornithologists Union.

Johnstone, R. E. and Storr, G. M. contributor Darnell, J. C. Volume 1. 1988, Volume 2. 2004, *Handbook of Western Australian Birds Vol. 1 & 2,* Western Australian Museum.

Morcombe, M. 2004 *Field Guide to Australian Birds*, Steve Parish Publishing.

North, A. J. *Nests and Eggs of Australia*, 1984, Oxford University Press, Melbourne.

Pizzey, G. and Knight, F. 1997, *Field Guide to the Birds of Australia*, Harper Collins.

Robson, C. 2002, *A Field Guide to the Birds of South-East Asia*, New Holland, United Kingdom.

Birding organizations in Western Australia

There is only one organization in Western Australia that concentrates purely on birds and that is the Western Australian Branch of BirdLife Australia which has its head office in Melbourne.

The WA office is located in the grounds of Bold Park. The address is as follows:

Western Australian Branch of BirdLife Australia
Peregrine House, 167 Perry Lakes Drive, Floreat WA 6014
Telephone (08) 93837749 Facsimile (08) 93878412
Email: mail@birdswa.com.au

The BirdLife Australia WA branch holds many birdwatching excursions. It also has regular talks held in the lecture hall close to the Perry Lakes Drive office. All members of the public are more than welcome.

The BirdLife Australia WA branch also publishes a quarterly newsletter highlighting meetings, sightings and various activities or reports on conservation or birds.

Acknowledgements

The vast majority of information contained in this publication is primarily based on 46 plus years of birding in Australia. However, I must acknowledge the phenomenal reference source of the various volumes of HANZAB (The Handbook of Australian and New Zealand Birds), for without this reference material, much of the information regarding subspecies could not have been written, although since that publication series, much new taxonomic information has become available, such is the ever-changing evolution of taxonomy nomenclature.

The ornithologist Ron Johnstone of the Western Australian Museum produced a very detailed twin volume series on the Birds of Western Australia. It was a useful source for me in terms of locating a few specific species, regarding the extent of their distribution. Ron has taken several of my spontaneous phone calls regarding such items as Barking Owl in the Pilbara etc. Ron is always on the ball and similar to me, as his memory is spot on for certain bird locations, as it's a subject we both love and it gets programmed in to the geographic memory bank. (Ron is second to none for that). Alan Burbidge, another fine ornithologist has answered in the past certain queries.

There are so many people who have assisted and are still assisting tirelessly with the Western Australian branch of Birdlife Australia. Many are the 'unsung heroes' and I think it appropriated that I thank those I know. They included: Bryce and Gail Wells; Perry and Alma de Rebeira. Perry managed to fight back from an almost certain death to be still with us and his wife Alma has been a brick through their journey they both have now been given an OAM this year a well deserved honour; Liz Walker; Rose and the late Alan Jones; John and Hazel Darnell who both have worked so hard to assist Ron Johnstone at the WA Museum; John Lawson and Lisa Richards my favourite Dryandra people; Shapelle McNee as said before, 'We all love you Shapelle'; George and Pam Agar; the tireless bird lister Frank O'Connor, who has and still is doing much to help the WA chapter of Bird Life Australia now chaired by the energetic Sue Mather who has done a sterling job as Chairperson like many before. And I acknowledge them. Those early years of Roger Jaensch and Doug Watkins who both are still actively involved in ornithological research; Ian Standring a good friend who gave advice on the early guide and always tests me out on 'What bird is this Simon?'; Michael and Irene Morcombe long time friends and multiple authors of many books; Ray Garstone a great bushman and birdwatcher; Wayne Zadow, a long standing friend of Ray Garstone and a very good birdwatcher in his own right, helping the south west people with birding; Alan Collins, simply a lovely young man, who gives freely of his great pelagic birding knowledge and now is ably assisted by John Graff, both of whom have corrected some of my seabird information; Brenda Newbey who has worked so hard to save one of our rarest birds the Ground Parrot; Leanne Quince who has always worked hard on my publications getting the graphics ready for print, alas due to the pressures of another book I lost out on her skills there but Toni Angelova of Graphics Source has done very well to take over and bring this rushed publication to print.

Thanks also to my long time friends Diane Beckingham and Rod Smith for doing a fantastic job with my testing grammar and spelling I really do appreciate the work you both have done. Diane would not give up and kept working till the very end finding so many missing details. There still will be faults with the book but it certainly will not be through their editing skills but more so the author's.

I sincerely thank the photographers who have helped with those images of species I lacked, namely Graeme Chapman one of our finest Australian bird photographers. Adrian Boyle who manages to get some of our rarest resident or vagrant birds, Alan Collins always so helpful to me, Brett Barrett for the Western Ground Parrot, Rohan Clarke the new breed of talented and very focused bird photographers who by the way also holds the second highest species sightings list for Australia. Rohan's identification skills regarding Albatross and other seabirds is of a very high standard as well.

To all the good people I took on the many tours I organized, you gave me some lasting wonderful memories. To those I have overlooked, I sincerely apologise.

To all I wish you happiness immersing yourselves in the solitude of the wonderful world of nature, away from the frenetic pace of this pressured world of ours.

Photograph by Martin Withers

Dedication to Martin Withers

I have dedicated this book to a wonderful wildlife photographer Martin Withers of the UK. Martin is sadly no longer with us and only recently passed away. Martin wrote to me less than a year before he died and it shows his acceptance of death. "Our world has been rocked rather by the news from my medical team that as well as my having Parkinson's Disease, I have also developed Motor Neurone Disease which means I will be dead within a year. A bit of a bummer but there is nothing to be done for it. At least I have a little time to get things sorted out for the benefit of Sally & Amy. I can't complain too much, I have had a wonderful life and been lucky enough to have seen many of the world's wonderful places and creatures. Spending time in the field with your good self was such a pleasure, and I have some wonderful memories of 'Oz' ".

I was privileged recently to see Martin in the UK, once when he could communicate and then again only days before his departure. He knew I was going to dedicate this book to him, as I wanted him, his wife Sally and daughter Amy to know that we all loved him. Martin gave me so much pleasure while I showed him and Sally around much of Australia when he was the UK tour leader for Hoskings Wildlife Tours. Martin worked closely with David Hoskings, another talented wildlife photographer, who in combination with his partner Jean run a very successful photographic agency.

Martin had a wonderful dry wit and I know his partner Sally and Amy his daughter will miss that very much. He never lost his cool on my tours and was always even tempered. His talent was not only through the lens of a camera but also with the pencil in hand, for he could sketch a landscape scene with real beauty in a very short space of time.

Martin will be missed by many, particularly Sally and Amy.

However Martin, you will always be with me for many years to come, particularly when I sit around the campfire, maybe in some remote part of the West Australian deserts. Do you know why ? Well on the back of that camp chair you left me, is hand sewn the words 'To you Martin'.

Crested Shrike-tit

The south west population of the Crested Shrike-tit may one day be split from the northern and eastern Shrike-tits as their distribution is more than 2,000 km away.

The western sub species has a white belly where as the eastern states birds have an all yellow belly.

Woodslane Press and Boiling Billy

Birds of Western Australia is just one of a growing series of illustrated books and outdoor guides from Sydney publishers Woodslane Press and Boiling Billy. To browse through other titles available from Woodslane, visit www.woodslanepress.com.au. If your local bookshop does not have stock of a Woodslane Press or Boiling Billy book, they can easily order it for you. In case of difficulty please contact our customer service team on 02 8445 2300 or info@woodslane.com.au.

Other books available include:

Australia's Great Western Deserts
$39.99
ISBN: 9781925868548

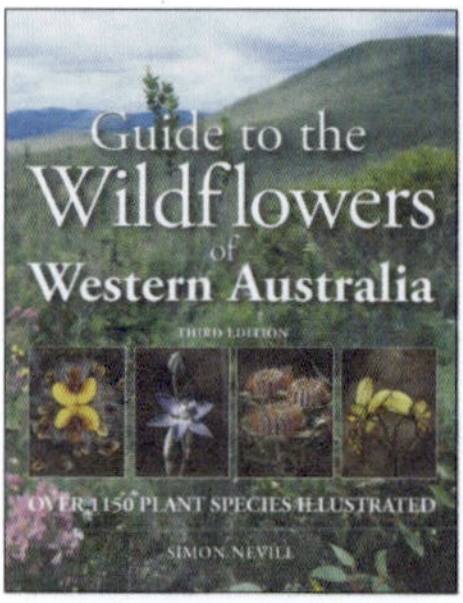

Guide to the Wildflowers of Western Australia
$34.99
ISBN: 9781925868036

Guide to the Wildlife of Perth
$39.99
ISBN: 9781921874246

Australian Geographic: Southwest Australia
$14.99
ISBN: 9781925868067

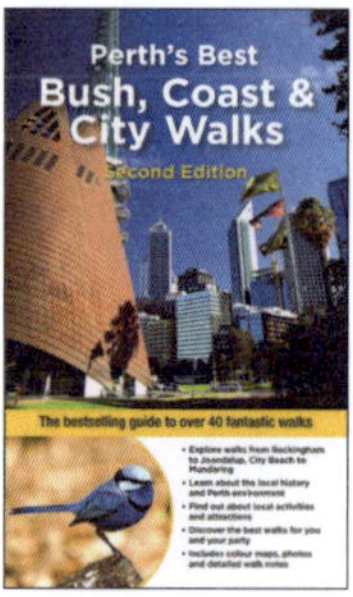

Perth's Best Bush, Coast & City Walks
$32.99
ISBN: 9781925868029

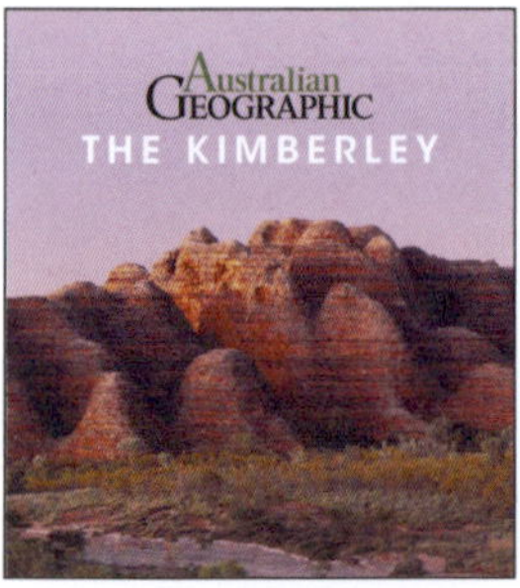

Australian Geographic: The Kimberley
$14.99
ISBN: 9781925403848

Best Walks of South West WA
$32.99
ISBN: 9781922800305

Best Walks of South Coast WA
$32.99
ISBN: 9781922800749

Compact Field Guide to Australian Birds
$34.99
ISBN: 9781922800732